SOC
Are you in?

We're Starting a Movement

Must-have resources for students and instructors

Affordable Solutions
For under $60, SOC presents all the content for your Intro Soc course through a set of print and digital resources!

Concise Chapters
Concepts and key terms are presented in a concise, engaging, and efficient format.

Online Study Tools
Students can practice or review with interactive online quizzing and printable flashcards at 4ltrpress.cengage.com/soc.

Superior Instructor Resources
You'll find chapter outlines, summaries, and key topics; links to news stories with discussion questions; and activities relevant to students' lives—all available online at 4ltrpress.cengage.com/soc.

Review Cards for Every Chapter
Perforated Chapter in Review cards help make study time more focused and efficient. Each card features key terms, concise review material, and helpful tables—all tied to the chapter key topics.

SOC 2009–2010 Edition
Nijole V. Benokraitis

Editor-in-Chief: Michelle Julet

Sr. Publisher: Linda Schreiber

Sr. Acquisitions Editor: Chris Caldeira

Director: Neil Marquardt

Managing Developmental Editor:
Jeremy Judson

Developmental Editor: Jamie Bryant,
B-books, Ltd.

Assistant Editor: Erin Parkins

Research Coordinator: Clara Goosman

Editorial Assistant: Rachael Krapf

Executive Marketing Manager:
Kimberly Kanakes

Executive Marketing Manager:
Kimberly Russell

Marketing Communications Manager:
Martha Pfeiffer

Production Director: Amy McGuire,
B-books, Ltd.

Media Editor: Lauren Keyes

Sr. Print Buyer: Judy Inouye

Production Service: B-books, Ltd.

Sr. Art Director: Caryl Gorska

Internal Designer: Ke Design

Cover Designer: Yvo Riezebos, Riezebos
Holzbaur Design Group

Cover Images: Corbis

Photography Manager: Deanna Ettinger

Photo Researcher: Charlotte Goldman

Library of Congress Control Number: 2009920119

SE ISBN-13: 978-0-495-60141-8
SE ISBN-10: 0-495-60141-1
IE ISBN-13: 978-0-495-60142-5
IE ISBN-10: 0-495-60142-X

Wadsworth Cengage Learning
10 Davis Drive
Belmont, CA 94002-3098
USA

Cengage Learning products are represented in Canada by Nelson Education, Ltd.

For your course and learning solutions, visit **academic.cengage.com**
Purchase any of our products at your local college store or at our preferred online store **www.ichapters.com**

Printed in the United States of America
1 2 3 4 5 6 7 12 11 10 09

BRIEF CONTENTS

CONTENTS

3 Culture 38

$*#@%**#@

4 Socialization 60

5 Social Interaction And Social Structure 80

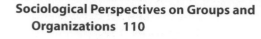

6 Groups, Organizations, and Social Institutions 98

7 Deviance, Crime, and the Criminal Justice System 116

8 Social Stratification 136

12 Work and the Economy 214

13 Families and Aging 232

14 Education 254

A sociological

imagination can inspire us to have more control over our lives.

what do you think?

No one is really unique even though everyone thinks she or he is.

1 2 3 4 5 6 7

strongly agree   strongly disagree

1

Thinking Like a Sociologist

Key Topics

In this chapter, we'll explore the following topics:

1 What Is Sociology?

2 What Is the Sociological Imagination?

3 Some Origins of Sociological Thinking

4 Contemporary Sociological Theories

In a nationwide survey of adult wireless subscribers, 80 percent said that using cell phones in a public place (like airports and restaurants) is "a major irritation" but 97 percent don't think they're part of the problem (Berger 2006). Why are "they" rude while "I'm" polite? And how do such perceptions shape our everyday behavior?

This chapter examines these and other questions. Let's begin by looking at what sociology is (and isn't) and how a "sociological imagination" can inspire us to have more control over our lives. We'll then look at how sociologists grapple with complex theoretical issues in explaining behavior.

> **sociology** the systematic study of social interaction at a variety of levels.

1 What Is Sociology?

Sociology is the systematic study of social interaction at a variety of levels. *Social interaction* is the process by which we act toward and react to people around us (see Chapter 5). When sociologists talk about the *systematic* study of social interaction, they mean that social behavior is regular and patterned and that it takes place between individuals, in small groups (such as families), large organizations (such as IBM), and entire societies (such as between the United States and other countries). But, you might protest, "I'm unique."

ARE YOU UNIQUE?

Yes and no. Each of us is unique in the sense that you and I are like no one else on earth. Even identical twins, who have the same physical characteristics and genetic matter, usually differ in personality and interests. One of my colleagues tells the story about his twin girls who received the same doll when they were three years old. One twin chattered that the doll's name was Lori, that she loved Lori, and would take good care of her. The second twin muttered, "Her name is Stupid," and flung the doll into a corner.

Despite some differences, identical twins, you, and I are like other people in most ways. Around the world, we experience grief when a loved one dies, participate in rituals that celebrate marriage or the birth of a child, and want to have healthy and happy lives. Some human behavior, like terrorist attacks, is unpredictable. For the most part, however, people conform to expected and acceptable behavior. From the time that we get up until we go to bed, we follow a variety of rules and customs about what we eat, how we drive, how we act in different social situations, and how we dress for work, classes, and leisure activities.

So what, you might shrug. Isn't it "obvious" that we dress differently for classes than for job interviews? Isn't all of this just plain old common sense?

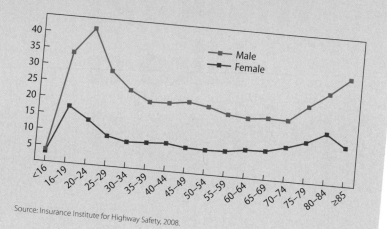

FIGURE 1.1
Are Teenagers the Most Dangerous Drivers?

Motor vehicle crash deaths per 100,000 people by age and gender, 2006

Source: Insurance Institute for Highway Safety, 2008.

ISN'T SOCIOLOGY JUST COMMON SENSE?

No. Sociology goes beyond what we call common sense in several ways:

- *Common sense often distorts reality.* If a teenage driver smashes into my car, I might conclude that "Everyone knows that teenagers are the most reckless drivers." In fact, auto crash deaths increase markedly at around age 70 (especially for men), almost equaling the auto crash death rate of male teenagers (see *Figure 1.1*). Thus, our common sense beliefs about teenage drivers being the most dangerous drivers aren't supported by "facts."

- *Common sense is often contradictory.* In terms of some of our favorite proverbs, should you "strike while the iron is hot" or "look before you leap"? And should we believe "out of sight, out of mind" or that "absence makes the heart grow fonder"?

- *Common sense perceptions change over time.* From the 1800s until at least 1950, U.S. Congressmen, some prominent sociologists, and influential journalists described European immigrants as feebleminded, criminal, immoral, and good for nothing (Carlson and Colburn 1972). These widespread common sense notions changed only after some sociologists started challenging such attitudes.

- *Much of our common sense is based on myths and misconceptions.* Common sense suggests that employers hire people who live nearby, especially for low-income jobs that don't require a costly commute. However, a sociological study of the working poor in Harlem, New York, found the opposite. Fast food employers preferred to hire people outside the community to decrease the costs of under-the-counter handouts to local friends and acquaintances (Newman 1999).

Thus, sociology goes well beyond common sense in understanding our social world. This is one of the reasons why a sociological perspective, and especially a "sociological imagination," is important.

2 What Is the Sociological Imagination?

how can we explain cultural variations? According to sociologist C. Wright Mills (1916-1962), our individual behavior is influenced by social factors—where and how others and we fit into the big picture. Mills (1959) described this intersection between individual lives and larger social influences as the **sociological imagination**. He emphasized the connection between personal troubles (biography) and structural (public and historical) issues. Mills noted, for example, that if only a few people are unemployed, that's a *personal* problem. If unemployment is widespread, it's a *public* problem because economic opportunities have collapsed and the problem requires solutions at the societal rather than at the individual level. In the words of a contemporary sociologist, the sociological imagination is "a means for many eye-opening experiences" because, among other things, it "empowers people to think about themselves, others, and what life is and could be in new and liberating ways" (Dandaneau 2001: 12).

The sociological imagination inspired generations of sociologists to examine the connections between one's life and the world and to apply sociology to attack many social problems using both micro- and macro-level approaches. Microsociology concentrates on the relationships between individual characteristics whereas macrosociology examines the relationships between institutional characteristics.

MICROSOCIOLOGY: HOW PEOPLE AFFECT OUR EVERYDAY LIVES

To some extent, we have many choices in our everyday lives. We decide, for example, where to shop, what to eat, and whether or not to buy a car. **Microsociology** focuses on small-scale patterns of individuals' social interaction in specific settings. In most of our relationships, we interact with others on a micro, or "small," level ("Aren't you feeling well?" or "What happened in class yesterday?"). Such everyday interactions involve what people think, say, or do on a daily basis.

MACROSOCIOLOGY: HOW SOCIAL STRUCTURE AFFECTS OUR EVERYDAY LIVES

Macrosociology focuses on large-scale patterns and processes that characterize society as a whole. Macro, or "large," approaches are especially useful in understanding some of the constraints—such as economic forces, social movements, and social and public policies—that limit many of our personal options on the micro level.

Microsociology and macrosociology differ conceptually, but they are interrelated. Consider the impact of technological innovations. Advances in medicine and other health-related improvements have led to a prolongation of life in industrialized countries, but because the average person can now expect to live into her or his 80s, poverty after retirement is more likely. On a micro level, researchers might analyze how older people interact with family members or others and how they cope with low incomes on a daily basis. On a macro level, sociologists might look at how medical costs deplete the savings of older people nationally and motivate them to reenter the workforce. Thus, understanding micro, macro, and micro-macro forces is one of the reasons why sociology is a powerful tool in understanding (and changing) our behavior and society at large (Ritzer 1992).

> **sociological imagination** the intersection between individual lives and larger social influences.
>
> **microsociology** the study of small-scale patterns of individuals' social interaction in specific settings.
>
> **macrosociology** the study of large-scale patterns and processes that characterize society as a whole.

NO WAY!

When I ask my students, "Would you marry someone you're not in love with?" most laugh, raise an eyebrow, or stare at me in disbelief. "Of course not!" they exclaim. In fact, the "open" courtship and dating systems common in Western nations, including the United States, are foreign to much of the world. In many African, Asian, Mediterranean, and Middle Eastern countries, marriages are arranged. In these societies, marriages forge bonds between families rather than individuals and preserve family continuity along religious and socioeconomic lines. Thus, love is not a prerequisite for marriage in societies that value the intergenerational and community relations of a kin group rather than an individual's choices (see Chapters 9 and 13).

Should colleges require all undergraduates to take a course on racial and ethnic relations? Why or why not?

WHY SOCIOLOGY IS IMPORTANT IN YOUR EVERYDAY LIFE

Sociology offers explanations that can greatly improve the quality of our everyday life. These explanations can influence or inform choices that range from personal decisions to social policies.

Making Informed Decisions

Knowing some sociology can help us make informed decisions that enrich the quality of our lives. In 1982, psychologist Carol Gilligan published an influential book which maintained that adolescent girls face a devastating drop in self-regard that boys don't experience. A decade later, clinical psychologist Mary Pipher's (1994) best seller contended, similarly, that teenage girls experience a fall in self-esteem from which many never recover. Both books, publicized by the popular press, generated considerable anxiety among parents, especially mothers, who worried about their daughters' emotional health.

Was the distress justified? No, because the conclusions were based on very small and nonrepresentative groups—Gilligan's on a private girls' school and Pipher's on a handful of troubled girls who sought counseling. In fact, well-designed studies since then have shown that the self-esteem scores of boys and girls are virtually identical (Barnett and Rivers 2004). Nonetheless, many academics and the mainstream press continue to promote the idea that girls have low self-esteem instead of worrying about popular misconceptions, knowing some-

thing about sociology and some basic research methods (which we'll address in the next chapter) can provide parents and teachers with valuable information.

Understanding Diversity

The racial and ethnic composition of the United States is becoming more diverse. By 2015, 62 percent of the U.S. population will be white, down from 76 percent in 1990 and 86 percent in 1950 (U.S. Census Bureau, 2008). As you'll see in later chapters, this racial/ethnic shift has already affected interpersonal relationships as well as education, politics, religion, and other spheres of social life.

There is also a rich diversity *within* ethnic groups. Latinos, for example, include persons of Mexican, Puerto Rican, Cuban, and Salvadoran descent—people who have different cultural backgrounds and reasons for immigrating to the United States or elsewhere. Asian Americans include people from locations as disparate as India, Manchuria, and Samoa. They follow different religions, speak different languages, and even use different alphabets.

Recognizing and understand diversity is one of the central themes in sociology. Our gender, social class, marital status, ethnicity, sexual orientation, and age—among other factors—shape our beliefs, behavior, and experiences. If, for example, you are a white, middle-class male who attends a private college, your experiences will be very different from those of a female Vietnamese immigrant who is struggling to pay expenses at a community college.

Almost all college and university sites feature "diversity." By their senior year, 41 percent of college students have taken an ethnic studies course, 38 percent have had a roommate of a different racial or ethnic background, and 40 percent say that promoting racial understanding is a "very important" or "essential" personal goal. However, 21 percent of these students report having little knowledge of or could "get along" with people from different races/cultures (Saenz and Barrera 2007).

Increasingly, nations around the world are intertwined through global political and economic ties. What happens in other societies often has a direct or indirect impact on contemporary U.S. life. Decisions in oil-producing countries, for example, affect gas prices, spur the development of "hybrid" cars that are less dependent on oil, and stimulate research on alternative sources of energy.

Evaluating Social and Public Policies

Sociology is also useful in evaluating social and public policies. For example, federal, state, and local governments spend over $30 billion each year to reduce illegal drug use. Such expensive efforts are often futile, however, because we have little reliable data on drug use, drug market economics, and enforcement activities (National Research Council 2001). Would you, the taxpayer, be better served if drug policies and budgets were based on accurate information?

PRACTICAL USES OF SOCIOLOGY

Sociology benefits us in several practical ways. The advantages range from learning to think critically to making career decisions.

Thinking Critically

Students develop a sociological imagination not just when they understand and can apply the concepts, but when they can think, speak, and write critically. Much of our thinking and decision-making is often impulsive and emotional. Critical thinking abilities enhance all learning knowledge, and problem solving (Paul and Elder 2007). "Critical sociological thinking" goes further because "students perceive and understand that their individual lives, choices, circumstances, and troubles are shaped by larger forces such as race, gender, social class, and social institutions" (Grauerholz and Bouma-Holtrop 2003: 493; see, also, Eckstein et al. 1995). (A *social institution*, which we'll examine in later chapters, is a set of widely shared beliefs and procedures that meet a society's basic needs).

Mate selection, for example, is not simply a matter of meeting and marrying one's "true love." Instead, whom we wed reflects larger social forces such as attending schools and having jobs that are often segregated by race and social class, laws that allow or forbid marriage between close relatives (such as uncles, aunts, and sometimes first cousins) or same-sex couples, and religious restrictions about inappropriate prospective spouses. In effect, then, most Americans select dating partners and marry people who are similar to themselves because filtering processes limit our choices.

Making Better Choices Amid Constraints

When there are widespread layoffs, people often blame themselves for being fired ("Maybe I should have worked harder"). Guilt, a feeling of inadequacy, and financial stresses can create family conflict and jeopardize interpersonal relationships. If you realize, however, that some of your personal problems reflect institutional constraints rather than individual failure, you'll be better equipped to handle some of life's "downers."

Besides helping us make better decisions, sociology encourages us to examine, critically and thoughtfully, the cultures and lives of people who are different from us. What's "natural" to you, for instance, may seem bizarre to someone from Europe, Asia, Africa, the Middle East, or Latin America (see Chapter 3). Recognizing the importance of cross-cultural variations, the U.S. military now provides its service members with training on Islamic culture, customs, and way of life.

Expanding Your Career Opportunities

A degree in sociology is a springboard for entering many jobs and professions. A national survey of the class of 2005 sociology majors found that, in fulltime and the largest job categories, 30 percent were in administrative support or management positions, 27 percent were employed in social service and counseling, and over 10 percent were in sales and marketing occupations (Spalter-Roth and Van Vooren 2008).

What skills do sociology majors feel they gain? Among graduating seniors, 73 percent felt they could develop evidence-based arguments, 67 percent said they could write a report that non-sociologists would understand, and 61 percent reported being able to interpret research findings (American Sociological Association 2006). In other cases, students major in sociology because they see it as a broad liberal arts base for professions such as law, education, medicine, social work, and counseling.

Even if you don't major in sociology, developing your sociological imagination can bring a depth and breadth of understanding to your workplace. Sociology courses help you learn to think abstractly and critically, formulate problems, ask appropriate questions, search for answers in the most reliable and up-to-date sources, organize material, and make effective oral presentations.

SOCIOLOGY AND OTHER SOCIAL SCIENCES: WHAT'S THE DIFFERENCE?

Many of the social sciences (disciplines that study human society) overlap. It is not unusual, then, for sociology textbooks (such as this one) to cite sources that include criminal justice, psychology, political science, and other social sciences. Despite the overlap, there are differences across the social sciences. For example,

- Sociologists laid the foundation for *criminology*, the scientific study of criminal behavior. *Criminal justice*, which attempts to control crime, often relies on sociological and criminological research.

- Most *anthropologists* study developing countries and may engage in archaeological exploration. Both anthropologists and sociologists consider how and why people behave as they do, but sociologists are more likely to focus on "modern" societies.

- *Economists* study the production, distribution, and consumption of goods and services. Sociologists tend to differ from economists by focusing on social relationships and do not generally assume that behavior is motivated solely by individual costs and benefits.

- For the most part, *historians* examine the past (such as the reign of a particular English king or queen) while sociologists tend to be more interested in contemporary behavior or attitudes.

- *Political scientists* focus on power relationships—how people vote, how laws are passed, and how governments exercise power. Sociologists study similar issues, but are more interested in understanding processes rather than events.

- *Psychologists* are usually interested in what occurs *within* people (such as emotions, perception, learning, personality and thinking), whereas sociologists focus on what occurs *between* people (such as interpersonal relationships, negotiating conflict, and the effect of social systems—like education and politics—on personal and group behavior).

- *Social workers* help people through an agency, organization, or nonprofit organization. In contrast,

How would different social scientists study the same phenomenon, such as homelessness? Criminologists might examine whether crime rates are higher among homeless people than those in the general population. Economists might measure the financial impact of services for the homeless in the nation and other countries. Political scientists might study whether and how government officials respond to homelessness. Psychologists might be more interested in how homelessness affects individuals' emotional and mental health. Social workers are most likely to try to provide needed services such as food, shelter, medical care, and jobs. Sociologists have been most interested in examining homelessness across gender, age, and social class and explaining how this social problem devastates families and communities.

According to sociologist Herbert Gans (2005), sociologists "study everything." There are currently 43 different subfields in sociology, and the number continues to rise. Why do sociologists' interests range across so many areas? Probably because the discipline's origins reflect the broad interests of its founders.

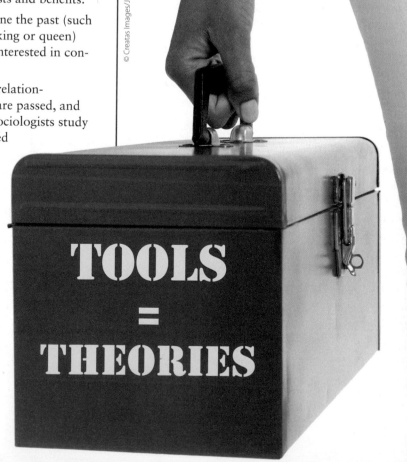

© Creatas Images/Jupiterimages

TOOLS = THEORIES

TRAFFIC TICKETS

© Mitchell Funk/Photographer's Choice/Getty Images

sociologists study people and generate research that social workers can then use to implement new policies, to challenge stereotypes, and to change existing services.

3 Some Origins of Sociological Thinking

During college, most of my classmates and I avoided taking theory courses (in all disciplines) as long as possible. "This stuff is boring, boring, boring," we'd complain, and "has nothing to do with the real world." Theorizing, in fact, is part of our everyday lives. Every time you try to explain why your family and friends behave as they do, for example, you are theorizing.

As people struggle to understand human behavior, they develop theories. A **theory** is a set of statements that explains why a phenomenon occurs. Theories produce knowledge, guide our research, help us analyze our findings, and, ideally, offer solutions for social problems. (Some sociologists differentiate between *theory* and *theoretical perspective*. Most use the terms interchangeably to explain how social phenomena are related to one another.)

Sociologist James White (2005: 170-171) describes theories as "tools" that don't profess to know "the truth" but "may need replacing" over time as our

© Visual Arts Library (London)/Alamy

understanding of society changes. Like hardware tools that change over the years, theories evolve over time to explain social phenomena. As you'll see shortly, sociological theories about behavior changed considerably after the rise of feminist perspectives during the late 1960s.

> **theory** a set of statements that explains why a phenomenon occurs.
>
> **empirical** information that is based on observations, experiments, or experiences rather than on ideology, religion, or intuition.

Sociological theories did not emerge overnight. Nineteenth-century thinkers grappled with some of the same questions that sociologists try to answer today: Why do people behave as they do? Why is society structured like it is? What holds society together? What pulls it apart? Of the many contributors to the development of sociology, some of the most influential were Auguste Comte, Harriet Martineau, Émile Durkheim, Karl Marx, Max Weber, Jane Addams, Georg Simmel, and W.E.B. Du Bois.

AUGUSTE COMTE

Auguste Comte (pronounced oh-gust KONT; 1798-1857) coined the term *sociology* and is often described as the "father of sociology." Comte's English translator, Harriet Martineau (1802-1876), fleshed out and publicized many of Comte's ideas. "We might, say, then, that sociology had parents of both sexes" (Adams and Sydie 2001: 32).

Comte believed that the study of society must be **empirical**. That is, information should be based on observations, experiments, or experiences rather than on ideology, religion, or intuition. He saw sociology as the scientific study of two aspects of society: social statics and social dynamics.

Father of Sociology—Auguste Comte

Social statics investigates how principles of social order explain a particular society as well as the interconnections between structures. *Social dynamics* explores how individuals and societies change over time. Comte's emphasis on social order and change within and across societies is still useful today because many sociologists examine the relationships between education and politics (social statics) as well as how their interconnections change over time (social dynamics).

HARRIET MARTINEAU

Harriet Martineau (1802-1876), an English author, published several dozen books covering a wide range of topics in social science, politics, literature, and history. Her translation and condensation of Auguste Comte's difficult material for popular consumption was largely responsible for the dissemination of Comte's work. She emphasized the importance of systematic data collection through observation and interviews and an objective analysis of records in explaining events and behavior and published the first methodological text for sociology (Adams and Sydie 2001).

Martineau, a feminist and strong opponent of slavery, denounced aspects of capitalism for being alienating and degrading. She was especially critical of machinery that resulted in injury and death, particularly of women and children. Martineau's suggestions for improving women's position in the workforce included education, nondiscriminatory employment, and training programs. She advocated women's admission into medical schools, emphasized the importance of physical fitness and exercise for girls and women, and investigated issues such as the care of infants, the rights of the aged, and the prevention of suicide and other social problems (Hoecker-Drysdale 1992). Her publications included thousands of articles and numerous books that criticized the injustices against women, slaves, children, the mentally ill, the poor, and prostitutes.

After a thirteen-month tour of the United States, Martineau described American women as being socialized to be subservient and dependent rather than equal marriage partners. She also criticized religious institutions for expecting women to be pious and passive rather than educating them in philosophy and politics. Most historians, who ridiculed such radical notions, dismissed her ideas.

ÉMILE DURKHEIM

Émile Durkheim (1858-1917), a French sociologist and writer, agreed with Comte that societies are characterized by unity and cohesion because its members are bound together by common interests and attitudes. According to Durkheim, however, Comte did not show that sociology could be scientific (and ignored Martineau's contributions on this subject).

Social Facts

To be scientific, Durkheim maintained, sociology must study **social facts**—aspects of social life, external to the individual, that can be measured. Sociologists can gauge *material facts* by examining such demographic characteristics as age, place of residence, and population size. They can determine *nonmaterial facts*, such as communication processes, by observing everyday behavior and how people relate to each other (see Chapters 3 to 6). Social facts also include *social currents* such as collective behavior and social movements (see Chapter 17).

Division of Labor

One of Durkheim's central questions was how people can be autonomous and individualistic while being integrated in society. **Social solidarity**, or social cohesiveness and harmony, according to Durkheim, is maintained by a **division of labor**—an interdependence of different tasks and occupations, characteristic of industrialized societies, that produce social unity and facilitate change.

As the division of labor becomes more specialized, people become increasingly more dependent on one another for specific goods and services. Currently, for example, many couples who are planning a wedding often contract and consult specific "providers" such as a photographer, florist, deejay, caterer, bartender, travel agent (for the honeymoon), and even a "wedding planner."

Social Integration

Durkheim, perhaps more than any of the other early pioneering theorists, showed the importance of combining theory with research. In his classic study, *Suicide*, Durkheim (1897) relied on extensive data collection to test his theory that suicide is related to social integration. For instance, Durkheim calculated the suicide rates of women and men, of the married and unmarried, and of Protestants, Catholics, and Jews.

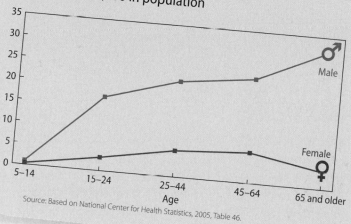

FIGURE 1.2
Suicide, by Sex and Age, 2002

Deaths per 100,000 in population

Male

Female

Source: Based on National Center for Health Statistics, 2005, Table 46.

He found that suicide rates reflected the degree to which individuals were integrated into family, group, and community life. Durkheim concluded that people who experience meaningful social relationships are less likely to commit suicide whereas those who feel lost, alone, helpless, or hopeless are more likely to commit suicide. That is, seemingly isolated individual acts are often the result of structural arrangements.

Are Durkheim's findings on social integration dated? No. Suicide has been one of the 11 leading causes of death in the United States since 1975. We typically read about the high suicide rates of teens. As *Figure 1.2* shows, however, the highest suicide rates are for people 65 years and older, especially white males (American Association of Suicidology 2006). Using Durkheim's analyses, the high suicide rates of older men may reflect being widowed and feeling alone, a sense of hopelessness due to terminal illnesses, and not being "connected" to support systems that women develop. Women are more likely than men to have close friends (especially other women), to maintain close ties with mothers and sisters, and to join community groups that provide assistance during troubling times (see Chapters 9 and 13).

KARL MARX

Karl Marx (1818-1883), a German social philosopher, is often described as the most influential social scientist who ever lived. Marx, like Comte and Durkheim, tried to explain changes in society that were taking place during the Industrial Revolution.

The Industrial Revolution began in England around 1780 and spread throughout Western Europe and the United States during the nineteenth century. A number of technological inventions—such as the spinning wheel, the steam engine, and large weaving looms—enabled the development of large-scale manufacturing and mining industries over a relatively short period of time. The extensive mechanization shifted agricultural and home-based work to factories in cities. As masses of people migrated from small farms to factories to find jobs, urbanization and capitalism grew rapidly.

Capitalism

Unlike his predecessors and contemporaries, Marx (1867/1967, 1964) maintained that economic issues produce divisiveness rather than social solidarity. According to Marx, the most important social changes reflected the development of **capitalism**, an economic system in which the ownership of the means of production—like land, factories, large sums of money, and machines—is in private hands. As a result, Marx saw industrial society as composed of three social classes:

- *capitalists*—the ruling elite who own the means of producing wealth (such as factories)

capitalism an economic system in which the ownership of the means of production—like land, factories, large sums of money, and machines—is in private hands.

- *petit bourgeoisie*—small business owners and owner workers who still have their own means of production but might end up in the proletariat because they are driven out by competition or their businesses fail

- *proletariat*—the masses of workers who depend on wages to survive, who have few resources, and who make up the working class

Class Conflict

Marx believed that society is divided into the haves (capitalists) and the have nots (proletariat). For Marx, capitalism is a class system where conflict between the classes is commonplace and where society is anything but cohesive. Instead, class antagonisms revolve around struggles between the capitalists who seek to increase their profits by exploiting workers and workers who resist but must give in because they depend on capitalists for jobs.

Marx felt that there was a close relationship between inequality, social conflict, and social class. Thus, he maintained, history is a series of class struggles between capitalists and workers. As wealth became more concentrated in the hands of a few capitalists, he thought, the ranks of an increasingly dissatisfied proletariat would swell, leading to bloody revolution and eventually a classless society. As you'll see in later chapters, some conflicts (such as the United States waging war on Iraq and many European countries' quotas on immigrants) reflect a struggle between the haves and the have nots. For the most part, however, the have not nations wield little power in resolving such class conflicts.

Alienation

In industrial capitalist systems, Marx (1844/1964) contended, **alienation**—the feeling of separation from one's group or society—is common across all social classes. Workers feel alienated because they don't own or control either the means of production or the product. Because meaningful labor is what makes us human, Marx maintained, our workplace has alienated us "from the essence of our humanness." In modern language, instead of collaborating, a capitalistic society encourages competition, backstabbing, and "looking out for number one."

According to Marx, capitalists are also alienated. They regard goods and services as important simply because they are sources of profit. Capitalists don't care who buys or sells their products, how the workers feel about the products they make, or whether buyers value the products or not. The major focus, for capitalists, is on increasing profits as much as possible rather than feeling "connected" to their products or the services they offer.

MAX WEBER

Max Weber (pronounced VAY-ber; 1864-1920) was a German sociologist, economist, legal scholar, historian, and politician. Weber rejected the Marxian view that economics was a major factor in explaining society. Instead, Weber focused on social organization, a subjective understanding of behavior, and a value-free sociology.

Social Organization

For Weber, economic factors were important, but ideas, religious values, ideologies, and charismatic leaders were just as crucial in shaping and changing societies. A complete understanding of society, according to Weber, must analyze the social organization and interrelationships between economic, political, and cultural institutions. In his *Protestant Ethic and the Spirit of Capitalism,* for example, Weber (1920) argued that the self-denial fostered by Calvinism supported the rise of capitalism, strengthened predestination, and shaped many of our current values about working hard (see Chapters 6 and 12).

Subjective Understanding

Unlike many of his predecessors, Weber stressed the differences, rather than the similarities, between the natural and the social sciences. Weber didn't dismiss "objective research," but he posited that an

© Comstock Images/Jupiterimages / © Hulton Archive/Getty Images

> Weber headed the first German institute of sociology and held a number of other prestigious university positions.

Is It Possible to Be a Value-Free Sociologist?

Max Weber was concerned about the popularity of professors who took political positions that pleased many of their students. He felt that these professors were behaving improperly because science, including sociology, must be "value free." Faculty must set their personal values aside to make a contribution to society. According to one sociologist who supports Weber's position, sociology's weakness is its tendency toward moralism and ideology.

> Many people become sociologists out of an impulse to reform society, fight injustice, and help people. Those sentiments are noble, but unless they are tempered by skepticism, discipline, and scientific detachment, they can be destructive. Especially when you are morally outraged and burning with a desire for action, you need to be cautious (Massey 2005: B12).

There is considerable disagreement, however, on whether sociologists can *really* be value free. Some argue that being value free is a myth because it's impossible for a scholar's attitudes and opinions to be totally divorced from her or his scholarship (Gouldner 1962). Many sociologists, after all, do research on topics that they consider significant and about which they have strong views. If you talk to your sociology instructor, for example, you'll probably find that she or he teaches and does research on topics in which she or he is intensely, and personally, interested.

Other sociologists maintain that one's values should be passionately partisan, frame research issues, and have an impact on improving society (Feagin 2001). That is, sociologists should not apologize for being subjective in their teaching and research.

understanding of society requires a "subjective" understanding of behavior. Such understanding, or *verstehen* (pronounced fer-SHTAY-en), requires knowing how people perceive the world in which they live. Weber described two types of *verstehen*. In *direct observational understanding*, the social scientist observes a person's facial expressions, gestures, and listens to his/her words. In *explanatory understanding*, the social scientist tries to grasp the intention and context of behavior.

If, in modern terms, a person bursts into tears (direct observational understanding), the observer knows what the person may be feeling (anger, sorrow, and so on). An explanatory understanding goes a step further by spelling out the reason for the behavior (rejection by a loved one, frustration when your computer crashes, humiliation if a boss yells at you in public).

value free separating one's personal values, opinions, ideology, and beliefs from scientific research.

Can sociologists be value free—especially when they have strong feelings about many societal issues? *Should* they be?

Value-Free Sociology

One of Weber's most lasting and controversial views was the notion that sociologists must be as objective, or "value free," as possible in analyzing society. A researcher who is **value free** is one who separates her or his personal values, opinions, ideology, and beliefs from scientific research.

During Weber's time, the government and other organizations demanded that university faculty teach the "right" ideas. Weber encouraged everyone to be involved as citizens, but he maintained that educators and scholars should be as dispassionate as possible about political and ideological positions. The task of the teacher, Weber argued, was to provide students with knowledge and scientific experience and not to "imprint" the teacher's personal political views. If educators introduce personal value judgments, according to Weber, a "full understanding of the facts ceases"

(Gerth and Mills 1946). The box "Is It Possible to Be a Value-Free Sociologist?" on page 13 examines this issue further.

JANE ADDAMS

Jane Addams (1860-1935) was a social worker who co-founded Hull House, one of the first settlement houses in Chicago that served as a community center for the neighborhood poor. An active reformer throughout her career, Jane Addams was a leader in the woman's suffrage movement and, in 1931, the first American woman to be awarded the Nobel Peace Prize for her advocacy of negotiating, rather than waging war, to settle disputes.

Sociologist Mary Jo Deegan (1986) describes Jane Addams as "the greatest woman sociologist of her day." She was ignored by her colleagues at the University of Chicago (the first sociology department established in the United States in 1892), however, because discrimination against women sociologists was "rampant" (p. 8).

Despite such discrimination, Addams published articles in numerous popular and scholarly

Near the end of his life, discouraged with the ongoing black discrimination in the United States, DuBois moved to Ghana.

journals as well as a number of books on the effects of social disorganization, immigration, and the everyday life of urban neighborhoods. Much of her work contributed to symbolic interactionism, an emerging school of thought. One of Addams' greatest intellectual legacies was her emphasis on applying knowledge to everyday problems. Her pioneering work in criminology included ecological maps of Chicago that were later credited to men (Moyer 2003).

W. E. B. DU BOIS

W. E. B. Du Bois (pronounced Do-BOICE; 1868-1963) was a prominent black sociologist, writer, editor, social reformer, and passionate orator. The author of almost two dozen books on Africans and black Americans, Du Bois spent most of his life responding to the critics and detractors of black life. He was the first African American to receive a Ph.D. from Harvard University, but once remarked, "I was in Harvard but not of it."

Du Bois helped found the National Association for the Advancement of Colored People (NAACP) and became editor of its journal, *Crisis*. He argued, unsuccessfully, that the NAACP should be led by African Americans, rather than whites. After a disagreement with the NAACP, which was committed to integration, Du Bois left. Shortly after that, he established the department of sociology at Atlanta University (now Clark Atlanta University), where he served as chair for ten years.

The problem of the twentieth century, he wrote, is the problem of the color line. Du Bois was certain that the race problem was one of ignorance and wanted to provide a "cure" for prejudice and discrimination. Such cures included black political power, civil rights, and providing blacks with a higher education rather than funneling blacks into technical schools.

These and other writings were unpopular at a time when Booker T. Washington, a well-known black educator, asked black people to be patient in demanding equal rights. As a result, Du

Bois was dismissed as a "radical" by his contemporaries and rediscovered by a new generation of black scholars only during the 1970s and 1980s. Among his many contributions, Du Bois examined the oppressive effects of race and class, described the numerous contributions of U.S. blacks to Western culture, and advocated women's rights. By the time he died, at age 95, Du Bois had authored 21 books, written over 100 essays and articles, and had played a key role in reshaping black-white relations in America (Du Bois 1986; Lewis 1993).

All of the early thinkers agreed that people are transformed by each other's actions, social patterns, and historical changes. Most importantly, these and other early contributors shaped contemporary sociological perspectives.

4 Contemporary Sociological Theries

how one defines "contemporary sociological theory" is somewhat arbitrary. The mid-twentieth century is a good starting point because "the late 1950s and 1960s have, in historical hindsight, been regarded as significant years of momentous changes in the social and cultural life of most Western societies" (Adams and Sydie 2001: 479). Some of the sociological perspectives had earlier origins, but all matured during this period. Like their predecessors, "modern" sociologists developed theoretical explanations that reflected their social and historical contexts such as the women's rights, gay rights, and anti-Vietnam war protests during the 1960s and 1970s; the impact of popular culture; and the increasing numbers of women who entered higher education and the labor force.

Sociologists typically use more than one theory in explaining human behavior and the theories often overlap. For greater clarity, we'll examine four perspectives separately—functionalism, conflict theory, feminist theories, and symbolic interactionism.

FUNCTIONALISM

Functionalism (also known as *structural functionalism*) maintains that society is a complex system of interdependent parts that work together to ensure a society's survival. Much of contemporary functionalism grew out of the work of Auguste Comte and Émile Durkheim, both of whom believed that human behavior is a result of social structures that promote order and integration in society. One of their contemporaries, English philosopher Herbert Spencer (1820-1903), used an organic analogy to explain the evolution of societies. To survive, Spencer (1862/1901) wrote, our vital organs—like the heart, lungs, kidneys, liver, and so on—must function together. Similarly, the parts of a society, like the parts of a body, work together to maintain the whole structure.

functionalism *(structural functionalism)* an approach that maintains that society is a complex system of interdependent parts that work together to ensure a society's survival.

Society Is a Social System

Prominent American sociologists, especially Talcott Parsons (1902-1979) and Robert K. Merton (1910-2003), developed these earlier ideas of structure and function. For these and other functionalists, a society is a system of major institutions such as government, religion, the economy, education, and the family.

Each institution or other social group has *structures*, or organized units, that are connected to each other and within which behavior occurs. Education structures such as colleges, for instance, are not only organized internally in terms of who does what and when but also depend on other structures such as government (to provide funding), business (to produce textbooks and construct buildings), and medical institutions (to ensure that students, staff, and faculty stay healthy).

"Sociologists typically use more than one theory in explaining human behavior and the theories often overlap."

dysfunctional social patterns that have a negative impact on a group or society.

manifest functions functions that are intended and recognized; they are present and clearly evident.

latent functions functions that are unintended and unrecognized; they are present but not immediately obvious.

conflict theory an approach that examines the ways in which groups disagree, struggle over power, and compete for scarce resources (such as property, wealth, and prestige).

Functions and Dysfunctions

Each structure fulfills certain *functions,* or purposes and activities, to meet different needs that contribute to a society's stability and survival (Merton, 1938). The purpose of education, for instance, is to transmit knowledge to the young, to teach them to be good citizens, and to prepare them for jobs (see Chapter 14).

Some social patterns are **dysfunctional** because they have a negative impact on a group or society. When one part of society isn't working, it affects all of the other parts by creating conflict, divisiveness, and social problems. Consider religion. In the United States, French, Portuguese, Spanish, and British missionaries were responsible for destroying much of the indigenous American Indian culture. The missionaries, determined to convert the "savages" to Christianity, eliminated many religious ceremonies and practices that they deemed "uncivilized" (Price 1981). More recently, religious intolerance has led to wars and terrorism (see Chapter 18 online).

Manifest and Latent Functions

There are two kinds of functions. **Manifest functions** are intended and recognized; they are present and clearly evident. **Latent functions** are unintended and unrecognized; they are present but not immediately obvious. Consider the marriage ceremony. The primary manifest function of the marriage ceremony is to publicize the formation of a new family unit and to legitimize sexual intercourse and childbirth (even though both occur outside of marriage in many industrialized countries). Its latent functions include the implicit communication of a "hands-off" message to suitors, providing the new couple with household goods and products through bridal shower and wedding gifts, and redefining family boundaries to include in-laws or stepfamily members.

Critical Evaluation

Functionalism is useful in seeing the "big picture" of interrelated structures and functions. According to some critics, however, functionalism is so focused on order and stability that it often ignores social change. For example, functionalists typically see high divorce rates as dysfunctional and as signaling the disintegration of the family rather than indicating a positive change (such as people leaving an unhappy situation).

A second criticism is that functionalism often ignores the inequality that a handful of powerful people create and maintain. Instead of challenging the status quo, some contend, functionalism simply describes it. Some critics have also charged that functionalism views society narrowly through white, male, middle-class lenses. According to some feminist scholars, for example, "functionalism tends to support a white middle-class family model emphasizing the economic activities of the male household head and domestic activities of his female subordinate" while ignoring non-traditional families, such as single-parent households (Lindsey 2005: 6).

CONFLICT THEORY

While functionalists emphasize order, stability, cohesion, and consensus, **conflict theory** examines the ways in which groups disagree, struggle over power, and compete for scarce resources (such as property, wealth, and prestige). In contrast to functionalists, conflict theorists see disagreement and the resulting changes in society as natural, inevitable, and even desirable.

Sources of Conflict

The conflict perspective has a long history. As you saw earlier, Karl Marx predicted that conflict would result from widespread economic inequality, and W. E. B. Du Bois denounced U.S. society for its ongoing racial discrimination that results in divisiveness. Since the 1960s, and as you'll see in later chapters, many sociologists—especially feminist and minority scholars—have emphasized that the key sources of economic inequity in any society also include race, ethnicity, gender, age, and sexual orientation.

Conflict theorists agree with functionalists that some societal arrangements are functional. But, conflict theorists ask, who benefits? And who loses? When corporations merge, workers in lower-end jobs are often the first to be laid off while the salaries and benefits of corporate executive officers (CEOs) soar and the value of stocks (usually held by upper middle and upper classes) increase. Thus, mergers might be functional for those at the upper end of the socioeconomic ladder but dysfunctional for those in the lower rungs.

Social Inequality

Unlike functionalists, conflict theorists see society not as cooperative and harmonious but as a system of widespread inequality. For conflict theorists, there is a continuous tension between the "haves" and the "have-nots," most of whom are children, women, minorities, and the poor.

Many conflict theorists focus on how those in power—typically white, wealthy, Anglo-Saxon, Protestant males (WASPs)—dominate political and economic decision making in U.S. society. This group controls a variety of institutions, such as education, criminal justice, and the media, and passes laws that benefit only small groups of people like themselves (see Chapters 8, 11, and 12).

Critical Evaluation

Conflict theory is important in explaining how societies create and cope with disagreements. However, some have criticized conflict theorists for overemphasizing competition and coercion at the expense of order and stability. Inequality exists and struggles over scarce resources occur, critics agree, but conflict theorists often ignore cooperation and harmony. Voters, for example, can boot dominant groups out of office and replace them with African Americans, Asians, Latinos, and women. Critics of conflict theory also point out that the have-nots can increase their power through negotiation, bargaining, lawsuits, and strikes.

Some critics also feel that conflict theory presents a negative view of human nature and neglects the importance of love and self-sacrifice, which are essential to family and other personal relationships. Because conflict perspectives examine institutional rather than personal choices and constraints, they don't give us insights on everyday individual behavior.

© Getty Images News / © Workbook Stock/Jupiter Images

FEMINIST THEORIES

feminist theories approaches that try to explain the social, economic, and political position of women in society with a view to freeing women from traditionally oppressive expectations, constraints, roles, and behavior.

Rebecca West, a British journalist and novelist once said, "I myself have never been able to find out precisely what feminism is; I only know that people call me a feminist whenever I express sentiments that differentiate me from a doormat." Feminist scholars agree with West and conflict theorists that much of society is characterized by tension and struggle between groups. They go a step further because **feminist theories** try to explain the social, economic, and political position of women in society with a view to freeing women from traditionally oppressive expectations, constraints, roles, and behavior. Thus, feminist perspectives maintain that women suffer injustice because of their sex and that people should be treated fairly and equally regardless of their race, ethnicity, national origin, age, religion, class, sexual orientation, disability, and other characteristics.

Focusing on Gender

According to many feminist scholars, women have historically been excluded from most sociological analyses (Smith 1987). Prior to the 1960s women's movement in the United States, very few sociologists published anything about gender roles, women's sexuality, fathers, or domestic violence. According to sociologist Myra Ferree (2005: B10), during the 1970s, "the Harvard social-science library could fit all its books on gender inequalities onto a single half-shelf." Since then, and because of feminist scholars, many researchers—both women and men—now routinely include gender as an important research variable on both micro and macro levels.

Listening to Many Voices

Feminist scholars contend that gender inequality is central to *all* behavior, from everyday interactions to organization structures and political and economic institutions. As a result, feminist theories encompass many perspectives. For example, *liberal feminism* emphasizes social and legal reform to create equal opportunities for women. *Radical feminism* sees male dominance as the major cause of women's inequality. *Global feminism* focuses on how the intersection of gender with race, social class, and colonization has exploited women in the developing world (see Lengermann and Niebrugge-Brantley 1992). Whether we identify ourselves as feminist or not, most of us are probably liberal feminists because we endorse equal opportunities for women and men in the workplace, politics, education, and other institutions.

Critical Evaluation

Feminist scholars have been effective in challenging discrimination in employment, among other practices, that have routinely excluded women who are not part of the "old boy network" (Wenneras and Wold 1997). One of the criticisms, however, is that many feminists are part of an "old girl network" that has not always welcomed different points of view from black, Asian American, Amer-

> "I myself have never been able to find out precisely what feminism is; I only know that people call me a feminist whenever I express sentiments that differentiate me from a doormat."
>
> –Rebecca West, British journalist

ican Indian, Muslim, Latina, lesbian, working-class, and disabled women (Almeida 1994; Lynn and Todoroff 1995; Jackson 1998).

Compared with other theorists, feminist scholars are more likely to embrace diversity. Some critics feel, however, that such inclusiveness dilutes agreement

Feminist theories encompass many perspectives.

© iStockphoto.com

on a variety of issues. If we want to eliminate gender inequality, for example, should feminist scholars argue that women should have the same rights and opportunities as men (liberal feminism) or that we should change institutional structures (radical feminism)?

A third criticism is that feminist perspectives tend to downplay social class inequality by focusing on low-income and minority women but not their male counterparts. Thus, some contend, feminist theories are not as gender-balanced as they claim. Some, including feminists, also question whether feminist scholars have lost their bearings by focusing on personal issues such as greater sexual freedom rather than broader social issues like poverty and wage inequality (Rowe-Finkbeiner 2004; Chesler 2006).

SYMBOLIC INTERACTIONISM

Symbolic interactionism (sometimes called *interactionism*) is a micro-level perspective that looks at individuals' everyday behavior through the communication of knowledge, ideas, beliefs, and attitudes. While functionalists and conflict theorists focus on structures and large systems, symbolic interactionists focus on *process* and keep the *person* at the center of their analysis.

There have been several influential symbolic interactionists whom we'll cover in later chapters. Stated briefly, George Herbert Mead's (1863-1931) proposal that the human mind and self arise in the process of social communication became the foundation of the symbolic interactionist schools of thought in sociology and social psychology. Herbert Blumer (1900-1987) coined the term symbolic interactionism in 1937, developed Mead's ideas, and emphasized that human beings interpret or "define" each other's actions instead of merely reacting to them, especially through symbols. Erving Goffman (1922-1982) contributed significantly to these earlier theories by examining human interaction in everyday situations ranging from jobs to funerals.

Constructing Meaning

Our actions are based on **interaction** in the sense that people take each other into account in their own behavior. Thus, we act differently in different social settings and continuously adjust our behavior, including our body language, as we interact (Goffman 1959; Blumer 1969). A woman's interactions with her husband are different from those with her children. And she will interact still differently when she is teaching a class

of students, talking to a colleague in the hall, or addressing an audience of colleagues at a professional conference.

For symbolic interactionists, society is *socially constructed* through human interpretation (O'Brien and Kollock 2001). The daughter who has batting practice with her dad will probably interpret her father's behavior as loving and involved. In contrast, she will see batting practice with her baseball coach as less personal and more goal oriented. In this sense, our interpretations of even the same behavior, such as batting practice, vary across situations and depending on the people with whom we interact.

> **symbolic interactionism** (*interactionism*) a micro-level perspective that looks at individuals' everyday behavior through the communication of knowledge, ideas, beliefs, and attitudes.
>
> **interaction** action in which people take each other into account in their own behavior.

We act differently in different situations.

WORK

PLAY

© Image Source/Jupiterimages

1 symbol ≠ 1 meaning

The American flag is one symbol, but it has different meanings for different groups.

Symbols and Shared Meanings

Symbolic interactionism looks at subjective, interpersonal meanings and at the ways in which we interact with and influence each other by communicating through *symbols*—words, gestures, or pictures that stand for something and that can have different meanings for different individuals.

After the 9/11 terrorist attacks, many Americans displayed the flag on buildings, bridges, homes, and cars to show their solidarity and pride in the United States. In contrast, some groups in Palestine and Pakistan burned the U.S. flag to show their contempt for American culture and policies. Thus, symbols are powerful forms of communication that show how people feel and interpret a situation.

To interact effectively, our symbols must have *shared meanings,* or agreed-upon definitions. One of the most important of these shared meanings is the *definition of the situation,* or the way we perceive reality and react to it. Relationships often end, for example, because partners define emotional closeness differently ("We broke up because Tom wanted sex. I wanted intimacy and conversation"). We typically learn our definitions of the situation through interaction with *significant others*—like parents, friends, relatives, and teachers—who play an important role in our socialization (as you'll see in Chapters 4 and 5).

Critical Evaluation

Unlike other theorists, symbolic interactionists show how people play an active role in shaping their lives on a micro level. One of the most common criticisms, however, is that symbolic interactionism ignores the impact of macro-level factors such as economic forces, social movements, and public policies on our everyday behavior. Because the United States (unlike many other Western countries) doesn't have paid maternity and paternity leaves, for example, only wealthy parents, and typically mothers rather than fathers, can afford to take time off from work to nurture and raise young children.

A related criticism is that interactionists often have an optimistic and unrealistic view of people's everyday choices. Most of us enjoy little flexibility in our daily lives because deeply embedded social arrangements and practices benefit those in power. For example, people are usually powerless when corporations transfer many jobs overseas or cut the pension funds of retired employees.

Some also feel that interaction theory is flawed because it ignores the irrational and unconscious aspects of human behavior (LaRossa and Reitzes 1993). People don't always consider the meaning of their actions or behave as reflectively as interactionists assume. Instead, we often act impulsively or say hurtful things without weighing the consequences of our actions or words.

For a summary of all these perspectives, see *Table 1.1.*

TABLE 1.1
Leading Contemporary Perspectives in Sociology

THEORETICAL PERSPECTIVE	FUNCTIONALIST	CONFLICT	FEMINIST	SYMBOLIC INTERACTIONIST
Level of Analysis	Macro	Macro	Macro and micro	Micro
Key Points	• Society is composed of interrelated, mutually dependent parts • Structures and functions maintain a society's or group's stability, cohesion, and continuity • Dysfunctional activities that threaten a society's or group's survival are controlled or eliminated	• Life is a continuous struggle between the "haves" and the "have nots" • People compete for limited resources that are controlled by a small number of powerful groups • Society is based on inequality in terms of ethnicity, race, social class, and gender	• Women experience widespread inequality in society because, as a group, they have little power • Gender, ethnicity, race, age, sexual orientation, and social class—rather than a person's intelligence and ability—explain many of our social interactions and lack of access to resources • Social change is possible only if we change our institutional structures and our day-to-day interactions	• People act on the basis of the meaning they attribute to others Meaning grows out of the social interaction that we have with others • People continuously reinterpret and reevaluate their knowledge and information in their everyday encounters
Key Questions	• What holds society together? How does it work? • What is the structure of society? • What functions does society perform? • How do structures and functions contribute to social stability?	• How are resources distributed in a society? • Who benefits when resources are limited? Who loses? • How do those in power protect their privileges? • When does conflict lead to social change?	• Do men and women experience social situations in the same way? • How does our everyday behavior reflect our gender, social class, age, race, ethnicity, sexual orientation, and other factors? • How do macro structures (such as the economy and the political system) shape our opportunities? • How can we change current structures through social activism?	• How does social interaction influence our behavior? • How do social interactions change across situations and between people? • Why does our behavior change because of our beliefs, attitudes, values, and roles? • How is "right" and "wrong" behavior defined, interpreted, reinforced, or discouraged?
Example	• A college education increases one's job opportunities and income	• Most low-income families cannot afford to pay for a college education	• Gender affects decisions about a major and which college to attend	• College students succeed or fail based on their degree of academic engagement

We are inundated

with *psychobabble*—psychological jargon about behaving and expressing one's feelings that has little scientific foundation.

what do you think?

Truly objective research is impossible.

1 2 3 4 5 6 7

strongly agree   strongly disagree

2

Examining Our Social World

Spring break is all about beer fests, wet-T-shirt contests, frolicking on the beach, and hooking up, right? Maybe not. A recent survey found that 70 percent of college students stay home with their parents and 84 percent of those who throng to vacation spots report consuming alcohol in moderation (The Nielsen Company 2008).

If you suspect that these numbers are too high or too low and wonder how the survey was conducted, you're thinking in terms of research, the focus of this chapter. We'll begin by considering what sociologists mean by social research, examine the scientific method, explore some of the major data collection methods, and end with a discussion of how ethical guidelines shape sociological research and why researchers sometimes violate these rules.

> **social research** research that examines human behavior.

1 What Is Social Research?

Social research examines human behavior. The process requires curiosity and imagination, but also knowing the rules and procedures that guide research aimed at describing and explaining why people behave as they do. Because knowledge is cumulative, social researchers continuously challenge the quality of findings and modify their research designs.

You'll recall that sociologists debate whether the discipline can or should be value free, especially in teaching (see Chapter 1). When it comes to research, they agree that the selection of a topic may be *subjective* (for example, choosing to examine immigration because of personal experiences or family background), but that the researcher should be *objective* in collecting, analyzing, and interpreting the data. That is, private beliefs, biases, and value judgments shouldn't intrude on the research process (Gray et al. 2007).

Researchers are skeptical creatures who question "commonsense" beliefs that can lead to misconceptions about society and human behavior. They are also skeptical of *psychobabble*—psychological jargon about behaving and expressing one's feelings that has

Key Topics

In this chapter, we'll explore the following topics:

scientific method the steps in the research process that include careful data collection, exact measurement, accurate recording and analysis of the findings, thoughtful interpretation of results, and, when appropriate, a generalization of the findings to a larger group.

variable a characteristic that can change in value or magnitude under different conditions.

hypothesis a statement of a relationship between two or more variables that researchers want to test.

independent variable a characteristic that determines or has an effect on the dependent variable.

dependent variable the outcome, which may be affected by the independent variable.

reliability the consistency with which the same measure produces similar results time after time.

SCIENTIFIC SOCIOLOGY

To discover patterns that explain behavior, sociologists rely on the **scientific method**, the steps in the research process that include careful data collection, exact measurement, accurate recording and analysis of the findings, thoughtful interpretation of results, and, when appropriate, a generalization of the findings to a larger group. Throughout this process, sociologists (like other social scientists) are interested in the relationships between variables.

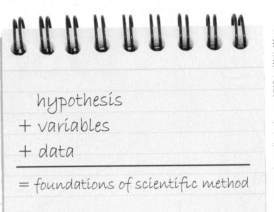

hypothesis
+ variables
+ data
—————————————
= foundations of scientific method

© Image Source/Jupiterimages / © Victor Watts/Alamy

little scientific foundation. One of the largest psychobabble growth industries is self-help books. Some of the information in these books may be sound, but much is simple-minded, biased, or generated by unlicensed, self-proclaimed "experts" who know less about people than you do. One of the best ways to defend against psychobabble is to understand the difference between personal opinions and social research.

2 The Scientific Method

hundreds of thousands of U.S. veterans get treatment for post-traumatic stress disorder (PTSD), using at least 12 different drugs and a variety of psychotherapies (such as cognitive restructuring and group therapy). According to the National Academy of Sciences (2007), however, *scientific* evidence shows that only one treatment may be effective in reducing PTSD symptoms. So, how could science help inform the treatment of U.S. veterans suffering from PTSD?

Variables and Hypotheses

A **variable** is a characteristic that can change in value or magnitude under different conditions. Variables can be attitudes, behaviors, or traits such as ethnicity, gender, and social class.

Scientists can simply ask a research question ("Have divorce rates increased or decreased?"), but they usually begin with a **hypothesis**, a statement of a relationship between two or more variables that they want to test ("Unemployment increases the risk of divorce"). In testing a hypothesis, sociologists predict a relationship between an **independent variable**, a characteristic that determines or has an effect, and the **dependent variable**, the outcome. In the previous example of a hypothesis, "unemployment" is the independent variable and "divorce" is the dependent variable.

Reliability and Validity

After asking a research question or stating one or more hypotheses, how do scientists measure the variables? Sociologists are always concerned about the reliability and validity of their measures. **Reliability** is the *consistency* with which the same measure produces similar results time after time. If, for example, you ask, "How old are you?" on two subsequent days and a respondent gives two different answers, such as 25 and 30, there's either something wrong with how you are asking the question or the respondent is lying. Respondents might lie, but scientists try to make sure that their measure is reliable.

Validity is the degree to which a measure is accurate and *really* measures what it claims to measure. Consider student course evaluations. The measures of a "good" professor often include items such as "The instructor is interesting" and "The instructor is fair." Because we don't really know what students mean by "interesting" and "fair," how accurate are such measures in differentiating between "good" and "bad" professors?

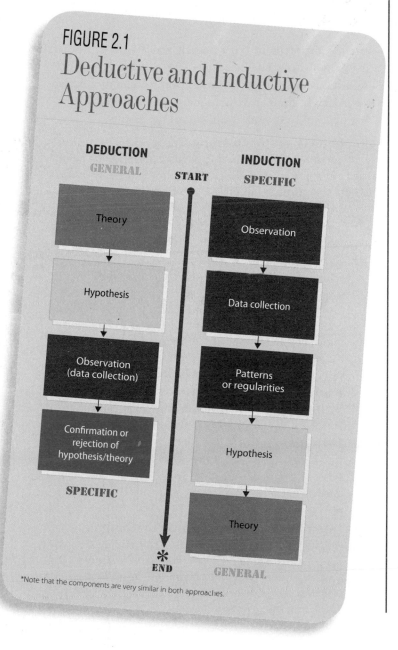

FIGURE 2.1
Deductive and Inductive Approaches

DEDUCTION
GENERAL

START

INDUCTION
SPECIFIC

Theory → Hypothesis → Observation (data collection) → Confirmation or rejection of hypothesis/theory

SPECIFIC

Observation → Data collection → Patterns or regularities → Hypothesis → Theory

END

GENERAL

*Note that the components are very similar in both approaches.

Deductive and Inductive Reasoning

Deduction and induction are two very different but equally valuable approaches for examining the relationship between variables. Stated generally, **deductive reasoning** begins with a theory, prediction, or general principle that is then tested through data collection. An alternative mode of inquiry, **inductive reasoning**, begins with a specific observation, followed by data collection and the development of a general conclusion or theory (see *Figure 2.1*).

Using a deductive approach, you might want to test a theory of academic success in which one of your hypotheses is the following: "Students who study in groups perform better on exams than those who study alone." You would collect relevant data, ultimately confirming or rejecting your hypotheses and theory. On the other hand, you might notice that your classmates who participate in study groups seem to get higher grades on exams than those who study alone. Using an inductive approach, you would collect data systematically, begin to detect patterns (or lack of them), formulate hypotheses, and end up with a theory that could be tested deductively. Most social science research involves both inductive and deductive reasoning.

Sampling

Early in the research process, sociologists decide what sampling procedures they will use. Ideally, researchers would like to study all the units of the population in which they are interested—say, all adolescents who use drugs. A **population** is any well-defined group of people (or things) about whom researchers want to know something. Obtaining information from populations is problematic, however. The population may be so large that it would be too expensive and time consuming to conduct the research. In other cases—such as that of all adolescents who use drugs—it may be impossible even to identify the population.

validity the degree to which a measure is accurate and really measures what it claims to measure.

deductive reasoning reasoning that begins with a theory, prediction, or general principle that is then tested through data collection.

inductive reasoning reasoning that begins with a specific observation, followed by data collection and the development of a general conclusion or theory.

population any well-defined group of people (or things) about whom researchers want to know something.

sample a group of people (or things) that are representative of the population researchers wish to study.

probability sample a sample for which each person (or thing, such as an e-mail address) has an equal chance of being selected because the selection is random.

nonprobability sample a sample for which little or no attempt is made to get a representative cross section of the population.

Researchers typically select a **sample**, a group of people (or things) that are representative of the population they wish to study. In obtaining a sample, researchers must decide whether to use probability or nonprobability sampling. A **probability sample** is one for which each person (or thing, such as an e-mail address) has an equal chance of being selected because the selection process is random. The most desirable feature of a probability sample is that the results can be generalized to the larger population. A **nonprobability sample** is any sample for which little or no attempt is made to get a representative cross section of the population. Instead, researchers use sampling criteria such as convenience or the availability of respondents or information.

Television news programs, newsmagazines, and entertainment shows often provide a toll-free number or a Web address and encourage viewers to vote on an issue (such as whether or not U.S. troops should leave Iraq). How representative are these voters? And how many enthusiasts skew the results by voting more than once? According to one observer, most Internet polls are "good for a few laughs" but are little more than "the latest in a long series of junk masquerading as indicators of public opinion because the participants aren't representative of everyone's opinion" (Wilt 1998: 23). But, you might think, if as many as 100,000 people respond to such a poll, isn't such a large number meaningful in reflecting how people think? No. Because the respondents are self-selected and don't comprise a random sample, they're not representative of the entire population.

THE RESEARCH PROCESS: A COOKBOOK

Hypotheses construction, establishing reliability and validity, using deductive and inductive reasoning, and sampling are some of the preliminary and often most

Are American Idol voters an example of a probability or nonprobability sample of the show's fans?

challenging steps in the research process. The process itself involves seven basic steps illustrated in *Figure 2.2,* which outlines the scientific method using an inductive approach beginning with an idea and ending with writing up (and sometimes publishing) the results.

1. **Choose a topic to study.** The topic can be general or very specific. Some sociologists begin with a new and unexplored idea; others extend or refine previous research findings. A topic can generate new information, replicate a previous study, or propose an intervention (such as implementing a new program for children in foster homes).

2. **Summarize the related research.** In what is often called a *review of the literature*, the sociologist summarizes the pertinent research, shows how her or his topic is related to previous and ongoing research, and indicates how the study will extend the body of knowledge in an area. Or, if the research is applied, the sociologist demonstrates that the proposed service or program will improve people's lives. In both cases, a summary of the related research makes clear the study's theoretical base and shows how the approach is systematic and cumulative in terms of past research.

3. **Formulate a hypothesis or ask a research question.** The sociologist next states a hypothesis or a research question. In either case, she or he needs to be sure that the measures of the variables are reliable and valid.

4. **Describe the data collection method(s).** In this step of the research process, the sociologist describes which method or combination of methods, sometimes called *methodology*, *procedure*, or *research design*, is best for testing the hypothesis or answering the research question. As you'll see shortly,

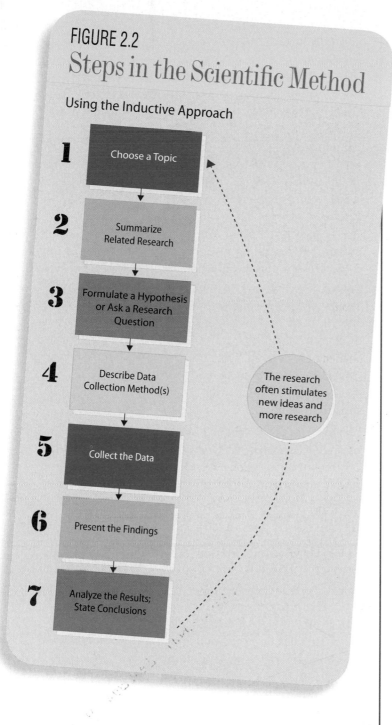

FIGURE 2.2

Steps in the Scientific Method

Using the Inductive Approach

1. Choose a Topic
2. Summarize Related Research
3. Formulate a Hypothesis or Ask a Research Question
4. Describe Data Collection Method(s)
5. Collect the Data
6. Present the Findings
7. Analyze the Results; State Conclusions

The research often stimulates new ideas and more research

each data collection technique has strengths and weaknesses that the researcher considers in deciding which is the most appropriate. The sociologist also describes the sampling technique, the sample size, and the characteristics of the respondents.

5. **Collect the data.** The actual data collection might rely on field work, surveys, experiments, or existing sources of information like Census Bureau data.

6. **Present the findings.** After coding (tabulating the results), running statistical tests, and analyzing the data, the sociologist presents the findings as clearly as possible. Because data can be interpreted in many

ways, it's up to the researcher to keep the summary of findings focused and understandable to the reader.

7. **Analyze and explain the results.** After presenting the data, the sociologist explains why these findings are important. This can be done in many ways. The researcher might show how the results provide new information, enrich our understanding of behavior or attitudes that researchers have examined previously, or refine existing theories or research approaches.

In drawing conclusions about the study, the sociologist typically discusses the study's *implications*. For instance, does a study of juvenile arrests suggest that new policies should be implemented, existing ones should be changed, or that current police practices may be affecting the results? That is, the researcher answers the question "So what?" by showing the importance and usefulness of the study.

QUALITATIVE AND QUANTITATIVE APPROACHES

The scientific method is important in both qualitative and quantitative approaches. In **qualitative research**, sociologists examine nonnumerical material that they then interpret. In a study of grandfathers who are raising their grandchildren, for example, the researcher tape-recorded in-depth interviews and then analyzed the responses to questions about issues like financial worries and daily parenting activities (Bullock 2005).

In **quantitative research**, sociologists focus on a numerical analysis of people's responses or specific characteristics, studying a wide range of attitudes, behaviors, and traits (homeowners versus renters). In one study, for example, the researchers analyzed data from a national survey of almost 7,000 respondents to examine the influence of grandparents who live with their children and grandchildren (Dunifon and Kowaleski-Jones 2007).

Which approach should a sociologist use? It depends on her or his purpose. Quantitative data can provide valuable information on characteristics such as national college graduation rates. Qualitative data, in contrast, might yield in-depth descriptions of why some college students drop out while others graduate. In many cases, sociologists use both approaches.

qualitative research research that examines nonnumerical material and interprets it.

quantitative research research that focuses on a numerical analysis of people's responses or specific characteristics.

CORRELATION AND CAUSATION

Sociologists know that some of the best predictors of divorce include marrying at a young age, having children before marriage, and experiencing domestic violence (see Chapter 13). This does not mean, however, that these and other characteristics *cause* divorce. Instead, divorce is more *likely* or *probable* in certain situations (when the marital partners are teenagers, for example).

Because nothing in life (except death) is certain, sociologists talk about *correlation*—the extent of the relationship between variables—rather than causation. They rarely use the term *cause* when interpreting their results because they cannot prove that a cause-and-effect relationship exists. Instead, sociologists "can only *suggest,* or at most *indicate*" a relationship between variables (Glenn 2001). Thus, a researcher would say that marrying at a young age is "more likely" to result in divorce, is "associated (or correlated)" with divorce, or "contributes to" rather than "causes" divorce.

3 Some Major Data Collection Methods

during the research process, sociologists typically use one or more of the following data collection techniques: surveys, secondary analysis, field research, content analysis, experiments, and evaluation research. Because each technique has benefits and limitations, a sociologist must consider which will provide the most accurate information given time and budget constraints.

SURVEYS

Researchers use **surveys** to systematically collect data from respondents using questionnaires, face-to-face or telephone interviews, or a combination of these. Questionnaires can be mailed, used during an interview, or self-administered (such as student course evaluations).

Telephone interviews are popular because they're a relatively inexpensive way to collect data. Researchers can obtain representative samples through *random-digit dialing*, which involves selecting area codes and exchanges (the next three numbers) followed by four random digits. In the procedure called *computer-assisted telephone interviewing* (CATI), the interviewer uses a computer to select random telephone numbers, reads the questions to the respondent from a computer screen, and then enters the replies in precoded spaces, saving time and expense by not having to reenter the data after the interview. Telephone surveys are becoming more challenging because of the increased use of cell phones. Because wireless carriers charge customers by the minute even for received calls, people may be less likely to complete a lengthy phone interview. Also, the Federal Communications Commission requires an interviewer to enter the number when calling a cell phone, which precludes the use of random-digit dialing (Thee 2007).

Advantages

Surveys are usually inexpensive and simple to administer and have a fast turnaround. Because the results are anonymous, respondents are generally willing to answer questions on sensitive topics such as income, sexual behavior, and the use of drugs (Hamby and Finkelhor 2001).

Face-to-face interviews have high response rates (up to 99 percent) compared with other data-collection techniques because they involve personal contact. In-depth interviews can provide rich detail about the social world and vivid descriptions of personal experiences, such as the emotional consequences of having a baby or coping with a death in the family. People are more likely to discuss such sensitive issues in an interview than via a mailed questionnaire or a phone survey (Weiss 2004).

Interviewers can also record a respondent's body language, facial expressions, and intonations, which can be useful in interpreting verbal responses. If a respondent does not understand a question or is reluctant to answer, the interviewer can clarify or probe. An observant interviewer can also gather information on variables such as social class by observing the respondent's home and neighborhood.

Disadvantages

One of the major limitations of surveys that use mailed questionnaires is a low response

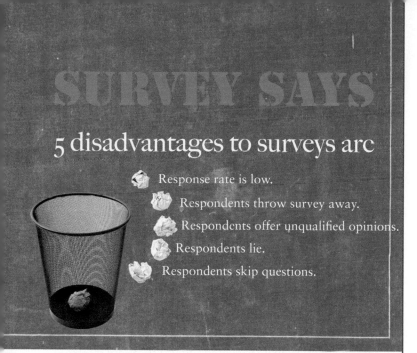

SURVEY SAYS

5 disadvantages to surveys arc

- Response rate is low.
- Respondents throw survey away.
- Respondents offer unqualified opinions.
- Respondents lie.
- Respondents skip questions.

rate, often only about 10 percent (Gray et al. 2007). If the questions are unclear, complicated, or perceived as offensive, a respondent may simply throw the questionnaire away. Also, from a third to a half of survey respondents offer opinions on subjects they know nothing about, such as fictitious legislation or nonexistent political figures (see Bishop et al. 1980).

Another problem with surveys is that people may skip or lie about questions that they feel are "too nosy." During the 2000 census, for example, 53 percent of the people who received the long form viewed questions about income as intrusive and ignored the items; 32 percent felt the same way about questions on physical or mental disabilities (Cohn 2000). If respondents lie or ignore questions, the research results will be invalid or limited (because the researcher has to exclude a key variable such as income). Also, surveys may be biased or unrepresentative of a larger population.

Unlike questionnaires and telephone surveys, face-to-face interviewing can be very expensive. Also, people may be less candid if the interviewer is of the opposite sex, and the interview environment itself can affect responses. For example, women may be less willing to admit they have been raped if questioned by men, by interviewers of different racial or ethnic backgrounds, or in the presence of family members (Herod 1993; Kane and Macaulay 1993).

Public opinion polls, common on television and in newspapers, often use surveys. Sociologist Joel Best (2001) suggests asking three basic questions about the numbers: (1) Who created these numbers? (2) Why were these numbers created? (3) How were these numbers created?

Asking a few other basic questions about a survey, like those in *Table 2.1*, will help you evaluate its credibility.

secondary analysis
examination of data that have been collected by someone else.

SECONDARY ANALYSIS

Sociologists also rely heavily on **secondary analysis**, which examines data that have been collected by someone else. The data may be historical materials (such as court proceedings), personal documents (such as letters and diaries), public records (such as state archives on births, marriages, and deaths), and official statistics (such as health information from the Centers for Disease Control and Prevention).

Advantages

In most cases, secondary analysis is convenient and inexpensive. Census Bureau data on topics such as income and employment are readily available at public college and university libraries and online. Also, many academic institutions buy tapes or CD-ROMs containing national, regional, and local data that can be used for faculty or student research or for class-related projects.

TABLE 2.1
Can I Trust These Numbers?

To determine a survey's credibility, ask:

- Who sponsored the survey? A government agency, a nonprofit partisan organization, a business, or a group that's lobbying for change?

- What is the purpose of the survey? To provide objective information, to promote an idea or a political candidate, or to get attention through sensationalism?

- How was the sample drawn? Randomly? Or was it a "self-selected opinion poll" (a SLOP, according to Tanur 1994)?

- How were the questions worded? Were they clear, objective, loaded, or biased? If the survey questions are not provided, why not?

- How did the researchers report their findings? Were they objective, or did they make value judgments?

Because secondary data are often *longitudinal* (collected at two or more points in time) rather than *cross-sectional* (collected at one point in time), researchers have the added advantage of being able to examine trends (such as voting patterns) over a number of years.

Disadvantages

Existing data sources may not have the information a researcher needs. For example, some of the statistics on remarriages and redivorces have been collected only since the early 1990s. Thus, it would be impossible for a researcher to make comparisons of these behaviors over decades. It may also be difficult to gain access to historical materials because the documents may be fragile, housed in only a few libraries in the country, or held in private collections.

If the existing sources don't include information the researcher is looking for, she or he may have to change the direction of the research or exclude some variables. If you wanted to find out why some mothers deliberately kill their infants, for example, you'd find very little national data (Meyer and Oberman 2001). Consequently, you'd have to rely on studies with small and nonrepresentative samples or collect national data yourself.

FIELD RESEARCH

In **field research**, sociologists collect data by systematically observing people in their natural surroundings. In *participant observation*, researchers interact with the people they are studying but do not reveal their identities as researchers. If you recorded interaction patterns between students and professors during your classes, you would be engaging in participant observation. In *nonparticipant observation*, researchers study phenomena without being part of the situation. For example, child psychologists, clinicians, and

Observational studies are much more complex and sophisticated than they appear to be to the general public or an inexperienced researcher.

sociologists often study young children in classrooms through one-way mirrors. (For a discussion of other variations in observational research, see Adler and Adler 1994.)

Sometimes field research can be dangerous. Sudhir Venkatesh, the son of Indian immigrants, grew up in a predominantly white suburb in Southern California. While a graduate student at the University of Chicago, he spent over six years studying the culture and respondents' of the Black Kings, a crack-selling gang in Chicago's inner city. In *Gang Leader for a Day* (2008), Venkatesh describes some of the difficulties in being a neutral observer while befriending community residents and especially J. T., the gang's leader.

Observational studies are usually highly structured and carefully designed research projects in which data are recorded and then converted to quantitative summaries. These studies may examine complex communication patterns, measure the frequency of acts (such as the number of head nods or angry statements), or note the duration of a particular behavior (such as the length of eye contact) (Stillars 1991). Thus, observational studies are much more complex and sophisticated than they appear to be to the general public or an inexperienced researcher.

Researchers sometimes combine participant and nonparticipant observation. Sociologist Elijah Anderson, for example, focused much of his research on households in West Philadelphia, an inner-city black community with high crime rates. As a nonparticipant observer, Anderson watched "decent" and "street" family members to learn how and why some poor residents take extraordinary measures to conform to mainstream values (such as maintaining a strong family life) while others engage in crime and violence (Anderson 1999). Over the years, Anderson's role of detached observer morphed into that of participant observer as he became personally involved with some of his research subjects. For example, he hired an ex-drug dealer as a part-time research assistant, found community members lawyers or jobs, encouraged his respondents to stay out of crime, and even loaned them money (Cose 1999).

Advantages

Unlike secondary analysis and most surveys, field research often provides in-depth understanding of attitudes and behavior. Observation is also more flexible than some other methods because the researcher can modify the research design, for example, by deciding to interview (rather than just observe) key people after the research has started.

Because observation doesn't disrupt the natural surroundings, the researcher doesn't directly influence the subjects. In *Nickel and Dimed*, a study based on participant observation, journalist Barbara Ehrenreich (2001) worked at several low-income jobs (such as waitressing in Florida and clerking in a Wal-Mart in Minnesota) to find out whether lower-wage American workers could actually get by on their earnings. Ehrenreich's conclusion that people can't get by on a $7-per-hour job supported what sociologists have been saying for decades. Because she is a best-selling journalist, however, Ehrenreich's project has become a well-known example of field research.

Disadvantages

If a researcher needs elaborate recording equipment, must travel far or often, or has to live in a different society or community for an extended period of time, observation can be expensive. Researchers who study other cultures must often learn a new language, a time-consuming task.

Doing field work in a country that's wracked by war may be dangerous for both researchers and their respondents. In a study of the major causes of death before and after the U.S. invasion of Iraq, for example, a team of American researchers feared the possibility of being killed or abducted. The researchers also had difficulty recruiting interviewers who were willing to travel to some of the most dangerous neighborhoods and convincing subjects to participate in a study that involved Americans (Guterman 2005).

A field researcher may also encounter barriers to collecting the desired data. Homeless and battered women's shelters, for example, are usually—and understandably—wary of researchers' intruding on their residents' privacy. Even if the researcher has access to such a group, it's often difficult to maintain objectivity while collecting and interpreting the data because of the researcher's emotional reactions such as anxiety, anger against perpetrators, and sympathy for subjects.

CONTENT ANALYSIS

content analysis data collection method that systematically examines some form of communication.

Content analysis is a data collection method that systematically examines some form of communication. This is an unobtrusive approach that a researcher can apply to virtually any form of written or oral communication: speeches, TV programs, newspaper articles, advertisements, office memos, songs, diaries, advice columns, poems, or e-mails, to mention just a few. The researcher develops categories for coding the material, sorts and analyzes the content of the material in terms of frequency, intensity, or other characteristics, and draws conclusions about the results.

As an example, look at the two greeting cards in *Figure 2.3* on the next page. What messages do they send about gender roles? Although these are only two cards, we could do a content analysis of all such cards at a particular store or of a large sample of cards from

Counting the homeless is an ongoing research problem. Communities must provide accurate numbers to qualify for federal funds, but census takers have difficulty getting such counts because safety rules prohibit them from entering private property (such as warehouses) or dark alleys. In one approach, the YWCA in Trenton, New Jersey, gave free haircuts at a fair for the homeless to attract the city's homeless people to a place where they could be counted (Jonsson, 2007). Is such field research unethical because it violates people's privacy? Or innovative in helping a community get more federal funding for the homeless?

© AP Images

a number of stores to determine whether these kinds of cards reinforce stereotypical gender role expectations.

Sociologists have used content analysis to study the content of sociology college textbooks, media coverage of heavy metal and rap music, and images of women and men in advertisements (Binder 1993; Shaw-Taylor and Benokraitis 1995; Larson and Hickman 2004). In other cases, content analysis has provided very practical information, such as identifying the most marketable sociology specialties in academic departments (Drentea and Xi 2003).

Advantages

A major advantage of content analysis is that it is usually inexpensive and sometimes less time-consuming than other data collection methods (especially surveys and field research). If, for example, you wanted to examine the content of television commercials aimed at the elderly, you wouldn't need fancy equipment, a travel budget, or a research staff.

A second advantage is that researchers can correct coding errors fairly easily by redoing the work. This is not the case with surveys. If you mail a questionnaire with poorly constructed items, it's too late. A third advantage is that content analysis is unobtrusive. Because researchers aren't dealing with human subjects, they don't need permission to do the research or worry about influencing the respondents' attitudes or behavior.

A final advantage is that content analysis often permits comparisons over time. In one study, the researchers examined the amount and intensity of violence in children's animated movies that were released between 1938 and 1999 (Yokota and Thompson 2000). Most other data collection methods wouldn't be able to analyze a phenomenon over a 62-year period.

Disadvantages

Content analysis can be very labor intensive, especially if a project is ambitious. In the research on children's animated movies, it took several years to code one or more of the major characters' words, expressions, and actions.

A related disadvantage is that the coding of material is very subjective. Having several researchers on a project can decrease the effects of subjectivity and thus increase the reliability of the analysis. Many content analyses are performed by only one researcher, however, decreasing the likelihood that the coding is objective.

Finally, content analysis often reflects social class biases. Because most books, articles, speeches, films and so forth are produced by people in upper socioeconomic levels, content analysis rarely captures the behavior or attitudes of working classes and the poor. Even when documents created by lower-class individuals or groups are available, it's difficult to determine whether or not the researchers reflect social class prejudices in their coding.

FIGURE 2.3
How Do Views of Female and Male Babies Differ?

As these cards illustrate, girls and boys are viewed differently from the time they are born. In these and other birth announcements, girls, but not boys, are typically described as "sweet" or "precious." Also, images usually show boys as active but girls as passive.

A sweet baby girl, so precious and new--

You Have a New Son!

EXPERIMENTS

Although experiments are more popular with physical scientists (such as medical researchers and psychologists) than among sociologists, they are an important data collection method. An **experiment** is a carefully controlled artificial situation that allows researchers to manipulate variables and measure the effects. In the classic experimental design, there are two equal-size groups that are very similar in characteristics such as sex, age, ethnicity or race, and education.

In the **experimental group**, the subjects are exposed to the independent variable. In the **control group**, they are not. Before the experiment, the researcher measures the dependent variable in both groups using a *pretest*. After the experimental group is exposed to the independent variable, the researcher measures both groups again using a *posttest*. If the researcher finds a difference in the measures of the dependent variable, she or he assumes that the independent variable is having an effect on the dependent variable.

Suppose, for example, that you wanted to find out whether practicing with study questions improves students' exam grades. You would provide the experimental group but not the control group with study questions before an exam. If the experimental group earned higher grades, you might conclude that, indeed, providing students with study questions for an exam improves grades.

Advantages

A major strength of experiments is that they come closer than any other data collection method to indicating a cause-and-effect relationship. Experiments are usually (but not always) less expensive and time-consuming than other data collection techniques (especially large surveys and multiyear field research). Regardless of where the research is conducted, there's often no need to purchase special equipment. A related strength is that because researchers recruit students or other volunteers (as in medical research) for their experiments, subjects are usually readily available and don't expect much, if any, monetary compensation (Babbie 2002).

A third advantage is that experiments can be replicated many times with different subjects. Such replication strengthens the researchers' confidence in the validity and reliability of the measures and the study's results. For example, doctors stopped prescribing hormone pills for menopausal women because better-designed experiments that replicated earlier studies found that the

pills increased the risk of breast cancer and strokes (Ioannidis 2005).

Disadvantages

A major disadvantage of laboratory experiments is that "the conditions so carefully controlled for the purpose of establishing causality may be so atypical or artificial that it becomes difficult to generalize from them to the outside world" (Gray et al. 2007: 280). Especially in laboratory settings, subjects know that they're being studied and might behave very differently than if they were in a natural setting (Babbie 2002).

A second drawback is that researchers must rely on volunteers, paid subjects, or a captive audience—such as college students who participate in faculty experiments. These students may resent being required to be in an experiment or might volunteer to do so (self select) to win favor with a professor.

A third disadvantage is the possibility that the conclusions drawn from experiments may not be accurate. Among other problems, attrition among the subjects may be very high, the members of experimental and control groups may communicate with each other about what's going on and behave differently as a result, and the presence of a researcher may affect the behavior of the people being studied (Cook and Campbell 1979; see, also, Chapter 6 on the Hawthorne effect).

Another limitation is that controlled laboratory settings are not suitable for studying large groups of people, the main concern of sociology. As you saw in Chapter 1, many sociologists examine macro-level issues (such as employment trends) that can't be observed in a laboratory setting.

EVALUATION RESEARCH

Evaluation research—which relies on all of the standard data collection techniques described so far—assesses the effectiveness of social programs in both the public and the private sectors. Many government and nonprofit agencies provide services that affect families and other

experiment a carefully controlled artificial situation that allows researchers to manipulate variables and measure the effects.

experimental group the group of subjects in an experiment who are exposed to the independent variable.

control group the group of subjects in an experiment who are not exposed to the independent variable.

evaluation research research that relies on all of the standard data collection techniques to assess the effectiveness of social programs in both the public and the private sectors.

groups both directly and indirectly. Examples include housing programs, programs to prevent teenage pregnancy, work-training programs, and drug-rehabilitation programs. Because local and state government budgets have been cut since the early 1980s, social service agencies have become increasingly concerned about the efficiency of their programs and often rely on evaluation research to streamline their services.

In evaluation research, sociologists use a variety of data collection methods such as examining an organization's records (secondary analysis and content analysis), using surveys to gauge employee and client satisfaction, and interviewing the staff and program recipients. Unlike the other data collection methods we've looked at, evaluation research is *applied* because it compares a program's achievements with its goals (Weiss 1998). The research findings are generally used to improve the operations of the program—changes that affect both the sponsoring agency and the people the program helps.

Advantages

Evaluation research is important because it examines actual efforts to deal with social problems. If the researchers rely on secondary analysis, the data collection costs are usually low. Also, the findings can be very valuable to program directors or agency heads in showing discrepancies between the original objectives and the program's actual functioning (Card et al. 1994).

Disadvantages

Evaluation research can be frustrating because politics plays an important role in determining what is evaluated, for whom the research is done, and how the results are used (or misused). Even though agency directors typically solicit evaluation research, they are not happy if the findings show that the program is excluding the neediest groups, that administrators are wasting money, or that caseworkers are making serious mistakes.

Also, practitioners rarely welcome the results of evaluation research if a sociologist concludes that a particular program isn't working. For example, since its inception in 1983, the DARE (Drug Abuse Resis-

I n a well-known 1973 experiment, psychologist Philip Zimbardo created a mock prison, using 21 undergraduate volunteers as prisoners and guards. The experiment was stopped after six days because some of the participants experienced intense negative reactions, and the study raised several ethical questions. For more information about this study, go to **Stanford Prison Experiment**, at www .prisonexp.org.

tance Education) program—which has relied on trained volunteers from local police departments who address grade-school children on the dangers of drug use—has been popular with schools, police departments, parents, and politicians across the country. When social scientists evaluated DARE in 1994, 1998, and 1999, however, they found that there was no significant difference in the drug use of students who had completed the DARE curriculum and those who hadn't done so. When the DARE funding was threatened, its promoters attacked the researchers' evaluations as "voodoo science" (Miller 2001).

RECAP OF RESEARCH APPROACHES

In summary, researchers have to weigh the benefits and limitations of each research approach in designing their studies (see *Table 2.2*). Often, they use a combination of methods because "many sociologists view the social world as a multi-faceted and multi-layered reality that reveals itself only in part with any single method" (Jacobs 2005: 4). For example, sociologists Brian Uzzi and Ryon Lancaster (2004), using qualitative approaches, conducted in-depth interviews with a handful of lawyers and clients and used content analysis to examine those results before embarking on a national survey of the social ties and pricing patterns in large U.S. law firms. Despite a researcher's commitment to objectivity, ethical debates and politically charged disagreements can influence sociological research.

© Richard Ross/Stone/Getty Images

4 Ethics, Politics, and Sociological Research

i n a scene in the movie *Kinsey*, government agents seize a box of materials being shipped by Dr. Alfred C. Kinsey, the pioneering sex researcher, and confiscate the contents because they are "obscene." Have such attitudes toward sex research changed? And, if so, what other ethical and political dilemmas do sociologists (and other social scientists) face in conducting their research?

ETHICAL RESEARCH

Because so much research relies on human subjects, the federal government and many professional organizations have devised codes of ethics to protect the participants. Common to most of these codes—regardless of the discipline or the type of research method used—are several minimal rules for professional conduct (see *Table 2.3* on page 37).

Some data collection methods are more susceptible to ethical violations than others. Surveys, secondary analysis, and content analysis are less vulnerable than field research and experiments because the researchers typically don't interact directly with subjects, interpret their behavior, or become personally involved with the respondents. In contrast, experiments and field research can raise ethical questions.

POLITICAL, RELIGIOUS, AND COMMUNITY PRESSURE

The legitimacy of social science research is especially likely to be challenged when studies focus on sensitive social, moral, and political issues. Research on teenage sexual behavior is valuable because it provides information that public health agencies and schools can circulate about sexually transmitted diseases, such as HIV, and contraception. Nonetheless, many local jurisdictions have refused to let social scientists study adolescent sexual behavior. Some parents feel that such research violates student privacy and might make a school district look bad (if a study reports a high incidence of drug use or sexual activity, for example). Moreover, some religious groups, school administrators, and politicians have opposed such studies because they feel the studies undermine traditional family values or make deviant behavior look normal. As a result, many social scientists don't do research on some of the most controversial issues because there is no funding or because a study would be opposed by a variety of groups that get considerable media attention (Carey 2004; Kempner et al. 2005).

Policy makers also play a major role in ignoring or suppressing research findings. In 2004, a team of social scientists who conducted house-to-house interviews in Iraq concluded that at least 100,000 civilians had been killed since the invasion by the United States in 2003. The deaths, especially of women and children, were due to air strikes, gunshot wounds, and a lack of sanitation and medical care for civilians with heart disease and other problems (Roberts et al. 2004). The Bush administration, the Department of Defense, politicians, and major U.S. newspapers ignored the study. A year later, President Bush announced that the war in Iraq had claimed the lives of 30,000 Iraqi citizens "more or less" (Guterman 2005; Vieth 2005).

Local and national advocacy groups also have a vested interest in ignoring research studies that might reduce their funding. For example, the National Alliance for the Mentally Ill embraces claims on the prevalence of mental disorders that are based on flawed

TABLE 2.2
Some Data Collection Methods in Sociological Research

METHOD	EXAMPLE	ADVANTAGES	DISADVANTAGES
Surveys	Sending questionnaires and/or interviewing students on why they succeeded in college or dropped out	Questionnaires are fairly inexpensive and simple to administer; interviews have high response rates; findings are often generalizable	Mailed questionnaires may have low response rates; respondents may be self-selected; interviews are usually expensive
Secondary analysis	Using data from the National Center for Education Statistics (or similar organizations) to examine why students drop out of college	Usually accessible, convenient and inexpensive; often longitudinal and historical	Information may be incomplete; some documents may be inaccessible; some data can't be collected over time
Field research	Observing first-year college students with high and low grade-point averages (GPAs) regarding their classroom participation and other activities	Flexible; offers deeper understanding of social behavior; usually inexpensive	Difficult to quantify and to maintain observer/subject boundaries; the observer may be biased or judgmental; findings are not generalizable
Content analysis	Comparing the transcripts of college graduates and dropouts on variables such as gender, race/ethnicity, and social class	Usually inexpensive; can recode errors easily; unobtrusive; permits comparisons over time	Can be labor-intensive; coding is often subjective (and may be distorted); may reflect social class biases
Experiments	Providing tutors to some students with low GPAs to find out if such resources increase college graduation rates	Usually inexpensive; plentiful supply of subjects; can be replicated	Volunteers and paid subjects aren't representative of a larger population; the laboratory setting is artificial
Evaluation research	Examining student records; interviewing administrators, faculty, and students; observing students in a variety of settings (such as classroom and extracurricular activities); and using surveys to determine students' employment and family responsibilities.	Usually inexpensive; valuable in real-life applications	Often political; findings might be rejected

community studies and that contend that half of Americans suffer from a mental disorder at some point, especially depression. In fact, better designed studies show that about 75 percent of depressive symptoms are normal and temporary (such as worrying about an upcoming exam or grieving the death of a family member or friend). If mental health organizations can convince politicians that mental illness is widespread, they can get more funding for mental health services. And pharmaceutical companies are eager to sell antidepressant drugs to alleviate depression. As a result, many mental health organizations and pharmaceutical companies benefit by labeling even normal depression as an "overwhelming problem" (Horwitz and Wakefield 2006: 23).

HOW COMMON IS SCIENTIFIC MISCONDUCT?

In a recent survey of almost 3,300 biomedical scientists, just 0.3 percent said that they had faked research results and 1.4 percent admitted to plagiarism. About a third, however, acknowledged that they had engaged in unethical behaviors such as not presenting data that contradicted their findings, violating some rules on protecting human subjects, cutting corners to complete a project, and changing the research design or study results to please the group that sponsored the research. The main reason given for the unethical behavior was the increasing pressure to publish, to win grants, and to earn promotions (Mello et al. 2005; De Vries et al. 2006).

One of the most serious ethical violations is using corporate funds to study a problem that the corporation or industry has created. A team of researchers at a prestigious Ivy League medical school concluded that 80 percent of lung cancer deaths could be prevented through the widespread use of computed tomography (CT) scans, implying that cigarettes are not as dangerous as physicians claim. It turned out that the $3.6 million CT study had been funded by the Foundation for Lung Cancer, which is underwritten by cigarette makers. The foundation claimed that it "had no control or influence over the research," but the editor in chief of the medical journal said that he would have never published the study if he had realized that the research had been sponsored by the tobacco industry (Henschke et al. 2006; Harris 2008).

Sociologists seem to be less likely to engage in unethical behavior than medical scientists. One of the reasons may be that sociological research is rarely supported by corporations (like tobacco and pharmaceutical compa-

nies) who have vested interests in getting the findings they want. Still, doing research on topics that politicians don't like may jeopardize a scholar's employment. For example, a tenured professor at a large state university came close to being fired because the state's governor disapproved of his research results, which showed that abstinence education doesn't decrease teenagers' sexual activity. In another case, a highly respected sociologist was uninvited to write a chapter on international migration for the United Nations because the White House feared that the research might present U.S. policies in a negative light (Bailey et al. 2002; Massey 2006).

TABLE 2.3
Some Basic Principles of Ethical Sociological Research

The American Sociological Association publishes ethical codes and guidelines for researchers. The key elements of these codes tell researchers that they …

- Must obtain all subjects' consent to participate in the research and their permission to quote from their responses and comments;
- May not exploit subjects or research assistants involved in the research for personal gain;
- Must never harm, humiliate, abuse, or coerce the participants in their studies, either physically or psychologically;
- Must honor all guarantees to participants of privacy, anonymity, and confidentiality;
- Must use the highest methodological standards and be as accurate as possible;
- Must describe the limitations and shortcomings of the research in their published reports;
- Must identify the sponsors who funded the research; and
- Must acknowledge the contributions of research assistants (usually underpaid and overworked graduate students) who participate in the research project.

Even people who pride

themselves on their individualism adhere to many cultural expectations.

what do you think?

Even though people say beauty is in the eye of the beholder, cultures have generally-accepted ideas of beauty.

1 2 3 4 5 6 7

strongly agree   strongly disagree

3 Culture

1 Culture and Society

Once when I returned a set of exams, a student who was unhappy with his grade blurted out an obscenity. A voice from the back of the classroom snapped, "You ain't got no culture, man!" The remark implied that refined people don't curse and that proper classroom behavior doesn't include using vulgar language.

As popularly used, *culture* often refers to appreciating the finer things in life, such as Shakespeare's sonnets, gourmet dining, and the opera. In contrast, sociologists use the term in a much broader sense: That is, **culture** refers to the learned and shared behaviors, beliefs, attitudes, values, and material objects that characterize a particular group or society. Thus, culture determines a people's total way of life.

Most human behavior is not random or haphazard. Among other things, culture influences what you eat; how you were raised and will raise your own children; if, when, and whom you'll marry; how you make and spend money; and what you read. Even people who pride themselves on their individualism adhere to many cultural expectations. The next time you're in class, for example, count how many students are *not* wearing jeans, T-shirts, sweatshirts, or sneakers—clothes that are the prevalent uniform for both college students and many adults in U.S. society.

A **society** is a group of people that has lived and worked together long enough to become an organized population and to think of themselves as a social unit (Linton 1936). Every society has a culture that guides people's interactions and behaviors. Society and culture are mutually dependent; neither can exist without the other. Because of this interdependence, social scientists sometimes use the terms *culture* and *society* interchangeably.

> **culture** the learned and shared behaviors, beliefs, attitudes, values, and material objects that characterize a particular group or society.
>
> **society** a group of people that has lived and worked together long enough to become an organized population and to think of themselves as a social unit.

Key Topics

In this chapter, we'll explore the following topics:

1 Culture and Society

2 The Building Blocks of Culture

3 Cultural Similarities

4 Cultural Diversity

5 Popular Culture

6 Cultural Change and Technology

7 Sociological Perspectives on Culture

SOME CHARACTERISTICS OF CULTURE

All human societies, despite their diversity, share some cultural characteristics and functions (Murdock 1940). We don't see culture directly, but it shapes our attitudes and behaviors.

1. **Culture is learned.** Culture is not innate but learned, and it shapes how we think, feel, and behave. If a child is born in one region of the world but raised in another, she or he will learn the customs, attitudes, and beliefs of the adopted culture.

2. **Culture is transmitted from one generation to the next.** We learn many customs, habits, and attitudes informally through interactions with parents, relatives, and friends and from the media. We also learn culture formally in settings such as schools, workplaces, and community organizations. Whether our learning is formal or informal, we don't have to reinvent the wheel through a process of trial and error. Instead, and because each generation transmits cultural information to the next one, culture is cumulative.

3. **Culture is shared.** Culture brings members of a society together. We have a sense of belonging because we share similar beliefs, values, and attitudes about what is right and wrong. Imagine the chaos if we did what we wanted (such as physically assaulting an annoying neighbor) or if we couldn't make numerous daily assumptions about other people's behavior (such as coming to work every day).

4. **Culture is adaptive and always changing.** Culture changes over time. New generations discard technological aspects of culture that are no longer practical, such as replacing typewriters with personal computers. Attitudes can also change over time. Compared with several generations ago, for example, many Americans now feel that premarital sex is acceptable or at least not wrong (see *Figure 3.1*).

To survive, people have to adapt to their surroundings. A few weeks after the 9/11 attacks, for example, President George Bush signed a new law—the USA PATRIOT Act (Uniting and Strengthening America by Providing Appropriate Tools to Intercept and Obstruct Terrorism). The purpose of the act was to make the country more secure by allowing government officials—especially the National Security Agency (NSA)—to track suspected terrorists by tapping phone lines, monitoring e-mail, inspecting bank accounts, and checking visits to Internet sites.

In late 2005, the *New York Times* reported that the NSA had been conducting electronic surveillance on millions of Americans without getting the necessary warrants. President Bush defended this monitoring of Americans' telephone and e-mail communications as

FIGURE 3.1
Is Premarital Sex Wrong?

Every year since 1969, the Gallup Institute has asked Americans the following question: "Do you think it is wrong for a man and a woman to have sexual relations before marriage, or not?" Here's how attitudes have changed over the years.

Source: Based on Saad 2001 and 2007.

Y-axis: Yes, premarital intercourse is wrong (percentage of respondents)

Year	1969	1987	1996	2007
Percentage	68	46	40	38

vital for preventing terrorist attacks. Critics charged that such eavesdropping was illegal and unconstitutional because it had been carried out without congressional approval or court oversight and violated citizens' civil rights ("Bush defends . . . " 2006; VandeHei and Eggen 2006). However, only a small number of Americans protested the surveillance because, among other reasons, most of us rarely question the government. Instead, we adapt to cultural changes that might include eavesdropping on our everyday communications (see "Domestic Spying . . . " 2006).

© Banana Stock/Jupiterimages

MATERIAL AND NONMATERIAL CULTURE

Culture is both material and nonmaterial (Ogburn 1922). **Material culture** consists of the tangible objects that members of a society make, use, and share. These creations include such diverse products as buildings, tools, music, weapons, jewelry, religious objects, and cell phones.

Nonmaterial culture includes the shared set of meanings that people in a society use to interpret and understand the world. Symbols, values, beliefs, sanctions, customs, and rules of behavior are elements of nonmaterial culture. (We'll examine these building blocks of culture shortly.) Some sociologists study the material aspects of culture, such as technology, but most are interested in nonmaterial culture, such as communication patterns, political opinions, and attitudes about spanking.

Material and nonmaterial culture influence each other. The automobile, for example, has changed every society that has adopted it. Among other things, cars have provided privacy during courting and dating, transported passengers relatively inexpensively, generated new laws (such as those requiring using seat belts), and created environmental problems.

2 The Building Blocks of Culture

those who are slender and attractive tend to make more money and to get promoted more often than those who are overweight or have average looks. For example:

- Good-looking people, especially men, earn 5 percent more per hour than average-looking people.
- Obese women, especially white women, earn 17 percent less than women of average weight.
- College students who view their instructors, especially their male professors, as good looking give them better course evaluations, which, in turn, generates more economic rewards, such as merit increases and promotions, for those professors (Hamermesh and Parker 2003; Engemann and Owyang 2005).

Such findings contradict the proverb "Beauty is in the eye of the beholder" (that someone or something is beautiful if the viewer perceives it so). How can we explain this contradiction? Why do most Americans agree on who is beautiful, average, or unattractive? Are the less attractive people victims of bias? Or do good-looking people develop self-confidence and social skills that enhance their economic opportunities?

To answer such questions, we must understand the building blocks of culture, especially symbols, language, values, and norms.

SYMBOLS

A **symbol** is anything that stands for something else and has a particular meaning for people who share a culture. In most societies, for example, a handshake represents friendship or courtesy, a wedding ring sends the message that a person isn't a potential dating partner, and a siren denotes an emergency. People influence each other through the use of symbols. A smile and a frown communicate different information and elicit different responses. Through symbols, we engage in *symbolic interaction* (see Chapter 1).

Symbols Take Many Forms

Written words are the most common symbols, but we also communicate by tattooing our bodies, getting breast implants, and purchasing goods and services that we feel might increase our social status. And gestures (such as raised fists, hugs, and stares) convey important messages about other people's feelings and attitudes.

Symbols Distinguish One Culture from Another

In Islamic societies, but not others, many women wear a veil to hide part of the face (from just below the eyes), or even cover the entire body (a *chador* or *burqa*). The veiling symbolizes religious beliefs, modesty, women's dress codes, and the segregation of women and men (see Chapter 8). Veiling also provides safety during crises. During the war in Iraq, for example, most women started wearing *burqas* because they feared being singled out and followed, kidnapped, or shot by insurgents (Spinner 2004).

Symbols Affect Cross-Cultural Views

Most Americans buy merchandise manufactured in China, Taiwan, Vietnam, or other countries without a second thought. In some countries, however, U.S. products such as Coca-Cola, Pepsi, hamburgers, and KFC signify American consumerism and cultural dominance, and are viewed as parasites that infiltrate and replace local beverages and foods, businesses, and traditional cultural values (Calvert 2001).

Symbols Can Unify or Divide a Society

Symbols can unify a country. Every Fourth of July, many Americans celebrate the anniversary of gaining independence from Britain with parades, firecrackers, barbecues, and numerous speeches by local and national politicians. All of this symbolic behavior signifies freedom and democracy, even though few Americans know the history of the struggle for independence. Many immigrants purposely choose July 4 as the date on which to be naturalized because it represents "the land of the free."

Symbols can also be divisive. For example, some white southerners fly the Confederate flag because they see it as a proud emblem of their Southern heritage. Others have abandoned the flag because it is used by racist groups, like the Ku Klux Klan, to symbolize slavery and white domination of African Americans.

Symbols Can Change Over Time

Symbols communicate information that varies across societies and may change over time. In 1986, for example, the International Red Cross changed its name to the International Red Cross and the Red Crescent Movement to encompass a number of Arab branches, and adopted a crescent emblem in addition to the well-known cross. Israel wanted to use a red Star of David, rejecting the cross as a Christian symbol and the crescent as an Islamic one. Red Cross officials offered a red diamond as the new shape, but some countries rejected the diamond because it represents bloody conflicts in many African countries that mine diamonds (Whitelaw 2000). The red cross and crescent continue to be the organization's emblems until the issue is resolved.

MAKE LOVE NOT WAR

S P O T

Left: The two emblems of the International Movement of the Red Cross and the Red Crescent. Right: The proposed red diamond emblem.

LANGUAGE

The 1994 Oregon Death with Dignity law allows physician-assisted suicide for terminally ill patients whose suffering is unbearable, who lose bodily functions, and who want control over their death. California, the only other state considering similar legislation, like Oregon, never mentions suicide in the bill. The choice of words makes a big difference in how people view doctor-assisted suicide. In a recent Gallup survey, 75 percent of Americans agreed that doctors should be allowed by law to "end the lives" of patients if a patient and his or her family request it. But when the question was worded to say that doctors should be permitted to "assist the patient to commit suicide," only 58 percent of the respondents agreed. Because of the stigma attached to suicide, we often use indirect language, like "aid in dying," "choice in dying," and "end-of-life options" (Nunberg 2007). Thus, language is important.

Language is a system of shared symbols that enables people to communicate with one another. In every society, children begin to grasp the essential structure of their language at a very early age and without any instruction. Babbling leads, rapidly, to uttering words and combinations of words. The average child knows approximately 900 words by the age of 2 and 8,000 words by age 6 (Hetherington et al. 2006).

Why Language Is Important

Language makes us human: It helps us interact with others, communicate information, express our feelings, and influence other people's attitudes and behavior. Language directs our thinking, controls our actions, shapes our expression of emotions, and gives us a sense of belonging to a group.

Language can also provoke anger and conflict. Recognizing this connection, two high schools in Connecticut started to fine students, up to $103, for cursing. The schools' officials hoped to decrease the fights that erupted when students used obscenities and vulgar language (Llana 2005). Language, thought, and behavior are indeed interrelated, which is especially evident when we consider gender and ethnicity.

> **language** a system of shared symbols that enables people to communicate with one another.

Language and Gender

Language has a profound influence on how we perceive and act toward women and men. Those who adhere to traditional usage of language contend that nouns such as *businessman, chairman, mailman,* and *mankind,* and pronouns such as *he* refer to both women and men and that women who object to such usage are too sensitive. Suppose, however, that all of your professors used only *she, her,* and *women* when they were referring to all people. Would the men in the class feel excluded?

Sometimes social problems are discussed in terms that refer to women, not men. For example, we often hear how many teens become pregnant each year but rarely how many boys or men impregnate teenage girls. Such language sends the message that girls, not boys or men, are the offenders.

Language, Race, and Ethnicity

Words—written and spoken—create and reinforce both positive and negative images about race and ethnicity. Someone might receive a *black mark,* and a *white lie* isn't really a lie. We *blackball* someone, *blacken* someone's reputation, view *blackguards* (villains) with dislike, and prosecute people who deal in the *black market.* In contrast, a *white knight* rescues people in distress, the *white hope* brings glory to a group, and the good guys wear *white hats.*

Racist or ethnic slurs, labels, and stereotypes demean and stigmatize people. Derogatory ethnic words abound: *honky, hebe, kike, spic, chink, jap, polack, wetback*, and many others. Common stereotypes hold that blacks are shiftless, Asians are uncommunicative, Irishmen are alcoholics, Jews are greedy, Latinos are lazy, and, most recently, that Arabs are terrorists. When black ministers denounce "the white man" or white politicians and others "accidentally" use racial slurs, they reinforce stereotypes and fan intergroup hostility (Attinasi 1994).

Self-ascribed racial epithets are as harmful as those imposed by outsiders. When Italians refer to themselves as *dagos* or African-Americans call each other *nigger*, they tacitly accept stereotypes about themselves and legitimize the general usage of such derogatory ethnic labels (Attinasi 1994).

Language and Social Change

Language is dynamic and changes over time. U.S. English is comprised of hundreds of thousands of words borrowed from many countries and from groups that were in the Americas before the colonists arrived (Carney 1997). *Table 3.1* provides a few examples of English words borrowed from other languages.

In response to cultural and technological changes, the *Oxford English Dictionary* now includes *hottie, daytrading, digital divide, e-book, baguette, chowhound, tweenager,* and *wazoo* among other new words ("New Words . . ." 2001). To some people's dismay and others' delight, some writers are now substituting *partner* for the traditional *spouse, wife,* or *husband.* Not long ago, we had plain old mail. Now we have *snail mail, e-mail, voicemail, fax mail,* and mail on our cell phones. Thus, as culture changes, so does its language.

VALUES

Values are the standards by which members of a particular culture define what is good or bad, moral or immoral, proper or improper, desirable or undesirable, beautiful or ugly. They are widely shared within a society

and provide *general guidelines* for everyday behavior rather than specific rules that apply to concrete situations. For example, when faculty catch students plagiarizing, students often plead innocence and blame the instructor ("*You* never told us exactly what *you* mean by plagiarism").

Major U.S. Values

Sociologist Robin Williams (1970: 452–500) has identified a number of core U.S. values. All are central to the American way of life because they are widespread, have endured over time, and reflect many people's intense feelings.

1. **Achievement and success.** U.S. culture stresses personal achievement, especially occupational success. Many Americans are captivated by people—including celebrities and affluent athletes—who flaunt their wealth and status.

2. **Activity and work.** Americans often seem to be in a hurry and want to "make things happen." They respect people who are focused and disciplined in their jobs and assume that hard work will be rewarded. Journalists and others often praise those who work past their retirement age.

3. **Morality.** The "typical" American thinks in terms of right or wrong, good or bad, ethical or unethical and focuses on how people *ought* to behave. When President Bush was campaigning in 2000, for instance, he emphasized the importance of creating "communities of character," where children are taught to value family, friendship, and respect for neighbors and other people's religious views (Bruni 2001).

TABLE 3.1
U.S. English Is a Mixed Salad

☐ Africa: apartheid, Kwanzaa, safari	☐ Spain: anchovy, bizarre
☐ Alaska and Siberia: husky, igloo, kayak	☐ Thailand: Siamese
☐ Bangladesh: bungalow, dinghy	☐ Turkey: baklava, caviar, kebob
☐ Hungary: coach, goulash, paprika	☐ France: bacon, police, ballet
☐ India: bandanna, cheetah, shampoo	☐ Japan: geisha, judo, sushi
☐ Iran and Afghanistan: bazaar, caravan, tiger	☐ Norway: iceberg, rig, walrus
☐ Israel: kosher, rabbi, Sabbath	☐ Mexico: avocado, chocolate, coyote
☐ Italy: fresco, spaghetti, piano	☐ Germany: strudel, vitamin, sauerkraut, kindergarten

To save the native tongue—Irish Gaelic—parts of Ireland's Galway County don't translate street signs into English (the majority language). Potential home buyers must submit to a rigorous oral test to see if they can speak Irish Gaelic because only fluent speakers are welcomed (O'Neill 2007). Is this discrimination? Or is it a legitimate way to protect a linguistic heritage?

4. **Humanitarianism.** U.S. society emphasizes concern for others, helpfulness, personal kindness, and offering comfort and support. During natural disasters—such as earthquakes, floods, fires, and famines—at home and abroad, many Americans are enormously generous, as when they contributed millions of dollars within a few days after Hurricanes Katrina and Rita in 2005.

5. **Efficiency and practicality.** Americans emphasize technological innovation, up-to-dateness, practicality, and getting things done. They are offended if described as "backward," "inefficient," "wasting time," or "useless." Many American colleges, in fact, now tout their programs or courses as being practical and useful instead of emphasizing knowledge and intellectual growth.

6. **Progress.** Americans focus on the future rather than the present or the past. The next time you walk down the aisle of a grocery store, note how many products are "new," "improved," and "better than ever." If Americans don't keep up with technological progress, they believe people in other countries will view them as "outmoded," "stagnant," and "computer-challenged." Many societies respect their elderly as founts of knowledge, but Americans tend to dismiss them as "old-fashioned" and "behind the times."

7. **Material comfort.** Americans consider it normal to want new products and services and to consume them heavily. Many work hard to pay for fancy new cars, large homes, and dream vacations (even if they can't afford them). Americans never have enough gadgets and always need more stuff. When the stuff accumulates, they buy stuff to stuff it into.

8. **Equality.** U.S. laws tell Americans that they all have been "created equal" and have the same legal rights, regardless of race, ethnicity, sex, religion, disability, age, or social class. Americans believe that if they work hard, apply themselves, and save money, they will be successful in the future

9. **Freedom.** Countless documents affirm freedom of speech, freedom of the press, and freedom of worship in the United States. Beginning with the colonists, immigrants have been drawn to this country to enjoy freedom from political persecution, economic problems, and religious intolerance.

10. **Conformity.** Individuals don't want to be seen as "strange," "peculiar," or "different." They conform because they want to be accepted, to get social approval from those they respect, and to be hired or promoted. Striving for success often means controlling one's impulses and biting one's tongue.

11. **Nationalism and patriotism.** Nationalism demands total and unquestioning allegiance to the country's symbols, rituals, and slogans. Any deviation is scorned as "un-American." Patriotism requires loyalty to institutions and symbols because they represent values such as democracy and respect for individuals.

12. **Democracy.** Democracy provides the average U.S. citizen with equal political rights and distributes power and decision making across several bodies (legislative, executive, and judicial branches). Democracy emphasizes equality, freedom, and faith in the people rather than giving power to a monarch, dictator, or emperor.

13. **Individuality.** American culture sets a high value on the development of each person. Thus we try to raise children to be responsible and self-respecting but also independent, creative, self-directed, self-motivated, and spontaneous.

Williams himself acknowledged troublesome patterns in the list because these core American values, as you may have noticed, don't always mesh with each other. For example, Americans value equality but are comfortable with enormous gaps in wealth and power and continue to discriminate against people because of their ethnicity, race, sexual orientation, gender, or age. We proclaim that we respect individuality but make hurtful comments about fat people and admit we don't like them (Mulvihill 2001). And we say we value responsibility but often blame television rather than parenting for children's bad behavior.

Values Are Emotion-Laden

Most of us are passionate about our values, which arouse strong emotions. Consider the nationwide outburst of patriotic behavior in the United States after the 9/11 attacks. When values such as democracy are threatened or assaulted, many people rally to protect them.

People who oppose widely shared cultural values, such as the freedom of expression, can be as fervent as those who defend them. Over the years, for example, small groups of parents and others have denounced books that they find objectionable (like *Heather Has Two Mommies* and *The Color Purple*) and sometimes succeed in removing them from libraries and schools (American Library Association 2007).

Values Vary Across Cultures and Change Over Time

As you'll see shortly, cultural values can change as a result of technological advances, the influx of immigration, and contact with outsiders. For example, the Japanese parliament recently passed a law making love of country a compulsory part of school curricula. The lawmakers and their numerous supporters hope that teaching patriotism will counteract the American-style emphasis on individualism and self-expression that they believe has undermined Japanese values of cooperation, self-discipline, responsibility, and respect for others (Wallace 2006).

In the United States, surveys of first-year college students show a shift in values. Between 1968 and 2006, for example, developing a meaningful philosophy of life plummeted in importance, while being rich became substantially more valued (Pryor et al. 2007). Consistent with our humanitarian values, however, a majority (70 percent) of the students surveyed in 2007 said that "helping others who are in difficulty" is essential or very important, but women were more likely to feel this way (see *Figure 3.2*).

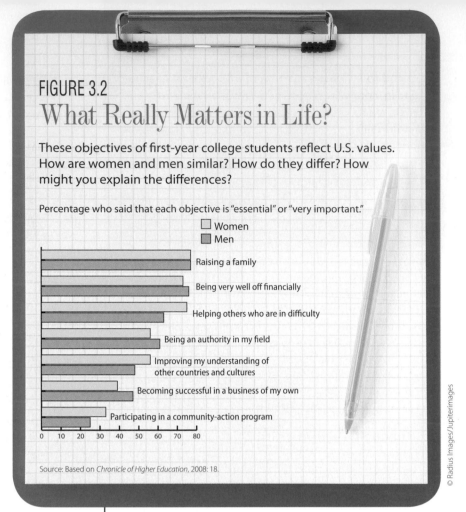

FIGURE 3.2
What Really Matters in Life?

These objectives of first-year college students reflect U.S. values. How are women and men similar? How do they differ? How might you explain the differences?

Percentage who said that each objective is "essential" or "very important."

☐ Women
■ Men

- Raising a family
- Being very well off financially
- Helping others who are in difficulty
- Being an authority in my field
- Improving my understanding of other countries and cultures
- Becoming successful in a business of my own
- Participating in a community-action program

0 10 20 30 40 50 60 70 80

Source: Based on *Chronicle of Higher Education*, 2008: 18.

NORMS

Values express general goals and broad guidelines for daily living, but **norms** are a society's specific rules of right and wrong behavior. Norms tell us what we *should, ought* and *must* do as well as what we *should not, ought not* and *must not* do: don't talk in church, stand in line, and so on. Norms are not universally applied to all groups, however. For example, a female professor's students sometimes comment on her clothes in their course evaluations and want her to "look like a professor." In contrast, her husband, also a professor, "has yet to hear a single student comment about his wardrobe." So, she changes her outfits every day and wears business-like clothes while her husband usually wears khaki pants (or jeans) and a polo or button-down shirt day in and day out (Johnston 2005). Thus, "her" and "his" norms differ.

Norms reflect values and, thus, are expectations shared by the members of the society at large or by the members of particular groups within a society. In the United States, where individualism is a basic value, young adults are expected to move out of their parents' home and to become independent and self-sufficient. In China, in contrast, where communal responsibility is a basic value, several generations live under the same roof, and children are expected to care for their aging parents.

Here are some general characteristics of norms:

- Most are *unwritten,* passed down orally from generation to generation (using the good dishes and tablecloth on special occasions).

- They are *instrumental* because they serve a specific purpose (getting rid of garbage that attracts roaches and rats).

- Some are *explicit* (save your money "for a rainy day") while others are *implicit* (being respectful during a wake or funeral).

- They *change* over time (it's now more acceptable than in the past to have a child out of wedlock but much less acceptable to smoke or to be overweight).

- Most are *conditional* because they apply in specific situations (slipping out of your smelly shoes may be fine at home but not on an airplane).

- Because they are situational, norms can be *rigid* ("You *must* turn in a term paper") or *flexible* ("You can have another week to finish the paper").

> **folkways** norms that members of a society (or a group within a society) look upon as not being critical and that may be broken without severe punishment.

Norms organize and regulate our behavior. We may not like many of these rules but they make our everyday lives more orderly and predictable. Imagine, for example, if your professor said, "This class meets from 2:00 to 5:00 on Tuesdays. If I'm a few hours late, just wait for me."

Sociologists differentiate three types of norms—folkways, mores, and laws—that vary because some rules are more important than others. As a result, a society punishes some wrongdoers more severely than others.

Folkways

About 69 percent of Americans feel that people are ruder than they were 20 or 30 years ago, citing in particular driving aggressively, using offensive language in public, and talking loudly on cell phones ("American Manners Poll" 2005). Rudeness violates **folkways,** norms that members of a society (or a group within a society) look upon as not being critical and that may be broken without severe punishment. Etiquette rules are good examples of folkways: Cover your mouth when you sneeze, say "please" and "thank you," and knock before entering someone's office.

We follow a variety of everyday conventions and customs unconsciously because we've internalized them since birth. We often don't realize that we conform to norms until someone violates one, such as picking one's nose in public or eating ice cream from the carton and then putting it back in the freezer at a friend's or relative's house.

Folkways vary from one country to another. Punctuality is important in many European countries but not in much of Latin America and the Mediterranean. Japan is notable for its gift giving among businesspeople, but this may be viewed as bribery in

At the 60th anniversary of the liberation of Auschwitz in 2005, Vice President Dick Cheney represented the United States. The other dignitaries wore formal overcoats, "gentlemen's" hats, and dress shoes. The vice president was dressed "in the kind of attire one typically wears to operate a snow blower:" a drab olive-green parka with a fur-trimmed hood, a knit cap, and hiking boots (Givhan 2005: C1). What does Mr. Cheney's attire say about how predominant casual dress has become in U.S. culture?

© AP Images

Western countries. Austrians, as well as Germans and the Swiss, consider chewing gum in public vulgar. There are also many differences in table manners: Europeans keep their hands above the table at all times, not in their laps; Koreans frown on sniffling or blowing one's nose at the table; and interrupting a speaker at the dinner table is a sign of disrespect in Turkey (Axtell et al. 1997).

Most folkways are firm ("You must be at work at 9:00 a.m."), but others are still evolving. Even before the Internet became widely accessible, for example, users started developing *netiquette,* do's and don'ts for online communication. These rules included the following: Don't write things in e-mail that you wouldn't say to a person's face, don't type in capitals because this indicates that you're shouting, don't "flame" people with harsh criticism or insults; and don't forward a personal message without the writer's permission.

The growth of technology and the increase in college enrollments, and thus in the average size of classes, have changed many folkways on campuses. In the past, a syllabus was typically a few pages long, presenting course requirements and deadlines for exams and papers. Now, many syllabi—some as long as 20 pages—look more like legal documents. They describe, often in great detail, a variety of rules on civil classroom behavior and Internet plagiarism as well as policies on disabilities, makeup exams, grading, late assignments, attendance, and laptop and cell phone use, even forbidding videos of professors that may appear on YouTube (Wasley 2008).

Mores

After some passengers complained, Southwest Airlines booted a woman off a flight for wearing a tight, low-cut shirt that showed considerable cleavage. The woman vowed to sue the airlines, but Southwest said that its rules can deny boarding to any passenger whose clothing is "lewd, obscene, or patently offensive." As this example illustrates, mores (pronounced "MOR-ays") are much stronger than folkways. **Mores** are norms that members of a society consider very important because they maintain moral and ethical behavior. Most people

believe that mores are crucial in upholding a decent and orderly way of life.

According to U.S. cultural mores, one must be sexually faithful to one's spouse or sexual partner and loyal to one's country and must not kill another person (except during war or in self-defense). Notice that while folkways emphasize *ought* to behavior, mores define *must* behavior.

Whereas folkways guide human conduct in the everyday areas of life, mores tend to focus on those aspects connected with sexuality, the family, or religion. The Ten Commandments are a good example of religious mores. Other mores include ethical guidelines (don't cheat, don't lie), expectations about interpersonal behavior (don't live on handouts), and rules about sexual partners (don't have sexual intercourse with children or family members).

Some mores (like the one forbidding incest) are nearly universal, but others vary across social groups and cultures. U.S. teenagers, for example, are more likely to have premarital sex than their counterparts in many other countries, especially Latin America and Africa. Even within the United States, teen sex varies considerably across ethnic and age groups (see Chapter 8).

Laws

The most rigid norms are **laws**, formal rules about behavior that are defined by a political authority that has the power to punish violators. Unlike folkways and most mores, laws are deliberate, formal, "precisely specified in written texts," and "enforced by a specialized bureaucracy," usually police and courts (Hechter and Opp 2001: xi).

Laws also vary over time and across societies. Ten states have allowed using marijuana for medical reasons, but the U.S. Supreme Court recently ruled that the federal government has the authority to override state laws. Judges in Iran can order public floggings of young men accused of drinking alcohol, distributing Western music CDs, or being alone with women who are not their relatives. Singapore, where chewing gum is now legal after a 12-year-ban, requires citizens who want to chew gum to submit their names and ID cards to the government (Anderson 2001; Knickerbocker 2005).

Laws impose social control by setting specific rules of conduct. Some laws, such as the U.S. Constitution, have existed for hundreds of years. In other cases, tradition and custom can become law over time.

Sanctions

Most people conform to norms because of **sanctions**, rewards for good or appropriate behavior and/or penalties for bad or inappropriate behavior. Children learn norms through both *positive* sanctions (praise, hugs, smiles, new toys) and *negative* sanctions (frowning, scolding, spanking, withdrawing love) (see Chapter 4).

Sanctions vary. When we violate folkways, the sanctions are relatively mild: gossip, ridicule, exclusion from a group. If you don't bathe or brush your teeth, you may not be invited to parties because others will see you as crude, but not sinful or evil. The sanctions for violating mores and some laws can be severe: loss of employment, expulsion from college, whipping, torture, banishment, imprisonment, and even execution. The sanctions are usually harsh because the bad behavior threatens the moral foundations of a society. The public floggings in Iran, you'll recall, punish offenders for their "immorality" and send a warning to others to follow the rules.

Sanctions aren't always consistent, despite universally held norms. In the United States, someone who can hire a good attorney might receive a lighter penalty for a serious crime. In some cultures, young girls and women may be lashed for engaging in premarital sex, but men aren't punished at all for the same behavior, even though they have also violated Islamic law (Agence France Presse 2001). Also, laws are sometimes enforced selectively or not at all. In the United States and elsewhere, for example, offenders who violate laws against discrimination, sexual harassment, and domestic violence are rarely prosecuted or punished (see Chapters 7 and 13).

IDEAL VERSUS REAL CULTURE

The **ideal culture** of a society comprises the beliefs, values, and norms that people say they hold or follow. In every culture, however, these standards differ from the society's **real culture**, or people's actual everyday behavior.

We don't always do what society says is ideal or what we profess to cherish. In 2008, for example, Eliot Spitzer, governor of New York, resigned after investigators found that he had spent as much as $80,000, over an extended period, with a high-priced prostitution service. Hypocrisy is fairly common in politics (see Chapter 11), but the scandal was especially shocking because, as New York's attorney general, Spitzer had broken up prostitution rings,

implemented the toughest and most comprehensive anti-prostitution laws in the nation, and increased the penalties for patronizing a prostitute. As a result, numerous human rights and women's groups had supported his campaign as governor (Bernstein 2008). Thus, ideal culture and our actual behavior are often contradictory.

> **sanctions** rewards for good or appropriate behavior and/or penalties for bad or inappropriate behavior.
>
> **ideal culture** the beliefs, values, and norms that people in a society say they hold or follow.
>
> **real culture** the actual everyday behavior of people in a society.
>
> **cultural universals** customs and practices that are common to all societies.

3 Cultural Similarities

You've seen that there's considerable diversity across societies in symbols, language, values, and norms. There are also striking uniformities, or cultural universals. Despite these uniformities, people who visit other countries often experience culture shock.

CULTURAL UNIVERSALS

Cultural universals are customs and practices that are common to all societies. Anthropologist George Murdock and his associates studied hundreds of societies and compiled a list of 88 categories that they found among all cultures (see *Table 3.2* on the next page).

There are many cultural universals, but specific behaviors vary across cultures, from one group to another in the same society, and across time. For example, all societies have food taboos, but specifics about what people ought and ought not eat differ across societies. About 75 percent of the world's people eat insects as part of their diet. In Thailand, locusts, crickets, silkworms, grasshoppers, ants, and other insects have become a big part of the Thai diet (Stolley and Taphaneeyapan 2002). South Korea has an estimated 6,000 restaurants that specialize in dog meat because it's easier to digest than many other meats. Traditionally, many restaurants in South Korea have served *poshintang*, or dog soup. Increasingly, however, as many South Koreans accept Western norms, some are treating dogs as pets and members of the family rather than as nutritional food sources (Demick 2002; Stolley

culture shock a sense of confusion, uncertainty, disorientation, or anxiety that accompanies exposure to an unfamiliar way of life or environment.

and Taphaneeyapan 2002; McLaughlin 2005).

CULTURE SHOCK

People who travel to other countries often experience **culture shock**, a sense of confusion, uncertainty, disorientation, or anxiety that accompanies exposure to an unfamiliar way of life or environment. Familiar cues about how to behave are missing or have a different meaning. Culture shock affects people differently, but the most stressful changes involve the type of food eaten, the type of clothes worn, punctuality, ideas about what offends people, the language spoken, differences in personal hygiene, the general pace of life, a lack of privacy, and concern about finances (Spradley and Phillips 1972; Pedersen 1995).

TABLE 3.2
Some Cultural Universals

athletics	folklore	inheritance rules	postnatal care
bodily adornments	food taboos	joking	property rights
community organization	funeral rites	kin terminology	puberty customs
cooking	games	language	religious rituals
courtship	gestures	magic	sexual restrictions
dancing	gift giving	marriage	status differentiation
division of labor	greetings	medicine	supernatural beings
dream interpretations	hair styles	mealtimes	trade
education	hospitality	modesty	visiting
ethics	housing	mourning	
etiquette	hygiene	music	
family feasts	incest taboos	names	

Source: Based on Murdock 1945.

To some degree, everyone is *culture bound*, having internalized cultural values. Even renewed contact with a once familiar culture can be unsettling. After living in the Midwest for 20 years, for example, one traveler encountered culture shock when she flew to Japan, her homeland:

> As soon as everyone is seated on the plane, the Japanese announcement welcoming us to the flight reminds me of the polite language I was taught as a child: always speak as though everything in the world were your fault. The bilingual announcements on the plane take twice as long in Japanese as in English because every Japanese announcement begins with a lengthy apology: "We apologize about how long it's taken to seat everyone and thank you for being so patient," "We apologize for the inconvenience you will no doubt experience in having to fill out the forms we are about to hand out." (Mori 1997: 5)

American students who have studied abroad often experience culture shock when they return, such as in readjusting to traffic jams, the pace of daily life, and the emphasis on work above personal life. For example, a student who studied in Ecuador said that she missed the sense of community she had felt: "Back in the United States, I noticed how separate and selfish people can be at times." (Goodkin 2005).

4 Cultural Diversity

there is considerable cultural variability *between* societies and sometimes *within* the same society. *Subcultures* and *countercultures* account for some of the complexity within a society.

SUBCULTURES

A **subculture** is a group or category of people whose distinctive ways of thinking, feeling, and acting differ somewhat from those of the larger society. A subculture is part of the larger, dominant culture but has particular values, beliefs, perspectives, lifestyle, or language. Members of subcultures often live in the same neighborhoods, associate with each other, have close personal relationships, and marry others who are similar to themselves.

One example of a subculture is college students. At residential campuses, most college students wear similar clothing, eat similar food, participate in similar recreational activities, and often date each other. Whether students live in campus housing or commute, they share a similar vocabulary that includes words like *syllabus*, *incomplete grade*, *dean's list*, and *core courses*. Many students are also members of other campus subcultures as well: sororities and fraternities, sports teams, clubs, or honor societies. In their home communities, college students are members of other subcultures, such as religious, political, and ethnic groups.

Whether we realize it or not, most of us are members of numerous subcultures. Subcultures reflect a variety of characteristics, interests, or activities:

- *Ethnicity* (Irish, Polish Americans, Vietnamese, Russians)
- *Religion* (Catholics, evangelical Christians, atheists, Mormons)
- *Politics* (Maine Republicans, Southern Democrats, independents, libertarians)
- *Sex and gender* (gay men, lesbians, transsexuals)
- *Age* (elderly widows, kindergarteners, middle schoolers)
- *Occupation* (surgeons, teachers, prostitutes, police officers, truck drivers)
- *Music and art* (jazz aficionados, opera buffs, art lovers)
- *Physical disability* (people who are deaf, quadriplegic, or blind)
- *Social class* (Boston Brahmin, working poor, middle class, jet set)
- *Recreation* (mountain bikers, bingo or poker players, motorcycle riders)

Some subcultures retreat from the dominant culture to preserve their beliefs and values. The Amish, for example, have created self-sustaining economic units, travel locally by horse and buggy, conduct religious services in their homes, make their own clothes, and generally shun modern conveniences such as electricity and phones. Despite their self-imposed isolation, the Amish have been affected by the dominant U.S. culture. They traditionally worked in agriculture, but as farming became less self-sustaining, many began small businesses that produce quilts, wood and leather products, and baked goods for tourists.

To fit in, members of most ethnic and religious subcultures adapt to the

> **subculture** a group or category of people whose distinctive ways of thinking, feeling, and acting differ somewhat from those of the larger society.

Some analysts describe the popular television program The Simpsons as countercultural because it ridicules the media's shallowness, portrays government in a cynical light, mocks indifferent or incompetent teachers and administrators, makes fun of stereotyping Asians who manage convenience stores, and scoffs at bungling and greedy law enforcement officers (Reeves 1999; Cantor 2001).

counterculture a group or category of people who deliberately oppose and consciously reject some of the basic beliefs, values, and norms of the dominant culture.

ethnocentrism the belief that one's culture and way of life are superior to those of other groups.

larger society. Some Chinese restaurants have changed their menus to accommodate the average American's taste for sweet-and-sour dishes but list more authentic food on a separate, Chinese-language menu. Fearing discrimination, some Muslim girls and women don't wear a *hijab*, or head scarf, to classes. And those who don't abide by religious laws (modest dress, a ban on alcohol, prayer five times a day, and limited interaction with the opposite sex) often feel pressure from their Muslim peers to conform to such orthodox practices (McMurtrie 2001).

In many instances, subcultures arise because of technological or other societal changes. With the emergence of the Internet, for example, subcultures arose that identified themselves as hackers, techies, or computer geeks.

COUNTERCULTURES

Unlike a subculture, a **counterculture** deliberately opposes and consciously rejects some of the basic beliefs, values, and norms of the dominant culture. Countercultures usually emerge when people believe they cannot achieve their goals within the existing society. As a result, such groups develop values and practices that run counter to those of the established society. Some countercultures are small and informal, but others have millions of members and are highly organized, like religious militants (see Chapter 15).

Most countercultures do not engage in illegal activities. During the 1960s, for example, social movements such as feminism, civil rights, and gay rights organized protests against mainstream views but stayed within the law. However, some countercultures are violent and extremist, such as the 888 active hate groups across the United States (see *Figure 3.3*), who intimidate ethnic groups and gays. There have been instances where counterculture

members have clearly violated laws: Some skinheads have murdered gays; antigovernment militia adherents bombed a federal building in Oklahoma, killing dozens of adults and children; and anti-abortion advocates have murdered physicians and bombed abortion clinics

ETHNOCENTRISM

When President Bush visited Queen Elizabeth II in England in 2003, he brought with him five of his personal chefs. The Queen was offended because she has a large staff of excellent cooks ("Bush's Cooks . . . " 2003). Was President Bush being ethnocentric?

Ethnocentrism is the belief that one's culture and way of life are superior to those of other groups. This attitude leads people to view other cultures as inferior, wrong, backward, immoral, or barbaric. Countries and people display their ethnocentrism in many ways. During the nineteenth and twentieth centuries, there was rampant anti-immigrant sentiment toward people coming from Ireland, Poland, and other European countries. The Chinese Exclusion Act of 1882 barred Chinese immigrants and the Immigration Act of 1924 used quotas to limit Italian and Jewish immigration.

FIGURE 3.3
Active Hate Groups in the United States: 2008

Source: Southern Poverty Law Center, www.splcenter.org (accessed October 22, 2008).

Because people internalize their culture and take it for granted, they may be hostile toward other cultures. Each group tends to see its way of life as the best and the most natural. Some of my black students argue that it's impossible for African Americans to be ethnocentric because they suffer much prejudice and discrimination. *Any* group can be ethnocentric, however (Rose 1997). An immigrant from Nigeria who assumes that all native-born African Americans are lazy and criminal is just as ethnocentric as a native-born African American who assumes that all Nigerians are arrogant and "uppity."

Ethnocentrism can sometimes be functional. Pride in one's country promotes loyalty and cultural unity. When children learn their country's national anthem and customs, they have a sense of belonging. Ethnocentrism also reinforces conformity and maintains stability. Members of a society become committed to their particular values and customs and transmit them to the next generation. As a result, life is (generally) orderly and predictable.

Ethnocentrism has its benefits, but it is usually dysfunctional. Viewing others as inferior generates hatred, discrimination, and conflict. Many of the recent wars, such as those in the former Yugoslavia and Rwanda (Africa), and the ongoing battles between Palestinians and Israelis reflect religious, ethnic, or political intolerance toward subgroups (see Chapter 15 and online Chapter 18). Thus, ethnocentrism discourages intergroup understanding and cooperation.

CULTURAL RELATIVISM

The opposite of ethnocentrism is **cultural relativism**, a belief that no culture is better than another and that a culture should be judged by its own standards. Most Japanese mothers stay home with their children, while many American mothers are employed outside the home. Is one practice better than another? No. Because Japanese fathers are expected to be the breadwinners, it's common for many Japanese women to be homemakers. In the United States, in contrast, many mothers are single heads of households who have to work to provide for themselves and their children. Also, economic recessions and the loss of many high-paying jobs in the United States have catapulted many middle-class women into the job market to help support their families (see Chapters 10, 12, and 13). Thus, Japanese and American parenting may be different, but one culture isn't better or worse than the other.

An appreciation of cultural relativism is practical and productive. Businesspeople and other travelers can overcome cultural barriers and improve communication if they understand and respect other cultures. However, respecting other cultures' customs and traditions does not require that we remain mute when countries violate human rights, practice female infanticide, or sell young girls for prostitution.

> **cultural relativism** the belief that no culture is better than another and that a culture should be judged by its own standards.
>
> **multiculturalism** (*cultural pluralism*) the coexistence of several cultures in the same geographic area, without any one culture dominating another.
>
> **popular culture** the beliefs, practices, activities, and products that are widely shared among a population in everyday life.

MULTICULTURALISM

Multiculturalism (sometimes called *cultural pluralism*) refers to the coexistence of several cultures in the same geographic area, without one culture dominating another. Many applaud multiculturalism because it encourages intracultural dialogue (for example, U.S. schools offering programs and courses in African American, Latino, Arabic, and Asian studies). Supporters hope that emphasizing multiculturalism—especially in academic institutions and the workplace—will decrease ethnocentrism, racism, sexism, and other forms of discrimination.

Despite its benefits, not everyone is enthusiastic about multiculturalism. Not learning the language of the country in which one lives and works, for instance, can be isolating and create on-the-job miscommunication, tension, and conflict. Some also feel that multiculturalism can destroy a country's national traditions, heritage, and identity because ethnic and religious subcultures may not support the dominant culture's values and beliefs (Watson 2000; Skerry 2002).

A major component of culture is popular culture. Popular culture has enormous significance in many contemporary societies.

5 Popular Culture

Popular culture refers to beliefs, practices, activities, and products that are widely shared within a population in everyday life. Popular culture includes television, music, magazines, radio, advertising, sports, hobbies, fads, fashions, and movies

The incidence of obesity among American children and teenagers has more than tripled, rising from 5 percent in the 1960s to almost 17 percent by 2006 (Ogden et al. 2008). There are many reasons for the increase, but physicians and researchers lay much of the blame on popular culture, especially the advertising industry. Marketers in the United States spend an estimated

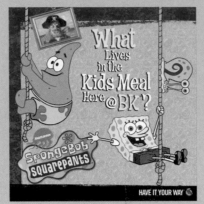

© PR/NewsFoto/Burger King Corporation

$10 billion a year to market products to children. Much of the advertising, particularly on television, uses cartoon characters like SpongeBob SquarePants and Scooby-Doo to sell sugary cereal, cookies, candy, and other high-calorie snacks (Institute of Medicine 2006; Strasburger et al. 2006). In contrast, many European countries forbid advertising on children's television programs.

mass media forms of communication designed to reach large numbers of people.

as well as the food we eat, the gossip we share, and the jokes we pass along to others. People produce and consume popular culture: They are not simply passive receptacles but influence popular culture by what they buy, how they spend their leisure time, and how they express themselves.

THE IMPACT OF POPULAR CULTURE

Popular culture can have positive and negative effects on our everyday lives. Most people do not believe everything they read or see on television but weigh the merit and credibility of much of the content. Further, in a national study of high school students, 46 percent of the teenagers identified a family member as their role model and another 16 percent chose a friend or a family friend rather than an entertainer or a sports figure (Hebert and Rivlin 2002). Thus, many young people aren't star-struck.

Most of us are highly influenced, nonetheless, by a popular culture that is largely controlled and manipulated to some extent by newspapers and magazines, television, movies, music, and ads (see Chapter 4). These **mass media**, or forms of communication designed to reach large numbers of people, have enormous power in shaping public perceptions and opinions. Let's look at a few examples.

Newspapers and Magazines

How accurate is the information we get from the mass media? Even though crime has decreased in the United States in recent years, newspapers and magazines have increased the amount of space they devote to covering violence. School violence is always a popular topic, even though schools are safer now than they were almost 20 years ago (Dinkes et al. 2007). In addition, the coverage can be deceptive. For example, a study of school violence concluded that six of the most influential newspapers (including the *New York Times, Washington Post,* and *Los Angeles Times*) portray violence in rural/suburban and urban schools differently. The newspapers were much less likely to print stories about shootings in urban school systems even though they occur twelve times more often than in rural schools. Violence in rural and suburban schools gets more coverage because many Americans have a stereotypical picture of rural life as peaceful and tranquil (Menifield et al. 2001).

Advertising and Commercials

The average American views 3,000 ads per day and at least 40,000 commercials on television per year, unless she or he has a service, such as TiVo, that skips commercials when recording television programs (Kilbourne 1999; Strasburger et al. 2006). We are constantly deluged with advertising in newspapers and magazines, on television and radio, in movie theaters, on billboards and the sides of buildings, on public transportation, and on the Internet. My Sunday newspaper comes in two big plastic bags, but about 80 percent of the content is advertising.

Many of my students, who claim that they "don't pay any attention to ads," come to class wearing branded apparel: Budweiser caps, Adidas sweatshirts, Old Navy T-shirts, or Nike footwear. A national study of people aged 15 to 26 years concluded that—regardless of gender, ethnicity, and educational level—exposure to alcohol advertising contributed to increased drinking. Those who saw more ads for alcoholic beverages tended to drink more and those who remembered the ads drank the most (Snyder et al. 2006).

Much mass media content is basically marketing. Many of the *Dr. Phil* shows plug his books and other products, as well as those of his wife and older son. Much of the content on television morning shows and MTV has become "a kind of sophisticated infomercial" (O'Donnell 2007:30). For example, a third of the content on morning shows (like the *Early Show* on CBS, the *Today Show* on NBC, and *Good Morning America* on ABC) is essentially selling something (a book, music, a movie, or another television program) that the corporation owns. One of the most lucrative alliances is between Hollywood and toy manufacturers. (See Chapter 4.)

And, according to some scholars, the U.S. mass media have reduced competition at home and have expanded cultural imperialism abroad.

CULTURAL IMPERIALISM

In **cultural imperialism**, the cultural values and products of one society influence or dominate those of another. Many countries complain that U.S. cultural imperialism displaces authentic local culture and results in cultural loss. The United States established the Internet, a global media network. The Internet uses mainly English and is heavily saturated with American advertising and popular culture (Louw 2001).

For example, a study of over 1,300 teenagers in 12 countries found that many enjoy American movies, television, celebrities, entertainers, and popular music. On the downside, images in U.S. mass media lead teens around the world to believe that Americans are extremely violent, criminal, and sexually immoral. U.S. news media, similarly, emphasize crime, corruption, sex, and violence far beyond what the ordinary American experiences on a daily basis (DeFleur and DeFleur 2003).

6 Cultural Change and Technology

"radio has no future" (Lord Kelvin, Scottish mathematician and physicist, 1897).

- "Everything that can be invented has been invented" (Charles H. Duell, U.S. commissioner of patents, 1899).

- "Television won't be able to hold on to any market it captures after the first six months. People will soon get tired of staring at a plywood box every night" (Darryl F. Zanuck, head of Hollywood's 20th Century-Fox studio, 1946).

- "There is no reason for any individual to have a computer in their home" (Kenneth Olsen, president and founder of Digital Equipment Corp., 1977).

Despite these predictions, radio, television, computers, and other new technologies have triggered major cultural changes around the world. This section examines why cultures persist, how and why they change, and what occurs when technology changes faster than cultural values, laws, and attitudes do.

> **cultural imperialism** the influence or domination of the cultural values and products of one society over those of another.
>
> **cultural integration** the consistency of various aspects of society that promotes order and stability.

CULTURAL PERSISTENCE: WHY CULTURES ARE STABLE

In many ways, culture is a conservative force. As you saw earlier, values, norms, and language are transmitted from generation to generation. Such **cultural integration**, or the consistency of various aspects of society, promotes order and stability. Even when new behaviors and beliefs emerge, they commonly adapt to existing ones. New immigrants, for example, may speak their native language at home and celebrate their own holy days, but they are expected to gradually absorb the new country's values, obey its civil and criminal codes, and adopt its national language. Life would be chaotic and unpredictable without such cultural integration.

CULTURAL DYNAMICS: WHY CULTURES CHANGE

Cultural stability is important, but all societies change over time. Some of the major reasons for cultural change include diffusion, invention and innovation, discovery, external pressures, and changes in the physical environment.

Diffusion

A culture may change due to *diffusion*, the process through which components of culture spread from one society to another. Such borrowing may have occurred so long ago that the members of a society consider their culture to be entirely their own creation. However, anthropologist Ralph Linton (1964) has estimated that 90 percent of the elements of any culture are a result of diffusion (see *Figure 3.4*).

Diffusion can be direct and conscious, occurring through trade, tourism, immigration, intermarriage, or the invasion of one country by another. Diffusion can also be indirect and largely unconscious, as in the Internet transmissions that zip around the world.

Invention and Innovation

Cultures change because people are continually finding new ways of doing things. Invention, the process of creating new things, brought about products such as toothpaste (invented in 3000 BC), eyeglasses (262 AD), flushable toilets (the sixteenth century), clothes dryers (early nineteenth century), can openers (1813), fax machines (1843—that's right, invented in 1843!), credit cards (1920s), sliced bread (1928), computer mouses (1964), Post-It notes (1980), and DVDs (1995).

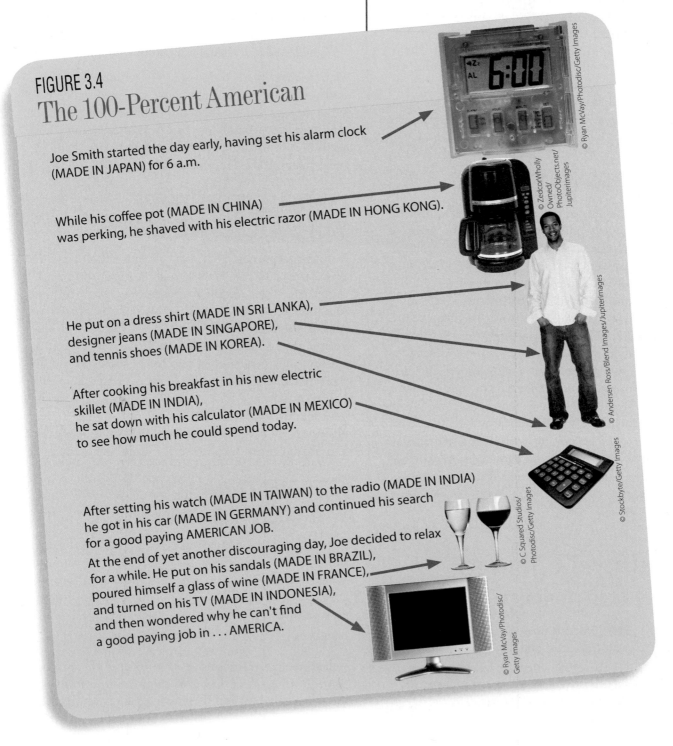

FIGURE 3.4
The 100-Percent American

Joe Smith started the day early, having set his alarm clock (MADE IN JAPAN) for 6 a.m.

While his coffee pot (MADE IN CHINA) was perking, he shaved with his electric razor (MADE IN HONG KONG).

He put on a dress shirt (MADE IN SRI LANKA), designer jeans (MADE IN SINGAPORE), and tennis shoes (MADE IN KOREA).

After cooking his breakfast in his new electric skillet (MADE IN INDIA), he sat down with his calculator (MADE IN MEXICO) to see how much he could spend today.

After setting his watch (MADE IN TAIWAN) to the radio (MADE IN INDIA) he got in his car (MADE IN GERMANY) and continued his search for a good paying AMERICAN JOB.

At the end of yet another discouraging day, Joe decided to relax for a while. He put on his sandals (MADE IN BRAZIL), poured himself a glass of wine (MADE IN FRANCE), and turned on his TV (MADE IN INDONESIA), and then wondered why he can't find a good paying job in . . . AMERICA.

© Ryan McVay/Photodisc/Getty Images

© ZedcorWholly Owned/PhotoObjects.net/Jupiterimages

© Andersen Ross/Blend Images/Jupiterimages

© Stockbyte/Getty Images

© C. Squared Studios/Photodisc/Getty Images

© Ryan McVay/Photodisc/Getty Images

Innovation—turning inventions into mass-market products—also sparks cultural changes. An innovator is someone determined to market an invention, even if it's someone else's good idea. For example, Henry Ford invented nothing new but "assembled into a car the discoveries of other men behind whom were centuries of work," an innovation that changed people's lives (Evans et al. 2006: 465).

Discovery

Like invention, *discovery* requires exploration and investigation and results in new products, insights, ideas, or behavior. The discovery of penicillin prolonged lives, which, in turn, meant that more grandparents (as well as great-grandparents) and grandchildren would get to know each other. However, longer lifespans also mean that children and grandchildren need to care for elderly family members over many years (see Chapter 13).

Discovery usually requires dedicated work and years of commitment, but some discoveries occur by chance. This is called the *serendipity effect*. For example, George de Mestral, a Swiss electrical engineer, was hiking through the woods. He was annoyed by burrs that clung to his clothing. Why were they so difficult to remove? A closer examination showed that the burrs had hook-like arms that locked into the open weave of his clothes. The discovery led de Mestral to invent a hook-and-loop fastener. His invention, Velcro—derived from the French words *velour* (velvet) and *crochet* (hooks)—can be now found on everything from clothing to spacecraft.

External Pressures

External pressure for cultural change can take various forms. In its most direct form—war, conquest, or colonization—the dominant group uses force or the threat of force to bring about cultural change in the other group. When the Soviet Union invaded and took over many small countries (such as Lithuania, Latvia, Estonia, Ukraine, Georgia, and Armenia) after World War II, it forbade citizens to speak their native languages, banned traditions and customs, and turned churches into warehouses.

Pressures for change can also be indirect. For example, some countries—such as Thailand, Vietnam, China, and Russia—have reduced their prostitution and international sex trafficking because of widespread criticism by the United Nations and some European countries (but not the United States). The United Nations has no power to intervene in a country's internal affairs but can embarrass nations by publicizing human rights violations (Farley 2001).

Changes in the Physical Environment

Changes in the physical environment also lead to cultural change. The potato blight in Ireland during the 1840s spurred massive emigration to the United States. Natural disasters, like earthquakes and floods, may cause death, destruction of homes, and other devastation. People may suffer depression because of the immense losses they have experienced (see Chapters 13 and 14). Disasters may also bring positive change. After major earthquakes in California, for instance, studies of architectural design resulted in new building codes that protected businesses and residences from collapsing during minor earthquakes (Bohannon 2005).

TECHNOLOGY AND CULTURAL LAG

Some parts of culture change more rapidly than others. **Cultural lag** refers to the gap when nonmaterial culture changes more slowly than material culture.

There are numerous examples of cultural lag in modern society, because ethical rules and government regulations haven't kept up with technological developments. Recently, for example, a graduate student posted a paper on a university Web site and found out, later, that an Internet term-paper mill was selling it without her knowledge or permission. Because intellectual property laws applying to Internet use are still being developed, this student was in court for several years before the Web site operator settled for an undisclosed amount (Foster 2006).

Cultural lag often creates uncertainty, ambiguity about what's right and wrong, conflicting values, and a feeling of helplessness. According to Naisbitt and his colleagues (1999: 3), we live in a "technologically intoxicated zone" where we both fear and worship technology, become obsessed with gadgets (like computers) even though they take up much of our time, don't deal with ethical issues raised by biotechnology (like the implications of cloning embryos), and rely on technology as quick fixes: "We want to believe that any given solution is only a purchase away."

Cultural lags have always existed and will continue in the future. Technology is necessary, but we can make conscious choices about how and when we use technological advances. Thus, cultural lags can be

> **cultural lag** the gap when nonmaterial culture changes more slowly than material culture.

viewed as opportunities for making positive changes that enhance culture.

7 Sociological Perspectives on Culture

What is the role of culture in modern society? And how does culture help us understand ourselves and the world around us? Functionalist, conflict, feminist, and interactionist scholars offer different answers to these and other questions about culture, but all provide important insights. (*Table 3.3* summarizes these perspectives).

FUNCTIONALISM

Functionalists focus on society as a system of interrelated parts (see Chapter 1). In their analysis of culture, similarly, functionalists emphasize the social bonds that attach people to society.

Key Issues

For functionalists, culture is a cement that binds society. As you saw earlier, norms and values shape our lives, provide guideposts for our everyday behavior, and promote stability. Especially in countries such as the United States that have high immigration rates, cultural norms and values help newcomers adjust to the society.

Functionalists also note that culture can be dysfunctional. For example, when subcultures such as the Amish refuse to immunize their children, some diseases can surge in the community (Brown 2005). Also, various countercultures (such as paramilitary groups) can create chaos by bombing federal buildings and killing or injuring hundreds of people.

Critical Evaluation

Functionalism is important in showing that shared norms and values create solidarity and stability in a culture. In emphasizing culture's meeting people's daily needs, however, functionalism often overlooks diversity and social change. For example, a number of influential functionalists argue that immigration should be restricted because it dilutes shared U.S. values, overlooking the many contributions that newcomers make to society (see Chapter 10).

CONFLICT THEORY

Unlike functionalists, conflict theorists argue that culture can generate considerable inequality. Because the rich and powerful determine economic, political, educational, and legal policies for their own benefit and control the mass media, the average American has little power in changing the status quo—whether it's low wages, a disastrous war, or corporate corruption (see Chapters 8 and 11).

Key Issues

According to conflict theorists, as a handful of powerful U.S. corporations increase their global influence, the rich get richer. The result, according to conflict theorists, is widespread inequality. The scientists who created the Internet opened it to the public. Recently, however, new laws have protected the interests of profit-making cable providers over those of individual users (Lessig 2001).

Conflict theorists also point out that technology benefits primarily the rich. For example, the C-Leg is a prosthetic leg that allows users to go down slopes or staircases, to run, or to stroll on a hilly path. Because the C-Leg costs about $50,000 and is typically not covered by medical insurance, only affluent people can take advantage of this technology (Austen 2002).

Critical Evaluation

Conflict theorists have provided important insights about why U.S. society (and others) suffers from widespread inequality. For example, values such as competitiveness benefit capitalists, who can threaten to fire workers if they aren't as productive as employers expect (see Chapter 12).

According to some critics, conflict theorists often emphasize divisiveness and don't appreciate how culture bonds people to a society. Capitalism may generate wide-

When media portrayals of women are absent or stereotypical, we get a distorted view of reality.

© Photodisc/Getty Images

spread inequality, but many people see even low-paid work as important if it helps them climb out of poverty.

FEMINIST THEORIES

Feminist scholars, who use both macro and micro approaches, agree with conflict theorists that material culture, especially, creates considerable inequality, but they focus on gender differences. Feminist scholars are also more likely than other theorists to examine multicultural variations across some groups.

Key Issues

Gender affects our cultural experiences. When media portrayals of women are absent or stereotypical, we get a distorted view of reality. For example, hip-hop and rap music that glorifies gang rape and violence against women degrades all women (see Chapter 9). Feminist scholars also emphasize that subcultures—for example, female students or single mothers—may experience our culture differently than their male counterparts do.

Critical Evaluation

Feminist analyses expand our understanding of cultural components that other theoretical perspectives ignore or gloss over. Like conflict theorists, however, feminist theorists often stress divisiveness rather than examining how culture integrates women and men into society. Another weakness is that feminist scholars typically focus on low-income and middle-class women as victims, overlooking the ways affluent women exploit these groups, both female and male.

SYMBOLIC INTERACTIONISM

Unlike functionalists and conflict theorists, symbolic interactionists examine culture through micro lenses. They are most interested in understanding how people interpret culture and transmit norms and values through social interaction.

Key Issues

Interactionists explain how culture influences our everyday lives. Language, you'll recall, shapes our views and behavior. Also, subcultures use different symbols to communicate their practices and beliefs. At Christmas, Christians often display nativity scenes of the birth of Jesus Christ and attend religious services. Jews celebrate Hanukkah by lighting eight candles on the menorah. African Americans light seven candles dedicated to particular principles (such as unity and self-determination) and celebrate with a *karamu*, or African feast. Ramadan, the ninth month of the Muslim year, includes prayer, fasting, and feasting on the final day.

Critical Evaluation

Micro approaches are useful in understanding what culture means to people and how these meanings differ across societies. However, symbolic interactionists don't address the linkages between culture and institutions. For example, it's important to recognize how language bonds people together, but interactionists say little about how organized groups (such as the English-only movement) try to maintain control over language or use language to promote their values.

TABLE 3.3
Sociological Explanations of Culture

THEORETICAL PERSPECTIVE	FUNCTIONALIST	CONFLICT	FEMINIST	SYMBOLIC INTERACTIONIST
Level Of Analysis	Macro	Macro	Macro and Micro	Micro
Key Points	• Similar beliefs bind people together and create stability. • Sharing core values unifies a society and promotes cultural solidarity.	• Culture benefits some groups at the expense of others. • As powerful economic monopolies increase worldwide, the rich get richer and the rest of us get poorer.	• Women and men often experience culture differently. • Cultural values and norms can increase inequality because of gender, race/ethnicity, and social class.	• Cultural symbols forge identities (that change over time). • Culture (such as norms and values) helps people merge into a society despite their differences.

Have you ever thought

about how you became the person you are today?

what do you think?

I'm always the same, no matter where I am or who's around.

1 2 3 4 5 6 7

strongly agree   strongly disagree

4

Socialization

Key Topics

In this chapter, we'll explore the following topics:

1 Socialization: Its Purpose and Importance

You've probably heard the expression that humans are born with a *tabula rasa*, meaning "blank slate." By as early as age 2, socialization has started to fill that slate. **Socialization** is the lifelong process of social interaction in which the individual acquires a social identity and ways of thinking, feeling, and acting that are essential for effective participation in a society. Thus, socialization transforms a naked, wet, and crying newborn into a person who becomes a social being.

> **socialization** the lifelong process of social interaction in which the individual acquires a social identity and ways of thinking, feeling, and acting that are essential for effective participation in a society.

WHAT IS THE PURPOSE OF SOCIALIZATION?

As you'll see, the process of socialization—from childhood to old age—can be relatively smooth or very bumpy, depending on factors such as age, gender, race/ethnicity, and social class. Generally, however, socialization has five key functions that range from providing us with a social identity to shaping that social identity so that it meshes with societal expectations.

Socialization Establishes Our Social Identity

Have you ever thought about how you became the person you are today? Sociology professors sometimes ask students to give twenty answers to the question "Who am I?" (Kuhn and McPartland 1954). How would you respond? You would probably include a variety of descriptions such as college student, single or married, female or male, your occupation, and family status (son or father, for example). All of your answers would reflect a sense of being someone, of your *self* (a concept we'll examine shortly). You are who you are largely because of socialization.

Socialization Teaches Us Role Taking

Why do you act differently in class than when chatting with your friends? Because we play different roles in different settings. A *role* is the behavior expected of a person in a particular social position (see Chapter 5). The way we interact with a parent is typically very different from the way we talk to an employer, a friend, or a professor. We all learn appropriate roles through the socialization process.

Socialization Controls Our Behavior

In learning appropriate roles, we absorb values and a variety of rules about how we should (and should not) interact in everyday situations. If we follow the rules, we're usually rewarded or at least accepted. If we break the rules, we may be punished. As a result, socialization controls our behavior.

We act in socially acceptable ways because we internalize societal values and beliefs. **Internalization** is the process of learning cultural behaviors and expectations so deeply that we assume they are correct and accept them without question. For example, several European nations provide generous employment leave for parents to care for preschool children, but fathers rarely take advantage of this benefit because most people, including women, assume that child rearing is a woman's responsibility (Forsberg 2005).

Despite our temperamental differences and personal quirks, the process of socialization teaches us to conform to societal expectations. This does *not* mean that we become programmed robots without the ability to make choices. Instead, many of our choices are limited by the cultural beliefs, values, and norms that we adopt through socialization. These limits, whether we agree with them or not, ensure that society is orderly and predictable.

Socialization Transmits Culture to the Next Generation

Socialization is the process of acquiring the culture in which we live. Each generation passes on the learned roles and rules to the next generation. The culture that is transmitted includes language, beliefs, values, norms, and symbols, as you saw in Chapter 3.

WHY IS SOCIALIZATION IMPORTANT?

Social isolation can be devastating. An example of its terrible effects is Genie. When Genie was 20 months old, her father decided she was "retarded" and locked her away in a back room, curtains drawn and the door shut. Genie was tied into a harness and placed in a wire mesh cage. She had almost no opportunity to overhear any conversation between others in the house. Genie's father frequently beat her with a wooden stick and never spoke to her. When she was discovered in 1970 at age 13, a psychiatrist described Genie as "unsocialized, primitive, and hardly human." Except for high-pitched whimpers, she never spoke. Genie had little bowel control, experienced rages, rubbed her face frantically, and tried to hurt herself. After living in a rehabilitation ward and a foster home, she learned to eat normally, was toilet-trained, and gradually developed a vocabulary, but her language use never progressed beyond that of a 3 or 4 year old (Curtiss 1977). (If you want to read more about Genie, see "The Civilizing of Genie," http://kccesl.tripod.com/genie.html.)

Since the 1930s, sociologists and psychologists have found that *institutionalization*, as in orphanages, has a negative impact on children's intellectual, physical, behavioral, social, and emotional development. And, the longer the institutionalization lasts, the greater the developmental problems, because the children experience little interaction with caregivers (MacLean 2003, provides a summary of the studies on this topic).

The research on children who are isolated (like Genie) or institutionalized (like orphans) shows that socialization is critical to our development. Talking, eating with utensils, and controlling our bowel movements do not come naturally. Instead, we learn to do all of these things beginning in infancy. When children are deprived of interaction with other people, they do not develop the characteristics that most of us see as normal and human.

2 Nature and Nurture

b iologists tend to focus on the role of heredity (or genetics) in human development. In contrast, most social scientists, including sociologists, underscore the role of learning, socialization, and culture. This difference of opinion is often called the *nature-nurture debate* (see *Table 4.1*). Many researchers think that heredity and environment overlap in shaping

TABLE 4.1
The Nature-Nurture Debate

NATURE — **Human development is . . .**
Innate
Biological, physiological
Due largely to heredity
Fairly fixed

NURTURE — **Human development is . . .**
Learned
Psychological, social, cultural
Due largely to environment
Fairly changeable

EMOTIONAL ATTACHMENT

the developing person, but there is still a tendency to emphasize either nature or nurture.

HOW IMPORTANT IS NATURE?

Some scientists propose that biological factors—especially the brain—play an important role in our development. A study of brain scans, for example, suggested that a region of the brain (the ventromedial prefrontal cortex) a few inches behind the bridge of the nose may predispose people to have a negative outlook on life. When activity in this region of the brain increases, people tend to be more anxious, irritable, angry, and unpleasant (Vedantam 2002). It's not clear, however, why many people have high brain activity yet remain calm, friendly, and pleasant.

How Biology Affects Behavior

How much impact does biology have on socialization? Some scientists point to unsuccessful attempts at sex reassignment to show that nature is more important than nurture. John Money, a highly respected medical psychologist at Johns Hopkins University Hospital, has published numerous articles and books maintaining that gender identity is not firm at birth but is determined as much by culture and nurture as by genes and hormones (Money and Ehrhardt 1972).

Recently, several scientists have challenged such conclusions. In 1963, twin boys were being circumcised. The penis of one of the infants, David Reimer, was accidentally burned off. Encouraged by John Money, the parents agreed to raise David

HARLOW STUDIES

In the early 1960s, psychologists Margaret and Harry Harlow (1962) conducted several studies on infant monkeys. In one group, a "mother" made of terrycloth provided no food, while the mother made of wire did so through an attached baby bottle containing milk. In another group, the terrycloth mother provided food but the wire mother did not. Regardless of which mother provided milk, when both groups of monkeys were frightened, they clung to the cloth mother. The Harlows concluded that warmth and comfort were more important to the infant monkeys than nourishment. Since then, some sociologists have cited the Harlow studies to argue that emotional attachment may be more critical than physical nourishment for human infants. Do you see any problems with sociologists' generalizing the results of animal studies to humans? Or not?

as "Brenda." The child's testicles were removed, and surgery to construct a vagina was planned. Money reported that the twins were growing into happy, well-adjusted children, setting a precedent for sex reassignment as the standard treatment for 15,000 newborns with similarly injured genitals (Colapinto 1997, 2001).

In the mid-1990s, a biologist and a psychiatrist followed up on Brenda's progress and concluded that the sex reassignment had not been successful. Almost from the beginning, Brenda refused to be treated like a girl. When her mother dressed her in frilly clothes as a toddler, Brenda tried to rip them off. She preferred to play with boys and stereotypical boys' toys such as machine guns. People in the community said that she "looks like a boy, talks like a boy." Brenda had no friends, and no one would play with her: "Every day I was picked on, every day I was teased, every day I was threatened" (Diamond and Sigmundson 1997: 300).

When she was 14, Brenda rebelled and stopped living as a girl: She refused to wear dresses, urinated standing up, turned down vaginal surgery, and decided she would either commit suicide or live as a male. When his father finally told David the true story of his birth and sex change, David recalls that "all of a sudden everything clicked. For the first time things made sense and I understood who and what I was" (Diamond and Sigmundson 1997: 300). David had a mastectomy (breast removal surgery) at the age of 14 and underwent several operations to reconstruct a penis. He was able to ejaculate but experienced little "erotic sensitivity." At age 25, he married an older woman and adopted her three children. He committed suicide in 2004 at the age of 38.

Most suicides have multiple motives that "come together in a perfect storm of misery." This was probably true of David: His brother died of an overdose of antidepressants in 2002 (after which David sank into a depression), his mother had suffered from depression all her life, and David experienced periodic bouts of unemployment. Most of all, some speculated, David committed suicide because of the "physical and mental torments he suffered in childhood that haunted him the rest of his life" (Colapinto 2004: 96). David's experience suggests to some scientists that nature outweighs nurture in shaping a person's gender identity.

There is considerable debate within the medical community about whether parents should allow surgery on their children who are born with ambiguous genitals (a topic we'll examine in Chapter 9). Some physicians argue that parents should be counseled until the child is old enough to decide whether to undergo surgery because sexual identity does not derive solely, or even primarily, from a person's genitals. Others contend that genital surgery helps establish a child's sexual identity at an early age and that it's irresponsible to do nothing (Weil 2006).

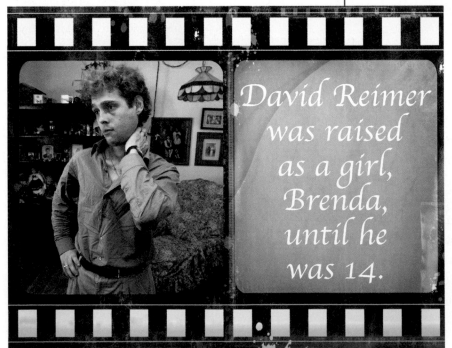

David Reimer was raised as a girl, Brenda, until he was 14.

Sociobiology and Socialization

Sociobiology is a theoretical approach that applies biological principles to explain the behavior of animals, including human beings. Sociobiologists argue, for example, that evolution and genetic factors (nature) can explain why men are generally more aggressive than women. To ensure that their genes will be passed on to offspring, males (human and animal) have to prevail over their rivals. The competition includes aggression, violence, weapons, and plain nastiness: "In the animal world, human and nonhuman, competition is often intense. Males typically threaten, bluff, and if necessary fight one another in their efforts to obtain access to females" (Barash 2002: B8).

Critics reject this sociobiological explanation of male violence. If men were innately aggressive, they would be equally violent across all societies. This is not the case. The proportion of women who have ever suffered physical violence by a male partner varies considerably: 80 percent in Vietnam, 61 percent in Peru, 20 percent in the United States, and 13 percent in Japan (Chelala 2002; Rennison 2003; World Health Organization 2005). Such variations presumably reflect cultural norms and practices and other environmental factors (nurture) rather than biology or genetics (nature) (Chesney-Lind and Pasko 2004).

HOW IMPORTANT IS NURTURE?

Most sociologists agree that nature affects human development. They maintain, however, that nurture is more significant than nature because socialization and culture shape even biological inputs.

How Behavior Affects Biology

Some research suggests that environment (nurture) influences children's genetic makeup (nature). A family history of alcoholism, for example, places a person at greater risk of developing alcohol problems. Serotonin, a brain transmitter, regulates mood, emotion, sleep, and appetite. When people abuse alcohol, serotonin's primary role as an inhibitor breaks down. As a result, alcoholics are often aggressive, depressed, and anxious (Schuckit 1999).

Children of alcoholics can inherit a malfunctioning serotonin transporter gene, become alcoholics themselves, and suffer from behavioral and emotional problems. Birth defects associated with prenatal alcohol exposure can occur in the first 3 to 8 weeks of pregnancy, before a woman even knows she's pregnant. A woman's single drinking "binge"—lasting four hours or more—can permanently damage the brain of an unborn child. A man who drinks heavily may have genetically damaged sperm that also leads to birth defects. Thus, the abuse of alcohol (an environmental factor) by either a woman or a man can have a devastating, irreversible, and lifelong negative impact on her or his child (Twitchell et al. 2001; Hetherington et al. 2006).

Adult behavior can influence a child's biological makeup in other ways. Physical, psychological, or sexual abuse can affect the developing brain during childhood, and the changes may trigger disorders such as depression in adulthood. An impaired

Abuse of alcohol can cause biological damage not only to drinkers, but also to their children's genetic makeup.

corpus callosum—the pathway integrating the two hemispheres of the brain—can result in dramatic shifts in mood and personality, especially for children who have suffered neglect and sexual abuse (Teicher 2000). Thus, childhood mistreatment can impair biological development.

Culture and Socialization

Sociologists often point to cross-cultural data to illustrate the importance of nurture. In a well-known study, anthropologist Margaret Mead (1935) observed three tribes—the Arapesh, Mundugumor, and Tchambuli—who lived fairly close to one another in New Guinea. She found three combinations of gender roles. Among the Arapesh, both men and women were nurturant with their children. The men were cooperative and sensitive, and they rarely engaged in warfare. The Mundugumor were just the opposite: Both men and women were competitive and aggressive, neither parent showed much tenderness toward their offspring, and both parents often used physical punishment to discipline their children. The Tchambuli reversed Western gender roles. The women were the economic providers, and the men took care of the children, sat around chatting, and spent a lot of time decorating themselves for tribal festivities. Mead concluded that attributes long considered either masculine (such as being aggressive) or feminine (such as being emotional) are culturally, not biologically, determined.

WHAT CAN WE CONCLUDE ABOUT THE NATURE-NURTURE DEBATE?

Scientists debate the relative importance of heredity and environment in human development, but both are essential in socialization. Studies of many pairs of identical twins have revealed that genetic factors seem to play a significant role in problems such as dyslexia and mental illness. Generally, however, at least half of the differences in twins' outcomes are due to environmental effects, such as the parents' social class (which affects children's educational opportunities), having friends who abuse alcohol and drugs, and cultural factors (Asian-American twins born in the United States, for example, are more likely than those of recent immigrants to suffer from obesity) (Guo 2005). In effect, then, nature and nurture interact in socialization processes and outcomes.

Margaret Mead (1901–1978) carried out a number of field studies in the Pacific. Many academics have described Mead as an influential feminist scholar because her work, especially her demonstration that gender roles and child-rearing practices differ in a variety of cultures, helped to break down stereotypes about what is natural (or innate).

3 Sociological Explanations of Socialization

a common misconception is that our social identity is carved in stone by about age 4. In reality, our attitudes and behavior can change over time:

One of the greatest thrills of being a woman of 70 is having the luxury to be open about what I really think. When I was younger, I was so afraid of hurting people or worried about what they would think of me that I kept my mouth shut. Now when I don't like something, I speak up. . . . And I feel better about myself now than I have at any other time in my life. (Belsky 1988: 65–66)

This woman's daughter and sister complained that she had suddenly become stubborn in her later years, but she had simply decided to be more frank. Theoretical explanations help us understand these and other developmental changes.

Functionalism provides a foundation for understanding the purposes of socialization described at the beginning of this chapter. For functionalists, socialization encourages conformity and maintains a society's stability by controlling disruptive behavior. However, functionalists don't tell us *how* socialization works on a micro level, and they treat socialization as a one-way process in which people adapt to culture rather than make choices and change society. Two important approaches that examine interpersonal relationships and other micro-level factors are social learning theories and symbolic interaction theories (*Table 4.2* summarizes these perspectives). Let's begin with social learning theories.

SOCIAL LEARNING THEORIES

The central notion of **social learning theories** is that people learn new attitudes, beliefs, and behaviors through social interaction, especially during childhood. We learn how to behave both directly and indirectly, for example, through observation and reinforcement. We can also learn how to act without actually performing the behavior.

© AP Images

© Photodisc/Getty Images

children whose parents drank at least monthly were three times more likely to buy alcohol. The researchers concluded that observing parental behavior may influence children to view smoking and drinking as appropriate and to adopt such practices later in life (Dalton et al. 2005).

Learning and Performing

Social learning theorists often make a distinction between *learning* and *performing* behavior. Children and adults can learn to do something through observation, but they don't always imitate the behavior. For example, children may see their friends cheat in school but don't do so themselves. Adults, similarly, may see their coworkers steal office supplies but buy their own.

Social learning theorists maintain that we behave in certain ways because of past rewards and punishments, modeling, and observation. We behave as we do, then, because our society teaches us what's appropriate and not (Bandura and Walters 1963; Mischel 1966; Lynn 1969).

Critical Evaluation

Social learning theories have contributed to our understanding of why we behave as we do, but much of their emphasis is on early socialization rather than on what occurs throughout life. Social learning theories also don't explain why reinforcement and modeling work for some children but not others, especially those in the same family. There may be personality differences, but if learning is as effective as social learning theorists maintain, siblings' attitudes and behavior should be more similar than different. This is often not the case, however, even with identical twins.

Direct and Indirect Learning

In most cases, our socialization is *direct*. Through reinforcement, we learn many behaviors by being rewarded or punished. A little girl who puts on her mother's makeup may be told she is cute, but her brother will be scolded ("Boys don't wear makeup!"). If a child is caught lying, she or he will be punished through a variety of sanctions, such as not being able to play with friends, not being allowed to watch television or use the computer, or having to do extra household chores.

Socialization also involves *indirect* reinforcement through *modeling* (imitating people who are important in our lives). In one study, for example, researchers observed preschoolers aged 2 to 6 who pretended to shop for a visiting friend of their Barbie or Ken doll. The pretend store stocked 133 miniature items, including meat, fruit, vegetables, snacks, cigarettes, beer, and wine. Overall, 28 percent of the children bought cigarettes, and 61 percent bought alcohol. The children whose parents smoked were almost four times more likely to buy cigarettes. The

---

TABLE 4.2
Key Elements of Socialization Theories

SOCIAL LEARNING THEORIES	SYMBOLIC INTERACTION THEORIES
• Social interaction is important in learning appropriate and inappropriate behavior. • Socialization relies on direct and indirect reinforcement.	• The self emerges through social interaction with significant others. • Socialization includes role taking and controlling the impression we give to others.
Example: Children learn how to behave when they are scolded or praised for specific behaviors.	*Example:* Children who are praised are more likely to develop a strong self-image than those who are always criticized.

Another weakness is that most social learning theories ignore factors such as birth order, which brings different advantages and disadvantages. Except for affluent families, larger families have fewer resources for the second, third, and later children. The first child may be disciplined more but may also be encouraged to pursue interests that could lead to higher education and a prestigious career. In contrast, later-born children may enjoy less parental attention and be less likely to receive financial support to pursue a college education (Zajonc and Markus 1975; Conley 2004).

SYMBOLIC INTERACTION THEORIES

Symbolic interaction theories have had a major impact in explaining social development. Sociologists Charles Horton Cooley (1864–1929), George Herbert Mead (1863–1931), and Erving Goffman (1922–1982) were especially influential in showing how social interaction shapes socialization.

Charles Horton Cooley: The Looking-Glass Self

After carefully observing the development of his young daughter, Charles Horton Cooley (1909/1983) concluded that children acquire a sense of who they are through their interactions with others, especially by imagining how others view them. The sense of self, then, is not innate but develops out of social relationships. Cooley called this social self the *reflected self*, or the **looking-glass self**, a self-image based on how we think others see us. He proposed that the looking-glass self develops in an ongoing process of three phases:

- *Phase 1: Perception.* We imagine how we appear to other people and how they *perceive* us ("She thinks I'm attractive" or "I bet he thinks I'm fat").

- *Phase 2: Interpretation of the perception.* We imagine how others *judge* us ("She's impressed with me" or "He's disgusted with the way I look").

- *Phase 3: Response.* We experience *self-feelings* based on what we regard to be others' judgments of us. If we think others see us in a favorable light, we may feel proud, happy, or self-confident ("I'm terrific"). If we think others see us in a negative light, we may feel angry, embarrassed, or insecure ("I'm pathetic").

Our interpretation of others' perceptions (Phase 2) may be totally wrong. The looking-glass self, remember, refers to how we *think* others see us rather than what they *actually* think of us. However, our perceptions of other people's views—whether we're right or wrong—mold our self-image, which then guides our behavior.

Cooley focused on how children acquire a sense of who they are through their interactions with others, but he noted that the process of forming a looking-glass self does not end in childhood. Instead, our self-concept may change over time because we reimagine ourselves as we think others see us—attractive or ugly, interesting or boring, intelligent or stupid, graceful or awkward, and so on.

Sometimes we're very aware of the process shaping our looking-glass self. Consider your own self-concept: If others keep telling you what a great student you are, you'll likely accept their judgments and see yourself as intelligent. In many settings, however—such as public places, large classrooms, and sports events—the looking-glass self is irrelevant because there is relatively little interpersonal interaction and the other people are strangers or aren't very significant in our lives.

George Herbert Mead: Development of the Self and Role Taking

Cooley described how an individual's sense of self emerges, but not how it develops. George Herbert Mead, one of Cooley's colleagues, took up this task. For Mead (1934), the most critical social interaction occurs in the family, the foundation of socialization.

Newborn infants lack a sense of **self**, an awareness of having a social identity. Gradually, they begin to differentiate themselves from their environment, distinguish the faces of caregivers and strangers, and develop a sense of self. We achieve self-awareness, according to Mead, when we learn to differentiate the *me* from the *I*—two parts of the self. The *I* is creative, imaginative, impulsive, spontaneous, nonconformist, self-centered, and sometimes unpredictable. The *me* which has been successfully socialized, is aware of the attitudes of others, has self-control, and has internalized social roles. Instead of impulsively and selfishly grabbing another child's toys, for example, as the *I* would do, the *me* asks for permission to use someone's toys and shares them with others.

For Mead, the *me* forms as children engage in **role taking**, learning to take the perspective of others. Children gradually acquire this ability beginning early in the socialization process, through three sequential phases or stages:

1. **Preparatory stage (roughly birth to 2 years).** An infant doesn't distinguish between the self and oth-

ers. The *I* is dominant, while the *me* is forming in the background. In this stage, children learn through imitation. They may mimic daddy's shaving or mommy's angry tone of voice without really understanding the parent's behavior. In this exploratory stage, children engage in behavior that they rarely associate with words or symbols, but they begin to understand cause and effect (for example, crying leads to being picked up). Gradually, as the child begins to recognize others' reactions (to form a looking-glass self), he or she builds the potential for a self.

2. **Play stage (roughly 2 to 6 years).** The child begins to use language and understand that words (like *dog* and *cat*) have a shared cultural meaning. Through play, children begin to learn role taking in two ways. First, they emulate the words and behavior of **significant others**, the people who are important in one's life, such as parents or other primary caregivers and siblings. The child learns that he or she has a self that is distinct from that of others, that others behave in many different ways, and that others expect her or him to behave in specific ways. In other words, the child learns social norms (see Chapter 3).

    In the play stage, the child moves beyond imitation and acts out imagined roles ("I'll be the mommy and you be the daddy"). The play stage involves relatively simple role taking because the child plays one role at a time and doesn't yet understand the relationships between roles. This stage is crucial, according to Mead, because the child *is learning to take the role of the other.* For the first time, the child tries to imagine how others behave or feel. The *me* grows stronger because the child is concerned about the judgments of significant others.

    Also in the play stage, children experience **anticipatory socialization**, the process of learning how to perform a role they don't yet occupy. By playing "mommy" or "daddy," children prepare themselves for eventually becoming parents. Anticipatory socialization continues in later years, for example, when expectant parents attend childbirth classes, job seekers practice their skills in mock interviews, and many high school students visit campuses and attend orientation meetings to prepare for college life.

3. **Game stage (roughly 6 years and older).** This stage involves acquiring the ability to understand connections between roles. The child must "not only take the role of

the other . . . but must assume the various roles of all participants in the game, and govern his action accordingly" (Mead 1964: 285). Mead used baseball to illustrate this stage. In baseball (or other organized games and activities), the child plays one role at a time (such as batter) but understands and anticipates the motivations and actions of other players (pitcher, shortstop, runners on bases) on both teams. As children grow older and interact with a wider range of people, they learn to respond to and fulfill a variety of social roles.

The game stage enables understanding and taking the role of the **generalized other**, people who do not have close ties to a child but who influence the child's internalization of society's norms and values. The generalized other, then, exerts control over the *I* and ensures some predictability in life. For sociologists, the development of the generalized other is a central feature of the socialization process because the *me* becomes an integral part of the self (see *Figure 4.1*).

Even after the generalized other has developed, the *me* never fully controls the *I*, even in adulthood. We sometimes break rules or act impulsively, even though we know better. Or, when we are frustrated or angry, we may criticize others for no reason.

Erving Goffman: Staging the Self in Everyday Life

Cooley and Mead examined how self-concept and role taking arise through interaction during early socialization. Erving Goffman (1959, 1969) extended these analyses by showing that we act differently in different settings throughout adulthood. Goffman proposed that social life mirrors the theater because we are like actors: We engage in "role performances," want to influence an "audience," and can have considerable control over the image that we project while we're "on stage," concepts explored in greater detail in Chapter 5.

In a process that Goffman called **impression management**, we provide information and cues to

significant others the people who are important in one's life, such as parents or other primary caregivers and siblings.

anticipatory socialization the process of learning how to perform a role one doesn't yet occupy.

generalized other a term used by George Herbert Mead to refer to people who do not have close ties to a child but who influence the child's internalization of society's norms and values.

impression management the process of providing information and cues to others to present oneself in a favorable light while downplaying or concealing one's less appealing qualities.

© C Squared Studios/Photodisc/Getty Images

FIGURE 4.1

Mead's Three Stages in Developing a Sense of Self

STAGE 1: **Preparatory Stage** (under age 2)

No distinction between self and others; the child is self-centered and self-absorbed

Learns through observation

STAGE 2: **Play Stage** (aged 2 to about 6)

Distinguishes between self and others

Imitates significant others (usually parents)

Learns role taking, assuming one role at a time, in "let's pretend" and other play that teaches **anticipatory socialization**

STAGE 3: **Game Stage** (aged 6 and older)

Understands and anticipates multiple roles

Connects to societal roles through the **generalized other**

others to present ourselves in a favorable light while downplaying or concealing our less appealing qualities. Being successful in this presentation of the self requires managing three types of expressive resources. First, we try to control the *setting,* the physical space, or "scene," where the interaction takes place and we play roles. In the classroom, "a professor may use items such as chalk, lecture notes, computers, videos, and desks to facilitate an engaging class and show that he or she is an excellent teacher" (Sandstrom et al. 2006: 105).

A second expressive resource that we try to control is *appearance*—features such as clothing and titles that convey information about our social status. When physicians or professors use the title "Doctor," for example, they are telling the "audience" to respond to them respectfully.

The third expressive resource is *manner*—the mood or style of behavior we display that sends important messages to the audience. For example, faculty members usually manage their manner in responding to students. When you e-mail a professor about a problem with your research project, the response is usually inviting ("I'll be happy to discuss your project") even though the professor may actually feel imposed upon

("Your project has nothing to do with this course; stop wasting my time"). Sometimes, no matter how much we try to manage setting, appearance, and manner, we slip up in our role performances. This leads to discomfort and embarrassment for everyone involved, but the audience is typically tactful:

Imagine that you are giving a presentation in one of your classes and, just as you make a key point, a large droplet of saliva sprays from your lips and lands near someone sitting in the front row. Although several of your classmates cannot help but notice it, none of them are likely to shout out, "Hey, you just about spit on someone in the first row!" Instead, they will probably show tact and act as if nothing unusual or embarrassing took place (Sandstrom et al. 2006: 109).

As we move from one situation to another, according to Goffman, we maintain self-control, for example, by avoiding emotional outbursts and altering our facial expressions and verbal tones. Thus, all of us engage in impression management almost every day.

Critical Evaluation

Symbolic interactionists have provided major insights about socialization, especially that of young children as they form a sense of self and learn role taking. Interactionists have also shown that fitting into our social world is a complex process that continues into and through adulthood. Like other theories, however, symbolic interactionism has its limitations. For example, it's not clear why some children have a more positive looking-glass self than others, even when the cues are consistently negative, as when some children who grow up in abusive homes and attend low-quality schools are successful later in life.

Some scholars have criticized the vagueness of Mead's essential concepts of self, me, and I. Because the concepts are imprecise, it is difficult, if not impossible, to measure them. Some sociologists have also challenged Mead's claim that children automatically pass through play and game stages as they get older. Instead, some critics contend, the extent of socialization depends on a child's social context, such as whether or not parents and other primary caregivers are actively involved in the child's upbringing and provide enriching interaction. Even then, some critics contend, some children never develop the role of the generalized other (see Ritzer 1992).

Other scholars have questioned the value of the concept of the generalized other in understanding early socialization processes. Because children interact in many social contexts (home, preschool, play groups), they do not simply assume the role of one generalized other, but many. That is, as children get older, they may have several **reference groups**, groups of people who shape an individual's self-image, behavior, values, and attitudes in different contexts (Merton and Rossi 1950; Shibutani 1986).

Another weakness is that interactionists credit people with more free will than they have. Individuals don't always have the power or ability to shape others' reactions or their social world. For example, impression management is a skill that is less common among lower socioeconomic groups than in the middle classes, whose members have internalized such expectations as part of fitting in. In many cases, also, how people react to us may reflect our membership in a particular group or social category (for example, due to one's sex, age, or race/ethnicity) that we can't change (Powers 2004).

A major criticism of Cooley, Mead, Goffman, and other interactionists, is that they tend to downplay or ignore macro-level and structural forces that affect our development (Ritzer 1992). As the next two sections show, socialization processes are diverse, especially in multicultural countries like the United States. Even in societies and communities that have similar characteristics, large-scale social structures (such as the economy, educational institutions, and popular culture) have a major impact on socialization. Let's begin by looking at socialization agents.

4 Primary Socialization Agents

babies waste no time in teaching adults to meet their needs:

> *By such responses as fretting, sounds of impatience or satisfaction, by facial expressions of pleasure, contentment, or alertness [the infant] . . . "tells" the parents when he wants to eat, when*

he will sleep, when he wants to be played with, picked up, or have his position changed. . . . From his behavior [parents] learn what he wants and what he will accept, what produces in him a state of well-being and good nature, and what will keep him from whining. (Rheingold 1969: 785–86)

reference groups groups of people who shape an individual's self-image, behavior, values, and attitudes in different contexts.

agents of socialization the individuals, groups, or institutions that teach us what we need to know to participate effectively in society.

As this quotation illustrates, infants socialize parents (and others) to be caretakers. The family, peer groups, teachers, and the media are some of the primary **agents of socialization**—the individuals, groups, or institutions that teach us what we need to know to participate effectively in society.

FAMILY

Parents, siblings, grandparents, and other family members play a critical role in our socialization. Parents, however, are the first and most influential socialization agents.

How Parents Socialize Children

The goal of socialization is to enable the child to regulate her or his own behavior and to make socially responsible decisions. In teaching their children social rules and roles, parents rely on several of the learning techniques discussed earlier in this chapter, such as reinforcement, to encourage desired behavior. Parents also manage many aspects of the environment that influence a child's social development: They choose the neighborhood where the child lives (which often determines what school a child attends), decorate the child's room in a masculine or feminine style, provide the child with particular toys and books, and arrange social events and other activities (such as sports, art, and music) to enrich the child's development.

Parenting Styles

Parents have a variety of parenting styles (see *Table 4.3*) that can affect socialization. In *authoritarian parenting*, parents tend to be harsh, unresponsive, and rigid and to use their power to control a child's behavior. *Authoritative parenting* is warm, responsive, and involved yet unobtrusive. Parents set reasonable limits and expect

TABLE 4.3
Parenting Styles

PARENTING STYLE	CHARACTERISTICS	EXAMPLE
Authoritarian	Very demanding, controlling, punitive	"You can't borrow the car because I said so."
Authoritative	Demanding, controlling, warm, supportive	"You can borrow the car, but be home by curfew."
Permissive	Not demanding, warm, indulgent, set few rules	"Borrow the car whenever you want."
Uninvolved	Neither supportive nor controlling	"I don't care what you do; I'm busy."

appropriately mature behavior from their children. *Permissive parenting* is lax: Parents set few rules but are usually warm and responsive. *Uninvolved parenting* is indifferent and neglectful. Parents focus on their own needs rather than those of the children, spend little time interacting with the children, and know little about their interests or whereabouts (Baumrind 1968, 1989; Maccoby and Martin 1983; Aunola and Nurmi 2005).

Healthy child development is most likely in authoritative homes, in which parents are consistent in combining warmth, monitoring, and discipline. Authoritative parenting styles tend to produce children who are self-reliant, achievement-oriented, and successful in school. Adolescents in households with authoritarian, permissive, or uninvolved parents tend to have poorer psychosocial development, lower school grades, lower self-reliance, and higher levels of delinquent behavior, and they are more likely to be swayed by harmful peer pressure (to use drugs and alcohol, for example) (National Center on Addiction and Substance Abuse 2003; Dorius et al. 2004; Eisenberg et al. 2005).

Such findings vary by social class and race/ethnicity, however. For many Latino and Asian immigrants, authoritarian parenting produces positive outcomes, such as academic success. This parenting style is also effective in safeguarding children who are growing up in low-income neighborhoods with high levels of crime and drug peddling (Brody et al. 2002; Pong et al. 2005).

Siblings

Siblings, like parents, play important roles in the socialization process. Children can bully and abuse their (usually) younger brothers and sisters through name-calling, ridiculing, destroying personal possessions, and even physical or sexual abuse. Such behavior can leave lasting emotional scars that lower a child's self-esteem, and physical abuse sends the message that violence is acceptable for resolving conflict now and in the future (Gelles 1997; Wiehe 1997).

Siblings can also be supportive: They can help younger brothers and sisters with their homework and protect them from neighborhood and school bullies. During adolescence and adulthood, siblings often act as confidantes and as role models on how to get along with people, and they often offer financial and emotional assistance during stressful times (Conger and Elder 1994; McCoy et al. 2002).

Grandparents

Grandparents are often the glue that keeps a family close. They may pass on family rituals and help their adult children by providing emotional support, encouragement, and day-to-day practical aid (such as baby-sitting). In addition, grandparents lend a hand during emergencies and crises, including illness and divorce (Szinovacz 1998 provides a good summary of the grandparenting role in various racial/ethnic families). No matter how strict they were with their own children, many grandparents

© Rubberball/Jupiterimages

are family mediators, advocates for their grandchildren's point of view, and emotional outlets who provide shoulders to cry on.

PLAY AND PEER GROUPS

Play is important in children's development because it provides pleasure, forms friendships, and builds communication, emotional, and social skills. Peer groups also play a significant role in our socialization. A **peer group** consists of people who are similar in age, social status, and interests. All of us are members of peer groups, but such groups are especially influential until about our mid-20s. After that, coworkers, spouses, children, and close friends are usually more important than peers in our everyday lives.

Play and Its Functions

As you saw earlier, George Herbert Mead believed that play and games were critical sources of socialization: We become more skilled in using language and symbols, learn role playing, and internalize roles that we don't necessarily enact (the generalized other). Play serves several other important functions.

First, play promotes cognitive development. Whether it's doing simple five-piece puzzles or tackling complex video games like *Sim City* and *Roller Coaster Tycoon*, play encourages children to think, formulate strategies, and budget and manage resources.

Second, play—especially when it's structured—keeps children out of trouble and enhances their social development. For example, a study of middle- and working-class 10-year-old children found that those who devoted more of their free time to structured and supervised activities—such as hobbies and sports—rather than just hanging out with their friends performed better academically, were emotionally better adjusted, and had fewer problems at school and at home. In effect, sports and hobbies provide children with constructive ways to channel their energy and intelligence (McHale 2001).

Some sociologists and other social scientists caution, however, that overprogramming childhood by creating too much structure (as opposed to allowing time for free play, which encourages creativity) can lead to stress. Children today tend to have more hectic schedules; those aged 12 and under have 6 hours a week of free play, compared with almost 10 hours in 1981. Many parents have no idea that their children are over-booked: Kids are three times more likely to feel time-deprived than their parents believe they are.

peer group any set of people who are similar in age, social status, and interests.

Some children don't mind. Others complain that their days are filled with "a mad scramble of sports, music lessons, prep courses, and then hours of homework" (Karasik 2000; Weiss 2001: 50).

Third, play can strengthen peer relationships. Beginning in elementary school, few things are more important to most children than being accepted by their peers. Even if children aren't popular, belonging to a friendship group enhances their psychological well-being and ability to cope with stress (Rubin et al. 1998; Scarlett et al. 2005).

Peers in Children's Socialization

Peer influence usually increases as children get older, and especially during the early teen years, peers often reinforce desirable behavior or skills in ways that enhance a child's self-image ("Wow, you're really good in math!"). In this sense, to use Cooley's concept, peers can help each other develop a positive looking-glass self.

Peers also serve as positive role models. Children acquire a wide array of information and knowledge by observing their peers. Even during the first days of school, children learn to imitate their peers at standing in line, raising their hands in class, and being quiet while the teacher is speaking. When children are assigned to classrooms in elementary school every fall, a peer grapevine quickly circulates information about how to behave ("If you don't do your homework even once, Ms. Hill will call your parents!"). Peers can also teach new skills. A child who's talented in art may help a schoolmate during art class, and one who's good in sports may teach others how to control a soccer ball or hold a baseball bat properly.

Not all peer influence is positive, however. Some cliques encourage high-risk behavior that includes sexual activity, smoking, drinking alcohol, and using other drugs. In most cases, however, peer acceptance (whether one is a jock, a computer geek, or a reporter on the yearbook staff) promotes friendship and a sense of fitting in. Thus, peer groups can be important sources of support for teens and increase their self-esteem, especially for adolescents who have poor relationships with their parents or feel insecure about their appearance or abilities (Strasburger and Wilson 2002).

TEACHERS AND SCHOOLS

Talk show host Oprah Winfrey often praises a fourth-grade teacher who recognized her abilities, encouraged her to read, and inspired her to excel academically. Have there been teachers, like Oprah's, who played a key role in shaping who you are today? Like family and peer groups, teachers and schools have a strong influence on our socialization.

The School's Role in Socialization

By the time children are 4 or 5, school fills an increasingly large portion of their lives. The primary purpose of the school is to instruct children and enhance their cognitive development. Schools do not simply transmit knowledge, they also teach children to think about the world in different ways. Because of the emphasis on multiculturalism, for example, children often learn about other cultures and customs (see Chapters 3 and 14). Even outside of classes, schools affect children's daily activities through homework assignments and participation in clubs and other extracurricular activities.

Because many parents are employed, schools have had to devote more time and resources to topics—such as sex education and drug-abuse prevention—that were once the sole responsibility of families. In many ways, then, schools play an increasing role in socialization.

Teachers' Impact on Children's Development

Teachers are among the most important socialization agents. From kindergarten through high school, teachers play numerous roles in the classroom—instructor, role model, evaluator, moral guide, and disciplinarian, to name just a few. Once children enter school, their relationships with their teachers are important for academic success. Kindergarten and elementary school teachers' reports of behavioral problems (such as unexcused absences) and poor grades often continue into the middle school years (Hamre and Pianta 2001).

POPULAR CULTURE AND THE MEDIA

Because of iPods, iPhones, instant messaging, YouTube, and social networking sites like Facebook, young people are rarely out of the reach of the electronic media. How does such technology affect socialization?

Electronic Media

Television can enhance children's socialization. For example, children aged 2 to 7 who spent a few hours a week watching educational programs such as *Sesame Street, Mister Rogers' Neighborhood, Reading Rainbow, Mr. Wizard's World,* and *3-2-1 Contact* had higher academic test scores 3 years later than those who watched many hours of entertainment-only programs and cartoons (Wright et al. 2001).

The American Academy of Pediatrics (2001: 424) advises parents to avoid television entirely for children younger than 2 years and to limit the viewing time of elementary school children to no more than 2 hours a day to encourage more interactive activities "that will promote proper brain development, such as talking, playing, singing, and reading together." Still, 68 percent of children under 2 view 2 to 3 hours of television daily, and 20 percent have a television in their bedroom, as do one-third of 3- to 6-year-olds (Garrison and Christakis 2005; Vandewater et al. 2007).

On average, seventh- to twelfth-graders spend almost 7 hours a day using electronic media compared with about 2 hours each with parents and friends, and less than 1 hour doing homework (Rideout et al. 2005). Generally, and even in middle-class homes, children and adolescents who use electronic media for recreation (rather than homework) score lower on math, language arts, and reading tests than their counterparts who use such media primarily for homework. High media usage can decrease academic success, but two-thirds of parents are especially concerned that the media contribute to young people's violent or sexual behaviors (Borzekowski and Robinson 2005; Rideout 2007).

Are such concerns justified? Some of the most consistent findings on the impact of television violence, particularly on children and teenagers, include the following:

- It desensitizes viewers to violence because half of the violent incidents show no pain and almost 75 percent of the offenders show no remorse and receive no punishment.
- It increases the fear of being victimized.
- It encourages high-risk behavior (sexual intercourse, drinking, smoking cigarettes and marijuana, cheating, stealing, cutting classes, and driving a car without permission).
- It increases the likelihood of learning aggressive and antisocial attitudes and behaviors, regardless of social class (Kunkel et al. 1999; Villani 2001; Anderson and Bushman 2002).

Many researchers agree that there are negative aspects of television viewing, such as obesity (because frequent television viewing replaces playing outside and encourages eating foods that are advertised on children's programs and are high in fat and sugar). They point out, however, that there is still no evidence that television viewing causes violence. For example, violent crimes in society have decreased despite the increased depiction of violence on television and other electronic media (Powell et al. 2007; Sternheimer 2007).

Like television, video games can provide enjoyment, entertainment, and relaxation. There is a growing consensus, however, that violent video games make violence seem normal. Playing violent video games like *Grand Theft Auto: Vice City, Halo,* and *Max Payne* can increase an individual's aggressive thoughts, feelings, and behavior both in laboratory settings and in real life. Violent video games may be more harmful than violent television and movies because they are interactive, engrossing, and teach aggression-related scripts for resolving conflict. Violent video games also encourage male-to-female violence because much of the violence depicted in them is directed at women (Anderson et al. 2003; Carnagey and Anderson 2005).

It's not clear, however, why violent video games affect people differently. Many young males enjoy playing such video games, for example, but aren't any more aggressive, vicious, or destructive than those who aren't video game enthusiasts (Williams and Skoric 2005).

Advertising and Commercials

The average young person under 18 views more than 3,000 ads every day on television, the Internet, and billboards and in magazines. Increasingly, advertisers are targeting children as early as possible. In the print media, for instance, young people see 45 percent more beer ads and 27 percent more ads for hard liquor

in teen magazines than adults do in their magazines (Strasburger et al. 2006).

We don't believe everything we read in ads, but they affect us nonetheless. Ads and commercials influence us throughout our lifetimes because they play on our insecurities and our search for quick fixes: "All advertisements speak the language of transformation. They tell consumers that their products will change their lives for the better if they buy a particular product" (Sturken and Cartwright 2001: 212–213). Ads and commercials tell us that a product (or a service) will make us happier, sexier, more popular, or more successful:

> *Mouthwash commercials are not about bad breath. They are about the need for social acceptance and, frequently, about the need to be sexually attractive. Beer commercials are almost always about a man's need to share the values of a peer group. An automobile commercial is usually about the need for autonomy or social status. ... Whatever problem you face (lack of self-esteem, lack of good taste, lack of attractiveness, lack of social acceptance), it can be solved, solved fast, and solved through a drug, a detergent, a machine or a salable technique. (Postman and Powers 1992: 123–124)*

A new form of advertising called advergaming, combines online games with advertising. Advergaming is growing rapidly. Sites like Wonka.com (Nestlé), Barbie.com (Mattel), and m-ms.com/us/fungames/games (M&Ms) attract millions of young children and provide marketers with an inexpensive way to "draw attention to their brand in a playful way, and for an extended period of time" (Moore 2006: 5).

5 Socialization Throughout Life

biological aging comes naturally, but social aging is quite a different matter. As we progress through the life course—from infancy to death—we are expected to act our age and to learn culturally approved norms, values, and roles.

INFANCY

Infants require what one sociologist calls "continuous coverage." That is, "They need to be talked to, listened to, cuddled, fed, cleaned, carried, rocked, burped, soothed, put to sleep, taken to the doctor and so on" (LaRossa 1986: 88). As you saw earlier, infants develop physically, cognitively, and socially as long as parents (and other primary caregivers) assume responsible roles.

Some scientists describe healthy infants' brains as "small computers" because of their enormous capacity for learning: "Babies are born with powerful programs already booted up and ready to run. Their experiences lead babies and young children to enrich, modify, revise, reshape, reorganize, and sometimes replace their initial [information]." As children explore their environment and interact with parents and others, they have "a succession of progressively more powerful and accurate programs" (Gopnik et al. 2001: 142–143).

CHILDHOOD

According to French historian Philippe Ariès (1962), childhood—viewed as a distinct stage of development—is a fairly recent phenomenon. In medieval Europe, children as young as 6 dressed like adults and took part in the same work and recreational activities as adults. In the U.S. colonies, child labor was nearly universal. At the beginning of the nineteenth century, American children began to spend more time playing than working, and adolescence became a new stage of life without adult responsibilities. More books for and about children were published, and people began to recognize children's individuality by giving them names that were different from those of their parents. In the

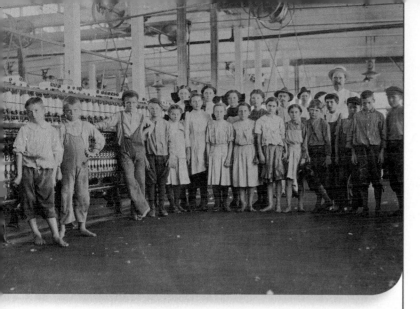

nineteenth century, people began for the first time to celebrate birthdays, especially those of children. There was also a marked decline in the use of corporal punishment, and physicians and others began to recognize the early onset of sexual feelings in children (Degler 1981; Demos 1986). At the turn of the twentieth century, children in the United States were first required to attend school until age 16, and then in 1938, the federal government passed its first law forbidding child labor.

Most children today enjoy happy and healthy lives, but many experience a rough socialization. About 1.5 million American children under age 18 have an incarcerated parent, a number that has increased by a third since 1991. Another 542,000 children are in the foster care system. And, every year, at least 1 million children experience mistreatment that includes neglect and physical and sexual abuse (Mumola 2000; U.S. Department of Health and Human Services 2007). Thus, for millions of U.S. children, the socialization process is shaky at best.

ADOLESCENCE AND YOUNG ADULTHOOD

Like the concept of childhood, that of the teenager didn't exist until recently; in fact, it is not yet universal. In many developing countries, children still work side by side with adults, marry at very young ages (even at 13), and experience a quick transition from childhood to adulthood. In industrial societies, adolescence is prolonged. Because an industrial society needs a highly trained and educated labor force, many young adults pursue their studies well into their late 20s and, consequently, postpone typical adult responsibilities such as marrying, having children, and getting a full-time job.

Many adolescents experience their first sense of total independence when they leave home after high school to find a job and their own housing or to attend college. Most young adults, especially women, leave the parental nest by age 23, but the proportion of those aged 25 to 34 who are living with their parents jumped from 9 percent in 1960 to nearly 17 percent in 2000 (Fields and Casper 2001). Some young adults live with their parents simply to enjoy the comforts of the parental nest. As one of my late-20s male students said, "I have lots of freedom without worrying about bills." Usually, however, macro-level factors push young adults to stay at their parents' home or move back in. Student loans, low wages, soaring house prices, credit card debt, and the desire to find "meaningful" work have made it harder for young adults, especially those raised in middle-class families, to maintain the lifestyles that their parents had provided.

ADULTHOOD

Socialization continues throughout adulthood, the period roughly between ages 21 and 65. Most adults adopt a series of new roles—work or career, marriage, parenthood, divorce, remarriage, buying a house, and experiencing the death of a parent or grandparent. We'll examine these transitions in later chapters. Here, let's consider two of the most important new roles in adulthood: work and parenthood.

Work Roles

The average American holds at least nine jobs from age 18 to age 34 alone (U.S. Department of Labor 2002). Such job changing means that our occupational socialization is an ongoing process. Learning work roles is difficult because each job has different demands and expectations. Even when there is formal job training, we must also learn the subtle rules that are implicit in many job settings. A supervisor who says, proudly and loudly, "I have an open door policy. Come in and chat whenever you want," may *really* mean, "I have an open door policy, but I really don't want to hear complaints." Even if we remain in the same job for many years, we must often acquire new skills, especially those related to computers, for example, or risk becoming labeled obsolete and being laid off.

As U.S. companies have transferred many jobs overseas, many workers have been fired or laid off. Being laid off can be stressful at any age. In the case of midlife men (generally defined as those aged 45 to 60), such changes are especially traumatic: They must learn new job-hunting skills and must often adjust to as

FIGURE 4.2
Yearly Cost of Raising Children

In 2007, middle-income families—those earning $45,800 to $77,100 a year—spent almost $11,000 *per year* for a child under 2. This amount does not include the costs of prenatal care or delivery.

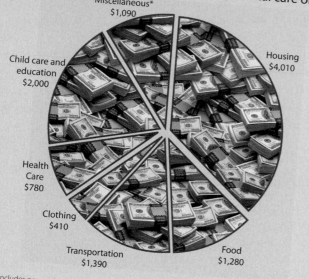

Miscellaneous*
$1,090

Child care and education
$2,000

Housing
$4,010

Health Care
$780

Clothing
$410

Transportation
$1,390

Food
$1,280

$10,960

*Includes personal care items, entertainment, and reading materials.
Source: Based on Lino 2008, Table EST1.

© Don Farrall/Photodisc/Getty Images/
© Dynamic Graphics/Creatas Images/Jupiterimages

much as a 70 percent decrease in their earnings. Instead of enjoying the security of a job in their midlife years, these men must undergo occupational socialization that assaults their self-image as good employees and family providers (see Chapter 12).

Parenting Roles

Like workplace roles, parenting does *not* come naturally. Most first-time parents muddle through by trial and error. Family sociologists often point out that we get more training for driving a car than for marriage and parenting. For example, most couples don't realize that raising children is expensive. Middle-income couples, with an average income of $61,000 a year, spend about 18 percent of their earnings on a child during the first 2 years (see *Figure 4.2*). Child-rearing costs are much higher for single parents and especially for low-income families if a child is disabled, chronically ill, or needs specialized care that welfare benefits don't cover (Lukemeyer et al. 2000; Lino 2008).

Arguments over finances and child-rearing strategies are two of the major reasons for divorce (see Chapter 13). All in all, adjusting to marital and parental roles during adulthood requires considerable patience, effort, and work.

LATER LIFE

Later life also requires adjusting to new situations. Because we live longer, we may spend 20 percent of our adult life in retirement. When retired people are unhappy, it's usually because of health or income problems rather than the loss of the worker role. If retirement benefits do not keep up with inflation or if a retiree is not covered by a pension plan, poverty can become an actuality soon after retirement (Toder 2005).

Most Americans aged 65 and over enjoy healthier lives than did past generations. Some older people move closer to their adult children and grandchildren. Those who can afford to do so pay from $300,000 to $500,000 for an upscale apartment or house in an age-restricted community that is equipped with modern kitchens and electronics, has medical facilities, and is close to shopping centers and golf courses (Wedner 2008).

Comedian Woody Allen once quipped, "It's not that I'm afraid to die. I just don't want to be there when it happens." Older people who are in poor health and experience continuous pain sometimes welcome death, but many Americans, including those who are over 65, avoid thinking about and preparing for new roles that can include frail health, dependence on others for caregiving, and, ultimately, death. Some clinicians say that death would be easier if our society implemented end-of-life conversations that included the elderly, family members, physicians, and other health care professionals (Larson and Tobin 2002; Ratner and Song 2002). For the most part, however, many aging people (and their families) avoid such discussions.

6 Resocialization

Socialization doesn't always occur in a linear or predictable fashion. Sometimes people need to learn totally new behavior patterns, or be resocialized. **Resocialization** is the process of unlearning old ways of doing things and adopting new attitudes, values, norms, and behavior. Much resocialization takes place in what sociologist Erving Goffman (1961) called **total institutions**—places, such as military boot camps, mental hospitals, prisons, concentration camps, and some religious orders, where people are isolated from the rest of society, stripped of their former identities, and required to conform to new rules and behavior.

Some resocialization is voluntary, such as when an American wife moves to the Middle East with her Iranian husband and must follow local customs about veiling, women's submissive roles, and staying out of public life. Other examples of voluntary resocialization include entering a religious group, seeking treatment in a drug-abuse rehabilitation facility, or serving in the military.

Resocialization can also be involuntary, as when children are sent to a foster home or a juvenile detention camp. Correctional institutions, especially, exemplify involuntary resocialization. Prisons have buildings that are physically separated from the rest of society. Whether built in rural or urban areas, prisons have high fences, barred windows, armed corrections officers at the entrance, and the staff has almost complete control over the prisoner. The inmates are strip-searched, deloused, and fingerprinted; their heads are shaved; they are issued a uniform; they are given a serial number; and they are told what to do and when to do it. Prisoners have almost no privacy, limited access to family and friends, and little communication with the outside world. The purpose of these practices and restrictions, according to Goffman, is to destroy any sense of individualism, autonomy, and past identity and to shape a more compliant person (we'll examine some of the effects of involuntary resocialization in Chapter 7).

As this chapter shows, socialization is a powerful force in shaping who we are. It does not produce robots, however. We remain creative creatures who adapt to new environments, change as we interact with others, and develop distinctive ways of achieving our goals.

resocialization the process of unlearning old ways of doing things and adopting new attitudes, values, norms, and behavior.

total institutions places where people are isolated from the rest of society, stripped of their former identities, and required to conform to new rules and behavior.

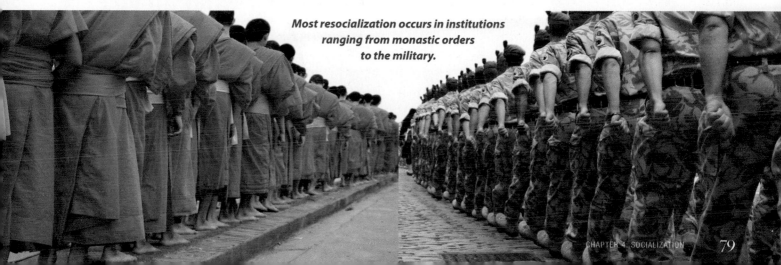

Most resocialization occurs in institutions ranging from monastic orders to the military.

All of us

conform to social structure.

what do you think?

I don't mind when people stand

close to me when talking.

1 2 3 4 5 6 7

strongly agree   strongly disagree

5

Social Interaction and Social Structure

Generally, African Americans have higher death rates at an early age and tend to receive lower-quality health care than whites, even when both groups are similar in family medical history, social class, and health insurance coverage (see Kreps 2006 for a summary of some of these studies). Health practitioners were puzzled by this racial difference until a groundbreaking study suggested that interaction might be a key factor. Regardless of race or ethnicity, physicians talked 43 percent more than their black patients but only 24 percent more than their white patients and the doctors' comments and voice tone were more positive with white patients (Johnson et al. 2004).

> **social interaction** the process by which we act toward and react to people around us.

Subsequent studies supported the finding that doctor-patient interaction patterns often differ with white and minority patients. Anticipating that doctors will be receptive to their questions, white patients often ask more questions, which in turn can elicit more medically useful information. In contrast, minority black patients may see comments about losing weight or reducing alcohol consumption as criticisms of their lifestyles and may not ask many questions to avoid appearing stupid or wasting the doctor's time (especially when a physician hurries from one patient to another) (Perloff et al. 2006).

The doctor-patient example illustrates four critical components of **social interaction**, the process by which we act toward and react to people around us (Maines 2001; Schwalbe 2001):

- Social interaction is central to all human social activity.
- People respond differently during social interaction, depending on what they think is at stake for them ("I may seem stupid if I ask a lot of questions").
- People influence each other's behavior through social interaction ("Mr. Smith doesn't have any questions, so I can move on to the next patient").
- Elements of social structure, such as race and ethnicity, affect all social interaction and can produce different personal outcomes.

Key Topics

In this chapter, we'll explore the following topics:

All of us, regardless of the situation, conform to a cultural social structure that shapes our roles, status, social interaction, and nonverbal and online communication.

1 Social Structure

in 2007, a Southwest Airlines attendant told a 23-year-old woman, after complaints from some passengers, that her revealing mini-skirt and tank top were inappropriate for a family airline. The woman was miffed because she had to cover herself with an airline blanket to avoid being escorted off the plane.

Whether you agree or disagree with airline policies about appropriate clothing and other things, such policies are part of social structure. **Social structure** is an organized pattern of behavior that governs people's relationships (Smelser 1988). Because social structure guides our actions, it gives us the feeling that life is orderly and predictable rather than haphazard or random. We're often not aware of the impact of social structure until we violate cultural rules, formal or informal, that dictate our daily behavior. People may sometimes resent the impact of social structure because it limits their personal choices. But, like it or not, social structure and social interaction—whether on an airplane or a doctor's office—shape our daily lives.

Every society has a social structure that encompasses essential components—statuses, roles, groups, organizations, and institutions (Smelser 1988). We'll examine groups, organizations, and institutions in later chapters. Let's take a closer look here at statuses and roles, two building blocks of our everyday lives.

2 Status

for most people, the word *status* signifies prestige: An executive, for example, has more status than a secretary, and a physician has a higher status than a nurse. For sociologists, **status** refers to a social position that an individual occupies in a society (Linton 1936). Thus, executive, secretary, physician, and nurse are all social statuses. Other statuses that are familiar to you include student, professor, musician, voter, sister, parent, police officer, and friend.

A status refers to *any* societal position within a culture. Statuses can be (and are) ranked, as you'll see in Chapter 8, but sociologists don't assume that one position is more prestigious than another. A mother, for example, is not more important than a father, and an adult is not more important than a child. Instead, all statuses are significant because they determine social identity, or who we are.

STATUS SET

Every person has many statuses at the same time (see *Figure 5.1*), and together they form her or his **status set**, a collection of social statuses that a person occupies at a given time (Merton 1968). Dionne, one of my students, is female, African American, 42 years old, divorced, mother of two, daughter, cousin, aunt, Baptist, Maryland voter, supervisor at a bank, volunteer at a soup kitchen, president of her homeowners association, country music fan, and stockholder. All of these socially defined positions (and perhaps others as well) make up Dionne's status set.

Status sets change throughout the life course. Dionne's statuses of child, single, and high school student changed to married, parent, and bank employee. Her status of married changed to divorced, and she added the status of college student. Because she will graduate next year and is considering remarrying and starting an after-school program, Dionne will add at least three more statuses to her status set and will also lose the statuses of divorced and college student. As Dionne ages, she will continue to gain new statuses (grandmother, retiree) and lose others (supervisor at a bank, or wife, if she is widowed).

Statuses are *relational,* or complementary, because they are connected to other statuses: A *husband* has a *wife,* a *real estate agent* has *customers,* and a *teacher* has *students.* No matter how many statuses you occupy, nearly every status is linked to that of one or more other people. These connections between statuses influence our behavior and relationships.

ASCRIBED AND ACHIEVED STATUS

Status sets include both ascribed and achieved statuses. An **ascribed status** is a social position that a person is born into. We can't control, change, or choose our ascribed statuses, which include sex (male or female), age, race, ethnicity, and family relationships. Your ascribed statuses, for example, might include *male, Latino,* and *brother.* Some argue that sex isn't really an ascribed status because people can have sex-change operations. However, a sex change operation doesn't change the fact that someone was born a male or a female (except in a minority of cases, as you'll see in Chapter 9). Thus, most of us are stuck with our ascribed statuses, like it or not.

An **achieved status**, in contrast, is a social position that a person attains through personal effort or assumes voluntarily. Your achieved statuses might include high school graduate, wife, and employee. Unlike our ascribed statuses, our achieved statuses can be controlled and changed. We have no choice about being a son or daughter (an ascribed status), but we have an option as to whether to become a parent (an achieved status).

Students sometimes argue that religion and social class are ascribed rather than achieved statuses. It's true that someone may be born into a family that practices a certain religion or one that is poor, middle class, or wealthy.

Because we can change these statuses through our own actions, however, neither religion nor social class is an ascribed status. A Catholic might convert to Judaism (or vice versa). And thousands of Americans born into poor or working-class families become millionaires because they work hard, save, and live modestly (Stanley and Danko 1996).

MASTER STATUS

An ascribed or achieved status can be a **master status** that determines a person's identity (Hughes 1945; Becker 1963). In most societies—including the United States—one's sex, age, physical ability, and race are master statuses because they are very visible. And when people meet each other, the first question is typically "What do you do for a living?" The answer helps to place a person's status in society ("She's a chemical engineer; she must be educated and successful").

A master status can be positive or negative. Many people admire Bill Gates, the founder of Microsoft, because he's a billionaire, a positive master status. Master statuses can also be negative. Instead of getting to know a person in a wheelchair, for example, we might stigmatize her or him as somehow imperfect and react to the disability rather than the person's accomplishments and engaging personality.

STATUS INCONSISTENCY

Because we hold many statuses, some clash. **Status inconsistency** refers to the conflict that arises from

achieved status a social position that a person attains through personal effort or assumes voluntarily.

master status an ascribed or achieved status that determines a person's identity.

status inconsistency the conflict or tension that arises from occupying social positions that are ranked differently.

FIGURE 5.1
Is This Your Status Set?

The status of college student is only one of your current statuses. What other statuses comprise your status set?

Fan

Brother or Sister

Son or Daughter

Parent or Grandparent

College Student

Significant Other or Spouse

Employer or Employee

Registered Voter

Consumer

Member of Religious Group

© Photodisc/Getty Images

occupying social positions that are ranked differently. Examples include a computer programmer who works as a bartender or a skilled welder who stocks shelves at Wal-Mart because neither can find a better job in a weak economy.

Such discrepancies often derail our interactions. In conversation with the computer scientist or the welder, what should we do? Ask how the lower-level job is going? Or avoid talking about work because it's a sensitive issue? We'll cover status inconsistency in more detail in Chapter 7. For now, you should be aware that you occupy many statuses, and some of them may clash now and in the future.

Leslie Visser graduated from Boston College with honors, was an accomplished athlete, and has been a successful sports analyst, journalist, and sportscaster for over 25 years. When describing Visser's achievement, writers typically focus on her gender: "The first woman to cover Monday Night Football" or "The first woman enshrined into the Pro Football Hall of Fame." Does Visser's master status diminish her achieved status?

3 Role

Each status is associated with one or more roles. A **role** is the behavior expected of a person who has a particular status. We occupy a status but play a role. In this sense, a role is the dynamic aspect of a status (Linton 1936).

College student is a status, but the role of a college student requires many *formal* behaviors such as going to class, reading, thinking, completing weekly assignments, writing papers, and taking exams. *Informal* behaviors may include joining a student club, befriending classmates, attending football games, and even abusing alcohol on weekends.

Like statuses, roles are relational (complementary). Playing the role of professor requires teaching, advising students, being present during office hours, responding to e-mail messages, and grading exams and assignments. Most professors are also expected to serve on committees, do research, publish articles and/or books, and perform services such as giving talks to community groups. Thus, the status of college student or profes-

sor involves numerous role requirements that proscribe who does what, where, when, and how.

Roles can be rigid or flexible. A person who occupies the status of secretary typically plays a role that is defined by rules about when to come to work, how to answer the phone, when to submit the necessary work and in what format, and how many sick days are allowed each year. A boss, on the other hand, usually enjoys considerable flexibility: She or he has more freedom to come in late or leave early, to determine which projects should be completed first, and to decide when to hire or fire employees.

Because roles are based on mutual obligations, they ensure that social relations are fairly orderly. We know what we are supposed to do and what others expect of us. If professors fail to meet their role obligations by missing many classes or coming to classes unprepared, students may respond by studying very little, cutting classes, and submitting negative course evaluations. If students miss classes, don't turn in the required work, or cheat, professors can fail them or assign low grades.

ROLE PERFORMANCE

Roles define how we are *expected* to behave in a particular status, but people vary considerably in their fulfillment of the responsibilities associated with their roles. Many college students succeed, while others fail; some professors inspire their students, while others put them to sleep. These differences reflect **role performance**, the *actual* behavior of a person who occupies a status. For example, a professor may vary her or his role by demanding more of graduate students than of undergraduates. An instructor may also relate differently to male students than to female students or may act differently with a teenager than with an older student who is anxious about returning to school.

ROLE SET

We occupy many statuses, and we play many roles associated with each status. A **role set** refers to the different roles attached to a single status. Every role set includes rights and responsibilities associated with people with different statuses and role sets. *Figure 5.2* illustrates six roles of a typical college student. Because a different set of norms governs each of these relationships, the student interacts differently with a classmate than with a reference librarian or a professor. All of these interactions, shaped by explicit or implicit rules, make up a student's role set.

These six roles reflect only one status—that of college student. If you think about other statuses that a college student may occupy (employee, son or daughter, parent, girlfriend or boyfriend, husband or wife), you can see that meeting the expectations of numerous role sets can create considerable role conflict and role strain.

ROLE CONFLICT AND ROLE STRAIN

Playing many roles often leads to **role conflict**, the frustrations and uncertainties a person experiences when confronted with the requirements of two or more statuses. College students who have a job, especially if it is a full-time one, often experience role conflict. The role conflict increases if the student has young children or cares for an aging parent.

Whereas role conflict arises from tensions *between* the roles of two or more statuses, **role strain** is the stress arising from incompatible demands among roles *within* a single status. Students experience role strain when several exams are scheduled on the same day or when three course papers have the same deadline. Faculty members experience role strain when teaching schedules conflict with presentations at professional conferences.

Almost all of us experience role strain because many inconsistencies are built into our roles. Employees expect supervisors to be friendly but also objective during job evaluations. Parents who want to be friends with their children often undermine their parental authority. And military chaplains report role strain in preaching about peace while blessing those about to go to war. *Table 5.1* on the next page presents examples of some of the factors that create role conflict and role strain.

role set the different roles attached to a single status.

role conflict the frustrations and uncertainties a person experiences when confronted with the requirements of two or more statuses.

role strain the stress arising from incompatible demands among roles within a single status.

COPING WITH ROLE CONFLICT AND ROLE STRAIN

Role conflict and role strain can produce tension, hostility, aggression, and stress-related physical problems such as insomnia, headaches, ulcers, eating disorders, anxiety attacks, chronic fatigue, nausea, weight loss or gain, and drug and alcohol abuse (Weber et al. 1997). Some role conflict and role strain may last only a few weeks but others are long-lived (working in a stressful or low-paying job, for instance).

To deal with role conflict and role strain some people *deny that there's a problem*. Employed mothers, especially those who are divorced, often become super moms—who provide home-cooked meals, attend their children's sports activities, and volunteer for a school's fund-raising campaign—even though they're exhausted, do laundry at midnight, and spend

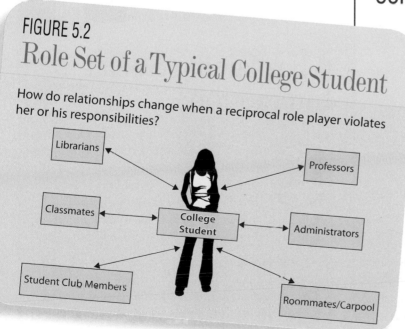

FIGURE 5.2
Role Set of a Typical College Student

How do relationships change when a reciprocal role player violates her or his responsibilities?

- Librarians
- Classmates
- Student Club Members
- College Student
- Professors
- Administrators
- Roommates/Carpool

TABLE 5.1
Why Do We Experience Role Conflict and Role Strain?

REASON	EXAMPLE
Because many people are over-extended, some roles are bound to conflict with others.	Students may study less than they want because employers demand that they work overtime or on weekends.
People have little or no training for many roles.	Parents are expected to live up to high standards and to turn out "perfect" kids even though they receive more training for driving a car than for parenting.
Some role expectations are unclear or contra-dictory.	Some employers pride themselves on having family-friendly policies but expect employees to work 12 hours a day, travel on weekends, and use vacation days to care for a sick child.
Highly demanding jobs often create difficulties at home.	Some jobs (such as being in the military, policing, firefighting, and serving on medical teams after a hurricane or flood) require people to be away from their families for extended periods of time or during crises.

Second, we can *set priorities*. If extracurricular activities interfere with studying, which is more important? Succeeding in college always requires making sacrifices, such as attending fewer parties and not seeing friends as often as we'd like.

Third, we can *compartmentalize* our roles. Many college students take courses in the morning, work part-time during the afternoon or evening, and devote part of the weekend to leisure activities. It's not always easy, but it's usually possible, to segregate our various roles.

Fourth, we can decide *not to take on more roles*. One of the most effective ways to avoid role conflict is to decline requests to do volunteer work, pressure from family or friends to take on unwanted tasks (such as babysitting), or pleas to become involved in college or community activities.

Finally, we can *exit* a role or status. In many cases, such as withdrawing from community activities and club offices, it's fairly easy to leave and can decrease role conflict and role strain considerably. Some role exits are painful and extend over a period of time, however. In the case of divorce, for instance, the process is usually spread over many years (sometimes decades), during which two people (and their children) often have to navigate through uncharted and troubled waters (Bohannon 1971).

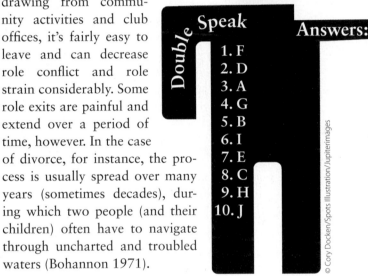

Double Speak

Answers:
1. F
2. D
3. A
4. G
5. B
6. I
7. E
8. C
9. H
10. J

© Cory Docken/Spots Illustration/Jupiterimages

4 Explaining Social Interaction

Statuses and roles are two critical components of social structure that shape our everyday relationships, but it is social interaction that provides the basis of these relationships and affects who you are, how you behave, and what you say. Social interaction seems natural and simple, but it's actually fairly complex:

Social interaction can be taxing on people's attention. Interactants must process what others are saying and doing, infer what they are feeling and thinking, and predict impending solutions.

their leisure time shopping with the kids. In effect, they're doing it all but neglecting their own needs and interests. Super moms may succeed over a number of years in dealing with role conflict and role strain. Eventually, however, they may become angry and resentful or experience health or emotional problems (Douglas and Michaels 2004). There are five other, more effective ways of minimizing role conflict and role strain.

First, we can reduce role conflict through *compromise* or *negotiation*. To decrease the conflict between work and family roles, many couples draw up schedules that call for the father to do more of the housework and child rearing.

All the while, they must plan their own utterances, monitor their actions, and confront the vast inner landscape of their thoughts, feelings, and bodily states. (Malle and Pearce 2001: 278)

Our social interaction begins in infancy and changes over time depending on the situation we're in and our sex, ethnicity, social class, marital status, and age, among other factors. For example, although many men enjoy compliments about their appearance, they rarely ask, "How do I look?" Women do so fairly frequently. And even if the man replies, "Fine," a woman may head for the closet to try on another outfit or worry about what's wrong with what she's wearing.

Why do we interact as we do? Why do our interaction patterns change over time? And why do people sometimes interpret the same words differently? Three micro-level perspectives—symbolic interactionism, social exchange theory, and feminist theories—provide answers to these and other questions. These perspectives offer distinctive contributions, but they have one characteristic in common—each explains how people communicate in their daily lives (*Table 5.2* on p. 90 summarizes these insights).

SYMBOLIC INTERACTIONISM

Symbolic interactionism, you'll recall, examines how people communicate knowledge, ideas, beliefs, and attitudes and how they interpret situations in everyday life (see Chapters 1 and 3). For interactionists, the most significant feature of all human communication is that people take each other and the context into account (Blumer 1969). When your professors ask, "How are you?" they expect a "Fine, thanks" and probably barely look at you. A doctor, on the other hand, usually looks you in the eye when asking "How are you?" and takes notes as soon as you start to reply. Thus, "How are you?" has different meanings in different social contexts and elicits different responses ("I'm doing okay" versus "I've been having a lot of headaches during the last few weeks").

How do we know how to react and what to expect from others? Interactionists say that much of the answer lies in understanding that people construct reality.

The Social Construction of Reality

What people perceive and understand as reality is a creation of the social interaction of individuals and groups. "Human reality is socially constructed reality" because people impose their subjective meanings on interactions to make sense of the world around them (Berger and Luckmann 1966: 172).

Our social interactions aren't isolated incidents. Instead, we produce, interpret, and share the reality of everyday life with others. This social construction of reality typically evolves through direct, face-to-face interaction, but the interaction can also be indirect, as in watching TV shows and movies or participating in online chat rooms. Moreover, word usage is important in shaping our perceptions of reality. Government officials, businesspeople, advertisers, politicians, educators, lobbyists, advocacy groups, and even social scientists use language deliberately to shape or change our perceptions of reality.

Doublespeak is "language that pretends to communicate but really doesn't. [It] makes the bad seem good, the negative appear positive, the unpleasant appear attractive or at least tolerable." There are several kinds of doublespeak: Euphemisms are inoffensive words or phrases that are used to avoid a harsh, unpleasant, or distasteful reality; gobbledygook (or bureaucratese) overwhelms the listener with big words and long sentences; and inflated language makes everyday things seem impressive and makes the simple seem complex (Lutz 1989: 1–6). Take the quiz above to see how much doublespeak you can decipher (answers on opposite page).

Social Interaction and Self-Fulfilling Prophecies

Our perceptions of reality shape our behavior. In an oft-cited statement, also known as the *Thomas Theorem*, sociologists W. I. Thomas and Dorothy Thomas (1928: 572) observed, "If men define situations as real, they are real in their consequences."

Carrying this idea further, sociologist Robert Merton (1948/1996) proposed that our definitions of reality can result in a **self-fulfilling prophecy:** If we define something as real and act upon it, it can, in fact, become real. For example, a national study of American adolescents in grades 7 to 12 found that students performed well in classrooms where teachers were empathetic, supportive, and promoted self-discipline. These students were less likely than those with less supportive teachers to use drugs, engage in violence, initiate sexual activity at an early age, and report a low level of emotional well-being (McNeely et al. 2002). Thus, teachers' positive expectations can elicit positive attitudes and behavior from students.

Our perceptions of reality shape our behavior, but how do people define that reality? For interactionists, two important methodological tools—ethnomethodology and dramaturgical analysis—help answer this question.

Ethnomethodology

A term coined by sociologist Harold Garfinkel (1967), **ethnomethodology** is the study of how people construct and learn to share definitions of reality that make everyday interactions possible.

For ethnomethodologists, reality is socially constructed because we base our interactions on common assumptions about what makes sense in specific situations, without thinking about the implicit rules of behavior that govern our interactions (Schutz 1967; Heritage 1984; Hilbert 1992).

People make sense of their everyday lives in two ways. First, by observing conversations, people can discover the general rules that we all use to interact meaningfully. For example, higher-status people are more likely to interrupt or ignore lower-status people (but not vice versa) (Wood 2003).

Second, people can understand interaction rules by breaking them. Over a number of years, Garfinkel instructed his students to purposely violate everyday interaction rules and then to analyze the results. In these exercises, some of his students went to a grocery store and insisted on paying more than was asked for a product. Others were instructed, in the course of an ordinary conversation and without indicating that any-

thing unusual was happening, "to bring their faces up to the subject's until their noses were almost touching" (Garfinkel 1967: 72). In these and other exercises during which students broke everyday interaction rules, they were penalized. Grocery clerks became hostile when the students insisted on paying more than the marked price for a product, and people backed off when their noses were almost touched: "Reports were filled with accounts of astonishment, bewilderment, shock, anxiety, embarrassment, and anger . . ." (Garfinkel 1967: 47).

Violating interaction rules, even unspoken ones, can trigger anger, hostility, and frustration. College students become upset if professors ignore teaching norms by using sarcasm and putdowns, coming to class unprepared or consistently late, and misplacing students' homework. Students also become distressed if instructors violate expected classroom interaction rules by constantly reading from the book or discouraging students from asking questions (Berkos et al. 2001).

Dramaturgical Analysis

Dramaturgical analysis is a technique that examines social interaction as if occurring on a stage where people play different roles and act out scenes for the "audiences" with whom they interact. According to sociologist Erving Goffman (1959, 1967), life is similar to a play in which each of us is an actor, and our social interaction is much like theater because each of us is always on stage and always performing. In our everyday performances, we present different versions of ourselves to people in different settings (audiences).

Because most of us try to present a positive image of ourselves, much social interaction involves *impression management,* a process of suppressing unfavorable traits and stressing favorable ones (see Chapter 4). To control information about ourselves, we often rely on props to convey or reinforce a particular image. For example, people may decorate their homes to create the impression that they're art collectors, and visitors (the audience) support this impression by admiring the artwork. Physicians, lawyers, and college professors may line their office walls with framed diplomas, medical certificates, or community awards to give the impression that they're competent, respected, and successful.

According to Goffman, the presentation of a performance involves front- and back-stage behaviors. The *front stage* is an area where an actual performance takes place. In front stage areas, such as living rooms or hotel restaurants, the setting is clean and the servers or hosts are typically polite and deferential to guests. The *back stage* is an area concealed from the audience, where

MENU

The chaotic and crowded back-stage conditions in many restaurant kitchens stand in stark contrast to the efficient and relaxed front-stage behavior of its servers.

© Luca Zampedri/Nonstock/Jupiterimages / © Thinkstock Images/Jupiterimages / © IStockphoto.com

people can relax. Bedrooms and restaurant kitchens are examples of back stages. After guests have left, the host and hostess may kick off their shoes and gossip about their company. In restaurants, cooks and servers may criticize the guests, use vulgar language, and yell at each other. Thus, the civility and decorum of the front stage may change to rudeness in the back stage.

Another example of front-stage and back-stage behavior involves faculty and students. Professors want to create the impression in the classroom that they're well prepared, knowledgeable, and hard working. In back-stage areas such as their offices, however, faculty members may complain to their colleagues that they're tired of teaching a particular course or grading terrible exams and papers, or they may confess that they don't always prepare for classes as well as they should. Students often plead with instructors for higher grades: "But I studied very hard" or "I'm an A student in my other courses." In back-stage areas such as dormitories, student lounges, or libraries, however, students may admit to their friends that they barely studied or that they have a low grade-point average.

SOCIAL EXCHANGE THEORY

The fundamental premise of **social exchange theory** is that a social interaction between two people is based on each person's trying to maximize rewards (or benefits)

> **social exchange theory** the perspective whose fundamental premise is that any social interaction between two people is based on each person's trying to maximize rewards (or benefits) and minimize punishments (or costs).

and minimize punishments (or costs). An interaction that elicits a reward, such as approval or a smile, is more likely to be repeated than an interaction that brings a cost, such as disapproval or criticism (Thibaut and Kelly 1959; Homans 1974; Blau 1986).

Interactions are most satisfying when there is a balance between giving and taking. A lack of reciprocity appears unfair and threatens the continuation of the relationship: "It indicates to the person receiving less that the other really does not care enough to maintain exchanges" (Cohen 1981: 46).

People bring various tangible and intangible resources to a relationship—such as money, status, intelligence, good looks, youth, power, talent, fame, and affection. Any of a person's resources can be traded for more, better, or different resources that another person possesses. Marriages between older men and young women, for example, often reflect an exchange of the man's power, money, and/or fame for the woman's youth, physical attractiveness, and ability to bear children.

Many of our cost-reward decisions are conscious and active, but others are passive or based on long-term negative interactions. For example, much of the research on domestic violence shows that women stay in abusive relationships because their self-esteem has eroded after

© Stockbyte/Getty Images

years of criticism and ridicule from both their parents and their spouses or partners ("You'll be lucky if anyone marries you," "You're dumb," "You're ugly," and so on). In effect, the abused women (and sometimes men) believe that they have nothing to exchange in a relationship, that they don't have the right to expect benefits (especially when the abuser blames the victim for provoking his anger), or that enduring an abusive relationship is better than being alone (Walker 2000; Bergen et al. 2005).

FEMINIST THEORIES

Feminist perspectives offer additional insights on social interaction. Three examples focus on power in relationships, emotional labor, and gender roles.

Interaction and Power

Many feminist scholars maintain that social interaction often involves a dominant-subordinate relationship. Sometimes authority and control are not obvious in an interaction, even though they strongly influence the way that people respond to each other, because people often manipulate those they say they love (see Chapter 13).

Interaction and Emotional Labor

Emotions are important components of social interaction. According to sociologist Arlie Hochschild, we learn *feeling rules* that shape the appropriate emotions for a given role or situation. For example, *emotional labor,* "the management of feeling to create a publicly observable facial and bodily display," is critical in many occupations that demand continuous interaction and masking of true feelings:

> The secretary who creates a cheerful office that announces her company as "friendly and dependable" and her boss as "up-and-coming," the waitress or waiter who creates an "atmosphere of pleasant dining," the tour guide or hotel receptionist who makes us feel welcome . . . all of them

must confront in some way or another the requirements of emotional labor. (Hochschild 1983: 11)

Hochschild maintains that because many women have jobs that deal with the public, they are usually more likely than men to perform work that requires emotional labor.

Interaction and Gender Roles

Women and men are more similar than different in their interactions. A study that recorded conversations of college students in the United States and Mexico found that women and men spoke about the same amount— 16,000 words per day. An analysis of studies published since the 1960s concluded that men are generally more talkative than women, but talkativeness depends on the particular situation. During decision-making tasks, men are more talkative than women, but when talking about themselves or interacting with children, women are more talkative than men (Leaper and Ayres 2007; Mehl et al. 2007).

Other studies support the findings that cultural norms and gender role expectations shape the sexes' communication patterns. Generally, women are socialized to be more comfortable talking about their feelings, while men are socialized to be dominant and take charge, especially in the workplace. Because women tend to use communication to develop and maintain relationships, talk is often an end in itself. It is a way to foster closeness and under-

TABLE 5.2

Sociological Explanations of Social Interaction

PERSPECTIVE	KEY POINTS
Symbolic Interactionist	• People create and define their reality through social interaction. • Our definitions of reality, which vary according to context, can lead to self-fulfilling prophecies.
Social Exchange	• Social interaction is based on a balancing of benefits and costs. • Relationships involve trading a variety of resources, such as money, youth, and good looks.
Feminist	• The sexes act similarly in many interactions but often differ in communication styles and speech patterns. • Men are more likely to use speech that's assertive (to achieve dominance and goals), while women are more likely to use language that connects with others.

standing. Women often ask questions that probe for a greater understanding of feelings and perceptions ("Do you think it was deliberate?" or "Were you glad it happened?"). Women are also much more likely than men to do conversational "maintenance work," such as asking questions that encourage conversation ("What happened at the meeting?") (Lakoff 1990; Robey et al. 1998).

Compared with female speech, men's speech often reflects conversational dominance, such as speaking more frequently and for longer periods of time. They also show dominance by interrupting others, reinterpreting the speaker's meaning, or rerouting the conversation. Men tend to express themselves in assertive, often absolutist, ways ("That approach won't work"). Compared with women, men's language is typically more forceful, direct, and authoritative rather than tentative. That is, male speech often exerts control, especially in defending their own ideas ("You have to provide more information to support your position") (Tannen 1990; Mulac 1998).

5 Nonverbal Communication

"We are what we repeatedly do," according to Greek philosopher Aristotle. In modern society, we often hear that "Actions speak louder than words," and "Don't tell me how to do this, show me!" Indeed, our bodies and nonverbal messages often communicate more clearly and loudly than our words.

Nonverbal communication consists of messages sent without using words. This silent language, a "language of behavior" that conveys our real feelings, can be more potent than our words (Hall 1959: 15). For example, sobbing sends a much stronger message than saying, "I feel very sad." Some of the most common nonverbal messages are silence, visual cues, touch, and personal space.

SILENCE

Silence transmits a variety of feelings and emotions: agreement, apathy, awe, confusion, disagreement, embarrassment, regret, respect, sadness, thoughtfulness, and fear, to name just a few. In various contexts, and at particular points in a conversation, silence means different things to different people. Sometimes silence saves us from embarrassing ourselves. Think, for example, about the times you fired off an angry e-mail message, regretted doing so an hour later, and cringed at the thought of getting the person's fuming response.

> **nonverbal communication** messages that are sent without using words.

In interpersonal relationships, the silent treatment can be devastating. Many of us have experienced the pain of getting this treatment from friends, family members, coworkers, or lovers. Not talking to people who are important to us builds up anger and hostility. Initially, the "offender" may work very hard to make the closemouthed person talk about a problem. Eventually, however, the target of the silent treatment will get fed up, give up, or end a relationship (Rosenberg 1993).

VISUAL CUES

Visual cues, another form of nonverbal communication, include gestures, facial expressions, and eye contact. Let's consider a few examples of how such body language works in our daily interactions.

Gestures

Most of us think we know what certain gestures mean—folding your arms across your chest indicates a closed, defensive attitude; leaning forward often shows interest; shrugging your shoulders signals indifference; narrowing your eyes and setting your jaw shows defiance; and smiling and nodding shows agreement. Finger-pointing is usually a gesture that directs attention outward, placing blame or responsibility on someone else. Baltimore Orioles first baseman Rafael Palmeiro and President Bill Clinton pointed fingers when denying, respectively, using steroids and having sexual relations with Monica Lewinski. The common "talk to the hand" gesture sends a stronger message: "Go away!" or "I'm not listening to you." Few gestures, however, clearly convey a meaning by themselves. Instead, they must be interpreted in context. Because of habit or hearing problems, for instance, a coworker may always lean forward when listening whether he or she is interested or not.

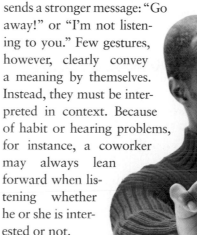

SOUTH AMERICA	UNITED STATES	JAPAN	FRANCE	GERMANY	OTHER COUNTRIES
not okay	okay	money	zero	vulgar gesture	better check first

The same gesture may have different meanings in different countries:

- In the United States, people use a firm handshake. In France, a quick handshake with only slight pressure is more acceptable.

- In Italy, Colombia, and China, people may wave good-bye by moving the palm and fingers back and forth, a gesture that typically means "come here" in the United States. In Malaysia, beckoning someone by moving the forefinger back and forth would be taken as an insult.

- Tapping one's elbow several times with the palm of one's hand indicates that someone is sneaky (in Holland), stupid (in Germany and Austria), or mean or stingy (in South America).

- Screwing a forefinger into one cheek signifies a dimple—a traditional sign of feminine beauty—and means, in Italy and Libya, "She is beautiful." The same gesture, in southern Spain, means that a man is effeminate and, in Germany, "You're crazy!"

- Fijians may fold their arms as a sign of respect when talking to another person. In Western countries, the same gesture usually means "I'm waiting," "I'm listening," or "This is pretty boring."

- In many Middle Eastern countries, people consider the shoes and the soles of one's feet to be unclean. As a result, stretching out one's legs with the feet pointing at someone or crossing one's legs so that a sole faces another person is considered rude (Morris 1994; Lynch and Hanson 1999; Jandt 2001).

Facial Expressions

Facial expressions often provide visual input that can be difficult to interpret. First, our facial expressions don't always show our true emotions. Parents, for example, tell their children "Don't you roll your eyes at me!" or "Look happy when Aunt Minnie hugs you." Thus, children learn that displaying their real feelings—especially when they're negative—is often unacceptable.

Second, faces can lie about feelings. Parents often know when children are lying because the children avoid eye contact, cry, swallow frequently, blink often, or start to sweat. Many adults, in contrast, monitor and control their facial expressions. They can deceive, successfully and over many years, because they have rehearsed the lies in their heads, are smooth talkers with a reputation for being trustworthy, or have gotten away with lying in the past (Ekman 1985; Sullivan 2001).

Third, our facial expressions don't always reflect how we feel because they change over time due to socialization. Consider smiling. Infants learn very quickly that a smile will get a parent's attention and a positive response (the parent smiles back) (Jones and Hong 2001). By kindergarten, most of us have learned to smile (and not smile) only in appropriate situations ("Don't smile during Uncle Fred's funeral, dear").

Finally, facial expressions are deceptive because of cultural variations. American businesspeople have grumbled that Germans are cool and aloof, while many German businesspeople have complained that Americans are excessively cheerful and hide their true feelings with grins and smiles. The Japanese, who believe that it's rude to display negative feelings in public, smile more than Americans do to disguise embarrassment, anger, and other negative emotions (Jandt 2001).

Eye Contact

We often communicate with our eyes. Are the eyes, as the proverb has it, the mirrors of the soul? Or do they hide our innermost feelings?

Eye contact serves several social purposes (Eisenberg and Smith 1971; Ekman and Friesen 1984; Siegman and Feldstein 1987). First, we get much information about other people by looking at their eyes. Eyes open wide show surprise, fear, or a flicker of interest. When we are angry, we stare in an unflinching manner. When we are sad or ashamed, our eyes may be cast down.

Second, eye contact is a potent stimulus. Speakers sometimes change their strategy when people in the audience don't look at them. Your instructors, for example, may speak louder, walk around the classroom, or revert to writing on a blackboard when they feel that students' lack of eye contact indicates boredom or daydreaming. In contrast, people usually evade the eyes of someone whose gaze is disturbingly unbroken and direct. When two people like one another, they establish eye contact more often and for longer durations than when there is tension in the relationship.

Finally, eye contact often varies across societies. In the United States, making eye contact is a key component of effective body language. Especially during job interviews, we're told, our eyes should signal attentiveness and interest, and "blinking, staring, or looking away whenever you begin speaking makes it hard for you to connect with your interviewer" (Bohannon 2000: 22). In some Asian cultures, like Japan, in contrast, students often avoid making eye contact with their instructors as a sign of respect (Jandt 2001).

TOUCH

Touch is another important form of nonverbal communication. Touching sends powerful messages about our feelings and our attitudes.

What Touching Communicates

Touching expresses our feelings toward another person. The sense of touch is activated long before birth:

The skin is the largest sense organ of the body ... [and] babies are surrounded and caressed by warm fluid and tissues from the beginning of fetal life. ... [After birth], the lips and hands have the most touch receptors; this may explain why newborns enjoy sucking their fingers (Klaus et al. 1995: 52).

Parents all over the world communicate with their infants through touch—stroking, holding, patting, rubbing, and cuddling.

Touching sends many messages, some positive (hugging, embracing, kissing, and holding hands) and some negative (hitting, shoving, pushing, and spanking). Other forms of touching are contradictory. One of my students, for example, left a boyfriend because "He said he loved me and trusted me but he gripped my arm

tightly every time I talked to another guy and pulled me away." The boyfriend's touch was controlling and threatening rather than affectionate. Between intimate partners, in addition, a decline in the amount of touching may signal that feelings are cooling off.

Gendered Touching

Whether touching is viewed as positive or negative depends on the situation and one's gender. In higher education, even when the faculty member is well liked, male and female students perceive her or his touching them differently. When female professors touch male students on the arm while talking to them, the students view the gesture as nurturing and friendly. When a male professor touches a female student on the arm, she may get nervous because she's afraid that the touching may escalate (Lannutti et al. 2001; Fogg 2005). (Unwanted touching, regardless of one's gender or sexual orientation, can lead to charges of sexual harassment, a topic we'll cover in later chapters.)

© Rod Aydelotte/Bloomberg News/Landov

Pictured here are President Bush and Saudi Crown Prince Abdullah at Bush's Texas ranch. Many U.S. journalists raised questions about two men holding hands. In response, Arab Americans were quick to point out that Arab society sees the outward display of affection between male friends as an expression of respect and trust. Thus, in that society, government officials and military officers often hold hands as they walk together or converse with one another.

Cross-Cultural Variations in Touching

As with other forms of nonverbal communication, the interpretation of touching varies from culture to culture. In some Middle Eastern countries, people don't offer anything to another with the left hand, the hand used to clean oneself after using the toilet. And among many Chinese and other Asian groups, hugging, back-slapping, and handshaking aren't as typical as they are in the United States, because such touching is seen as too intimate (Lynch and Hanson 1999).

According to one scholar, compared with those in some other societies, people in the United States are "touch-deprived" because they have one of the lowest rates of casual touch in the world:

> If you are talking to a friend in a coffee shop in the United States, you might touch each other once or twice an hour. If you were British and in a London coffee shop, you probably won't touch each other at all. But if you were French and in a Parisian café, you might touch each other a hundred times in an hour! (Jandt 2001: 117)

PERSONAL SPACE

The distance that people establish between themselves when they interact is an important aspect of nonverbal communication. Personal space plays a significant role in our everyday nonverbal interactions, reflects power and status, and varies across societies.

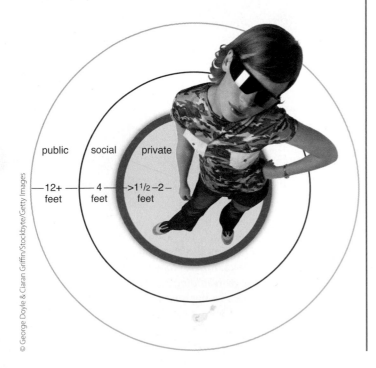

public   social   private

—12+—   —4—   —>1½–2—
feet   feet   feet

© George Doyle & Ciaran Griffin/Stockbyte/Getty Images

Some Characteristics of Personal Space

Our living space is public or private. In the public sphere, which is usually formal, we have clearly delineated spaces: "This is your locker," "That's her office," or "You're parking in my spot." We usually decorate our public spaces with businesslike artifacts such as awards or framed photos of an institution's accomplishments.

Private spaces send a different message. They convey informality and a relaxed feeling. Private spaces—such as homes, apartments, and dorm rooms—are often mini-museums that reflect people's interests, hobbies, and personalities. We can tell a lot about people by simply looking around their rooms or homes: the newspapers and books they read, the music they enjoy, whether they're neat or sloppy, and so on.

People sometimes invade our personal space by standing too close to us in a line, leaning against us, using all of the shared armrest in tight airline seats, or plopping their feet on the chair next to us in a classroom. Not all space intrusions are physical encroachments. Loud cell phone conversations are a good example of an auditory intrusion into our personal space.

For example, in intimate relationships (such as between family members), our personal space is often 2 feet or less because our social interactions are comfortable and affectionate and we are comfortable with close physical proximity. In contrast, in public situations (such as someone speaking to a large audience or faculty in large classrooms), the personal space is often 12 feet or more because the speaker, who has a higher social status than the audience, has a formal relationship with the listeners and avoids close physical contact (Hall 1966).

Space and Power

Space signifies who has privilege, status, and power (Chapman and Hockey 1999; Falah and Nagel 2005). Wealthy people can afford enormous apartments in the city or houses in the suburbs, while the poor are crowded into the most undesirable sections of a city or town or in trailer parks. Executives, including college presidents and deans, have large offices (even entire suites), while faculty members often share office space, even though many of their discussions with students are confidential. Secretaries' space is typically open to the public and crowded with file cabinets and computers. Generally, the higher the socioeconomic status, the greater the consumption of space, including large cars, reserved parking spaces, private dining areas, first-class airline seats, and luxurious skyboxes at sports stadiums.

Cross-Cultural Variations and Space

Cultural values and norms dictate how we use space. Americans not only stand in line but also have strict queuing rules: "Hey, the end of the line is back there" and "That guy is trying to cut into the line." In contrast, "along with the Italians and Spaniards, the French are among the least queue-conscious in Europe" (Jandt 2001: 109). In these countries and others, people routinely push into the front of a group waiting for taxis, food, and tickets, and nobody objects. North Americans maintain personal space in an elevator by moving to the corners or to the back. In contrast, an Arab male may stand right next to or touching another man even when no one else is in the elevator. Because most Arab men don't share the American concept of personal space in public places and in private conversations, they consider it offensive if the other man steps or leans away. In many Middle Eastern countries, however, there are stringent rules about women and men separating themselves spatially during religious ceremonies and about women avoiding any physical contact with men in public places (Jandt 2001; Office of the Deputy Chief 2006; see, also, Chapter 3).

6 Online Communication

during the latter part of the twentieth century, the Internet changed many people's interactions and communication processes significantly. We have replaced many of our face-to-face activities with a virtual reality, where we obtain services and exchange information, ideas, and greetings in *cyberspace*, an online world of computer networks. This final section addresses two questions: Who's online and why? And is online interaction beneficial or harmful?

WHO'S ONLINE AND WHY?

The percentage of adult Americans connected to the Internet increased from practically zero in 1994 to over 75 percent in 2007 (Pew Internet & American Life Project Surveys 2007). E-mail is still the most common form of online communication, but instant messaging (IM) and text messaging are very popular.

Variations by Sex, Age, Ethnicity, and Social Class

Equal numbers of women and men (73 percent) use the Internet. There has been an enormous growth in Internet use across most age groups, but only 35 percent of Americans aged 65 and older are online compared with 70 percent of those aged 50 to 64 and at least 85 percent of those aged 49 and younger. Many older Americans say that they don't need the Internet, while others live on fixed incomes and can't afford either computers or monthly charges for Web service (Fox 2006; Pew Internet & American Life Project Surveys 2007).

Asian Americans are the most wired group in the United States. About 90 percent are Internet users compared with 80 percent of English-speaking Latinos, 75 percent of whites, and 59 percent of African Americans (Pew Internet & American Life Project Surveys 2007). Asian Americans are more likely to use the Internet because many parents in this group are professionals who can afford computers and online service and encourage their children to use the Internet for education, a major avenue of upward mobility (see Chapters 10 and 14). Latinos with lower educational levels and little English proficiency are the least likely to be connected to the Internet (Fox and Livingston 2007).

The biggest digital divide is that between social classes. The higher a family's income, the greater the likelihood that its members are Internet users (see *Figure 5.3*). Affluent families can buy home computers, pay for Internet service, and send their children to schools that provide computer-related instruction. Thus, the children from the poorest families are the most likely to lack technological skills.

What We Do (and Don't Do) Online

On an average day, about 135 million Americans go online. Of more than 70 activities they participate in, the most common are sending e-mail, finding information, and reading the news (Pew Internet & American Life Project Surveys 2008). Fully 45 percent of users say that the Internet has played a crucial role in making a major financial decision, dealing with a major health matter, buying a car, choosing a college, or getting additional career training (Horrigan and Rainie 2006). Thus, the Internet has an important impact on many Americans' lives.

Women's and men's online activities tend to differ. Both use search engines (like Google) heavily, but women are more likely to use e-mail to contact friends and family members for news and advice. Men, in contrast, are more likely to engage in financial activities (such as buying or selling stock), to research products (such as cars or appliances), and to get information on their hobbies or recreational activities (Fallows 2005).

Such differences aren't surprising. Gender roles have changed, but most women still have the primary responsibility for child care and domestic tasks, while men are more likely to make major financial decisions and, even when they're married and have children, have more time to pursue leisure interests, including Internet activities (see Chapter 13).

IS ONLINE COMMUNICATION BENEFICIAL OR HARMFUL?

During the late 1990s, a number of scholars predicted that the Internet would replace close offline relationships, diminish people's involvement in interpersonal and community relationships, and isolate them socially. Contrary to such gloomy forecasts, the Internet has increased interaction, especially with people that we care about. In addition, millions of Americans use the Internet to plan religious activities, arrange neighborhood gatherings, petition local politicians, and contact civic organizations. All of these activities have strengthened rather than diluted interpersonal and community ties (Boase et al. 2006).

However, a major disadvantage of cyberspace communication is that it is more impersonal than face-to-face interaction. According to one faculty member, for instance, e-mail has made teaching more isolating and less satisfying:

> . . . outside the classroom, my interactions with students have mostly evaporated. Students used to drop by my office all the time, and I miss that. They used to come by during office hours to ask a question about a point raised in class or to clarify the requirements for an assignment. After the business part of our conversation was over, we'd chat. (Connolly 2001: B5)

Cyberspace has provided us with a convenient and efficient communication tool, but it is limited by constraints. Consider, for example, the benefits and costs related to its convenience, how we communicate, work and leisure, and interpersonal relationships.

Convenience

A major advantage of online communication is its convenience. We can send or respond to messages at our leisure, without worrying about the schedules of people who live in different time zones, and we can save much time by attaching documents to e-mail messages rather than using regular mail. On the negative side, being online makes us vulnerable to *spam* (unsolicited commercial e-mail sent to a large number of addresses), viruses that can destroy material on a hard drive, and hoaxes—all of which interfere with online communication.

How We Communicate

Online interaction lets us think before we speak. During face-to-face and phone discussions, we sometimes regret having been impulsive or rude. Spontaneous e-mail messages can get us into trouble, but most of us learn fairly quickly that we can compose our thoughts and formulate our ideas more clearly and persuasively in an e-mail than during

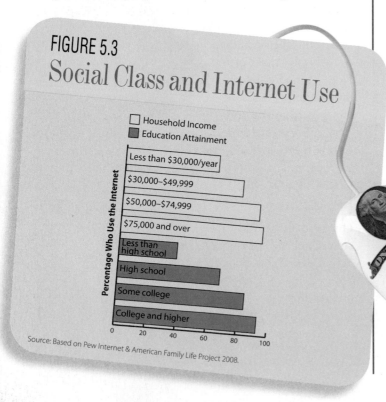

FIGURE 5.3
Social Class and Internet Use

Household Income
Education Attainment

Less than $30,000/year
$30,000–$49,999
$50,000–$74,999
$75,000 and over
Less than high school
High school
Some college
College and higher

Percentage Who Use the Internet

0   20   40   60   80   100

Source: Based on Pew Internet & American Family Life Project 2008.

Even the most neutral [electronic] exchanges can spark problems.

many face-to-face interactions. In Goffman's terms, we can manage the impressions that we give off to others.

On the negative side, even the most neutral exchanges can spark problems. To avoid misunderstandings, e-mail users often use *emoticons* (like smiley faces or frowning ones) to convey voice inflections and facial expressions. Still, because we don't see the other person's body language, miscommunication is common. A simple and matter-of-fact directive from an instructor, like "I want you to rewrite this assignment and resubmit it," would usually be perceived as normal by a student in a face-to-face encounter but may be interpreted as abrupt in an e-mail. Also, people sometimes forward private e-mails to others without the writer's consent, causing embarrassment or anger.

Work and Leisure

In 2008, 30 percent of Americans *telecommuted* (worked from home) at least two days a week, up from only 9 percent in 1995, but down from 32 percent in 2006, and the vast majority earned $75,000 or more a year (Saad 2008). Telecommuters cite numerous benefits: They spend less money on clothes and gasoline, and less time in traffic; have a more flexible work schedule; spend more time with children instead of using child care; and can attend many of their children's school functions. In addition, the country's almost 4 million telecommuters collectively save about 840 million gallons of gas and 14 million tons of carbon dioxide emissions a year. That's equal to taking 2 million vehicles off the road (Heubeck 2005; Jones 2006; Consumer Electronics Association 2007).

On the negative side, telecommuting often blurs the line between home life and work. Increasingly, many of us use our personal time to catch up with work-related assignments and responsibilities. Further, 60 percent of Americans work during vacations. Even at the office, the average employee has added two extra hours of work a day just to check and respond to work-related e-mail, and one out of five Americans feels that e-mail cuts into their leisure time because a coworker, supervisor, or client can contact them at any time and usually expects a quick response (Alvarez 2005).

Relationships and Privacy

Online communication promotes new friendships and romantic relationships. However, about 66 percent of Internet users feel that using online dating services is dangerous because participants provide considerable personal information that may lead to cyber-stalking and harassment. Even so, 11 percent of American adults have used an online dating site (Madden and Lenhart 2006).

A major cost of online interaction is jeopardizing our privacy because e-mail is neither anonymous nor confidential. About 60 percent of companies monitor and preserve their employees' e-mail messages and can use them to discipline or fire people. It's also becoming increasingly common for employers to search networking sites before deciding whether to make a job offer to a graduating college student. E-mails don't disappear after being deleted but may last indefinitely in cyberspace. People often say things in e-mails they would never say in person, especially when they're flirting, angry, frustrated, or tired. The best ways to avoid such problems are to write each e-mail message as though it were a formal letter, to read each message aloud to catch hostile language, to refrain from sending love notes or other confidential material, and to never send anything you wouldn't want your mom to read (Weinstock 2004; Dang and Tucker 2005).

One in three American teens who use the Internet, most of them girls, have experienced cyber-bullying which involves a wide range of annoying and potentially menacing activities such as receiving threatening messages, having an embarrassing picture posted, or having rumors about them spread online (Lenhart 2007).

© Image Source/Jupiterimages

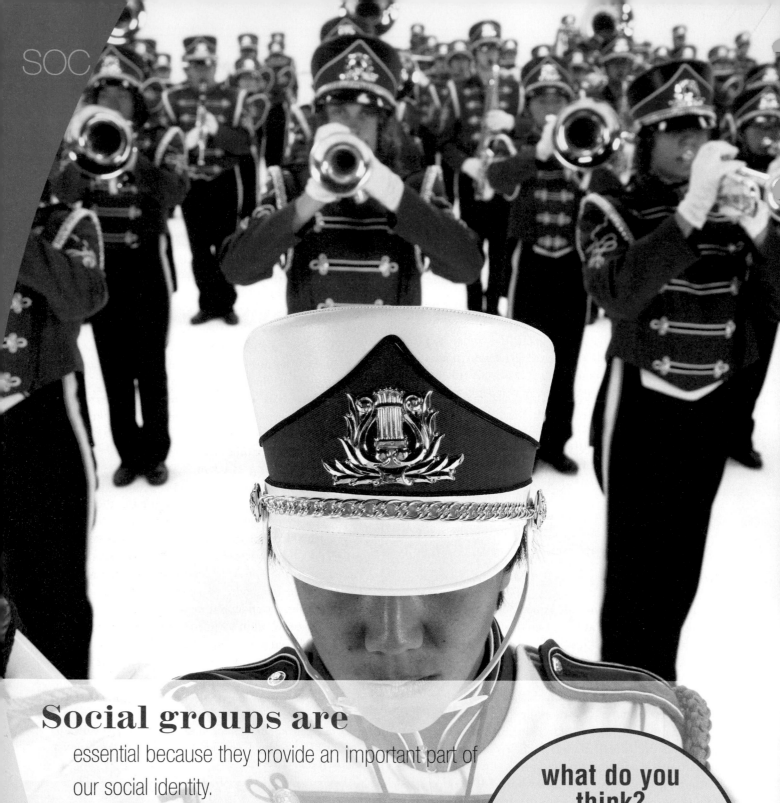

Social groups are

essential because they provide an important part of
our social identity.

what do you think?

It's easy to resist pressure to
follow the crowd.

1 2 3 4 5 6 7

strongly agree   strongly disagree

6

Groups, Organizations, and Social Institutions

Key Topics

In this chapter, we'll explore the following topics:

1 Social Groups

every Saturday night, from 75 to 100 drivers meet at a parking lot in Baltimore to show off and discuss their customized, souped-up cars. The drivers are typically males, represent a variety of racial/ethnic backgrounds, and range in age from the late teens to early 40s. Most pour their paychecks into making their cars faster, paying for wild paint jobs, and installing television sets and stereo systems. They hang out until about 2 A.M., even though these get-togethers sometimes create conflict at home: "My wife complains that I put Saturday night before her" (Mitchell 2002: A6). These car enthusiasts are a social group.

> **social group** two or more people who interact with one another and who share a common identity and a sense of belonging or "we-ness."

WHAT IS A SOCIAL GROUP?

A **social group** consists of two or more people who interact with one another and who share a common identity and a sense of belonging or "we-ness." Friends, families, work groups, religious congregations, clubs, athletic teams, and organizations are all examples of social groups. Each of us is a member of many groups simultaneously. Some groups are highly organized and stable (local political parties); others are fluid and temporary

hot rodders = social group

Why are these hot rod fans in California an example of a social group?

(high school classmates). Interaction, especially face-to-face interaction, is the key ingredient in creating and maintaining groups.

Social groups are essential because they provide an important part of our social identity and help us understand the behavior of other people in our society. Some groups, nonetheless, are more binding and significant than others because they shape our social and moral beliefs. Our most important social groups are primary and secondary groups, in-groups and out-groups, and reference groups.

PRIMARY GROUPS AND SECONDARY GROUPS

Suppose your car's battery died this morning, your sociology professor gave a pop quiz for which you hadn't studied, your computer's hard drive crashed, and your microwave stopped working. Whom might you call to vent? Your answer reflects the difference between primary and secondary groups.

Primary Groups

A **primary group** is a relatively small group of people who engage in intimate face-to-face interaction over an extended period of time. For sociologist Charles Horton Cooley (1909/1983: 24), the most significant primary groups were "the family, the play-group of children, and the neighborhood or community of elders" because they are "first" and central in shaping a person's social and moral development.

Primary groups are our emotional glue. We call members of our primary group to share good news or to gripe. Primary group members are typically understanding, supportive, and tolerant when we're in a bad mood or selfish. They have a powerful influence on our social identity because

we interact with them on a regular and intimate basis over many years, usually throughout our lives. Because primary group members genuinely care about each other, they contribute to one another's personal development, security, and well-being. Our family and close friends, for example, stick with us through good and bad, and we feel comfortable in being ourselves in their presence.

Secondary Groups

A **secondary group** is a large, usually formal, impersonal, and temporary collection of people who pursue a specific goal or activity. Your sociology class is a good example of a secondary group. You might have a few friends in class, but students typically interact infrequently and formally. When the semester (or quarter) is over and you've accomplished your goal of passing the course, you may not see each other again (especially if you're attending a large college or university). And you certainly wouldn't call your professor if you needed a sympathetic ear at the end of a bad day. Other examples of secondary groups include political parties, labor unions, and employees of a company.

Unlike primary groups, secondary groups are usually highly structured: There are many rules and regulations, people know (or care) little about each other personally, relationships are formal, and members are expected to fulfill particular functions. Whereas primary groups meet our *expressive* (emotional) needs, secondary groups fulfill *instrumental* (task-oriented) needs. Once a task or activity is completed—whether it's earning a grade, turning in a committee report, or building a bridge—secondary groups usually split up, and go on to become part of other secondary groups.

Table 6.1 summarizes the characteristics of primary and secondary groups. These characteristics represent **ideal types**, general traits that describe a social phenomenon rather than every case. Ideal types provide composite pictures of how social phenomena differ rather than specific descriptions of reality. Because primary and secondary groups are ideal types, their characteristics can vary. Thus, members of primary groups may sometimes devote themselves to meeting instrumental needs (such as creating a family-owned business), and members of secondary groups (such as military units and athletic teams) can develop a lasting personal closeness.

EMOTIONAL GLUE

© Lew Robertson/Corbis

IN-GROUPS AND OUT-GROUPS

All good people agree,
And all good people say,
All nice people, like Us, are We
And everyone else is They.
—*from* We and They, *by Rudyard Kipling*

Rudyard Kipling's poem "We and They" captures the essence of in-groups and out-groups. Members of an **in-group** share a sense of identity and "we-ness" that typically excludes and devalues outsiders. **Out-groups** consist of people who are viewed and treated negatively because they are seen as having values, beliefs, and other characteristics different from those of the in-group. For example, "we" vegetarians are healthier than "you" meat eaters, "we" computer nerds are smarter than "you" fraternity and sorority "types," and so on.

Based on ascribed or achieved statuses, almost everyone sees others as members of in-groups and out-groups. From ancient times to the present, people in various parts of the world have made "we" and "they" distinctions based on race or ethnicity, gender, religion, age, social class, and other social and biological criteria (Coser 1956; Tajfel 1982; Hinkle and Schopler 1986).

Out-group members are usually aware of being outsiders. For example, overweight people are often viewed by others as lazy, sloppy, unmotivated, and undisciplined. Many overweight people have internalized such negative attitudes and see themselves as part of an out-group. In a national study of overweight people, nearly 40 percent had negative attitudes about fat people (including themselves), a third would rather experience a divorce than be obese, 15 percent would give up 10 years or more of life to be thinner, and 4 percent would trade blindness for obesity (Schwartz et al. 2006).

In-groups and out-groups affect our feelings about ourselves and others, our social identity, and our life outcomes. In-groups can be a positive influence in promoting individuals' sense of self-worth and belonging and increasing group solidarity and cohesion, but they can also create conflict and provoke inhumane actions,

in-groups people who share a sense of identity and "we-ness" that typically excludes and devalues outsiders.

out-groups people who are viewed and treated negatively because they are seen as having values, beliefs, and other characteristics different from those of an in-group.

TABLE 6.1
Characteristics of Primary and Secondary Groups

	CHARACTERISTICS OF A PRIMARY GROUP	CHARACTERISTICS OF A SECONDARY GROUP
Interaction	• Face to face • The group is usually small	• Face to face or indirect • The group is usually large
Communication	• Communication is emotional, personal, and satisfying	• Communication is emotionally neutral and impersonal
Relationships	• Intimate, warm, and informal • Usually long-term • Valued for their own sake (expressive)	• Typically remote, cool, and formal • Usually short-term • Goal-oriented (instrumental)
Individual Conformity	• Individuals are relatively free to stray from norms and rules	• Individuals are expected to adhere to rules and regulations
Membership	• Members are not easily replaced	• Members are easily replaced
Examples	• Family, close friends, girlfriends and boyfriends, self-help groups, street gangs	• College classes, political parties, professional associations, religious organizations

Rapper Puff Daddy (Sean Combs) performs during The Source Hip Hop Awards in Los Angeles. Documentary filmmaker Byron Hurt has criticized hip-hop music as violent, homophobic (showing a hatred of homosexuals), and misogynist (showing a hatred of women). Do you think that Hurt is an important member of a reference group for all or only some rap artists and fans?

reference group a collection of people who shape our behavior, values, and attitudes.

even war. For example, in-group/out-group hostilities have fueled the Palestinian-Israeli battles over the occupation of the West Bank, the eviction of white farmers in South Africa, and ongoing civil wars in some African nations.

REFERENCE GROUPS

Besides belonging to primary groups, secondary groups, in-groups, and out-groups, we also have reference groups. A **reference group** is a collection of people who shape our behavior, values, and attitudes (Merton and Rossi 1950). Reference groups influence who we are, what we do, and who we'd like to be in the future. Unlike primary groups, however, reference groups rarely provide personal support or face-to-face interaction over time.

Reference groups might be groups with which we already associate, like a college club or a recreational athletic team. They can also be groups that we admire and aspire to be part of, such as successful financial advisors or country club members. Each person has many reference groups. Your sociology professor, for example, may be a member of several professional sociological associations, a golf enthusiast, a parent, and a homeowner. Identification with each of these groups influences her or his everyday attitudes and actions.

Like in-groups and primary groups, reference groups can have a strong impact on our self-identity,

self-esteem, and sense of belonging because they shape our current and future attitudes and behavior. We often add or drop reference groups throughout the life course. If you aspire to move up the occupational ladder, for example, your reference group may change from entry-level employees to managers, vice-presidents, and CEOs.

GROUP CONFORMITY

Most Americans see themselves as rugged individualists who have a mind of their own and don't bow to group pressure (see the discussion of values in Chapter 3). A number of studies have shown, however, that most Americans are profoundly influenced by group pressure. Four of the best-known of these are studies by Solomon Asch, Stanley Milgram, Philip Zimbardo, and Irving Janis.

Asch's Research

In a now-classic study of group influence, social psychologist Solomon Asch (1952) told subjects that they were taking part in an experiment on visual judgment. After arranging six to eight male undergraduates around a table, Asch showed them the line drawn on card 1 and asked them to match the line to one of three lines on card 2 (see *Figure 6.1*). The correct answer, clearly, is line C.

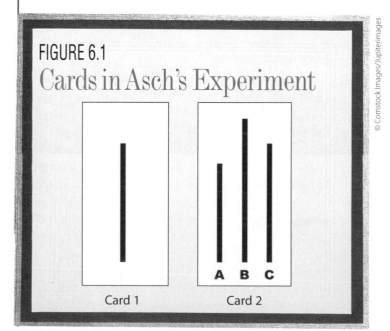

FIGURE 6.1
Cards in Asch's Experiment

A   B   C

Card 1   Card 2

All but one of the participants in each group—and that one usually sat in the next-to-last seat—were Asch's confederates, or accomplices. In the first test with each group, all the confederates selected the correct matching line. In the other tests, each of the confederates, one by one, deliberately chose an incorrect line. Thus, the nonconfederate participant faced a situation in which seven other group members had unanimously agreed on a wrong answer. Averaged over all of the trials, 37 percent of the nonconfederate participants ended up agreeing with the false judgments of the group. When those participants were asked to judge the length of the lines alone, away from the influence of the group, they made errors only 1 percent of the time.

Asch's research demonstrated the power of groups over individuals. Even when we know that something is clearly wrong, we may go along with the group to avoid ridicule or exclusion. Remember that these experiments were done in a laboratory and with groups whose members didn't know each other. A group's influence on members' judgments and behavior can be even stronger when it is a real-life primary group or a reference group.

The learner in Milgram's experiment was instructed to fake pain and fear.

Milgram's Research

In a well-known laboratory experiment on obedience conducted by psychologist Stanley Milgram (1963, 1965), 40 volunteers were asked to administer electric shocks to other study participants. In each experimental trial, one participant was a "teacher" and the other a "learner," one of Milgram's accomplices. The teachers were businessmen, professionals, and blue-collar workers. The learner was strapped into a realistic-looking chair that supposedly regulated electric currents. The learner was not actually receiving a shock but was told to fake pain and fear. The teacher read aloud pairs of words that the learner had to memorize. Whenever the learner didn't answer correctly, the teacher was told to apply an electric shock from a low of 15 volts to a high of 450 volts. When the learners shrieked in pain, the majority of the teachers, although distressed, obeyed the study supervisor and administered the shocks when told to do so.

Milgram's study was controversial. Ordering electric shocks raised numerous ethical questions about the subjects suffering extreme emotional stress (see Chapter 2). However, the results showed that an astonishingly large proportion of the subjects inflicted pain on others to obey an authority figure. Many social scientists used Milgram's findings to explain the Abu Ghraib prison in Iraq where U.S. soldiers abused and humiliated Iraqi prisoners because "I was just following orders."

Zimbardo's Research

The Stanford Prison Experiment conducted by social psychologist Philip Zimbardo also underscores the power of groups on human behavior (Haney et al. 1973; Zimbardo 1975). Zimbardo advertised in a local newspaper for volunteers for an experiment on prison life and selected 24 young men, most of them college students, as participants.

On a Sunday morning in August 1971, nine young men were "arrested" in their homes by Palo Alto police as neighbors watched. The "prisoners" were booked and transported to a mock prison that Zimbardo and his colleagues had constructed in the basement of the psychology building at Stanford University. The prisoners were searched, issued an identification number, and outfitted in a dresslike shirt and heavy ankle chains. Those assigned to be "guards" were given uniforms, billy clubs, whistles, and reflective sunglasses. The guards were told that that their job was to maintain control of the prisoners but not to use violence. All the young men quickly assumed the roles of either obedient and docile prisoners or autocratic and controlling guards. The guards became increasingly more cruel and demanding. The prisoners complied with dehumanizing demands (such as eating filthy sausages) to gain the guards' approval and bowed to their authority. (For a slide show of this experiment, go to http://www.prisonexp.org.)

Zimbardo's study was scheduled to run for two weeks but was terminated after six days because the guards became increasingly aggressive and humiliated the prisoners. Among other things, they forced the prisoners to clean out toilet bowls with their bare hands, locked them in a closet, and made them stand at attention for hours. Instead of simply walking out or rebelling, the prisoners became withdrawn and depressed. Zimbardo ended the experiment because of the prisoners' stressed reactions.

The experiment raised numerous ethical questions about the harmful treatment of participants. It demonstrated, however, the powerful effect of group conformity: People exercise authority, even to the point of hurting others or themselves, or submit to authority if there is group pressure to conform (Zimbardo et al. 2000).

Janis's Research

Sometimes intelligent people, even those in positions of high responsibility, make disastrous and irrational decisions. Why? Social psychologist Irving Janis (1972: 9) cautioned presidents and other heads of state to be wary of **groupthink**, a tendency of in-group members to conform without critically testing, analyzing, and evaluating ideas, that results in a narrow view of an issue. To preserve friendly relations and to remain loyal to the group, Janis argued, individuals don't raise controversial issues, question weak arguments, or probe "softheaded thinking." As a result, influential leaders often make decisions—based on their advisors' consensus—that turn out to be political and economic fiascoes.

A United States Senate study (2004) on intelligence agencies provides an example of groupthink. The report concluded that U.S. leaders' decision to invade Iraq in 2003 was based on a groupthink dynamic that relied on unproven and inaccurate assumptions, inadequate

> *Sometimes intelligent people, even those in positions of high responsibility, make disastrous and irrational decisions.*

or misleading sources, and a dismissal of conflicting information, which showed that Iraq had no weapons of mass destruction.

Janis and other researchers have focused on high-level decision-making, but groupthink is common in all kinds of groups—student clubs, PTAs, search committees, professional organizations, and juries, for example. If group members are aware of the negative characteristics of groupthink, they can avoid some of the pitfalls by setting up and following democratic discussion and voting processes, hammering out disagreements, and seeking advice from informed and objective people outside the group.

SOCIAL NETWORKS

Groups exist within the context of larger social units, such as social networks. A **social network** is a web of social ties that links an individual to others. A social network may involve as few as three people or as many as millions.

Some of our social networks, such as our primary and secondary groups, may be tightly knit, involve interactions on a daily basis, and have clear boundaries about who belongs and who doesn't. In other cases, our social networks connect us to large numbers of people whom we don't know personally and with whom we interact rarely or indirectly, and the group's boundaries are fluid or unclear. Examples of such distant networks include members of the American Sociological Association and *New York Times* readers.

The Internet, especially, has generated numerous interlocking social networks that unite people across many states and countries who share similar concerns, interests, or characteristics that include gender, education, occupation, sexual orientation, age, religion, and/or ethnicity. MySpace, for example, is a very popular social network that includes millions of international subscribers with a broad range of characteristics.

2 Formal Organizations

from the time that you turn on the faucet to brush your teeth until you sit down in your first class, you've participated in dozens of formal organizations. A **formal organization** is a complex and structured secondary group that has been deliberately created to achieve specific goals in an efficient manner. We depend on a variety of formal organizations to provide goods and services in a stable and predictable way, including companies that supply clean water for brushing our teeth, numerous food producers who stock our favorite breakfast items, and garment industries and retailers that produce and sell the clothes we wear.

CHARACTERISTICS OF FORMAL ORGANIZATIONS

Formal organizations share some common characteristics:

- Social statuses and roles are organized around shared expectations and goals.
- Norms governing social relationships among members specify rights, duties, and sanctions.
- A formal hierarchy includes leaders or individuals who are in charge.

Two of the most widespread and important types of formal organizations in the United States are voluntary associations and bureaucracies.

VOLUNTARY ASSOCIATIONS

A **voluntary association** is a formal organization created by people who share a common set of interests and who are not paid for their participation. Unlike bureaucracies, as you'll see shortly, voluntary associations vary quite a bit in organizational structure. Some small voluntary organizations, such as book clubs, don't have a formal hierarchy, a rigid set of rules or regulations, or even specific goals. Others, like investment clubs, have a constitution, written regulations about the duties and rights of officers and members, minutes, and an agenda that's circulated before each meeting.

Helping others is a strong American value (see Chapter 3). Whether it's an earthquake in Mexico, a drought in Somalia, or a cyclone in Burma, Americans generously send money, food, and personnel. Every year, millions of Americans—many of whom have full-time jobs—are volunteers in a variety of local, regional, and national organizations. These volunteers save government agencies billions of dollars every year by providing needed services (see *Table 6.2*).

In the United States, there are more than 1.2 million charities, social welfare organizations, and religious congregations that are voluntary associations (Independent Sector 2006). If local voluntary associations that focus on education, self-

help, and environmental issues were included in the count, the numbers of groups and members would be staggering. Overall, there are four types of voluntary associations: groups that have a charitable or altruistic purpose (like the United Way), groups that focus on political issues (like the Green Party), groups that organize around occupations (like the National Association of Realtors), and groups that have a recreational orientation (like the Beer Can Collectors of America).

Who is most likely to join voluntary associations? It depends on the organization. Charitable/altruistic and recreational associations attract people from a variety of social classes, age categories, and ethnic/racial backgrounds. Political and occupation-related associations, on the other hand, are more common among people from higher socioeconomic backgrounds, who have money to fund the groups' activities and time to attend local, national, and regional conferences.

BUREAUCRACIES

When I ask my students to describe a bureaucracy, their answers often include adjectives like "bungling," "impossible," "buck-passing," "frustrating," "hellish," and worse. Such reactions are understandable because, for most of us, dealing with a bureaucracy is synonymous with red tape, slow-moving lines, long and tedious forms, and rude employees.

For sociologists, a **bureaucracy** is a formal organization that is designed to accomplish goals and tasks

> **voluntary association** a formal organization created by people who share a common set of interests and who are not paid for their participation.
>
> **bureaucracy** a formal organization that is designed to accomplish goals and tasks through the efforts of a large number of people in the most efficient and rational way possible.

TABLE 6.2
Volunteering in the United States, 2007

Percentage of adults who volunteered	27
Total number of volunteers aged 16 and older	62 million
Average annual hours per volunteer	52
Percentage of volunteers who worked full time	34
Estimated hourly value of volunteer time	$19.51 per hour
Total dollar value of volunteer time	$295 billion

Sources: Based on Corporation for National and Community Service 2007; Independent Sector 2008; U.S. Department of Labor 2008.

through the efforts of a large number of people in the most efficient and rational way possible. Bureaucracies are not a modern invention but existed thousands of years ago in ancient Egypt, China, and Africa. Some bureaucracies function more smoothly than others, and some formal organizations are more bureaucratic than others. Whether they are relatively small (such as a 500-bed hospital) or huge (such as Medicare, a U.S. government agency that provides health insurance for millions of Americans aged 65 and older), bureaucracies share some common characteristics.

Ideal Characteristics of Bureaucracies

Max Weber (1925/1947) identified six key characteristics of the ideal type of bureaucracy. Remember that ideal types describe distinctive and abstract traits rather than those that fit any specific organization. In Weber's model, then, the following characteristics describe what an efficient and productive bureaucracy *should* be like:

- *High degree of division of labor and specialization.* Individuals who work in the bureaucracy perform very specific tasks.

- *Hierarchy of authority.* Workers are arranged in a hierarchy where each person is supervised by someone in a higher position. The resulting pyramids—often presented in organizational charts—show who has authority over whom and who is responsible to whom. Thus, there is a chain of command, stretching from top to bottom, that coordinates decision making. *Figure 6.2* shows a simplified organizational chart of a small state university of 5,000 students. If this chart included secretaries, food services, plant maintenance, libraries, human resources, student housing, and numerous other offices and staff, you can see that the bureaucratic structure would be much more complex. The shaded boxes represent the flow of authority from the top (governor of the state) to the bottom (students in a particular academic department).

- *Explicit written rules and regulations.* Detailed written rules and regulations cover almost every possible kind of situation and problem that might arise. They address a variety of issues, including hiring, firing, salary scales, rules for sick pay and absences, and everyday operations. If a person has a question, all she or he has to do is look up the answer.

- *Impersonality.* There is no place in a bureaucracy for personal likes or dislikes or emotional whims. Instead, employees are expected to follow the rules, to get the work done, and to behave professionally. An imper-

sonal workplace in which all employees are treated equally minimizes conflict and favoritism and increases efficiency.

- *Qualifications-based employment.* People are hired based on objective criteria such as skills, education, experience, and scores on standardized examinations. If workers perform well and have the necessary credentials and technical competence, they'll move up the career ladder.

- *Separation of work and ownership.* Neither managers nor employees own the offices they work in, the desks they sit at, the machinery they use, or the products that they make, invent, or design.

For Weber, all of these characteristics produce an efficient and rational bureaucracy. A "rational matter-of-factness," he maintained, dehumanized bureaucracies but also made them more productive by "eliminating from official business love, hatred, and all purely personal, irrational, and emotional elements which escape calculation" (Weber 1946: 216). Weber viewed bureaucracies as superior to other forms of organization because they're more efficient and more predictable, but he also worried that bureaucracies could become "iron cages" because people become trapped in them, "their basic humanity denied."

To avoid medical and health insurance bureaucracies, more people are turning to convenient care clinics—usually staffed by nurse practitioners—that offer low-cost treatment for more than 25 common conditions, such as strep throat and ear infections, and provide health screening tests, immunizations, and physicals. The clinics are located in drugstores and large retail stores like Wal-Mart. There is no need to make an appointment and minimal (if any) waiting; the patients are in and out in about 15 minutes, and most of the clinics are open during evenings and weekends.

© Burke/Triolo/Brand X Pictures/Jupiterimages

FIGURE 6.2
The Bureaucratic Structure of a Small Public University

alienation a feeling of isolation, meaninglessness, and powerlessness that may affect workers in a bureaucracy.

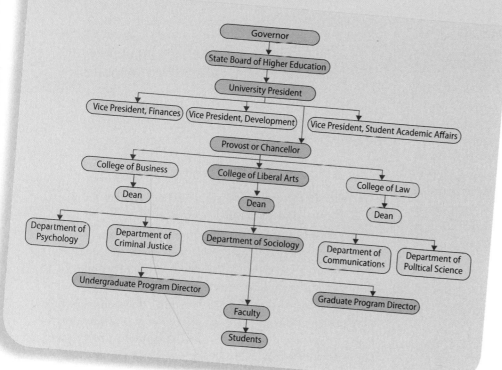

Shortcomings of Bureaucracies

A survey of federal workers found that 40 percent believed that many of their managers accepted shoddy work from lazy or incompetent employees and 40 percent rated the morale of their agency as low. Also, a third said that their organization did a bad job of attracting and retaining talented workers, and a third reported being dissatisfied with their prospects for advancement (Light 2002).

Such responses explain, in part, why many Americans see the federal government as plodding and unresponsive, but what about other bureaucracies? Weber described the ideal characteristics of a productive and efficient bureaucracy, but what's the reality? As you read through the following list of problems, think about the ones you've experienced while working in a bureaucracy or dealing with one:

- **Weak reward systems** reduce the motivation to do a good job and are thus a major source of inefficiency and lack of innovation (Barton 1980). Besides low wages or salaries, weak reward systems include few or no health benefits, little recognition, unsafe equipment and work environments, and few incentives to be creative.

- **Rigid rules** discourage creativity. An employee in a bureaucracy who suggests a way to decrease

red tape, for example, will probably be criticized for being lazy or not following the rules, rather than being praised for being innovative.

- Rigid rules create and reinforce **bureaucratic ritualism**, a preoccupation with rules and regulations rather than the organization's objectives (Merton 1968). Instead of questioning whether all the red tape is really necessary, bureaucrats are usually more concerned that people follow established procedures because "we've always done it this way."

- Rigid rules and ritualism often lead to **alienation**, a feeling of isolation, meaninglessness, and powerlessness. Workers who don't feel connected to the organization's goals or who believe that they have little control may experience indifference to or hostility toward their jobs. When people are reduced to a "small cog," they feel dehumanized and see little opportunity for advancement. Alienation—at all levels—may result in high turnover, tardiness, absenteeism, stealing, sabotage, stress, health problems, and, in some cases, whistle-blowing (reporting organizational misconduct to legal authorities).

- **Communication problems** are common in bureaucracies. Because communication typically flows down rather than up the hierarchy, employees (including many managers) below the highest echelons rarely know what's going on. Supervisors and their subordinates may be reluctant to discuss problems or offer suggestions for fear of being criticized or fired. Those at the top often make ill-informed decisions based on incomplete or inaccurate information, as illustrated by groupthink (Blau and Meyer 1987; Fletcher and Taplin 2002).

- **Parkinson's Law** is the idea that work expands to fill the time available for its completion (Parkinson 1962). This means that, even if employees finish an assigned task before the deadline, they'll look busy and act as though they're still working on the task to safeguard their jobs or to avoid getting another assignment. Parkinson's Law also explains a bureaucracy's tendency to keep getting larger. That is, bosses may not crack down on idle workers because higher productivity may reduce the possibility of increasing the budget or hiring more people.

- A related idea, the *Peter Principle*, proposes that workers are promoted until they reach their level of incompetence (Peter and Hull 1969). In many bureaucracies, employees who perform well are promoted to the next level, usually into administrative positions. Thus, project directors become managers, teachers become principals, and nurses become hospital administrators. Sooner or later, those promoted find themselves in positions for which they lack sufficient knowledge, capabilities, training, competence, or experience.

- The **iron law of oligarchy** is the tendency of a bureaucracy to become increasingly dominated by a small group of people (Michels 1911/1949). A handful of people can control and rule a bureaucracy because the top officials and leaders monopolize information and resources. As a result, those at the top maintain their power and privilege.

- The cumulative effect of these and other bureaucratic dysfunctions can result in *dehumanization* because weak reward systems, ritualism, rigid rules, and the other shortcomings limit organizational creativity and freedom. In fact, sociologist George Ritzer argues that some of the problems and characteristics that dominate bureaucracies are increasingly shaping U.S. society. As the "The McDonaldization of Society" box shows, Americans' everyday life is becoming more automated, rigid, and impersonal.

Once established, bureaucracies are almost impossible to eliminate. For example, the U.S. Department of Education is a huge bureaucracy that continues to grow even though it has done little to improve the public educational system and even restricts college opportunities for many low-income students (see Chapter 14). However, most bureaucracies continue to function, largely because of the internal development of informal groups.

THE INFORMAL SIDE OF BUREAUCRACY

Informal networks develop at all levels of a bureaucracy, from the very bottom to the very top. At the top, a CEO or a few members of the board of directors run the organization. They often make decisions infor-

mally, and the board then rubber-stamps those decisions. And, because of personal ties and connections, the top officials at large corporations often know each other, consult with each other, and serve on each other's boards. For example, George J. Mitchell, former U.S. Senator, serves on the board of directors of Disney, two large and influential law firms in Washington, DC, a large insurance company, FedEx Corporation, Xerox Corporation, Casella Waste Systems, Staples, and Starwood Hotels & Resorts. Such interlocking ties suggest that a small group of people holds much of the informal decision-making power across many organizations.

Historic Studies of Informal Structures

What about informal structures at the lower ranks? For many years, and well into the 1930s, experts concentrated on organizational efficiency based on the principles of *scientific management* developed by Frederick Winslow Taylor, a mechanical engineer who wanted to improve industrial productivity. Taylor (1911/1967) considered workers—especially those in factories and on assembly lines—mere adjuncts to machines, assumed that people do not like to work, and maintained that employees must be prodded by the promise of financial incentives, close supervision, and clear goals that they can attain with little effort. Taylor believed that the best results would come from a partnership between a trained and qualified management and a cooperative workforce, but that management had to enforce the cooperation because most workers are incapable of handling even the simplest tasks.

A number of extensive studies conducted by industrial psychologists and sociologists between 1927 and 1932 shook up Taylor's views. The research teams studied employees at the Western Electric Company's Hawthorne plant in Chicago. The *Hawthorne studies*, as they're often called, found that informal groups were critical to the organization's functioning (Roethlisberger and Dickson 1939/1942; Mayo 1945; Landsberger 1958).

Informal social groups can promote an organization's goals if they collaborate, are cohesive, and motivate each other (Barnard 1938). They can also resist an organization's goals and formal rules. In one of the Hawthorne studies, Roethlisberger and Dickson (1939/1942) spent six months observing a group of 14 men who wired telephone switchboards in what the company called the "bank-wiring room."

The bank-wiring room work group consisted of nine wiremen, three solderers, and two inspectors. Although management offered financial incentives for higher productivity, the work group pressured people to limit out-

The McDonaldization of Society

According to sociologist George Ritzer (1996: 1), the organizational principles that underlie McDonald's, the well-known fast-food chain, are beginning to dominate "more and more sectors of American society as well as of the rest of the world." McDonaldization has four components: efficiency, calculability, predictability, and control.

1 *Efficiency* means that consumers have a quick way of getting meals. In a society where people rush from one place to another and where both parents are likely to work or single parents are pressed for time, McDonald's (and similar franchises) offers an efficient way to satisfy hunger and avoid much "fuss and mess."

Like their customers, McDonald's workers function efficiently: The menu is limited, the registers are automated, and employees perform their tasks rapidly and easily.

2 *Calculability* emphasizes the quantitative aspects of the products sold (portion size and cost) and the time it takes to get the products. Customers often feel that they are getting a lot of food for what appears to be a nominal sum of money and that a trip to a fast-food restaurant will take less time than eating at home. In reality, soft drinks are sold at a 600 percent markup because most of the space in the cup is taken up by ice. It costs at least twice as much and takes twice as long to get to the restaurant, stand in line, and pick up the food than to prepare a similar meal at home.

3 *Predictability* means that products and services will be the same over time and in all locales. "There is great comfort in knowing that McDonald's offers no surprises" (Ritzer: 10). The Quarter Pounder is the same regardless where customers order. Outside of the United States, "homesick American tourists in far-off countries can take comfort in the knowledge that they will likely run into those familiar golden arches and the restaurant they have become so accustomed to" (Ritzer: 81).

Workers, like customers, behave in predictable ways: "There are, for example, six steps to window service: greet the customer, take the order, assemble the order, present the order, receive payment, thank the customer and ask for repeat business" (Ritzer: 81).

4 *Control* means that technology shapes behavior. McDonald's controls customers subtly by offering limited menus and uncomfortable seats that encourage diners to eat quickly and leave. "Consumers know that they are supposed to line up, move to the counter, order their food, pay, carry the food to an available table, eat, gather up their debris, deposit it in the trash receptacle, and return to their cars. People are moved along in this system not by a conveyor belt, but by the unwritten, but universally known, norms for eating in a fast-food restaurant" (Ritzer: 105).

McDonald's controls employees more openly and directly. Workers are trained to do a limited number of things in precisely the same way, managers and inspectors make sure that subordinates toe the line, and McDonald's has steadily replaced human beings with technologies that include soft-drink dispensers that shut off when the glass is full and a machine that rings and lifts the french fries out of the oil when they are crisp. Such technology increases the corporation's control over workers because employees don't have to use their own judgment or need many skills to prepare and serve the food.

Efficiency, calculability, predictability, and control reflect a *rational* system (remember Weber?) that increases a bureaucracy's efficiency. On the other hand, Ritzer contends, McDonaldization reflects the "irrationality of rationality" because the results can be harmful. For example, the huge farms that now produce "uniform potatoes to create those predictable french fries" of the same size rely on the extensive use of chemicals that then pollute water supplies. And the enormous amount of nonbiodegradeable trash that McDonald's produces wastes our money because we—and not McDonald's—pay for landfills.

According to Ritzer (2008: 11, 119), McDonald's is such a powerful model that many businesses have cloned its four dimensions of efficiency, calculability, predictability, and control. Examples include "McDentists" and "McDoctors," drive-in clinics designed to deal quickly with minor dental and medical problems; "McChild" care centers like KinderCare; and McPaper" newspapers like *USA Today.*

put. The wiremen, for instance, developed the following norms that controlled the group's behavior:

1. You should not turn out too much work. If you do, you are a "rate-buster" and a "speed king."

2. You should not turn out too little work. If you do, you are a "chiseler."

3. You should not tell a supervisor anything that will be detrimental to a co-worker. If you do, you are a "squealer."

4. You should not "act officious." If you are an inspector, for example, you should not act like one.

If an individual violated any of these norms, the other workers used "binging" (striking a person on the shoulder) as a form of punishment.

Why did the men control productivity rather than take advantage of the management's promise of higher wages? They feared that if they produced at a high level, they would be required to produce at that level regularly. They also worried that high productivity would lead to layoffs because supervisors would conclude that fewer workers could achieve the same output. In addition, the bank-wiring room men experienced high morale and job satisfaction because they felt they had some control over their work. In contradiction to Taylor's scientific management perspective, then, the Hawthorne studies uncovered the importance of informal structures and concluded that there is an important relationship between formal and informal organization.

Modern Work Teams

Since the Hawthorne studies, many managers have recognized that informal groups pervade organizations and affect workers' productivity. As a result, numerous organizations have implemented alternative management strategies.

Today, *self-managing work teams* are the dominant model in most large organizations (Cloke and Goldsmith 2002; Neider and Schriesheim 2005). Contrary to Taylor's notion that workers do and managers think, self-managing work teams, sometimes referred to as *postbureaucratic organizations,* involve groups of 10 to 15 people who take on the duties of their former supervisors. Instead of being told what to do by a boss, self-managing workers gather and interpret information, act on the information, and take collective responsibility for their actions.

Typically, a self-managing work team is responsible for completing a specific, well-defined job func-

tion. The team has the authority to make essential decisions to complete the job—whether it's designing a new refrigerator, or handling a food service for a university. Members of a self-managing team order the materials they need and coordinate with other groups within and outside the organization. Because they are not held back by unresponsive managers, self-managing teams theoretically become committed to the organization and its success (Wellins et al. 1991).

How effective are self-managing work teams? The data are mixed. A number of case studies show that well-designed and well-executed team initiatives improve employee morale and the quality of products and services with relatively small costs. The initiatives that fail are marked by low support from management and union leaders, management's interfering with the group's work, or ineffective team leaders who discourage a team's creativity or autonomy (Walton and Hackman 1986; Manz and Sims 1987; Barker 1993).

3 Sociological Perspectives on Groups and Organizations

how do organizations work? And how do social groups affect massive bureaucracies? In answering such questions, functionalism, conflict theory, feminist theories, and symbolic interactionism provide different insights. Taken together, these perspectives provide a multifaceted understanding of groups and organizations (*Table 6.3* on p. 114 summarizes these perspectives).

FUNCTIONALISM: COOPERATION WORKS

Much of what you've read about social groups and formal organizations so far in this chapter has come from functionalist perspectives, which emphasize that groups and organizations are composed of interrelated, mutually dependent parts. As Weber pointed out, when a bureaucracy operates rationally and efficiently, workers cooperate and maintain the organization's stability and continuity.

Garment work is traditionally piecework, with a worker making only one part of a garment. With piecework, the more you produce, the more you earn.

© Nel Beckerman/Stone/Getty Images

Today, many companies still apply Taylor's basic principles of scientific management, especially in industries where the work is largely routine. For example, for workers whose job requires sewing, the labor is broken down into numerous specific tasks, supervisors closely monitor each task, and the more pieces a worker finishes, the higher the pay. In many manufacturing plants, where wages aren't necessarily based on piecework, workers and bosses must cooperate across a number of sequential steps to turn out a final product, whether it's an automobile or a can of soup.

Functionalists also note that organizations (and especially bureaucracies) can be dysfunctional. Workers may be alienated because of weak reward systems, favoritism, and incompetent supervisors. Such dissatisfaction can result in absenteeism, low morale and productivity, and quitting.

Critical Evaluation

Functionalism has been useful in understanding how groups and organizations fulfill essential functions such as motivating people to get work done and achieving a common goal. For critics, especially conflict theorists, functionalists exaggerate harmony and tend to gloss over dysfunctions such as worker dissatisfaction and alienation.

A related issue is whether informal social networks improve worker morale and control as much as func-

tionalists claim. After all, a tedious and monotonous job is tedious and monotonous regardless of the degree of informal coworker interaction.

CONFLICT THEORY: SOME BENEFIT MORE THAN OTHERS

Conflict theorists contend that organizations are based on vast differences in power and control. In addition to gender and ethnicity, which we'll examine shortly, social class is important in determining individuals' places in organizations. In many companies, those at the higher levels are more comfortable hiring and promoting others similar to themselves. Phrases like "fitting the mold" and "having common interests" are often code words for belonging to the in-group, most often consisting of higher-income white males who share similar recreational activities (like golf or tennis) and know where to shop for acceptable attire (like pin-striped suits).

Inequality in income, status, and other rewards means that owners and managers can easily exploit workers. Those at the top dictate to those at the middle and the bottom. Because organizations serve elites, conflict theorists argue, they are usually undemocratic and ignore workers' needs and interests. As a result, workers have little input in deciding how their work is structured. In fact, formal organizations, especially large corporations, cause harm to workers, consumers, communities, and the environment because they are driven by a relentless pursuit of power and profit (Bakan 2004).

Even in organizations where most of the workers have considerable status and prestige, elites can control the agenda. For example, more than half of the 889 scientists at the Environmental Protection Agency (EPA) who responded to a survey said that they had experienced considerable internal interference with their work, such as being forced to misrepresent findings or to exclude material that showed that the EPA was ignoring environmental laws. Of the respondents, 76 percent held either a master's degree or a doctorate, 83 percent were in high-level government positions, and 37 respondents served in key positions just below the top presidential appointees. The report concluded that the EPA was "under siege from political pressure": "On numerous issues—ranging from mercury pollution to groundwater contamination to climate change—political appointees of the George W. Bush administration have edited scientific documents,

Some organizations pay to have their products incorporated into the story lines of television programs because many Americans have technology (such as TiVo) that enables them to skip commercials. In one episode of The Office, Michael Scott, the branch manager, introduced casual Fridays and kept proclaiming "I love my new Levi's!" For conflict theorists, how does such product placement (or stealth advertising, according to some critics) manipulate viewers?

glass ceiling a collection of attitudinal or organizational biases in the workplace that prevent women from advancing to leadership positions.

manipulated scientific assessment, and generally sought to undermine the science behind dozens of EPA regulations" (Union of Concerned Scientists 2008: 2). Thus, top bureaucrats can stifle information that contradicts their views.

Critical Evaluation

Among its contributions, conflict theory has underscored the inequality within groups and organizations that saps workers' motivation and limits their economic success. Critics, however, fault conflict theory for assuming that greater equality always leads to a more successful and productive organization. As you saw earlier, even a model of nontraditional organization such as self-managing work teams can fail because of ineffective team leaders or management interference with teams.

In addition, do conflict theorists focus too much on organizational deficiencies? There are, after all, many formal organizations, including some bureaucracies, that are efficient and fair.

FEMINIST THEORIES: MEN BENEFIT MORE THAN WOMEN

Feminist scholars agree with functionalists that organizations can be valuable vehicles for attaining common goals. They also agree with conflict theorists that those with power protect their self-interests. Feminist analyses differ from those of functionalists and conflict theorists in two ways, however: (1) feminists emphasize that across all social classes, women (and especially minority women) consistently fare worse than men, and (2) they rely on both macro and micro data to explain why gender is a major variable affecting roles in formal organizations, especially leadership roles.

Women have enjoyed increased success in organizations since the 1980s, but rarely at the highest levels. For example, women hold more than half of all management and professional positions in the United States, but make up less than 2 percent of *Fortune* 500 CEOs, 15 percent of board directors at *Fortune* 500 companies, and 17 percent of board directors at *Fortune* 100 companies (Catalyst 2007; Alliance for Board Diversity 2008).

Large numbers of women still hit a **glass ceiling**, a collection of attitudinal or organizational biases in the workplace that prevent them from advancing to leadership positions. Organizational barriers reflect, in large part, stereotypes about gender roles. When researchers interviewed senior-level U.S. executives, they found that *both* men and women tended to describe women as better at stereotypically feminine "caretaking" skills, such as supporting and encouraging others, and to assert that men excel at stereotypically masculine "taking charge" skills, such as influencing superiors and problem solving. Furthermore, when women act in ways that are consistent with gender stereotypes, they are viewed as less competent leaders (too soft); when they act in ways that are inconsistent with gender stereotypes, they're considered unfeminine (too tough). Thus, a "double bind—that nagging sense that whatever

you do, you can do no right"—means that women face higher standards and receive lower rewards than men and have to prove that they can lead, over and over again (Sabattini et al. 2007; Foust-Cummings et al. 2008).

Critical Evaluation

Feminist theories have expanded the conflict perspective on groups and organizations by showing that many talented women are still treated like outsiders. One weakness of feminist theories, however, is that much of the emphasis is still on white and black women in professional and managerial positions, even though Latinas and Asian American women comprise a large segment of these workers and those in the general labor force.

A second limitation is that even when feminist scholars say that both sexes suffer from organizational stereotypes, they offer little data on how such stereotypes affect men. In addition, some female leaders are playing a major role in reshaping organizational structures because they have good communication skills, know how to increase morale and productivity, and are comfortable with diversity—all assets in today's global economy (Helgesen 2008). The question, then, is whether some feminist scholars are spending too much time focusing on women's ongoing struggles rather than their progress.

SYMBOLIC INTERACTIONISM: PEOPLE DEFINE AND SHAPE THEIR SITUATIONS

While functionalist, conflict, and (some) feminist theorists examine groups and organizations on a macro level, symbolic interactionists focus on interpersonal relationships. Interactionists emphasize that individuals' perception and definition of a situation shape group dynamics and, consequently, organizations. Group leaders or members can create or reinforce conformity (as shown in the Asch and Milgram studies). Informal groups can also determine what goes on in an organization by refusing to obey the rules and implementing their own (as in the Hawthorne studies). Thus, according to symbolic interactionists, individuals make choices, change rules, and mold their own

identities instead of being passive members of a group who are manipulated (Kivisto and Pittman 2001).

Critical Evaluation

Symbolic interactionism has made important contributions in describing how members of groups and organizations interpret the world around them and, as a result, affect what goes on. Despite these contributions, the interactionists' perspective has several weaknesses. Some scholars have wondered whether results of some of the influential studies (especially Milgram's) reflected the participants' responses to a laboratory situation where they obeyed the experimenter's instructions rather than the way they would have behaved in normal circumstances (Blass 2000).

Because symbolic interactionism is a micro-level theory, it ignores macro-level factors that result in the exploitation of workers and consumers. About 54 percent of young people aged 8 to 21 buy products—whether they need them or not—because of television ads, a macro-level factor (Martin 2006). In addition, and in contrast to interactionists' claims, most people have little control in shaping the situations in which they find themselves. Instead, formal organizations (such as the U.S. government and large companies) often invade the privacy of citizens, workers, or consumers by collecting information and monitoring people's behavior.

We've looked at the general characteristics of social groups and formal organizations, both of which are part of larger structures that sociologists call *social institutions*. The concluding section introduces you to this important sociological concept and lays the foundation for the discussion of social institutions in later chapters.

Sam Walton, the founder of Wal-Mart, kept in touch with his employees, called them "associates," and promised them "limitless opportunities" in the company. As a result, the employees, despite their low wages, worked hard because they felt that they had a stake in the "Wal-Mart family" (Frank 2006).

TABLE 6.3
Sociological Perspectives on Groups and Organizations

THEORETICAL PERSPECTIVE	LEVEL OF ANALYSIS	MAIN POINTS	KEY QUESTIONS
Functionalist	Macro	Organizations are made up of interrelated parts and rules and regulations that produce cooperation in meeting a common goal.	• Why are some organizations more effective than others? • How do dysfunctions prevent organizations from being rational and effective?
Conflict	Macro	Organizations promote inequality that benefits elites, not workers.	• Who controls an organization's resources and decision making? • How do those with power protect their interests and privileges?
Feminist	Macro and micro	Organizations tend not to recognize or reward talented women and regularly exclude them from decision-making processes.	• Why do many women hit a glass ceiling? • How do gender stereotypes affect women in groups and organizations?
Symbolic Interactionist	Micro	People aren't puppets but can determine what goes on in a group or organization.	• Why do people ignore or change an organization's rules? • How do members of social groups influence workplace behavior?

social institution an organized and established social system that meets one or more of a society's basic needs.

4 Social Institutions

Sociologists define a **social institution** (or simply, *institution*) as an organized and established social system that meets one or more of a society's basic needs. Some social institutions are almost universal because they ensure a society's survival and practically all members of a society participate. According to functionalists, there are five major social institutions worldwide:

• The *family* replaces the members of a society through procreation, socializes children, and legitimizes sexual activity between adults.

• The *economy* organizes a society's development, production, distribution, and consumption of goods and services.

• *Political institutions* maintain law and order, passes legislation, and forms military groups for internal and external defense.

• *Education* helps to socialize children and transmits knowledge and provides information and training for jobs and other work-related activities.

• *Religion* encompasses beliefs and practices that offer a sense of meaning and purpose to life.

Besides these core universal institutions, others that have emerged include law, science, medicine, sports, health care, criminal justice, and the military.

WHY SOCIAL INSTITUTIONS ARE IMPORTANT

Social institutions are abstractions, but they have an organized purpose, weave together norms and values, and, consequently, guide human behavior. Although no two families are exactly alike, the institution of the family reflects broadly shared cultural agreements about what a family should be and do. In the same vein, no two banks are exactly alike, but the economy as an institution creates and perpetuates a variety of rules that usually make banking predictable and fairly stable.

HOW SOCIAL INSTITUTIONS ARE INTERCONNECTED

The various social institutions in a society are linked to one another. Through taxes, the economy provides revenues for the political institution, education, and those families that need financial support. Education, in turn, trains children (and adults) in the general and specific skills required for productive participation in the economy. The family instills many of the values of hard work and success that educational institutions reinforce. Families and schools could not survive without the goods provided by the economy, and the economy needs workers who have been socialized by the family and trained by schools to enter the labor force in adulthood. And, in many communities, criminal justice, religious, legal, and political institutions have formed coalitions to decrease substance abuse and other social problems.

Interconnections between institutions can also be dysfunctional. Consider the linkages between five social institutions—the economy, the political system, medicine, education, and the media. Media organizations (such as television networks, magazines, and newspapers) feature ads from companies that make drugs and medical devices and want to target consumers directly. Whether their health problems are real or imagined, consumers then request particular brands from their doctors. Also, the pharmaceutical industry spends $12 billion a year marketing to doctors. More than 87,000 sales representatives shower nearly 95 percent of physicians with free samples of drugs (that influence usage and sales), lavish dinners, tickets to sports events, and cruises and vacations during so-called

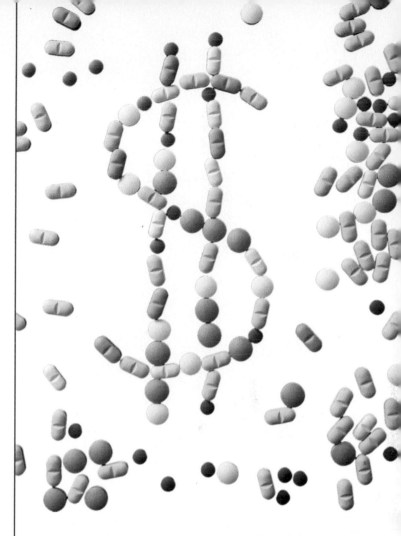

medical seminars. Drug industry lobbyists earn up to $600,000 a year for persuading legislators to support bills that increase the costs of prescription drugs and limit the availability of generic drugs. Legislators, like physicians, are rewarded with vacations, golfing trips, and other gifts. And a team of researchers recently found that pharmaceutical companies have conducted research on their own drugs and paid prominent academic scientists to put their names on the resulting papers, which were then published in prestigious medical journals (Campbell et al. 2007; Ross et al. 2008). In 2005, the University of Michigan health systems banned the delivery of free lunches by pharmaceutical sales representatives. The University estimated that the meals had been worth about $2.5 million per year (Saul 2007).

As you've seen in this chapter, social groups, formal organizations, and institutions shape our behavior. Nonetheless, we're not robots who succumb to bureaucracies. Instead, and despite numerous constraints, we can make better decisions if we understand how groups, organizations, and institutions can help or hinder us in our everyday lives.

5 major soc. instit:
- family
- economy
- political system
- education
- religion

© Marcelo Wain/IStockphoto.com

© Radius images/Jupiterimages

All of us violate

some of society's rules.

what do you think?

There would be less crime if the punishment was more severe.

1 2 3 4 5 6 7

strongly agree   strongly disagree

7

Deviance, Crime, and the Criminal Justice System

Key Topics

In this chapter, we'll explore the following topics:

At one time or another, all of us violate some of society's rules. This chapter explores such violations. We begin by examining the characteristics of deviance and crime, discuss several theoretical explanations of why people deviate, and end the chapter by looking at institutional attempts to control and change rule-breaking behavior. First, however, take the brief quiz on page 118 to see how much you know about deviance and crime in the United States.

> **deviance** behavior that violates expected rules or norms.

1 What Is Deviance?

have you ever driven above the speed limit? Cheated on an exam? Hit someone? All are examples of **deviance**, behavior that violates expected rules or norms. The word *deviance* has a derogatory connotation, but sociologists don't make such value judgments. Instead, they are interested in understanding and explaining behavior—sometimes positive but usually negative—that violates social rules.

POSITIVE AND NEGATIVE DEVIANCE

Most people behave within the boundaries of accepted standards that a group or society considers normal. When we violate these expectations, we're often quick to admit fault, to apologize, and to promise to improve. Deviant acts—both positive and negative—fall outside of the range of normal behavior.

Positive deviance is behavior that overconforms to social expectations. Some students show positive deviance by getting consistently high grades. *Negative deviance* is behavior that falls below social expectations because people are not aware of the rules, reject them, or place a higher priority on different norms. Students with consistently low grades are usually not studying enough because they are involved in other activities that they give a higher priority (such as jobs, family care, or socializing with friends). Our reactions to negative deviance are fairly standard ("If you don't study, you'll fail"), but our reactions to positive deviance may be mixed. For example, a high achiever—whether in class or on the job—may be admired or may be accused of sucking

© Nicolas Belton/iStockphoto

stigma a negative label that devalues a person and changes her or his self-concept and social identity.

up or being a brown noser because she or he performs above the group average.

KEY CHARACTERISTICS OF DEVIANCE

Deviance is universal because it exists in every society. Sociologists emphasize, however, that the key characteristics of deviance can vary quite a bit over time, from group to group, and from culture to culture (Sumner, 1906; Schur 1968):

- *Deviance can be a condition or belief rather than a behavior.* People don't necessarily have to do something to be considered deviant. We can be branded as outsiders simply because of our appearance (very tall or very short, obese or very thin). And, in many societies, skin color or sexual orientation can result in being ostracized. We may also be deemed deviant because of our beliefs, such as being a Muslim in a predominantly Christian neighborhood or vice versa (Becker 1963; see, also, Chapter 15).

- *Deviance is accompanied by social stigmas.* A **stigma** is a negative label that devalues a person and changes her or his self-concept and social identity. Stigmatized individuals may respond to being discredited in many ways: They may alter their appearance (for example, through cosmetic surgery), associate with others like themselves who accept them (as in gangs), hide information about some aspect of their deviance (for example, an ex-con who does not reveal that status), or divert attention from a stigma by excelling in some area (such as music or sports) (Goffman 1963).

- *Deviance varies across and within societies.* What is appropriate or tolerated in one society may be deviant in another. For example, in the United States and many other Western countries, people who practice witchcraft are usually viewed as oddballs but tolerated. In some African nations, in contrast, children and women who are branded as witches are cast out by relatives or exiled.

- *Deviance is formal or informal.* Formal deviance is behavior that violates laws. A major example is crime, a

topic that we'll examine shortly. In contrast, *informal deviance* is behavior that breaks customary practices. Body piercing and tattoos are currently popular, but most employers have a negative view of both practices. Even when people are no longer formally deviant, they may describe themselves as informally deviant (as a recovering alcoholic or an ex-con).

- *Perceptions of deviance can change over time.* Many behaviors that were acceptable in the past are now seen as deviant. Only during the 1980s and 1990s did U.S. laws define date rape, marital rape, stalking, and child abuse as crimes. And, recently, smoking—widely accepted in the past—has been prohibited in practically all U.S. restaurants, airports, workplaces, and even some public parks.

On the other hand, most Americans now shrug off behaviors that were stigmatized in the past. Cohabitation and out-of-wedlock births, seen as sinful and immoral only a few decades ago, are now widespread and accepted by most Americans. Many states have legalized some forms of gambling (lotteries and slot machines) that were previously illegal. And, instead of stigmatizing alcoholism and other drug abuse as a personal failing, insurance companies and every state government now provide counseling and rehabilitation programs.

WHO DEFINES DEVIANCE?

Because deviance is culturally relative and the standards change over time, who decides what's right or wrong? Those who have authority or power. During our early

---

HOW MUCH DO YOU KNOW ABOUT U.S. DEVIANCE AND CRIME?

True or False?

1. Most crime victims are women.
2. Crime rates have decreased during the last decade.
3. People are more likely to be arrested for a drug violation than for driving while drunk.
4. Serious crimes such as murder and assault are more common than less serious crimes like illegal gambling and prostitution.
5. People in affluent neighborhoods are more likely to experience burglary than those in poor neighborhoods.
6. About 15 percent of all prison inmates are women.
7. Death penalties deter crime.
8. Compared with other nations, the United States has a low number of people in prisons.

The answers for #2 and #3 are true; the rest are false. You'll see why as you read this chapter.

More than 1,000 women in Ghana live in exile in witch camps. Most are poor widows or older women who are blamed for outbreaks of disease in their villages and the illness or death of relatives or neighbors.

© Markus Matzel/Peter Arnold Inc.

years, parents and teachers define acceptable and unacceptable behavior in our everyday interactions and behavior. As we get older, we learn that legal institutions also define what is deviant, for example, not allowing us to drive until age 16 or to purchase or consume liquor until age 21. Often, however, children get mixed messages about what is normal and what is deviant. For example, many K–12 teachers complain that their students use foul language (even in the first grade), but many parents themselves curse or excuse their children's profanity (Strauss 2005).

2 What Is Crime?

What comes to mind when you hear the word *crime?* Most of us typically imagine a murder or a violent physical attack. In fact, such offenses constitute a minority of deviant behavior. **Crime** is a violation of societal norms and rules for which punishment is specified by public law. Many sociologists are **criminologists,** researchers who use scientific methods to study the nature, extent, cause, and control of criminal behavior. Measuring crime may seem straightforward, but the task is not as simple as it appears.

MEASURING CRIME

One newspaper announces that U.S. crimes rates are falling, but another proclaims that they are rising. Which is correct? It depends, among other factors, on the newspaper's data source. There are several primary sources of crime statistics, but two of the most important are official data and victim surveys. The best known and most widely cited source of official statistics is the FBI's Uniform Crime Report (UCR), which includes crimes reported to the police and arrests made each year.

Because the UCR has been published since 1930, its statistics are useful in examining trends over time. However, the UCR doesn't include federal offenses such as corporate crime, kidnapping, and Internet crimes, or anything on the roughly 60 percent of all crimes committed in the United States that go unreported to the police (Hart and Rennison 2003). Also, many crimes don't result in arrests (see *Figure 7.1* on the next page): There are no witnesses or the police don't investigate many offenses because they must devote most of their limited resources to violent crimes rather than property crimes. Because of these and other problems, many criminologists also use victim surveys to measure the extent of crime.

© Nicholas Belton/iStockphoto.com

Are any of these people deviant? (Why or why not?)

© BananaStock/Jupiterimages / © Image Source/Jupiterimages / © Seth Kushner/Workbook Stock/Jupiterimages / © Image Source/Jupiterimages /

A **victim survey** is a method of gathering data that involves interviewing people about their experiences as crime victims. Every year, the U.S. Department of Justice sponsors the most widely used survey—the National Crime Victimization Survey (NCVS). Because the response rates are at least 90 percent, the NCVS offers a more accurate picture of offenses than does the UCR. Also, the NCVS measures include both reported and unreported crime and are not affected by police discretion in deciding whether or not to arrest an offender. Still, some crimes are underreported because people may not be totally honest. Some don't want to admit having been victims—especially when the perpetrator is a family member or friend. Others may be too embarrassed to tell the interviewer that they were victimized while drunk or engaged in drug sales (Hagan 2008).

HOW MUCH CRIME IS THERE?

No one knows the extent of U.S. crime because even the best sources (like the UCR and NCVS) are only estimates. Still, of the almost 15 million arrests in 2006, there were more arrests for property crimes (19 percent) and drug abuse violations (13 percent) than violent crimes (4 percent). Violent crimes are most likely to be covered by the media and, consequently, inspire the greatest fear, but Americans are much more likely to be victimized by theft or burglary than to be murdered, raped, robbed, or assaulted with a deadly weapon. In fact, 88 percent of serious crimes are property crimes (such as auto theft and burglary) rather than physical attacks (Federal Bureau of Investigation 2007).

Offenses that are the least likely to be reported—such as illicit drug use, prostitution, drunkenness, and illegal gambling—are often referred to as *public order crimes,* or *victimless crimes.* **Victimless crimes** are acts that violate laws but involve individuals who don't consider themselves victims. For example, prostitutes and sexual exploiters argue that they are simply providing services to people who want them and substance abusers claim that they aren't hurting anyone but themselves. Some contend that this term is misleading because victimless crimes often lead to property and violent

crimes, as when addicts commit burglary, rob people at gunpoint, and engage in identity theft to get money for drugs. Also, families can become victims, as when alcoholism results in auto fatalities, domestic violence, and higher divorce rates. In a recent national survey, 70 percent of the respondents said that a family member's drug abuse had a negative effect on the emotional or mental health of at least one other family member, and 39 percent experienced financial problems because they went into debt to pay for treatment for an addict who had no health insurance coverage (Saad 2006).

VICTIMS AND OFFENDERS

In 2006, many national newspapers headlined a story about two older women, aged 73 and 75, who befriended two homeless men, aged 50 and 73, took out $2.2 million life insurance policies on them, and then killed them. Such reports skew our perceptions of criminality because most offenders are men and people aged 65 and older are rarely either offenders or victims.

Who Are the Victims?

Most crime victims are men, African Americans, people under age 25, and those who are poor and live in urban areas. Except for rape and assaults by intimate partners

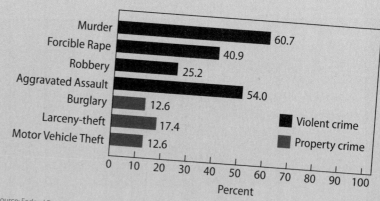

FIGURE 7.1

Percentage of Crimes Cleared by Arrest, 2006

Nationwide, only 44 percent of violent crimes and 16 percent of property crimes are cleared by arrest (someone has been arrested and turned over to the courts for prosecution). So, are arrest rates accurate measures of the prevalence of these crimes?

Murder — 60.7
Forcible Rape — 40.9
Robbery — 25.2
Aggravated Assault — 54.0
Burglary — 12.6
Larceny-theft — 17.4
Motor Vehicle Theft — 12.6

■ Violent crime
■ Property crime

Percent

Source: Federal Bureau of Investigation 2007.

(such as spouses, ex-spouses, and boyfriends), men are almost twice as likely as women to be victims of violent crimes, such as robbery and assault with a deadly weapon. About half of all homicide victims are black men, a majority of them between the ages of 17 and 29 (Harrell 2007).

Older people are much less likely to be victimized than their younger counterparts. Those aged 65 and older experience nonfatal violent crime (such as robbery) only 5 percent as frequently as younger people do, but the crimes are especially traumatic because older people are usually physically unable to escape or to fight back and it takes them longer to recover from a crime both physically and emotionally. Scam artists are especially likely to target older people because they tend to be trusting of others, often live alone, and many have accumulated savings (Alt and Wells 2004; Klaus 2005).

Who Are the Offenders?

Most offenders have never been caught, but arrest rates show patterns by age, gender, race and ethnicity, and social class. Generally, the older Americans get, the less likely they are to engage in crime. Those aged 29 and under constitute 58 percent of all arrestees. In contrast, 22 percent of arrestees are 40 or older, and many of the crimes involve drug abuse violations and driving under the influence (Fox and Zawitz 2004; Federal Bureau of Investigation 2007).

Of all those arrested in the United States, 76 percent are men. Men make up 82 percent of persons arrested for violent crime and 68 percent of those arrested for property crime. Compared with women, men are ten times more likely to commit murder. In other crimes, 95 percent of carjackers, 90 percent of those who rob, and 92 percent of those arrested for weapons violations and sexual offenses are men (Rennison 2003; Klaus 2006; Federal Bureau of Investigation 2007).

Almost 70 percent of those arrested are white, 28 per-

Men are more likely to be victims of violent crime than women.

cent are black, and the remainder is American Indian or Asian American. (The FBI doesn't provide data on Latinos because it collects information on race but not ethnicity.) African Americans make up about 13 percent of the general population but account for 29 percent of violent crimes. Across all U.S. racial groups, young males (aged 18 to 24) have the highest arrest rates—making up about 46 percent of all those arrested (Federal Bureau of Investigation 2007).

Social class also affects crime rates. Offense rates are higher in poor inner-city areas than in suburban and wealthier neighborhoods, and prison statistics consistently show that those incarcerated have low educational levels. For example, 68 percent of state prisoners don't have a high school diploma, while only 11 percent have attended college or have a college degree (Harlow 2003; Federal Bureau of Investigation 2007).

Crime rates are higher in low-income areas, but does this mean that poor people are more deviant? Because police devote more resources to poor neighborhoods, those at the lower end of the socioeconomic ladder are more likely to be caught, arrested, prosecuted, and incarcerated. Middle-class criminals are more elusive. For example, the theft of *intellectual property* (IP) is a crime that includes offenses such as software piracy, bootlegging musical recordings and movies, selling company trade secrets, and copyright violations. Between 1994 and 2002, less than 1 percent of more than 1 million IP theft suspects were caught, but of those prosecuted, 50 percent were college educated (Motivans 2004).

Most of us don't commit crimes but rather conform to laws and cultural expectations. We usually conform, however, not because we're innately good, but because of social control.

3 Controlling Deviance and Crime

Social control refers to the techniques and strategies that regulate people's behavior in society. The purpose of social control is to eliminate, or at least reduce, deviance. Ensuring people's conformity to group or societal expectations involves both informal and formal social control.

social control the techniques and strategies that regulate people's behavior in society.

© Comstock Images/Jupiterimages

sanctions punishments or rewards for obeying or violating a norm.

INFORMAL AND FORMAL SOCIAL CONTROL

Most conformity comes from the internalization of norms during the powerful process of socialization. Because we genuinely care about the opinions of family members, friends, and teachers, we try to live up to their expectations (see Chapter 4). In this sense, we conform because of *informal social controls* that we learn and internalize during childhood.

In addition to informal mechanisms, *formal social control* regulates social behavior. Unlike informal social control, formal social control exists outside of the individual. For example, a college dean might threaten a student with suspension or expulsion, a supervisor might submit a negative evaluation of an employee, or a business might install security cameras. Many formal social control agents, such as police and judges, spend considerable effort trying to prevent or decrease deviance.

POSITIVE AND NEGATIVE SANCTIONS

Most of us conform because of **sanctions**, punishments or rewards for obeying or violating a norm. *Positive sanctions* are rewards for desirable behavior and include a variety of facial expressions (such as smiling), body language (such as hugging), comments (such as "Congratulations!"), and other forms of recognition (such as good grades, trophies, and promotions). Positive sanctions are very effective because they increase our self-confidence, self-esteem, and motivation, especially when the rewards are deserved.

Negative sanctions are punishments that convey disapproval for violating a norm. Negative sanctions range from mild and informal expressions (such as frowns and gossip) to more severe and formal reactions (such as fines, arrests, and incarcerations). Some American Indian reservations have revived the traditional penalty of banishment to deal with gangs and drugs. The troublemakers can be ordered off the reservation and stripped of their tribal membership (Snell 2007).

Growing marijuana is moving to the suburbs. Growers pay at least $750,000 for a house in an affluent subdivision, gut the interiors, and install sophisticated artificial lighting and watering systems. Neighbors are rarely aware of what's going on because they assume their communities are safe from crime and rarely know the others who live on their street (Blankenstein and Barboza 2007; Ritter 2007).

The ultimate formal negative sanction in most societies is execution.

The next sections examine four important sociological perspectives—functionalism, conflict theory, feminist theories, and symbolic interactionism—that help us understand why people are deviant (*Table 7.1* on p. 130 summarizes these theories).

4 Functionalist Perspectives on Deviance and Crime

for functionalists, deviance and crime are normal parts of the social structure. Functionalists don't endorse undesirable behavior, but they view deviance and crime as both functional and dysfunctional.

HOW DEVIANCE AND CRIME CAN BE BOTH DYSFUNCTIONAL AND FUNCTIONAL

Remember that, for sociologists, *functional* doesn't mean "good" and *dysfunctional* doesn't mean "bad." Instead, sociologists simply look at the positive and negative effects of behaviors on groups and societies. Crime and deviance are dysfunctional when they:

- *Create tension and insecurity.* Crime makes people uneasy. Any violation of norms—a babysitter who cancels at the last minute or the theft of your laptop computer—makes life unpredictable and increases anxiety.

- *Erode trust in personal and formal relationships.* Crimes such as date rape and stalking make us suspicious of other people. Of the almost 4 million Americans who experienced identity theft in 2004, nearly a third said that they had problems in obtaining banking services or credit cards because financial institutions didn't trust them (Baum 2006).

- *Damage confidence in institutions.* After the scandals involving Enron and other corporations, one of the founders of Intel, a high-tech company, said that he was "embarrassed and ashamed" to be a corporate executive (Hochberg 2002: A17). Millions of people, even those who didn't lose money, worry that their retirement funds may disappear in the future as a result of similar corporate scams.

- *Are costly.* Besides personal costs to victims (such as fear, emotional trauma, and physical injury), deviance is expensive. All of us pay higher prices for consumer goods and services (such as auto and property insurance), as well as taxes for prosecuting criminals, and for building and maintaining prisons.

Deviance and crime can also be functional because they provide a number of societal benefits (Durkheim 1893/1964; Sagarin 1975; Erikson 1966):

- *Affirm cultural norms and values.* Negative reactions, such as expelling a college student who's caught cheating or incarcerating a bank robber, assert a society's rules and its values about being honest and law-abiding.

- *Provide temporary safety valves.* Some deviance is accepted under certain conditions, as when people "blow off steam." Typically, a community and its police will tolerate noise, underage drinking, loud music, and obnoxious behavior during college students' spring break because it's a short-lived nuisance.

- *Create social unity.* Some deviant behavior, especially when there is a common enemy, unites a group, community, or society. Following the 9/11 terrorist attacks, most Americans experienced feelings of greater solidarity, despite differences in age, political views, social class, and race and ethnicity.

- *Improve the economy.* Deviance and crime can benefit a community financially. Some towns welcome new prisons because they generate jobs and stimulate the economy. Unlike a manufacturing plant that might close down or move to another country, a prison is usually stable. A prison also pays property taxes and boosts businesses such as motels and restaurants that cater to visiting family members and friends of prisoners (Kilborn 2001; Staples 2004).

- *Trigger social change.* Crime and deviance are warning signs that a system isn't working properly. In response, new laws and rules, such as hate crime legislation and campus sex crime prevention acts, may be established (Carr 2005). Also, states may get tougher about enforcing laws, such as those against drunk driving, and impose harsher sentences for murderers and child molesters.

anomie the condition in which people are unsure of how to behave because of absent, conflicting, or confusing social norms.

ANOMIE AND SOCIAL STRAIN

Functionalists have offered a variety of explanations for deviance and crime. Two of the most influential analyses are anomie and strain theories, both of which seek to explain why so many people commit crimes and engage in deviant behavior even though they share many of the same goals and values as people who conform to social norms.

Durkheim's Concept of Anomie

Émile Durkheim (1893/1964; 1897/1951) introduced the term **anomie** to describe the condition in which people are unsure of how to behave because of absent, conflicting, or confusing social norms. During periods of rapid social change, such as industrialization in Durkheim's time, societal rules may break down. As many young people moved to the city to look for jobs in the nineteenth century, norms about proper behavior that existed in the countryside crumbled. Even today, as you'll see in Chapter 16, many urban newcomers experience anonymity and miss the neighborliness that was common at home.

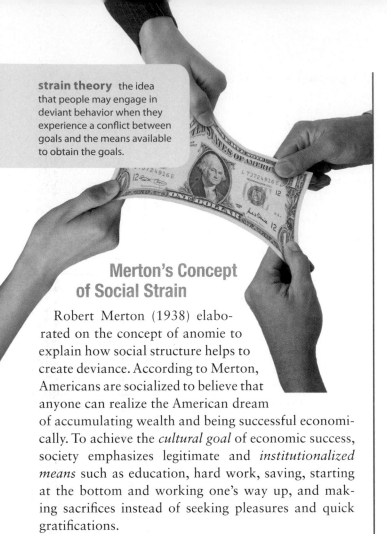

© John Lund/Blend Images/Jupiterimages

strain theory the idea that people may engage in deviant behavior when they experience a conflict between goals and the means available to obtain the goals.

Merton's Concept of Social Strain

Robert Merton (1938) elaborated on the concept of anomie to explain how social structure helps to create deviance. According to Merton, Americans are socialized to believe that anyone can realize the American dream of accumulating wealth and being successful economically. To achieve the *cultural goal* of economic success, society emphasizes legitimate and *institutionalized means* such as education, hard work, saving, starting at the bottom and working one's way up, and making sacrifices instead of seeking pleasures and quick gratifications.

In a highly stratified society like the United States, however, many people don't have access to institutionalized means for financial success. Families may be locked into poverty, and parents may lose their jobs or be unable to afford their children's college education. Thus, the stage is set for anomie—a feeling that one is

being denied a chance to become prosperous and successful—which, in turn, ignites anxiety and anger.

How do people respond? Merton's **strain theory** posits that people may engage in deviant behavior when they experience a conflict between goals and the means available to obtain the goals. Not all people turn to deviance in resolving social strain (see *Figure 7.2*). Most of us *conform* by working harder and longer to become successful. The fact that you're reading this textbook shows that, using Merton's language, your mode of adaptation to strain is conformity—one of achieving success through an institutionalized means such as a college education.

Merton's other four modes of adaptation reflect deviance. *Innovation* occurs when people have endorsed the cultural goal of economic success but turn to illegitimate means, especially crime, to achieve their goal. For many people living in inner-city ghettos, education, hard work, and deferred gratification are often unachievable. For others, crime is a quick way to become richer. For innovators, "it's not how you play the game but whether you win or lose."

In *ritualism,* people don't expect to get rich but get the necessary education and experience to obtain or retain their jobs. In many bureaucratic positions, employees don't make decisions and can't move up the ladder, but their jobs are relatively secure. Ritualists aren't criminals, according to Merton, but they are deviant because they have given up on becoming financially successful. Instead, they do what they are told and "go along to get along."

In *retreatism,* people have rejected both the goals and the means for success. Merton used examples like vagrants, alcoholics, and drug addicts, some of whom have given up because they feel it's impossible to succeed.

In *rebellion,* people feel so alienated that they want to change the social structure entirely by substituting new goals and means for the original ones. Merton used the example of revolutionaries who want to replace capitalism with socialism or communism. A contemporary example is paramilitary groups around the United States that oppose the government.

CRITICAL EVALUATION

A major contribution of functionalism is showing how social structure, and not just individual attitudes and behavior, produces deviance and crime. Functionalism also helps us understand current and emerging forms of deviance. For instance, cheating

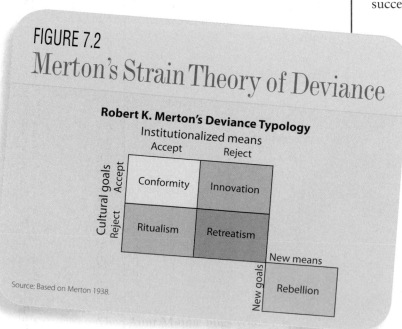

FIGURE 7.2
Merton's Strain Theory of Deviance

Robert K. Merton's Deviance Typology

	Institutionalized means	
Cultural goals	**Accept**	**Reject**
Accept	Conformity	Innovation
Reject	Ritualism	Retreatism

	New means
New goals	Rebellion

Source: Based on Merton 1938.

has always existed, but the number of college students who admit to cheating increased from 11 percent in 1963 to 70 percent in 2005. Perhaps students are simply more honest today than in the past in admitting they cheat. From a functionalist perspective, however, cheating has probably increased because students feel more pressure to do well to get into graduate or professional schools and to secure a good job. Because of such pressure, faculty members sometimes don't report cheating and may even encourage dishonesty by accepting master's theses and doctoral dissertations that they suspect or even know have been plagiarized (McCabe 2005; Wasley 2006).

Functionalist theories also have weaknesses. The concepts of anomie and strain theory are limited because they overlook the fact that not everyone in the United States embraces financial success as a major goal in life. Functionalist theories also don't explain why women's crime rates are much lower than men's (especially since women have fewer legitimate opportunities for financial success) and why people commit some crimes (such as murdering an intimate partner or setting fires just for kicks) that have nothing to do with being successful (Anderson and Dyson 2002; Williams and McShane 2004).

The most consistent criticism is that functionalism typically focuses on lower-class deviance and crime. Conflict theorists have filled this gap by examining middle- and upper-class crime.

5 Conflict Perspectives on Deviance and Crime

functionalists, interested in the behavior of individuals, ask, "Why do some people commit crimes and others do not?" In contrast, most conflict theorists, focusing on who makes the laws, ask, "Why are some acts defined as criminal while others are not?" (Akers 1997).

CAPITALISM, POWER, SOCIAL INEQUALITY, AND CRIME

For conflict theorists, the most powerful groups in society control the law, which defines what's deviant and who will be punished. Because the law embodies the values of those who create it, those outside the power structure (such as youth and members of disadvantaged groups) are more likely to be prosecuted and less likely to have the resources to fight the criminal justice sys-

tem when they feel that they are treated unjustly (Chambliss 1969; Chambliss and Seidman 1982).

Conflict theorists often focus on white-collar crime to show that most laws protect the interests of the few rather than the many (Sutherland 1949). **White-collar crime** refers to illegal activities committed by high-status individuals in the course of their occupation. There is a wide range of white-collar crimes—from thefts in business enterprises to high-tech crimes such as Internet fraud.

According to Dutch criminologist Willem Bonger (1916/1969), capitalism is the root cause of criminal behavior because it breeds egoism, or placing one's interests above those of others: Workers act selfishly and engage in crime because their poverty blunts concern for others; capitalists, competing for profits, are greedy parasites who exploit workers and break the law to become richer. A number of influential American sociologists have argued, like Bonger, that there is a strong association between capitalism and white-collar crime. Others have expanded Bonger's perspective and maintain that only the behaviors that injure the economic interests or challenge the political power of the dominant class are punished (Vold 1958; Turk 1969, 1976; Quinney 1980). Four of the most common types of white-collar crimes are occupational crimes, corporate crimes, cybercrimes, and organized crime.

Occupational crimes are illegal activities committed in the workplace by individuals acting solely in their own personal interest. Many middle- and upper-class criminals are in positions of trust, have many opportunities for theft, and use rationalizations for committing their crimes (Cressey 1953). For example, an executive secretary for a top banker in Goldman Sachs embezzled $5 million over several years. She managed the boss's expense account and routinely wired blocks of his money to her banks in Greece. Her justification was that she was not being paid enough (Sorkin 2002).

Corporate crimes (also known as *organizational crimes*) are illegal acts committed by executives to benefit themselves and their companies. Corporate crimes include a vast array of illegal activities such

white-collar crime illegal activities committed by high-status individuals in the course of their occupation.

occupational crimes crimes committed in the workplace by individuals acting solely in their own personal interest.

corporate crimes white-collar crimes committed by executives to benefit themselves and their companies (also known as *organizational crimes*).

cybercrime white-collar crimes that are conducted online.

organized crime activities of individuals and groups that supply illegal goods and services for profit.

as conspiracies to stifle free market competition, price-fixing, tax evasion, and false advertising. The target of the crime can be the general public, the environment, or even a company's own workers. Often, corporate offenders commit multiple crimes such as stock fraud, insider trading, and perjury. One of these corporate criminals was the former chief executive of WorldCom, a telecommunications company, who was indicted for orchestrating an $11 billion fraud that led to the company's bankruptcy and billions of dollars of loss to employees and stockholders (Belson 2005). In 2005 and 2006, about a dozen U.S. CEOs, convicted of corporate fraud, were sentenced to prison. Jeffrey Skilling, a former Enron CEO, defrauded investors of billions of dollars but claimed that he was innocent. In contrast, when executives in South Korea's Hyundai Corporation were simply accused of corruption in 2006, they apologized publicly during a news conference.

And, most recently, the Securities and Exchange Commission (SEC) has been investigating the former Countrywide Financial Corporation Chairman and CEO who sold $10 million of his stock two weeks before the mortgage market began collapsing and, shortly after that, the Countrywide stock plunged from $45/share to less than $5/share (Kristof and Reckard 2008).

Cybercrime, illegal activities that are conducted online, is a new category of white-collar offenses. These high-tech crimes include defrauding consumers with bogus financial investments, embezzling, being paid to recommend stock on chat rooms, and stealing business data. Other offenses include sabotaging computer systems, hacking (gaining unauthorized access to computers), and stealing confidential information. In 2007 alone, the Internet Crime Complaint Center (2007) received almost 207,000 complaints, an all time high, amounting to nearly $240 million in reported losses. The most common offenses were Internet auction frauds, undelivered merchandise or payments, and credit/debit card fraud. In addition, every year nearly 10 million Americans experience identity theft (Dash 2006; Kane and Wall 2006).

Organized crime refers to activities of individuals and groups that supply illegal goods and services for profit. Organized crime includes drug distribution, loan-sharking (lending money at illegal rates), prostitution, illegal gambling, pornography, theft rings, hijacking

cargo, and laundering illegal money through legitimate enterprises. The *Godfather* films and the popular television show *The Sopranos* romanticized organized crime and presented the perpetrators as primarily Italian males. Italian-run organized crime exists, but most organized crime groups consist of Latinos, Asian Americans, and blacks. Some Russian and other Eastern European groups have been operating on U.S. soil since the 1970s. And, in some cases, women are active in organized crime. In 1997, for example, Heidi Fleiss—known as the "Hollywood Madam" because she had numerous famous and wealthy clients—was convicted of running a prostitution ring and tax evasion. She served 21 months in a federal prison but complained, bitterly, that none of her clients were prosecuted.

LAW ENFORCEMENT AND SOCIAL CLASS

Why do so many people commit white-collar crimes? Because they can, according to conflict theorists. First, most white-collar crimes are *not criminalized* (Turk 1969; Lilly et al. 1995). Because "greed, dubious bookkeeping, and suspiciously timed trading" are unethical but not necessarily criminal, the law is often vague, and prosecutors often find it difficult to demonstrate "not only that an action violated a specific law but also that the executive intentionally committed the bad act" (Sasseen 2005: 60).

Second, there is *minimal enforcement* and few penalties. For example, top officials at the Xerox Corporation agreed to pay $22 million to settle accusations of accounting fraud involving a scam that made $1.4 billion in profits (Norris 2003). Thus, Xerox kept almost 98 percent of the ill-gotten profits. If, using a comparable percentage, you made $35,000 a year illegally and had to pay a fine of $550, what would be the financial incentive for honesty?

Third, white-collar crimes thrive because of *privilege and corruption*. The common cultural background shared by judges and many white-collar defendants leads to greater leniency for these offenders than for street criminals. The federal government continues to award business worth billions to companies that repeatedly break the laws against polluting the air and water, cause deaths and injuries to their workers because of unsafe equipment, and even defraud the government (Silverstein 2002).

Finally, law enforcement has been inadequate due to a *redistribution of resources*. Prosecutions for white-collar crimes have dropped 28 percent since 9/11 because many FBI agents were reassigned to homeland security investigations and the prosecution of undocumented immigrants. The FBI had 2,385 agents fighting white-collar crime in 2000. By 2004, the number had dropped to 1,882 (Marks 2006).

CRITICAL EVALUATION

Conflict theories have been useful in highlighting the linkages between power and social class that may lead to criminal laws that benefit those at the top. Another strength is that conflict theory identifies biases present in the criminal justice system, such as the allocation of more resources to prosecute traditional rather than white-collar offenders.

Critics, however, point to several weaknesses. First, some contend that conflict theory exaggerates the importance of capitalism in explaining white-collar crime. Other capitalistic societies, like Japan, have much lower crime rates because of greater social solidarity and control of deviant behavior through shaming and restricting individual freedom (Leonardsen 2004). In effect, critics say, capitalism is not a major reason for crime.

A second criticism is that conflict theory deemphasizes the crimes committed by the poor. The costs of murder, rape, and other street crimes exceed $1 trillion a year in legal expenses, victim injuries and wage losses, and crime prevention activities (Anderson 1999). Thus, some critics maintain, low-income people are just as deviant as the rich, and their crimes are costly to society. A related problem is that conflict theories tend to ignore the fact that many affluent people, including corporate executives, don't always get away with their crimes. In 2005 alone, the cases pursued by the FBI resulted in 497 indictments and 317 convictions of corporate criminals. And, in 2005 and 2006, federal grand juries indicted four congressmen for embezzlement and other crimes (Federal Bureau of Investigation 2005; Pershing 2008).

Third, some critics say that conflict theory ignores the ways crime is functional for society as a whole. As you saw earlier, deviance provides jobs and affirms law-abiding cultural norms and values. Finally, many contend, the most influential conflict theories focus almost entirely on men (Moyer 2001; Belknap 2007). Feminist theories have filled this gap.

6 Feminist Perspectives on Deviance and Crime

for much of its history, sociology focused almost entirely on male offenders. As in most other academic disciplines, nearly all sociologists were men who saw women as not worthy of much analytical attention or assumed that explanations of male behavior were equally applicable to females (Flavin 2001; Simpson and Gibbs 2006; Belknap 2007). To remedy such omissions, feminist scholars have concentrated on girls and women as victims and offenders.

WOMEN AS VICTIMS

In 2008, the residents of a small town near Vienna, Austria, were stunned to learn that Josef Fritzl, 73, had kept one of his daughters imprisoned for 24 years in a basement dungeon, repeatedly raping her and fathering her seven children. Three were raised above ground by him and his wife, one died in infancy, and three were raised in the cramped, windowless basement. One of the renters of the three-story building that Fritzl owned suspected that something was going on, but said nothing because he was afraid of losing his apartment. Neighbors described Josef Fritzl, an electrical engineer, as a "man of stature" in the community who dressed well and raised seven well-behaved biological children with his wife. When arrested, Fritzl said he always knew that what he was doing was wrong, but was happy about the children his daughter bore him and rationalized his criminal behavior by insisting that that he was good to his incestuous offspring because he brought them toys and books (Landler 2008).

The Fritzl case is a recent example of the physical and sexual victimization of women and children. As you saw earlier, many of the serious crimes (like murder and robbery) are committed by men against other men, but women and girls are commonly the victims of sexual assault, rape, intimate partner violence, stalking, sexual exploitation, female infanticide, and other crimes that degrade women and deny them basic human rights.

Feminist scholars offer several explanations for women's victimization. In patriarchal societies, including the United States, men historically have dominated the government and the legal system. Because women have less access to the structures of power, they are at a disadvantage in creating and implementing laws that are more sympathetic to female victims. For example, of the 193 countries in the world, only 104 have made rape a crime, and the existing laws are rarely enforced (Price and Sokoloff 2004; United Nations Development Fund for Women 2007). Such inequity diminishes women's control over their lives and increases their invisibility as victims.

A related reason for female victimization is the effect of culture on gender roles. Many girls and women have been socialized to be victims of male violence because of societal images of women as weaker, less intelligent, and less valued than men: "Girls are rewarded for passivity and feminine behavior, whereas boys are rewarded for aggressiveness and masculine behavior. These stereotypes are often reaffirmed in the media where strong, independent female characters are rare, but violent, controlling male characters are abundant" (Belknap 2007: 243). In effect, then, both sexes internalize the belief that male victimization of females is normal.

But, you might be thinking, there are many strong women who reject sex stereotypes and could therefore escape or avoid victimization. Feminist theorists point to several problems with such assumptions. One is that low-income women often believe that leaving an abuser could result in greater economic hardship for themselves and the children. Even among women with college degrees (as you'll see in Chapter 13), many have internalized beliefs that male violence is okay, that they bring the battering on themselves ("I shouldn't have disagreed with him"), or that trying to escape will result in being killed. In addition, because patriarchal societies often don't enforce laws that punish male offenders who commit violence against women, many victims feel that they don't have the resources or options to escape.

WOMEN AS OFFENDERS

Because men are more likely than women to commit crimes and to commit more serious crimes, it's not clear why female arrests have increased. Some analysts propose that girls and women are becoming more violent because of a breakdown of family, religion, and community; a rise in inadequate schooling that results in high dropout rates; greater assertiveness; and the pervasive violence in much of today's entertainment. Others maintain that the higher arrest rates are a byproduct of policy changes, such as more aggressive policing, and the greater likelihood that parents and school officials will call the police to deal with girls' unruly behavior (Alder and Worrall 2004; Prothrow-Stith and Spivak 2005; Steffensmeier et al. 2005).

Some of the explanations of female crime parallel those of women's victimization. According to some feminist criminologists, women who commit crimes experience mistreatment that begins in early childhood: Girls are more likely than boys to be victims of family-related sexual abuse, the assaults start at a young age, and the abuse lasts longer. These factors can lead to suicide as well as to delinquent or criminal offenses such as running away from home, truancy, drug abuse, and prostitution (Chesney-Lind and Pasko 2004; Snyder and Sickmund 2006).

Holding money exchanged for sex, the child of a prostituted woman stands in the doorway of a brothel in Phnom Penh, Cambodia. Children, especially girls who are raised in brothels, are highly vulnerable to sexual exploitation. Many of the customers are Western men.

Other feminists emphasize patriarchy and women's limited economic opportunities. Because of their marginalization in the economy, women may resort to criminal activities—especially shoplifting, petty theft, and prostitution—to survive financially or to support a family. Women's offenses are highest in cities, where women's economic oppression and poverty are greatest. To decrease female (and male) crime, feminist scholars propose providing women with greater equality in employment and job training so that they can get a bigger piece of the economic pie (Simon and Landis 1991; Radosh 1993; Heimer et al. 2006).

CRITICAL EVALUATION

Feminist sociologists have been at the forefront of studying women offenders and victims: "The bottom line is that gender shapes human behavior in all arenas, and crime and victimization are no exceptions" (Heimer and Kruttschnitt 2006: 1). If it had not been for feminist scholars, there would probably still be little awareness of crimes such as date rape, stalking, domestic violence, and the international sex trafficking of women and girls. An estimated 61 percent of rapes and sexual assaults are not reported, but rapes have decreased by almost 85 percent since 1980. The reasons for these lower rates are unclear, but much of the decline may be due to feminist activists and scholars who have taught women to avoid drugs and unsafe situations and, more importantly, who have worked for the passage of tougher rape and domestic violence legislation (Britton 2003; Schulz 2004; Federal Bureau of Investigation 2007).

Some critics contend that feminist analyses have not gone far enough in showing how women's experiences as victims and offenders differ due to social class, race, ethnicity, and sexual orientation. Others maintain that because concepts such as patriarchy are difficult to measure, feminist research has yet to show, specifically, how patriarchy produces crime and affects women. Another criticism is that most feminist analysis emphasizes direct male violence against women (such as partner abuse and rape) and street crime but says little about women's white-collar crimes (Belknap 2001; Moyer 2001; Friedrichs 2004).

Feminist scholars analyze deviance and crime on both micro and macro levels (see Heimer and Kruttschnitt 2006). A fourth theoretical perspective, symbolic interactionism, relies almost exclusively on micro-level approaches.

7 Symbolic Interaction Perspectives on Deviance and Crime

the proportion of U.S. adolescents aged 12 to 17 who admitted using illicit street drugs (like marijuana and crack) dropped from 12 percent in 2002 to 10 percent in 2006. During the same period, abuse of prescription drugs (like pain relievers and stimulants) grew from 5 percent to 6 percent. The latter percentages may seem low, but they represent nearly 19 percent of teens. There are many reasons for illegal drug use, but 42 percent of teens say that they feel pressure from friends to abuse prescription drugs and that doing so is an important part of fitting in (Office of National Drug Control Policy 2007; Substance Abuse and Mental Health Services Administration 2007).

Such research supports symbolic interactionists' claim that people learn deviant behavior in their everyday lives. Symbolic interactionists offer many theories, but two of the best known are differential association theory and labeling theory.

DIFFERENTIAL ASSOCIATION THEORY

Dale "Rooster" Bogle's family in Oregon is an example of crime running in families. Even though Rooster, the father, served time in prison for theft, he taught his sons and daughters to survive by stealing:

> By the time the boys were 10 years old they were breaking into liquor stores for their dad or stealing tractor-trailer trucks, hundreds of them. The girls turned to petty crimes to support their drug addictions. . . . By official count, 28 in the Bogle clan have been arrested and convicted, including several of Rooster's grandchildren. . . . "Rooster raised us to be outlaws," said Tracey Bogle, the youngest of Rooster's children. "What you're raised with you grow to become. You don't escape." Tracey, 29, is serving a 15-year sentence for kidnapping, rape, assault, robbery

and burglary. . . . He committed the crimes with one of his older brothers (Butterfield 2002: 1).

Tracey's comment, "What you're raised with you grow to become," illustrates differential association theory. Sociologist Edwin Sutherland coined the term **differential association** to indicate that people learn deviance through interaction, especially with significant others like family members and friends. Through such interaction, people learn techniques for committing criminal behavior, and the values, motives, rationalizations, and attitudes that reinforce such behavior. People become deviant, according to Sutherland, if they have more contact with significant others who violate laws than with those who are law-abiding (Sutherland and Cressey 1970). Thus, in Rooster's clan, almost everyone wound up in prison because they grew up in an environment that taught deviance rather than conformity.

Sutherland emphasized that differential association doesn't occur overnight. Instead, people are most likely to engage in crime if they are exposed to deviant values (1) early in life, (2) frequently, (3) over a long period of time, and (4) from important people (parents, siblings, close friends, important business associates). When your parents complain that you're hanging out with the wrong crowd or when they encourage you to participate in college activities, they're showing an intuitive understanding of differential association theory.

Considerable research supports differential association theory. Almost 47 percent of state prisoners have a parent or other close relative who has also been incarcerated. Parents who abstain from cigarettes and illegal drugs, drink responsibly, and provide loving support and communication are less likely to raise children who use and abuse tobacco, alcohol, or drugs. Even before age 13, children who associate with peers who commit crimes are more likely to do so themselves because they pick up attitudes and values from their friends (Wiig and Widom 2003; *Family Matters* 2005; Conway and McCord 2005). Thus, according to differential association theory, we are products of our socialization.

LABELING THEORIES

Have you ever been accused of something wrong that you didn't do? What about getting credit for something that you and others know you didn't deserve? In either case, did people start treating you differently? The reac-

TABLE 7.1
Sociological Explanations of Deviance and Crime

THEORETICAL PERSPECTIVE	LEVEL OF ANALYSIS	KEY POINTS
Functionalist	Macro	• Anomie increases the likelihood of deviance.
		• Crime occurs when people experience blocked opportunities to achieve the culturally approved goal of economic success.
Conflict	Macro	• Laws protect the interests of the few (primarily those in the upper classes) rather than the rights of the many.
		• Law enforcement is rarely directed at the illegal activities of the powerful.
Feminist	Macro and Micro	• Crimes committed by women reflect their general oppression due to social, economic, and political inequality.
		• Many women are criminal offenders or victims because of culturally organized beliefs and practices that are sexist and patriarchal.
Symbolic Interactionist	Micro	• People learn deviant and criminal behavior from others—like parents and friends—who are important in their everyday lives.
		• If people are labeled or stigmatized as deviant, they are likely to develop deviant self-concepts and engage in criminal behavior.

tions of others are the crux of **labeling theory**, which holds that society's reaction to behavior is a major factor in defining oneself or others as deviant. In some of the earliest studies, researchers found that teenagers who were caught in misbehavior were tagged as delinquents. Such tagging changed the child's self-concept and resulted in more deviance and criminal behavior (Tannenbaum 1938). During the 1950s and 1960s, two influential sociologists—Howard Becker and Edwin Lemert—extended labeling theory.

Becker: Deviance Is in the Eyes of the Beholder

According to Howard Becker (1963), being a deviant or a criminal depends on how others react: "Deviance is *not* a quality of the act the person commits, but rather a consequence of the application by others of rules and sanctions to an 'offender.' The deviant is one to whom that label has successfully been applied; *deviant behavior is behavior that people so label*" (p. 9).

Some people are never caught or prosecuted for crimes they commit, and thus are not labeled as deviant. In other cases, people may be innocent of breaking laws (such as cheating on our taxes) but falsely accused and stigmatized. In effect, then, deviance is in the eye of the beholder because societal reaction, rather than an act, labels people as law-abiding or deviant. Moreover, labeling can lead to secondary deviance.

Lemert: Primary and Secondary Deviance

Edwin Lemert (1951, 1967) expanded on the effects of societal reactions by differentiating between primary and secondary deviance. **Primary deviance** is the initial violation of a norm or law. Primary deviance can range from relatively minor offenses, such as not attending a family member's funeral, to serious offenses, such as stealing and murder.

Even if people aren't guilty of primary deviance, labeling can result in **secondary deviance**, rule-breaking behavior that people adopt in response to the reactions of others. A teenager who is caught trying marijuana may be labeled "a druggie." Because the individual is rejected by others, he or she may accept the "druggie" label, associate with drug users, and become involved in a drug-using subculture. According to Lemert, a single deviant act will rarely result in secondary deviance. The more times that a person is labeled, however, the higher the probability that she or he will accept the label and engage in deviant behavior.

There's considerable evidence that labeling impacts people's lives. For example, nearly half of all Americans who experience severe health problems such as depression, schizophrenia, and eating disorders never seek treatment because they fear being stigmatized as mentally ill. Also, ex-inmates have a very difficult time reentering society because of numerous barriers such as employers' refusing to hire people with criminal records and landlords' unwillingness to rent apartments to people who have been released from prison (U.S. Department of Health and Human Services 1999; Hirsch et al. 2002).

labeling theory a perspective that holds that society's reaction to behavior is a major factor in defining oneself or others as deviant.

primary deviance the initial violation of a norm or law.

secondary deviance rule-breaking behavior that people adopt in response to the reactions of others.

CRITICAL EVALUATION

Symbolic interactionists' theories are important in understanding the everyday processes that contribute to deviant and criminal behavior. Differential association theory explains how social interaction increases a person's likelihood of engaging in deviant behavior. Labeling theory, by emphasizing the importance of social reactions, shows the dangers of categorizing people negatively because the stigmas can lead to criminal careers.

Despite these contributions, symbolic interactionists' theories have several weaknesses. Differential association theory doesn't explain impulsive crimes of rage (such as domestic murder) committed by people who have grown up in law-abiding families. This theory also ignores the possibility that deviant values and behaviors can be unlearned, as when young children encounter teachers or other adults whom they respect (Anderson and Dyson 2002; Williams and McShane 2004).

Critics have faulted labeling theory on a number of other points: The theory exaggerates the importance of judgments in altering a person's self-concept, doesn't explain why crime rates are higher in the South than in other parts of the United States or at particular times of the year (such as before holidays), and doesn't tell us why people commit crimes. Some critics also contend that social reactions are the *result* rather than the *cause* of deviant behavior and that social control agents (such as the police) are most likely to label people who have committed serious crimes or who have a long criminal record (Schurman-Kauflin 2000; Benson 2002).

Conflict theorists criticize symbolic interactionists for ignoring structural factors—such as poverty and low-paid jobs—that create or reinforce deviance and crime. And,

criminal justice system
the government agencies—including the police, courts, and prisons—that are charged with enforcing laws, passing judgment on offenders, and changing criminal behavior.

according to feminist criminologists, symbolic interactionists' theories are limited because, with a few exceptions, they still focus almost exclusively on male criminality (Currie 1985; Belknap 2007).

8 The Criminal Justice System and Social Control

Social institutions like the family, education, and religion try to maintain social control over moral misbehavior, whereas the criminal justice system has the *legal* power to control crime and punish offenders. The **criminal justice system** refers to government agencies—including the police, courts, and prisons—that are charged with enforcing laws, passing judgment on offenders, and changing criminal behavior.

The U.S. criminal justice system is extensive and expensive. In 2003, the last year for which data are available, the United States spent a record $185 billion on police, prisons, and courts. This is $38 billion more than in 1999 and represents a 148 percent increase since 1982 (Hughes 2006). These costs don't include indirect expenses such as providing victims with medical treatment and counseling or businesses that lose productive workers who quit because they are so traumatized by being victims of a crime that they can no longer cope with the everyday responsibilities of a job. Because of the high costs, the criminal justice system relies on three major approaches in controlling crime: prevention and intervention, punishment, and rehabilitation. Punishment, as you'll see, is the least effective approach.

PREVENTION AND INTERVENTION

Because most crime first occurs in adolescence (or earlier), criminal justice agencies often focus many of their prevention and intervention efforts on juveniles and their families. The most common sources of prevention and intervention are social service agencies, community outreach programs, and the police.

Social Service Agencies and Community Outreach Programs

Numerous organizations—comprised of law enforcement professionals, social workers, and nonprofit groups—try to prevent crime. A typical criminal career that spans one person's juvenile and adult years costs society about $1.5 million. Because many crimes involve substance abuse, all states have implemented a variety of intervention programs. For all age groups, the average annual cost per person for treatment of alcohol or drug abuse is $1,443 in an outpatient facility, $3,840 for residential treatment, and $7,415 for outpatient methadone treatment. Such costs are a bargain compared with paying, on average, about $24,000 a year per prison inmate. The cost is much higher if we include expenses for foster care for children who have drug-addicted or incarcerated parents. Because of high prison costs, the U.S. Department of Justice, among other federal agencies, funds numerous initiatives that try to decrease youth violence and delinquency, drug dealing, rape, robbery, prostitution, and domestic violence (Burns et al. 2003; Office of Applied Studies 2004; Stephen 2004).

How well do prevention and intervention programs work? The results are mixed, but a national study of adolescents aged 12 to 17 found that the grades of those who participated in a 5-month drug treatment program improved by about 30 percent, often to a level of B or better. And, in Baltimore, Maryland—one of the cities with the highest drug abuse and crime rates in the nation—a study found that, 1 year after people had entered public residential treatment, their heroin use dropped by 69 percent, cocaine use by 48 percent, and criminal activity by 64 percent (Sugg 2002; Spiess 2003).

Police

The primary role of the police is to enforce society's laws. A more controversial issue is whether the police can prevent crimes. Police can head off some crimes by cruising high-risk areas in patrol cars or by having more officers on foot patrols. Concentrating on *hot spots*, areas of high criminal activity, reduces crime only temporarily, however, because criminals simply move to

$1,443 Outpatient

$3,840 Residential

$7,415 Outpatient residential

$24,000 Incarceration

© Scott Hancock/Rubberball/Jupiterimages / © Porcorex/iStockphoto

other parts of the city (U.S. Department of Justice 1996; Braga 2003). Also, police can't prevent most crimes because they have no control over macro-level factors that lead to criminal behavior—poverty, unemployment, low educational opportunities, and neighborhood deterioration, for example.

PUNISHING CRIME

Those who support a **crime control model** believe that crimes rates increase when offenders don't fear apprehension or punishment. This perspective emphasizes protecting society and supports a tough approach toward criminals in terms of sentencing, imprisonment, and capital punishment.

Sentencing and Incarceration

After a defendant has been found guilty of a criminal offense or has pleaded guilty, a judge (and sometimes a jury) imposes a *sentence*, a penalty. A sentence can be a fine, probation (supervision), incarceration, or capital punishment. About 31 percent of those sentenced get probation; the others serve a term in a local jail or a state or federal prison. Many people question the fairness of sentencing because those convicted of similar crimes can receive widely different prison sentences depending, among other factors, on race/ethnicity, social class,

variations in state sentencing laws, and how juries and judges evaluate the seriousness of a crime (Durose and Langan 2004; Whelan 2007).

> **crime control model** an approach that holds that crimes rates increase when offenders don't fear apprehension or punishment.

Violent crimes have dropped 25 percent since 1987, but harsher sentencing for lesser crimes has resulted in higher incarceration rates. In 2008, for the first time in history, one in every 100 American adults was in prison (see *Figure 7.3*). The United States has less than 5 percent of the world's population, but almost a quarter of the planet's prisoners. Besides the sheer number of inmates, the United States is also the global leader in inmates per capita (751 per 100,000 people), ahead of nations like Russia (627 per 100,000 people) and England (151 per 100,000 people) (Pew Center on the States 2008).

Capital Punishment

The numbers have fluctuated over the years, but almost 70 percent of Americans support *capital punishment* (the death penalty), and this percentage has been fairly consistent since 1937. Since 1990, 108 nations have abolished the death penalty or suspended executions. In 2007, three states—California, Florida, and Texas—accounted for 44 percent of the total death row population in the United States (Newport 2007; Death Penalty Information Center 2008).

Of the 1,047 Americans executed between 1976 and 2006, 57 percent were white, 34 percent were black, 7 percent were Latino, and 2 percent were from other racial/ethnic groups. Many Americans believe that executions deter crime, but there is no evidence that this is the case. For example, the South, which accounts for almost 82 percent of executions, also has the highest murder rates. Families of victims, especially, maintain that speeding up executions would decrease homicide rates. Others argue that doing so would punish the innocent because 216 death row inmates (63

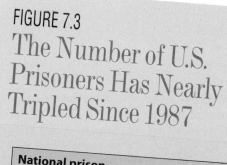

FIGURE 7.3

The Number of U.S. Prisoners Has Nearly Tripled Since 1987

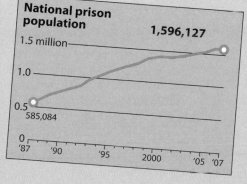

National prison population

1,596,127

1.5 million

1.0

0.5 — 585,084

0

'87   '90   '95   2000   '05 '07

Source: Based on Pew Center on the States 2008: 5.

percent of them African Americans) have recently been released after DNA evidence showed that they had not committed the crime of which they were convicted (Death Penalty Information Center 2008; Innocence Project 2008).

It's tempting to assume that the more people behind bars, the less crime there will be. Despite the surge of national and state spending on correctional facilities, *recidivism* (being arrested for committing another offense after being released) has barely changed since 1980: More than half of released prisoners are back behind bars within 3 years (Pew Center on the States 2008).

REHABILITATION

Rehabilitation, a third approach to controlling deviance, maintains that appropriate treatment can change offenders into productive, law-abiding citizens. Advocates argue that public assistance, educational opportunities, job training, and crisis intervention programs can reduce recidivism. According to a criminology professor who served 11 years in a federal prison, "If we really want to lower recidivism rates, prisoners should be released with Social Security cards, current drivers' licenses, and sufficient gate money to cover rent and food for three months. They should be provided, upon their request, with professional services (employment assistance, personal and family counseling, drug and alcohol treatment programs, and medical services)" (Ross and Richards 2002: 177).

Are rehabilitation programs successful? Yes, especially if they provide employ-

> **It's tempting to assume that the more people behind bars, the less crime there will be.**

ment after release. Other effective rehabilitation efforts have offered training for a trade while in prison, earning a high school or college degree while in prison or after release, and services that address several needs (such as housing, employment, and medical services) rather than just one (like counseling for drug abuse) (Jacobson 2005; Lowenkamp and Latessa 2005).

One of the most successful training programs involves released prisoners in the food industry. A criminologist founded the Delancey Street Restaurant in San Francisco, with a staff made up entirely of ex-inmates. The training not only provides skills in the food industry but also help in obtaining GEDs, preparing résumés, tackling business problems, and "smiling at diners no matter how annoying they are" (Cohen 2004: A18). Some of the ex-inmates who participated in this program have moved on to private business, one was elected to the San Francisco Board of Supervisors, and another headed the city's housing agency. Other communities have implemented the Delancey Street Restaurant model with similar success. The ex-felons, who earn certificates in culinary arts and licenses as food managers, are in great demand at restaurants (Lund 2004; Clayton 2005).

Willie "O" Pete Williams smiles as he answers a question during a news conference at the office of the Georgia Innocence Project in Atlanta, in January 2007. The Georgia Innocence Project was able to use DNA evidence to free Williams from prison after spending 21 years behind bars for rape.

© AP Images

Learning Your Way

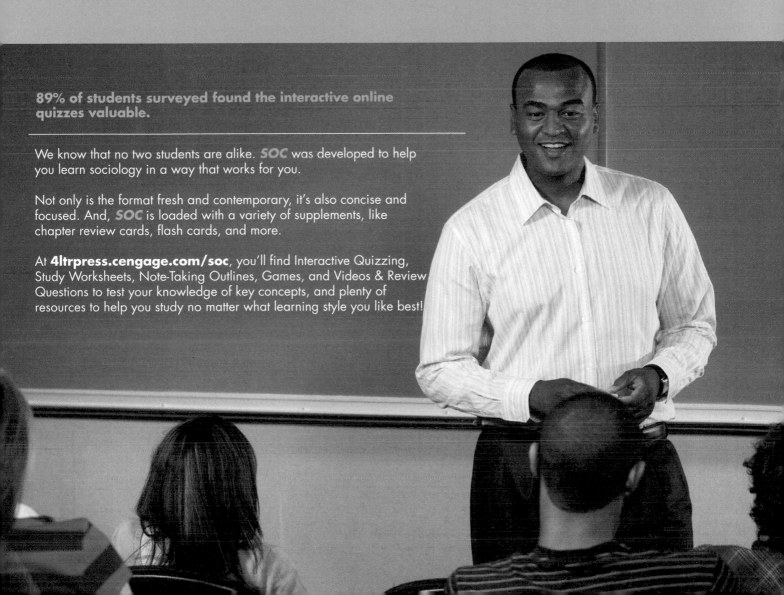

89% of students surveyed found the interactive online quizzes valuable.

We know that no two students are alike. *SOC* was developed to help you learn sociology in a way that works for you.

Not only is the format fresh and contemporary, it's also concise and focused. And, *SOC* is loaded with a variety of supplements, like chapter review cards, flash cards, and more.

At **4ltrpress.cengage.com/soc**, you'll find Interactive Quizzing, Study Worksheets, Note-Taking Outlines, Games, and Videos & Review Questions to test your knowledge of key concepts, and plenty of resources to help you study no matter what learning style you like best!

Having resources
can mean the difference between life and death.

what do you think?

Individuals can change their social class.

1 2 3 4 5 6 7

strongly agree   strongly disagree

8

Social Stratification

Having resources can mean the difference between life and death. This chapter examines stratification, considers why people move up or down the economic ladder, discusses some of the sociological theories that explain why there are haves and have-nots, and looks at global inequality.

> **social stratification** the hierarchical ranking of people in a society who have different access to valued resources, such as property, prestige, power, and status.
>
> **open stratification system** a system that is based on individual achievement and allows movement up or down.
>
> **closed stratification system** a system in which movement from one social position to another is limited by ascribed statuses such as gender, skin color, and family background.

1 What Is Social Stratification?

Social stratification is the hierarchical ranking of people who have different access to valued resources of a society, such as property, prestige, power, and status. All societies are stratified, but some more than others. An **open stratification system** is based on individual achievement and allows movement up or down. In a **closed stratification system**, movement from one social position to another is limited by ascribed statuses such as one's sex, skin color, and family background. Closed stratification systems are considerably more fixed than open ones, but no stratification system is completely open or completely closed.

CLOSED STRATIFICATION SYSTEMS

Let's first look at two closed stratification systems: slavery and castes. Both exist today. In *slavery*, an extreme form of inequality, people own others as property and have almost total control over their lives. The United Nations banned slavery worldwide in 1948, but it persists in many countries. In *chattel slavery*, for example, people are bought and sold as commodities, sometimes multiple times. Chattel slaves are often abducted from their homes, inherited, or given as gifts.

India's Dalits, like those pictured here, still perform unpleasant tasks such as burning corpses.

Members of educated classes, such as this Brahmin couple at an elegant wedding ceremony, enjoy high social status and respect.

social class a category of people who have a similar standing or rank in a society based on wealth, education, power, prestige, and other valued resources.

wealth the money and other economic assets that a person or family owns, including property and income.

Such practices take place in many countries in the Middle East, Africa, the Balkans, and Asia (U.S. Department of State 2006; Lampman 2007).

Castes, a second type of closed stratification system, are social categories based on heredity. Because social status is ascribed at birth, caste members are severely restricted in their choice of occupations, residence, and social relationships. A good example is India, where a caste system existed for more than 3,000 years. At the top were the Brahmins (priest, scholars, and educated class), followed by Kshatriyas (kings and warriors), Vaishyas (merchants and farmers), and Shudras (peasants, laborers, and craftspeople). On the bottom rung were the Dalits (formerly referred to as "untouchables") who were very poor and performed the most menial and unpleasant jobs, such as collecting waste and cleaning streets. Fearing being "polluted," persons of higher castes would not interact with the Dalits (Mendelsohn and Vicziany 1998).

India outlawed the caste system in 1949. Caste barriers have broken down in large cities, but divisions are deeply embedded and social distinctions still persist. People identify castes through their surnames and socialize with and marry within their own castes. In more than half the classrooms across India, Dalit children are often forced to sit in the back and to eat separately and are bullied, assaulted, and humiliated. To discourage such discrimination, the government has passed laws to provide lower castes with more educational opportunities (Neelakantan 2008).

OPEN STRATIFICATION SYSTEMS

In open stratification systems, social classes are relatively fluid because they are based on achieved rather than ascribed statuses. A **social class** is a category of people who have a similar standing or rank in a society based on wealth, education, power, prestige, and other valued resources. Theoretically, people in open stratification systems can move from one class to another. As you'll see throughout much of this chapter, however, the more resources someone has at birth, the greater her or his chance of moving into a higher social class. In effect, then, open stratification systems are not as open as many people believe.

2 Dimensions of Stratification

In the nineteenth century, the English novelist Jane Austen wrote, "A large income is the best recipe for happiness I ever heard of." More than a hundred years later, the late American entertainer Sophie Tucker quipped, "I've been rich and I've been poor. Rich is better." Income is a critical factor of stratification, but it's not the only one. Instead, sociologists use a multidimensional approach that includes wealth, prestige, and power.

WEALTH

Wealth is the money and other economic assets that a person or family owns, including property and income. *Property* comes in many forms, such as buildings, land, stocks and bonds, retirement savings, and personal possessions such as furniture, jewelry, and works of art. *Income* is money a person receives regularly, usually in the form of wages or a salary but also as rents, interest on savings accounts, dividends on stock, royalties, or the proceeds from a business.

Income and wealth are different in several important ways. Unlike income, wealth is *cumulative*. Wealth increases over time, especially through investment, whereas income is usually spent on everyday expenses.

Second, because wealth is accumulated over time, much of it can be *passed on to the next generation*. With an estimated $65 billion income in 2008, Bill Gates, co-founder of Microsoft Corporation, is one of the wealthiest people in the world. If each of Gates' three children inherits only 1 percent of his fortune, she or he will get at least $650 million! Third, *wealth produces income* ("it takes money to make money"). For example, a person with substantial stock portfolios can collect several million a year from the dividends.

U.S. wealth and income inequality is staggering. The top 1 percent of U.S. households own 34 percent of all wealth and 17 percent of all income; the bottom 40 percent have only 0.2 percent of all wealth and 12 percent of all income. More employed Americans have stock because of retirement plans, such as a 401(k), but overall, fewer than half owns any stock at all. The richest 20 percent own 90 percent of all stock. Because the cost of housing and utilities has surged, middle-class families report being worse off today than just 5 years ago (Allegretto and Gray 2007; Wolff 2007).

In 2008, there were 1,125 billionaires worldwide: 40 percent were Americans who owned 28 percent of global income. Of the 469 richest Americans, 77 were new billionaires. Because the median U.S. household income in 2007 was $50,233, it's difficult for most of us to imagine having an annual income of a million, much less a billion dollars ("Billionaires 2008" 2008; DeNavas-Walt et al. 2008).

The U.S. economy has expanded since the late 1960s, but income inequality has increased. Between 1970 and 2007, the percentage of total income going to the top 5 percent and top 20 percent of households rose, but it dropped for the other groups (see *Table 8.1*). The income discrepancy would be much greater for 80 percent of Americans if only one member of the household worked (Kennickell 2006; Aron-Dine and Shapiro 2007).

> **prestige** respect, recognition, or regard attached to social positions.

PRESTIGE

A second dimension of social stratification is **prestige**—respect, recognition, or regard attached to social positions. Prestige is based on many criteria, including wealth, family background, fame, leadership, power, occupation, and accomplishments. For example, every college convocation acknowledges students who graduate cum laude, magna cum laude, and summa cum laude.

We typically evaluate others according to the kind of work they do. *Table 8.2* on the next page presents a sample of prestige scores for U.S. occupations. Although, theoretically, the scores range from 0 to 100, no occupation is actually worthless or perfect. As a result, the prestige scores typically range from 20 to 86. In general, nonmanual occupations (such as lawyer and optometrist) typically rank higher in prestige than manual jobs (such as bus driver and garbage collector). Studies of occupational prestige in 57 other countries have found results similar to these for the United States (Hauser and Featherman 1977; Treiman 1977).

If you examine *Table 8.2*, you'll notice several characteristics shared by the most prestigious occupations:

- They *require more formal education* (such as college or postgraduate degrees) and/or extensive training. Physicians (86) require from 23 to 28 years of school and then fulfill internship and residency requirements after receiving a medical degree.

- They *pay more*, even though there are some exceptions. A realtor (49) or a truck driver (30) may earn more than a registered nurse (66), but registered nurses are likely to earn more over a lifetime because they have steady employment, good health benefits, retirement programs, and more opportunities to find other jobs during layoffs or career changes.

- They *are seen as more socially important*. An elementary school teacher (64) may earn less than the school's janitor (22), but the teacher's job is more prestigious because of teachers' contributions to reading, writing, and thinking skills.

TABLE 8.1
A Widening U.S. Income Gap, 1970–2007

The percentage of the total U.S. income that went to households at various levels.

	1970	1990	2007
The top 5 percent	16.6	18.6	21.2
The top 20 percent	43.3	46.6	49.7
The second 20 percent	24.5	24.0	23.4
The third 20 percent	17.4	15.9	14.8
The fourth 20 percent	10.8	9.6	8.7
The bottom 20 percent	4.1	3.9	3.4

Source: Based on DeNavas-Walt and Cleveland 2002, Table A-2; and DeNavas-Walt et al. 2008, Table 2.

power the ability of individuals or groups to achieve goals, control events, and maintain influence over others despite opposition.

- They *involve more abstract thought and mental activity*. An architect (73) must use far more imagination in designing a building than a carpenter (39) who performs very specific tasks.

- They *offer greater self-expression, autonomy, and freedom from supervision*. A dentist (72) has considerably more freedom in performing her or his job than a dental hygienist (52). In effect, then, higher-prestige occupations provide more privileges.

POWER

A third important dimension of social stratification is **power**, the ability of individuals or groups to achieve goals, control events, and maintain influence over others despite opposition. In every society, power is based on social class, but there are other sources of power. One is custom or tradition; for example, the chief of a tribe may have total authority based on tradition. Another source of power is being charismatic or eloquent or having other traits that inspire large groups of people. Leaders like Mahatma Gandhi and Martin Luther King Jr. motivated millions of people to demand change, peace, and social justice. Power is also tied to particular occupations: Your professor has the power to give you an A or an F, and a police officer can stop you and give you a ticket.

Sociologist C. Wright Mills (1956) coined the term *power elite* to describe a small and tightly knit group of white men—especially corporation heads, political leaders, and high-ranking military officers—who make all the important decisions in U.S. society. More recently, sociologists Richard Zweigenhaft and William Domhoff (1998) have agreed that a socially cohesive and very wealthy group of men continues to be a "ruling class" that dominates much of the American economy and government. (We'll examine power elites in Chapters 11 and 12.)

A person's ranking may be about equal in terms of wealth, prestige, and power. A Supreme Court justice,

TABLE 8.2
Prestige Scores for Selected Occupations in the United States

Do you agree with these rankings? Are there occupations that you think should rank higher or lower? If so, why?

HIGHER-PRESTIGE JOBS	MEDIUM-PRESTIGE JOBS	LOWER-PRESTIGE JOBS
86 Physician	59 Police officer	39 Carpenter
85 Supreme Court judge	58 Actor	36 Child-care worker
78 Lawyer	55 Radio/TV announcer	36 Hairdresser
74 College professor	54 Librarian	35 Assembly-line worker
74 Computer systems analyst	53 Firefighter	33 Cashier in supermarket
73 Architect	52 Dental hygienist	32 Bus driver
72 Dentist	52 Social worker	31 Auto body repairperson
69 Member of the clergy	51 Electrician	30 Truck driver
67 Optometrist	49 Funeral director	28 Garbage collector
66 Registered nurse	49 Realtor	28 Waiter/waitress
66 High school teacher	48 Manager of a supermarket	25 Bartender
65 Accountant	47 Mail carrier	23 Cleaner, private home
64 Elementary school teacher	46 Secretary	22 Janitor
62 Veterinarian	45 Plumber	22 Telephone solicitor
61 Airline pilot	43 Bank teller	21 Filling station attendant

Sources: Based on Nakao and Treas 1992; and J. Davis et al. 2005.

for instance, is usually affluent, enjoys a great deal of prestige, and wields considerable power. In many cases, however, there can be *status inconsistency*, the condition in which a person ranks differently on various stratification dimensions. Consider funeral directors. Their prestige is relatively low, but most have higher incomes than college professors, who are among the most educated people in U.S. society and have relatively high prestige (see *Table 8.2*). You'll recall that status inconsistency can be stressful and lead to frustration and depression (see Chapter 5).

We've looked at stratification systems and the dimensions of stratification, but how, specifically, do people in different social classes behave? And how does social class affect our behavior?

3 Social Class in America

a good indicator of social class is **socioeconomic status (SES)**, an overall ranking of a person's position in the class hierarchy based on income, education, and occupation. How do sociologists measure social class? Some ask residents to identify the social classes in their communities (the reputational approach), some ask people to place themselves in one of a number of classes (the subjective approach), but most use SES indicators (the objective approach).

Because there are different ways of measuring social class, sociologists don't agree on the number of social classes in the United States. However, there is consensus that there are four general social classes—upper, middle, working, and lower. Except for the working class, most sociologists often divide these classes further into more specific strata.

Sociologists Dennis Gilbert and Joseph Kahl (1993) developed a teardrop model of the American class structure based, primarily, on income and occupation (see *Figure 8.1*). They caution that social class is a complicated concept and that specifying the dividing lines between classes is as much art as science, but the model provides a good overview of the U.S. social class structure. Besides income, education,

and occupation, social classes also differ in values, power, prestige, social networks, and *lifestyles* (tastes, preferences, and ways of living).

THE UPPER CLASS

You saw earlier that the upper class controls a vastly disproportionate amount of the total U.S. wealth. There are actually two upper classes in the United States: the upper-upper class and the lower-upper class.

The Upper-Upper Class

Upper-upper class members rarely appear on the lists of wealthiest individuals published by *Forbes* or *Business Week*. Because they value their privacy, some upper-upper class members refuse to be listed even in

> **socioeconomic status (SES)** an overall ranking of a person's position in the class hierarchy based on income, education, and occupation.

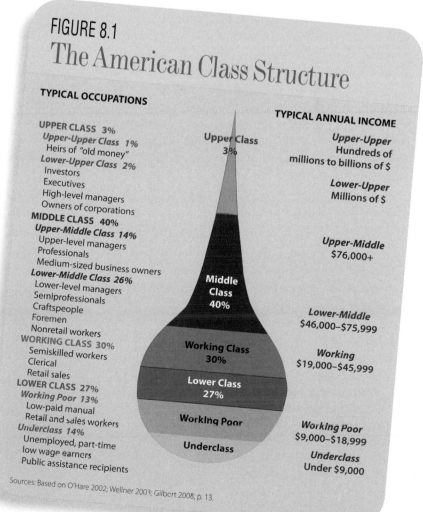

FIGURE 8.1
The American Class Structure

TYPICAL OCCUPATIONS

UPPER CLASS 3%
Upper-Upper Class 1%
  Heirs of "old money"
Lower-Upper Class 2%
  Investors
  Executives
  High-level managers
  Owners of corporations
MIDDLE CLASS 40%
Upper-Middle Class 14%
  Upper-level managers
  Professionals
  Medium-sized business owners
Lower-Middle Class 26%
  Lower-level managers
  Semiprofessionals
  Craftspeople
  Foremen
  Nonretail workers
WORKING CLASS 30%
  Semiskilled workers
  Clerical
  Retail sales
LOWER CLASS 27%
Working Poor 13%
  Low-paid manual
  Retail and sales workers
Underclass 14%
  Unemployed, part-time
  low wage earners
  Public assistance recipients

Upper Class 3%
Middle Class 40%
Working Class 30%
Lower Class 27%
Working Poor
Underclass

TYPICAL ANNUAL INCOME

Upper-Upper Hundreds of millions to billions of $

Lower-Upper Millions of $

Upper-Middle $76,000+

Lower-Middle $46,000–$75,999

Working $19,000–$45,999

Working Poor $9,000–$18,999

Underclass Under $9,000

Sources: Based on O'Hare 2002; Wellner 2003; Gilbert 2008, p. 13.

The Social Register, an inventory of America's social elite that has been published since 1887. Whether the income originally came from legal or illegal activities (Phillips 2002), most upper-upper class members have enormous wealth that has been passed down through generations. Some of the "old money" families include the Vanderbilts, DuPonts, Rockefellers, and Fords.

Members of the upper-upper class send their children to private schools and Ivy League universities to strengthen social relationships with their own kind and to meet marriage partners from their own class. Exclusive schooling and social clubs also provide connections to other upper-upper class members with similar backgrounds and lifestyles. Not everyone in the upper-upper class knows everyone else, but "everybody knows somebody who knows someone in other areas of the country" (Domhoff 1986: 49–50; see, also, Kendall 2002).

Having an inherited fortune brings with it power (McNamee and Miller 1998). Upper-upper class white males, especially, shape the economic and political climate through a variety of mechanisms: dominating the upper levels of business and finance, holding top political positions in the federal government, underwriting thousands of tax-free foundations including think tanks and research institutes that formulate national policies, and shaping public opinion through the mass media (Zweigenhaft and Domhoff 2006).

The Lower-Upper Class

The lower-upper class—which is much more diverse than the upper-upper class—is the *nouveau riche,* those with "new money." Some, like the Kennedys, amassed fortunes several generations ago, but many of the newly rich—like Donald Trump, Bill Gates, and the late President Ronald Reagan—worked for their income rather than inherited it. Besides business entrepreneurs, the lower-upper class also includes high-level managers of huge corporations, self-made millionaires, and some highly paid athletes and actors, but their lifestyles differ considerably.

Some lower-upper class members flaunt their newly earned wealth. Sociologist

Thorstein Veblen (1899/1953) coined the term **conspicuous consumption** to refer to lavish spending on goods and services to display one's social status and enhance one's prestige. For example, lower-upper class people may have personal chefs, take exotic vacations (often in private planes), and send their children to $4,000-an-hour math tutors and summer camps at French chateaus. Such conspicuous consumption sends the message "I'm very, very rich."

Because they lack the "right" ancestry, have not attended the "right" schools, and have usually made their money by working for it, lower-uppers are not accepted into "old money" circles that have strong feelings of in-group solidarity. Still, lower-upper class members engage in lifestyles and rituals that try to parallel those of the upper-upper class. For example, they have their own exclusive clubs, live in affluent neighborhoods segregated from adjacent poorer ones, and vacation in exclusive locations.

THE MIDDLE CLASS

Most Americans describe themselves as middle class, including 41 percent of those with incomes below $20,000 and 33 percent of those with incomes above $150,000 (Pew Research Center 2008). There are several strata in the middle class, but sociologists often distinguish between the *upper middle-class* and the *lower-middle class.*

The Upper-Middle Class

Upper-middle class members, although rich, live on earned income rather than accumulated wealth. Their salaries are high enough to provide economic stability and sizeable savings, but usually because both spouses have careers.

Status symbols—cars, clothes, vacations, and so on—are not limited to the wealthy. In a recent national poll, 81 percent of Americans said they felt social pressure to buy high-priced goods, even if it meant going into debt, to convey a certain image within their social class (Steinhauer 2005). Many upper-middle class members, especially, use status symbols to show that they've made it, by buying luxurious cars (such as a Jaguar X-type sedan that starts at $30,000), upscale kitchen appliances, designer handbags, and expensive jewelry.

The occupations of this group—mainly professional and managerial—usually require advanced degrees in business, law, medicine, and the physical sciences. People in this class include corporate executives and managers (but not those at the top), high government officials, business owners, physicians, and successful lawyers and stockbrokers. Many of these occupations have considerable on-the-job autonomy and freedom from supervision, but more than half of upper-middle class members work 50 or more hours per week, compared with 28 percent of the general population (Gardyn 2001).

Upper-middle class members are likely to own homes in fashionable communities that are segregated by race and social class and to send their children to private or public schools that have a good reputation. Parents enroll their children in a variety of activities (ballet, sports, music); send them to summer camps or programs that offer instruction in science, computer science, or the arts; and expect their children to achieve or surpass their own economic and educational levels.

The upper-middle class exercises a large and growing influence in American society. Many of its members vote, volunteer, make campaign donations, run for local offices, and participate in fund-raising activities, especially for their children's schools. Members of this class often spend their leisure time attending arts performances, including classical music concerts, plays, the opera, and the ballet.

The Lower-Middle Class

The lower-middle class is even more diverse than the upper-middle class and is comprised of people in non-manual and semiprofessional occupations. Non-manual jobs include office staff, low-level managers, owners of small businesses, medical and dental technicians, legal and medical secretaries, police officers, sales workers (such as insurance salespeople and real estate agents), and some highly skilled blue-collar workers (like building contractors). Examples of semiprofessional occupations are nursing, social work, and teaching. Almost all of these jobs require training beyond high school and many, especially the semiprofessional occupations, require a college degree. Most

families in the lower-middle class rely on two incomes to maintain a comfortable standard of living.

Unlike upper-middle class jobs, those in the lower-middle class have less autonomy and freedom from supervision and there is little chance for advancement. People are more likely to follow orders than to give them. Many lower-middle class jobs are relatively secure, but the workers worry about taxes, inflation, and layoffs as workplaces become more computerized. Except for some retirement funds, most have only modest savings to cover emergencies. Many buy used or inexpensive late-model cars, eat out fairly regularly at middle-income restaurants, and take occasional vacations, but they rarely have the income to buy luxury products without going deeply into debt.

The lower-middle class is much less active politically than the upper-middle class, but many of its members vote and work as volunteers for local candidates or on community projects. Women are often involved in raising funds for schools, and men (and some women) devote much of their leisure time to organizing and supervising sports programs offered by local recreation departments.

THE WORKING CLASS

The working class consists of skilled and semiskilled laborers, factory employees, and other blue-collar workers in manual occupations. They are construction workers, assembly-line workers, truck drivers, auto mechanics, repair personnel, bartenders, and skilled craft workers like carpenters and electricians. Most of the jobs are blue collar, but some—like clerks and retail sales workers in the service sector—are white collar.

© RubberBall/Jupiterimages

working poor people who work at least 27 weeks a year but receive such low wages that they live in or near poverty.

underclass people who are persistently poor and seldom employed, segregated residentially, and relatively isolated from the rest of the population.

People who fill working-class jobs often have a high school degree but no college education. There are few opportunities for advancement, and many of the positions provide little or no opportunity for advancement. In a weak economy, especially, the bargaining power of working-class people diminishes. Large numbers see their work as boring and routine—a source of income to survive on rather than a means of attaining personal fulfillment. Most of the semiskilled jobs require little training, are mechanized, and closely supervised.

Working-class members who purchase homes, including mobile homes, may experience foreclosure because of delinquent payments. Many in the working class use credit cards to pay off bills each month but then can barely pay the monthly minimum. Debts become overwhelming when borrowers who live from paycheck to paycheck suffer setbacks such as divorce, illness, or job loss (Mayer 2002).

Compared with those in the middle class, people in the working class are much more likely to watch television as their primary recreational activity rather than attend cultural events or get involved in organizational activities. They are less likely to vote because they often feel politically powerless. Despite low civic participation, most working-class people are very involved with friends and family members.

THE LOWER CLASS

People in the lower class are at the bottom of the economic ladder because they have little education, few occupational skills, work in minimum wage jobs, or are often unemployed. Even though most of the lower class is poor, sociologists often distinguish between the *working poor* and the *underclass*.

The Working Poor

About 61 percent of low-wage workers—11 percent of all Americans—earn less than $7.73 an hour. Most comprise the **working poor**, people who work at least 27 weeks a year but receive such low wages that they live in or near poverty. In 2005, almost half of American families headed by a married couple with at least one full-time, full-year worker were among the working poor (Holzer 2007).

Some sociologists describe the working poor as "the excluded class" who are often treated as "waste" because they typically "fill the most undesirable jobs in restaurant kitchens or as nighttime cleaners of downtown buildings" (Perrucci and Wysong 1999: 28). Several sociologists have recently proposed that the working poor also encompass the "missing class" of "near poor whose incomes place them above the poverty line, but well below the middle class":

> *Near-poor women work in clothing stores, minimarts, and child-care centers; they clean subway cars on the night shift. Near-poor men often hold down more than one job, working days as aides for the mentally impaired and nights as security guards. (Newman and Chen 2007: B10)*

Up to half of the people who visit food pantries and soup kitchens in some states are from the working poor (Ryan 2008). Many attribute their situation to bad luck or fate and feel powerless over their economic insecurity. Like the underclass beneath them, the working poor are generally alienated from political processes.

The Underclass

The **underclass**, which occupies the bottom rung of the U.S. social ladder, consists of people who are persistently poor, segregated residentially, and relatively isolated from the rest of the population. Most rarely work; they are chronically unemployed or drift in and out of jobs. Social scientists commonly use the term *underclass* to describe inner-city minorities, but it applies to people of any race or ethnicity who are locked in destitution and have little chance of moving out of abject poverty. They may work erratically or at part-time jobs, but their lack of skills, low educational levels, and, in many cases, disabilities make it difficult for them to find regular, full-time jobs (Beeghley 2000; Gilbert 2008).

For sociologist William Julius Wilson (1996), being in the underclass or one of "the ghetto poor" is the result of joblessness rather than poverty, because the underclass grew when most middle-class people—both white and black—moved to the suburbs, the number of factory jobs declined, and the poor became more isolated. Others argue that the underclass is locked in a "culture of poverty" that has more to do with personal values than with structural factors like unemployment. Most social scientists agree, however, that the underclass experiences a wide range of social problems—

crime, welfare, family dissolution, drug abuse, poor health, and domestic violence.

Our social class position, more than any other single variable, affects just about all aspects of our lives. Max Weber referred to the consequences of social stratification as **life chances**, the extent to which people have positive experiences and can secure the good things in life (such as food, housing, education, and good health) because they have economic resources. The United States is the richest nation in the world (see Chapter 12), but the gap between the affluent and the poor is increasing.

4 Poverty in America

there is more poverty in the United States today than 35 years ago: over 37 million people were poor in 2007, compared with 23 million in 1973 (DeNavas-Walt et al. 2008). What do sociologists mean by poverty? Who are the poor? And why?

WHAT IS POVERTY?

There are two ways to define poverty: absolute and relative. **Absolute poverty** is not having enough money to afford the most basic necessities of life, such as food, clothing, and shelter ("what I need"). **Relative poverty** is not having enough money to maintain an average standard of living ("what I want").

The **poverty line** is the minimal level of income that the federal government considers necessary for basic subsistence. To determine the poverty line, the Department of Agriculture (DOA) estimates the annual cost of food that meets minimum nutritional guidelines and then multiplies this figure by three to cover the minimum cost of clothing, housing, health care, and other necessities. Anyone whose income is below this line is considered officially poor and is eligible for government assistance (such as food stamps).

The poverty line, which in 2007 was $21,027 for a family of four (two adults and two children), is adjusted every year to reflect cost-of-living increases. If a family makes a dollar more than the poverty line figure, they are not officially categorized as poor. Also, many people earn considerably less than the poverty threshold. In 2007, for example, 43 percent of poor families—a group the Census Bureau refers to as "severely poor"—earned less than half of the poverty threshold (DeNavas-Walt et al. 2008).

Some feel that the official definition of poverty is inflated. They argue, for example, that poverty levels—which were developed in the mid-1960s—do not include the value of such noncash benefits as food stamps, medical services (like Medicare and Medicaid), and public housing subsidies.

Others claim that the poverty line is unrealistically low because it ignores many current needs of poor people. For example, single mothers require affordable child care so that they can work and pay for transportation costs to child-care centers and jobs. These critics also contend that a poor person who lives in a metropolitan area or in a state such as California and Massachusetts needs more money to survive—often three to four times more than someone who lives in the rural South—primarily because of higher housing costs (Jolliffe 2006).

WHO ARE THE POOR?

Poverty isn't random. Both historically and currently, the poor share some common characteristics that include age, gender, family structure, and race and ethnicity. Children make up only 25 percent of the U.S. population but 36 percent (over 13 million) of the poor. Among older Americans, people 65 and over make up 13 percent of the total population but about 10 percent of the poor (DeNavas-Walt et al. 2008). Although it is still too high, the poverty rate among older Americans is at an all-time low because government programs for the elderly, like Medicare and Medicaid, have kept up with the rate of inflation; in contrast, many programs for children living in poverty have been reduced or eliminated since 1980.

Of all people 18 and older who are poor, 57 percent are women (DeNavas-Walt et al. 2008). Researcher Diana Pearce (1978) coined the term the **feminization of poverty** to describe the higher likelihood that female heads of households will be poor. Because of increases in divorce and unmarried childbearing, single-mother

life chances the extent to which people have positive experiences and can secure the good things in life because they have economic resources.

absolute poverty not having enough money to afford the most basic necessities of life.

relative poverty not having enough money to maintain an average standard of living.

poverty line the minimal level of income that the federal government considers necessary for basic subsistence.

feminization of poverty the higher likelihood that female heads of households will be poor.

families are at least five times more likely to be poor than are married-couple families, and they are disproportionately represented among the long-term poor, especially when biological and divorced fathers don't support their children (Grall 2005). Besides marital status, other major reasons for the feminization of poverty are low-paying jobs and wage discrimination, topics we'll discuss in more detail in Chapters 9 and 12.

African Americans, American Indians, and Latinos are twice as likely to be poor as whites, Asian Americans, and Pacific Islanders (DeNavas-Walt et al. 2006). These poverty differences have changed little since 1971, largely because high school dropout rates are especially high among blacks, Latinos, and American Indians due to class differences (see Chapter 14).

WHY ARE PEOPLE POOR?

About 69 percent of Americans say that wealth in this country should be more evenly distributed, but an equal percentage also believe that it's possible to start out poor and get rich through hard work (University of Connecticut . . . 2007). There are two general explanations for why people are poor: One blames the poor; the other emphasizes societal factors.

Blaming the Poor: Individual Characteristics

A majority of Americans assert that, since this is a land of opportunity, anyone who doesn't get ahead just isn't working hard enough: "Almost anyone can get rich if they put their mind to it" (Jones 2007). Some researchers see poor people as "deficient" and innately inferior due to genetic factors that produce lower intelligence and cognitive abilities (see Herrnstein and Murray 1994).

Proponents of an influential view that sees a *culture of poverty* contend that the poor are deficient: They share certain values, beliefs, and attitudes about life that differ from those who are not poor, are more permissive in raising their children, and are more likely to seek immediate gratification instead of planning for the future (Lewis 1966; Banfield 1974). The assertion that these values, beliefs, and attitudes are transmitted from generation to generation implies that the poor create their own problems through a self-perpetuating cycle of poverty ("like father, like son").

Blaming Society: Structural Characteristics

In contrast to blaming the poor, most sociologists assert that a society's organization creates and sustains poverty. In a classic article on poverty, sociologist Herbert Gans (1971) maintained that poverty and inequality have many functions:

- the poor ensure that society's dirty yet necessary work gets done (such as dishwashing and cleaning bedpans in hospitals);

SIX BIGGEST MYTHS ABOUT THE POOR

MYTH 1. PEOPLE ARE POOR BECAUSE THEY ARE LAZY AND REFUSE TO WORK. Of poor people 16 years of age and older, 12 percent work full time year-round, and another 25 percent work part time (Mead 2008).

MYTH 2. MOST POOR PEOPLE ARE MINORITIES. Almost 44 percent of people living in poverty are white. In proportions, however, 8 percent of whites live in poverty compared with 25 percent of African Americans, 22 percent of Latinos, and 10 percent of Asian Americans (DeNavas-Walt et al. 2008).

MYTH 3. MOST POOR PEOPLE LIVE IN INNER CITIES. A large number (43 percent) of the poor live in inner cities, but the rest live in urban areas outside of inner cities, the suburbs, small towns, and rural communities (DeNavas-Walt et al. 2008).

MYTH 4. MOST OF THE POOR ARE SINGLE MOTHERS. Of all families living in poverty, 53 percent are single mothers and their children, but 37 percent of married-couple families and 9 percent of father-headed households are poor (DeNavas-Walt et al. 2008).

MYTH 5. MOST OF THE POOR ARE ELDERLY. About 10 percent of people 65 years and older are poor, but 36 percent of the poor are children under age 18 (DeNavas-Walt et al. 2008).

MYTH 6. THE POOR GET SPECIAL ADVANTAGES. The poor pay more for goods and services than do wealthier people. Supermarket chains and discount stores rarely locate in low-income communities, and because the poor have limited access to banks or other financial institutions, they must often rely on "check-cashing stores" that charge high rates for cashing checks or borrowing money (Jeffery 2006).

- the poor subsidize the middle and upper classes by working for low wages;

- the poor buy goods and services that would otherwise be rejected (such as day-old bread, used cars, and the services of old, retired, or incompetent professionals); and

- the poor absorb the costs of societal change and community growth (as when they are pushed out of their homes by urban renewal and construction of expressways, parks, and stadiums).

Thus, according to Gans, poverty persists in the United States because many people benefit from the consequences.

Which perspective is more accurate—blaming the poor or blaming society? Some people are poor because they're lazy, focus on the present, and would rather get a handout than a job. However, researchers find little support for the argument that poverty is transmitted from generation to generation. Instead, almost 8 out of 10 Americans receive public assistance for only 2 years. Most people are poor because of economic conditions (especially low wages), job loss, physical or mental disabilities, or an inability to afford health insurance, which, in turn, can result in acute health problems that interfere with employment. And, as more businesses relocate to the suburbs, poor minorities—especially those concentrated in inner cities—are unlikely to hear about employment possibilities or to have the transportation to get to even low-skill jobs (Nichols 2006; Bernstein et al. 2007).

HOMELESSNESS

One of the most devastating consequences of poverty is homelessness. The uncounted homeless include people who live in automobiles, have makeshift housing (such as boxes and boxcars), or stay with relatives for short periods. An estimated 3.5 million people (about 1 percent of Americans, a third of whom are children) are likely to experience homelessness in a given year (National Coalition for the Homeless 2007a).

Single men comprise 51 percent of the homeless, families with children 30 percent, single women 17

© iStockphoto.com

Almost one-third of America's homeless are families with children. Because there aren't enough shelters, many live in their cars.

percent, and unaccompanied youth 2 percent. The homeless population is estimated to be 42 percent African-America, 39 percent white, 13 percent Latino, 4 percent American Indian, and 2 percent Asian. In cities, about 16 percent of homeless people are considered mentally ill, 26 percent are substance abusers, 13 percent are employed, and 9 percent are veterans (United States Conference of Mayors, 2006; U.S. Department of Housing and Urban Development, 2007).

Homelessness is due to a combination of factors, some of which are beyond the control of individuals. Poverty, lack of education or marketable skills, low-paying jobs, unemployment, domestic violence, substance abuse, the inability of relatives and friends to help during crises, a shortage of affordable housing, and a decline in public assistance are among the most common reasons for homelessness. Young mothers with very young children are especially likely to become homeless. The homeless also include teenage runaways escaping from family violence or incest (Cintrón-Vélez 2002; Burt et al. 2004; National Coalition for the Homeless 2007b).

> **social mobility** a person's ability to move up or down the class hierarchy.
>
> **horizontal mobility** moving from one position to another at the same class level.

5 Social Mobility

most Americans believe in **social mobility**, a person's ability to move up or down the class hierarchy. The movement results from a variety of factors. Before considering these factors, let's look at the types of social mobility that sociologists typically examine.

TYPES OF SOCIAL MOBILITY

There are different types of social mobility: horizontal and vertical mobility and intergenerational and intragenerational mobility.

Horizontal mobility means moving from one position to another at the same class level, or making a lateral move. Tracey, a nurse, might move from the pediatrics to the obstetrics department, but her salary won't change much because the position is similar in responsibilities and qualifications. Sociologists are generally not very interested in horizontal mobility because it involves little

J.K. Rowling worked as a teacher but was divorced and living on public assistance when she wrote Harry Potter and the Philosopher's Stone *during her daughter's naps. The book was published in 1997; by 2007, Rowling, one of the world's richest women and a billionaire, was considerably wealthier than the Queen of England.*

vertical mobility moving up or down the class hierarchy.

intragenerational mobility moving up or down the class hierarchy over a lifetime.

intergenerational mobility moving up or down the class hierarchy relative to the position of one's parents.

change in one's social class. **Vertical mobility** refers to moving up or down the class hierarchy. If Tracey wants a higher salary, she may decide to undergo more training and become a physician's assistant (PA). Working under the supervision of a doctor, a PA examines, diagnoses, and treats patients. If Tracey is laid off and can't find another job, she might experience downward mobility. Vertical mobility can be intragenerational and intergenerational.

Intragenerational mobility refers to moving up or down the class hierarchy over a lifetime. If Tracey begins as a nurse's assistant, becomes a registered nurse, and then a PA, she experiences intragenerational mobility. **Intergenerational mobility** is moving up or down the class hierarchy relative to the position of one's parents. It is a change in social class that occurs across two or more generations. If Tracey's parents were blue-collar workers, her upward movement to the middle-class would be an example of intergenerational mobility.

Intragenerational and intergenerational mobility can be downward or upward. One individual who experienced upward intergenerational mobility was Sam Walton, the founder of Wal-Mart. He worked in his father's store while attending high school, graduated from the University of Missouri, and opened his first store, Ben Franklin, with the help of a loan from his father-in-law. When he died in 1992, at the age of 74, Walton was the world's second-richest man, behind Bill Gates. If Walton had lost his fortune and wound up on public assistance, he would have experienced downward intragenerational and intergenerational mobility.

When social mobility occurs, most moves are short. A child from a working-class family, for example, is more likely to move up to the lower-middle class than to jump to the lower-upper class. The same is true of downward mobility: Someone from the middle class is more likely to slide into the working class than to drop to the underclass.

WHAT AFFECTS SOCIAL MOBILITY?

According to many sociologists, vertical social mobility doesn't always reflect people's talents, intelligence, or hard work. Instead, such social mobility depends on structural, demographic, and individual factors.

Structural Factors

Macro-level variables, over which we have little or no control as individuals, affect social mobility in many ways. First, *changes in the economy* spur upward or downward mobility. During an economic boom, the number of jobs increases, and many people, including those on public assistance, have an opportunity to move up. Second, the *number of available positions* in particular occupations changes over time. The need for agricultural workers dwindled significantly in the United States during the last 100 years, while the demand for clerical, technical, and professional workers mushroomed. As a result, an expanding occupational structure created more room in the middle of the social class hierarchy. Third, *immigration* stimulates social mobility. Because most immigrants take low-paying jobs, groups that are already there advance to higher positions. Such upward mobility continues as poor immigrants take the least desirable jobs, allowing others to move into higher-status occupations (Haskins 2007b).

Demographic Factors

Demographic factors, which are usually interrelated, also affect social mobility. Three of the most important are education, gender, and race and ethnicity.

Education is a critical factor in social mobility. Especially when the economy is slumping, people with college and graduate degrees fare better than those with a high school education or less. Those who don't graduate from high school often face long and frequent bouts of unemployment, must get by with temporary employment, and may move down the socioeconomic ladder (Haskins 2007a).

In terms of *gender,* women's massive entry into the labor force during the 1980s increased family income considerably as well as fostering single women's upward mobility. For both sexes, but especially women, whom one marries also affects upward or downward mobility.

Regarding *race and ethnicity,* African Americans and Latinos, especially those from low-income backgrounds, usually experience little upward mobility. Black and Latino middle classes have grown since the 1970s, but both groups still lag significantly behind whites in median family income. A major reason is that white parents have more wealth that they can pass down to their children (Sernau 2001; Isaacs 2007a, 2007b).

Individual Factors

When researchers look at the combined effect of structural and demographic factors such as immigration, SES, race, and ethnicity, they account for about half of the reasons for upward or downward social mobility. Individual factors, although more difficult to measure, also affect social mobility.

Because family background is a critical factor in determining social class, the best way to be upwardly mobile is to choose the right parents. Consider that only 7 of the 44 U.S. presidents came from the lower-middle class or below. Abraham Lincoln, although born in a log cabin, had a father who was one of the wealthiest people in his community. And about a third of the students with low grades at Ivy League universities wouldn't be there if their parents weren't celebrities, well-known politicians, or others who donated at least $25 million to the school (Golden 2006).

Parents tend to *socialize* their children to assume an expected class position. Our class upbringing influences interests and activities that determine what French sociologist Pierre Bourdieu (1984) called *habitus*—the habits of speech and lifestyle that determine where one will feel comfortable and knowledgeable. Upper-class parents emphasize flexibility, autonomy, and creativity because they expect their children to step into positions that require such characteristics. Poor and working-class parents stress obedience, honesty, and appearance—the marks of respectability that many employers expect.

Connections and chance are especially important factors in upward mobility. Many people get jobs by word-of-mouth rather than by searching in newspaper ads or Jobs.com. In a candid autobiographical sketch, sociologist S. M. Miller (2001: 1, 3) described his rise in academia, where connections and networking, rather than merit, "made the difference." He concluded that

© AP Images

Many children raised in poor families are upwardly mobile. Dr. Benjamin S. Carson is a world-renowned physician who is currently Director of Pediatric Neurosurgery at the prestigious Johns Hopkins Hospital in Baltimore, Maryland. Benjamin was 8 years old when his mother, who married when she was only 13, divorced his father. Mrs. Carson sometimes worked three jobs at a time to provide for Benjamin and his older brother, Curtis. When the boys' grades fell, Mrs. Carson limited their watching television and wouldn't let them play until they had finished their homework. She required her sons to read two library books a week and to give her written reports, even though she could barely understand the reports because she had left school after third grade. Since the founding of the Carson Scholars Fund in 1994, Dr. Carson and his wife have distributed 3,400 scholarships to graduating high school seniors to attend college. In 2008, President Bush awarded Dr. Carson a Medal of Freedom, the nation's highest civilian award, for his "groundbreaking contributions to medicine and his inspiring efforts to help America's youth fulfill their potential" (Nitkin 2008: 7B).

many employment decisions are based not on a person's individual abilities but on the "social capital of connections—the inequitable distribution of access to people who can help you get a good job."

6 Why There Are Haves and Have-Nots

hy are societies stratified? *Table 8.3* on p. 151 summarizes the key points of functionalist, conflict, feminist, and symbolic interactionist theories. Let's begin by looking at a long-standing debate between functionalists and conflict theorists on why there are haves and have-nots.

FUNCTIONALIST PERSPECTIVES: STRATIFICATION BENEFITS SOCIETY

Functionalists see stratification as both necessary and inevitable because class provides each individual a place in the social world and motivates people to contribute to the maintenance of society. Without a system of unequal rewards, functionalists argue, many important jobs wouldn't be performed.

The Davis-Moore Thesis

Sociologists Kingsley Davis and Wilbert Moore (1945) developed the most influential functionalist perspective on social stratification that persists today. The **Davis-Moore thesis**, as it is commonly called, asserts that social stratification has beneficial consequences for a society's operation. The key arguments of the Davis-Moore thesis can be summarized as follows:

1. **Every society must fill a wide variety of positions and ensure that people accomplish important tasks.** In the United States, we need teachers, doctors, farmers, trash collectors, engineers, secretaries, plumbers, police officers, and so on.

2. **Some positions are more important than others for a society's survival.** Doctors, for example, provide more critical services to ensure a society's survival than do lawyers, engineers, or bank tellers.

3. **The most qualified people must fill the most important positions.** Some jobs require more skill, training, or intelligence than others because they are more demanding, and it's often difficult to replace the workers doing such jobs. Brain surgeons and pilots, for example, must have many years of training and aren't replaced as easily as nurses or flight attendants, whose training is much shorter and less rigorous.

4. **Society must offer greater rewards to motivate the most qualified people to fill the most important positions.** People won't undergo many years of education or training unless they are rewarded by money, power, status, and/or prestige. If doctors and nurses earned the same salaries, there wouldn't be much incentive for doctors to spend so many years earning a medical degree.

According to the Davis-Moore thesis and other functionalists' perspectives, then, stratification and inequality are necessary to motivate people to work hard and to succeed. In open class systems, functionalists claim, inequality reflects the existence of a **meritocracy**, a belief that individuals are rewarded for what they do and how well rather than on the basis of their ascribed status.

Critical Evaluation

Melvin Tumin (1953) was the first sociologist to challenge Davis and Moore's functionalist perspective. First, he argued, societies don't always reward the positions that are the most important for its members' survival. For example, by age 29, golfer Tiger Woods was earning almost $100 million a year. Most doctors don't earn such sums during their entire working lives. And, in 2007, the average annual salary of U.S. elementary and high school teachers, who play critical roles in society, was only about $50,000.

If the highest-paid professional athletes, actors, and pop musicians went on strike, many of us would probably barely notice. If, on the other hand, garbage collectors, secretaries, teachers, doctors and nurses, truck drivers, and mail carriers refused to work, society would grind to a halt. Thus, according to Tumin, there's little association between earnings and the jobs that keep a society going.

Second, Tumin argued, Davis and Moore overlook the many ways that stratification limits the discovery of talent. Where wealth is differentially distributed, for instance, access to education, especially higher education, depends on the wealth of one's parents. As a result, large segments of the population are likely to be deprived of the chance to even discover what their talents are, and society loses.

Third, Tumin criticized Davis and Moore for ignoring the critical role of inheritance. In upper social classes, sons and daughters don't have to worry about jobs because their inherited wealth guarantees a lifetime income. Even if class differences were eliminated today, it would take low-income families an average of six generations (about 120 years) to reduce the wealth gap that has accumulated over hundreds of years (Isaacs 2007c).

CONFLICT PERSPECTIVES: STRATIFICATION HURTS SOCIETY

Like Tumin, conflict theorists maintain that social stratification is dysfunctional because it harms individuals and societies. In the mid-nineteenth century, Karl Marx spent much of his life in the midst of the enormous and tumultuous changes taking place during the Industrial Revolution in England (see Chapter 1). His analysis (1934) of social class and inequality has had a profound influence on modern sociology, especially conflict theory.

Capitalism Benefits the Rich

Marx was aware that a diversity of classes can exist at any one time, but he predicted that capitalist societies would ultimately be reduced to two social classes: the capitalist class, or bourgeoisie, and the working class, or proletariat. The **bourgeoisie**, those who own the means of production, such as factories, land, banks, and other sources of income, can amass wealth and power. The **proletariat**, workers who sell their labor for wages, earn barely enough to keep themselves and their families alive.

As the numbers of oppressed and alienated workers increased, Marx said, the proletariat would overthrow the bourgeoisie. "A century after his death, it is apparent that Marx was a better sociologist than he was a prophet" (Gilbert 2008: 7), because revolutions have not occurred in capitalist countries.

Conflict theorists agree with Marx, however, that social stratification benefits the rich at the expense of the workers. A major example is **corporate welfare**, an array of direct subsidies, tax breaks, and assistance that the government has created for businesses (see Chapter 11). For example, despite ongoing accusations that Wal-Mart exploits its workers and forces smaller retailers out of business, between 2004 and

bourgeoisie those who own the means of production and can amass wealth and power.

proletariat workers who sell their labor for wages.

corporate welfare an array of direct subsidies, tax breaks, and assistance that the government has created for businesses.

TABLE 8.3
Sociological Explanations of Social Stratification

PERSPECTIVE	LEVEL OF ANALYSIS	KEY POINTS
Functionalist	Macro	• Fills social positions that are necessary for a society's survival • Motivates people to succeed and ensures that the most qualified people will fill the most important positions
Conflict	Macro	• Encourages workers' exploitation and promotes the interests of the rich and powerful • Ignores a wealth of talent among the poor
Feminist	Macro and micro	• Constructs numerous barriers in patriarchal societies that limit women's achieving wealth, status, and prestige • Requires most women, not men, to juggle domestic and employment responsibilities that impede upward mobility
Symbolic Interactionist	Micro	• Shapes stratification through socialization, everyday interaction, and group membership • Reflects social class identification through symbols, especially products that signify social status

2007 alone, 15 state and local governments provided Wal-Mart with more than $200 million in property tax breaks, job training funds, and grants that came from tax revenues (Mattera and Purinton 2004; "Wal-Mart Subsidy Watch" 2007).

Critical Evaluation

Some scholars have criticized Marxian and later conflict theories for several reasons. First, even though the concentration of corporate wealth has increased during the last 100 years, the polarization of classes and impoverishment of workers that Marx expected in industrialized countries has not come about. In fact, many people's income has increased considerably since Marx's day, there's an abundance of manufactured goods in capitalist societies, and welfare programs dilute widespread unrest (Glassman 2000).

Second, some question whether people always act primarily out of economic self-interest. For example, Chuck Feeney, the founder of Duty Free Shoppers, has already given $4 billion of his fortune to charities and has instructed his board to donate the rest (another $4 billion) by 2016 (Roosevelt 2008).

Third, functionalists, especially, criticize conflict theorists for underrating the ability of individuals to be upwardly mobile. That is, if people really want to succeed, they can do so by working hard and making sacrifices.

FEMINIST PERSPECTIVES: WOMEN ARE ALMOST ALWAYS AT THE BOTTOM

For feminist scholars, functionalist and conflict theories are limited because they typically focus on men in describing and analyzing social stratification and social class. As a result, women are largely invisible.

Patriarchy Benefits Most Men, not Women

Patriarchy, according to feminist sociologists, undermines the upward mobility of even the most talented women. For example, of the world's top 20 billionaires, only one is a woman, and she inherited her wealth from her father, the founder of the cosmetics company L'Oréal. Of the 39 women among the richest CEOs on the *Forbes 400* list, the majority also inherited their wealth from their fathers or husbands ("Cash Countesses" 2007; "Billionaires 2008" 2008). And, among CEOs, the best-paid men earn six times more than the best-paid women and their bonuses are three times higher (Dobrzynski 2008).

Entertainment is the only industry where women have achieved a significant degree of economic success; of the 100 top-paid U.S. celebrities, about 30 are women (Goldman et al. 2007). Many (like singers Madonna and Céline Dion) are well-known, but only a few, including Oprah Winfrey, also enjoy some measure of political influence and prestige, not just name recognition.

Feminist theorists contend that in a patriarchal system men shape the social stratification system because they control a disproportionate share of wealth, prestige, and power. On a micro level, individual women may have more power than men if they earn the major portion of a household's income (Lorber 2005). On a macro level, however, the feminization of poverty often results in women's downward mobility. In addition, women often have to overcome economic and educational inequities as well as juggle domestic and workplace responsibilities (see Chapters 9, 12, and 13).

Critical Evaluation

Feminist sociologists point out that patriarchy affects both social stratification and social class, but they typically focus on poor women (see Kendall 2002). A related problem is that many feminist scholars examine women's and men's income differences and the effects of social stratification on both sexes but say little about social mobility, especially for middle and working-class women.

SYMBOLIC INTERACTIONIST PERSPECTIVES: PEOPLE CREATE AND SHAPE STRATIFICATION

The major contribution of symbolic interactionism is showing us how people create and reinforce stratification. For symbolic interactionists, individuals aren't necessarily helpless victims of macro-level forces (like economic shifts), but can shape and change their social realities.

People Have a Great Deal of Control Over Their Own Lives

People in upper, middle, working, and lower classes interact and socialize their children differently because of family background, education level, and income. People in different social classes also acquire and use symbols differently. For example, lower-upper class members may engage in conspicuous consumption to show off their new wealth (by buying a yacht, for example), while people in the middle class may go into debt to give an appearance of being richer than they are.

Critical Evaluation

Symbolic interactionists' theories are important in understanding the everyday processes that underlie social stratification, but there are several weaknesses. First, the theories don't explain why—despite the same family background and socialization—one child in a family is upwardly mobile, while another plunges into poverty.

Second, it's not clear why some people (across social classes and racial/ethnic groups and in both sexes) are more obsessed than others with status symbols and reputation. For example, some billionaires live modestly and avoid the media, while others, such as billionaire Donald Trump, constantly show off their wealth and even do television commercials to increase their billions.

Finally, conflict theorists, especially, fault symbolic interactionists for ignoring structural factors—including meritocracy and inherited wealth—that create and reinforce inequality. For example, many talented and intelligent teenagers drop out of high school because of limited family income and other resources (see Chapter 14).

7 Inequality across Societies

So far, we've focused on U.S. stratification. All societies are stratified but the inequality varies a great deal. Let's look at some of these variations. In northern Afghanistan, Abdul Ghaffer weaves the carpets that his family has made for generations. It takes three family members, working from dawn until past midnight, to earn about $60 a month weaving 3 square meters of carpet. Asked if he feels his labor should be better paid, Ghaffer says yes, "but if I raise my price, the carpet dealer will go to someone else who charges less. I have to survive" (Baldauf 2001:7).

Ghaffer and his family are very poor but better off than billions of others around the world. Many people describe our planet as a global village in which societies are connected economically, politically, and socially because of technology. This is true for only a handful at the top of the class hierarchy. Most people live in very different, separate, and unequal worlds.

LIVING WORLDS APART

The richest 2 percent of adults in the world own more than half of global wealth. In contrast, the bottom half of the adult population owns barely 1 percent of global wealth. The top 5 percent of individuals in the world receive about one-third of total world income, compared with only 5 percent of world income for the poorest 40 percent (Davies et al. 2006; Milanovic 2006). Such differences in individual wealth are astounding, but what about variations across countries?

The World Bank (2008) describes global stratification in terms of high-income, middle-income, and low-income countries. There are many poor people in high-income countries and some very affluent people in low-income countries, but *Table 8.4* provides an overview of the enormous economic disparities across countries.

High-income countries have a developed industrial economy and an annual gross national income (GNI) of almost $37,000 per person. The combined population of these countries is only 15 percent of the world's population but enjoys 78 percent of the world's income (see *Table 8.4*). The 60 high-income countries include the United States and Canada in North America; Great Britain, France, Germany, Portugal, Norway, Switzerland, and other industrialized nations in Western Europe; Japan in Asia; Australia and New Zealand in Oceania; and Israel. To a greater extent than other countries, high-income countries have access to health services, safe water, and higher education.

Middle-income countries, which comprise about 48 percent of the world's population, have a developing industrial economy and a considerably lower GPI per person than high-income countries (see *Table 8.4*). The 96 middle-income countries include most of the European countries that broke away from the Soviet Union in 1991, Russia, Mexico, Latin America, several countries in northern and southern Africa (including Algeria, Libya, and Angola), and many countries in the Middle East (including Turkey, Iran, Iraq, and Saudi Arabia). Most people in middle-income countries have less access to education, health services, food, and amenities like electricity, automobiles, and telephones than people in high-income countries, but their standard of living is higher than in low-income countries.

Low-income countries are the least industrialized and are largely agricultural. Most people in these countries are peasant farmers or live in villages. The 53 nations in this group, comprising 40 percent of the world's population, have an average GNI of $649 per person (see *Table 8.4*). These countries are primarily in Central and East Africa, parts of Asia (Pakistan, India, and China), and Indonesia (Java, Sumatra, and Borneo). People in low-income countries experience a low standard of living, are impoverished, and have little access to health services, education, and safe water.

TABLE 8.4
Global Economic Inequality, 2006

	NUMBER OF COUNTRIES	POPULATION	PERCENTAGE OF WORLD POPULATION	TOTAL GNI	PERCENTAGE OF WORLD GNI	GNI PER CAPITA
Low income	53	2.4 billion	40	$1.6 million	3	$ 649
Middle income	96	3.1 billion	48	$9.4 million	19	$ 3,053
High income	60	1 billion	15	$37.7 billion	78	$36,608
World	209	6.5 billion		$48 billion		$ 7,448
United States		300 million	5	$13 billion	28	$44,710

Note: GNI is the Gross National Income.

Source: Based on material in World Bank 2008.

In 2006, of the world's 6.5 billion people, almost half lived on less than $2 a day and 1.2 billion—almost a fifth—on less than $1 a day, what economists classify as extreme poverty. About 93 percent of those in extreme poverty live in three regions: East Asia, South Asia, and sub-Saharan Africa. In some countries, such as Haiti, where 75 percent of the population earn less than $2 a day, the most destitute feed their children with patties made of mud, oil, and sugar. Those who survive starvation experience stunted physical growth, permanent mental retardation, lower intelligence levels, and a reduced capacity to learn by age 2. Such problems create a vicious cycle: Hunger in childhood can lead to permanently dulled minds, an inability to pursue a livelihood, and having children with a diminished mental capacity (Martens 2005; World Food Program 2006; Lacey 2008).

WHY IS INEQUALITY UNIVERSAL?

There are many theories that try to explain why inequality is universal, but three of the most influential have been modernization theory, dependency theory, and world-system theory.

Modernization theory claims that low-income countries are poor because of traditional attitudes about technology and institutions. These countries don't have attitudes and values that lead to experimentation and the use of modern technology, but instead adhere to traditional customs that isolate them from competing in a global economy. In effect, modernization theory blames poor nations for their poverty and other problems. After the key foundations of modernity and capitalism are in place, this perspective maintains, low-income countries will prosper.

Dependency theory contends that the main reason why low-income countries are underdeveloped is because they are pawns that high-income countries exploit and dominate. Rich nations wield an enormous amount of power by exporting jobs overseas, manipulating foreign aid, draining less powerful countries of their resources, penetrating other countries with multinational corporations, and coercing national governments to comply with their interests (by not passing environmental laws, for example). In effect, according to dependency theorists, high-income countries benefit because the poor provide cheap labor and don't protest even though they work in hazardous conditions and earn under $1 a day.

More recently, *world-system theory,* similar to dependency theory, argues that "the economic realities of the world system help rich countries stay rich while poor countries stay poor" (Bradshaw and Wallace 1996: 44). That is, those countries that dominate the world economy (like the United States) influence the economies of low-income countries because the workers depend on external markets for jobs. High-income countries can extract raw materials (like diamonds) with little cost. They can also set the prices for the agricultural products that low-income countries export regardless of market prices, forcing many small farmers to abandon their fields because they can't pay for labor, fertilizer, and other costs (Carl 2002).

None of these perspectives explain inequality across *all* societies. Instead, global inequality is due to a combination of factors, including a country's values and customs, and exploitation by high-income countries.

During the last few years, we've heard about the booming economies of India and China, but not everyone has profited. In India, some people have become enormously wealthy, while others remain very poor. A tent city in India houses many of the workers who earn about $1.30 a day building new office towers for the affluent nearby (left). In China, 10 percent of urban households control 40 percent of urban wealth. A man in Shanghai begs as wealthier residents pass by (right).

Gender and gender roles

are social creations.

what do you think?

Sexuality is determined biologically.

1 2 3 4 5 6 7

strongly agree   strongly disagree

9

Gender and Sexuality

Key Topics

In this chapter, we'll explore the following topics:

This chapter describes how gender, sex, and sexuality shape our lives. We'll examine sexual orientation, some current controversies, and sociological explanations of gender inequality and sexuality. Let's begin by looking at whether women and men are as different as some self-help writers claim.

> **sex** the biological characteristics with which we are born.
>
> **gender** learned attitudes and behaviors that characterize people of one sex or the other.

1 Female-Male Similarities and Differences

many people use the terms *sex* and *gender* interchangeably, but they have distinct meanings. The terms are related, but sex is a biological designation, whereas gender and gender roles are social creations.

SEX AND GENDER

Sex refers to the biological characteristics with which we are born—chromosomes, anatomy, hormones, and other physical and physiological attributes. These attributes *influence* our behavior (such as shaving beards and wearing bras), but do not *determine* how we think, feel, and act. Whether we see ourselves and others as feminine or masculine depends on gender, a considerably more complex concept than sex.

Gender refers to learned attitudes and behaviors that characterize people of one sex or the other. Gender is based on social and cultural expectations rather than on physical traits. Thus, most people are *born* either male or female, but we *learn* to be women or men because we associate conventional behavior patterns with each sex. In many societies, for example, women are expected to look young and attractive as long as possible and men are expected to amass as much income and wealth as possible.

GENDER IDENTITY AND GENDER ROLES

Children develop a **gender identity**, or a perception of themselves as either masculine or feminine, early in life. Many Mexican baby girls but not boys have pierced ears, for example, and hairstyles and clothing for American toddlers differ by sex. Gender identity, which typically corresponds to a person's biological sex, is learned in early childhood and usually remains relatively fixed throughout life.

About 90 percent of U.S. women and men believe that women are more emotional than men. In fact, both sexes experience emotions such as anger, happiness, and sadness just as deeply. What differs is *how* women and men express their emotions because men are more likely to suppress their feelings whereas women tend to show their emotions more openly (Newport 2001; Simon and Nath 2004).

Such differences are due to **gender roles**—the characteristics, attitudes, feelings, and behaviors that society expects of females and males. A review of research on the differences between the sexes conducted over the last 20 years concluded that females and males are much more alike than different on a number of characteristics, including cognitive abilities, verbal and nonverbal communication, leadership traits, and self-esteem (Hyde 2005). Still, sociologists often describe our social roles as *gendered*, in that males and females are often treated differently because of their sex (Howard and Hollander 1997).

In the United States, people are more likely now than in the past to pursue jobs and other activities based on their abil-ity and interests rather than their sex. For the most part, however, American society still has fairly rigid gender roles and widespread **gender stereotypes**, expectations about how people will look, act, think, and feel based on their sex.

gender identity a perception of oneself as either masculine or feminine.

gender roles the characteristics, attitudes, feelings, and behaviors that society expects of females and males.

gender stereotypes expectations about how people will look, act, think, and feel based on their sex.

PIctured left, Ana Carolina Reston, a 21-year-old Brazilian model, died from complications resulting from anorexia. She was 5 feet 8 inches tall and weighed 88 pounds. Below, a model on the runway of the summer 2007 fashion show of Guy Laroche in Paris.

Anorexia nervosa is a potentially life-threatening eating disorder characterized by self-starvation and excessive weight loss. In the United States, eating disorders affect nearly 10 million women and 1 million men (National Eating Disorders Association 2008).

2 Contemporary Gender Stratification and Inequality

Since 1893, the U.S. Postal Service has published hundreds of commemorative postage stamps that honor a place, event, or person, but only one out of five people honored in this way has been a woman. The Postal Service defends its selections with the argument that the vast majority of stamp collectors are males who prefer to buy stamps about "guy topics." Others argue that such marginalization sends a message that women and their contributions are unimportant (Long 2006).

The much lower number of commemoratives honoring women reflects **sexism**, an attitude or behavior that discriminates against one sex, usually women, based on the assumed superiority of the other sex. Much sexism, such as that of the Postal Service, is subtle. Because most people have internalized beliefs that women are inferior to men, we don't even notice women's exclusion (Benokraitis 1997).

Considerable sexism is still blatant, however, because it's visible, intentional, and easily documented. According to numerous women bloggers, for example, cybersexism (including stalking, death threats, and hate speech) is prevalent, and women whose user names indicate their sex are 25 times more likely than men to experience online harassment (Valenti 2007).

Boys and men also experience sexism. One of my students wrote the following during an online discussion of gender roles:

> *Some parents live their dreams through their sons by forcing them to be in sports. I disagree with this but want my [9-year-old] son to be "all boy." He's the worst player on the basketball team at school and wanted to take dance lessons, including ballet. I assured him that this was not going to happen. I'm going to enroll him in soccer and see if he does better.*

Is this mother suppressing her son's natural dancing talent? We'll never know because she, like many parents, expects her son to fulfill gender roles that meet with society's approval.

Sexism is widespread due to **gender stratification**—people's unequal access to wealth, power, status, prestige, opportunity, and other valued resources as a result of their sex. We'll examine some of this stratification in greater depth in later chapters. For now, let's look briefly at gender inequality in the family, education, the workplace, and politics.

> **sexism** an attitude or behavior that discriminates against one sex, usually females, based on the assumed superiority of the other sex.
>
> **gender stratification** people's unequal access to wealth, power, status, prestige, and other valued resources as a result of their sex.

At an early age, sex-appropriate children's toys prepare girls and boys for future adult roles. Do the toys reflect children's own choices or gender socialization by parents, teachers, other adults, and the media?

The Butler Café is becoming very popular in Japan. It's a place for young women to unwind from the stresses of the outside world, including pressure to conform and marry. The only men present are butlers who respond in less than 6 seconds if a customer rings a golden bell located on each table.

Credit: © Yoshikazu Tsuno/AFP/Getty Images

GENDER AND FAMILY LIFE

Americans have more choices today, but is family life less gendered than in the past? Probably not as much as we think. If Americans were free to do either, half of women and 68 percent of men would prefer to work outside the home rather than stay home to take care of their house and family. Few adults have this either-or choice, however. Instead, nearly three out of four married mothers are in the labor force, compared with two out of four in 1970 (Saad 2007; U.S. Census Bureau 2008).

Having a husband creates an extra 7 hours of housework a week for a woman, but many couples say that they share domestic tasks more equally than their parents or grandparents did. Still, there continues to be a division of labor along traditional gender lines; for example, 68 percent of wives do the laundry (Newport 2008; Stafford 2008).

Men are doing more in the home, but women shoulder twice as much child care and housework as men. Women's domestic work declined from 50 hours in 1965 to about 26 hours in 2004, while men's increased from 12 to 16 hours. These changes are due partly to men's greater efforts and partly to a decrease in the amount that women do because of having full-time employment (Bianchi et al. 2006; England 2006).

GENDER AND EDUCATION

Many teachers and schools send gendered messages to children that follow them from preschool to college. When children enter kindergarten, they perform similarly on both reading and mathematics tests. By the third grade, however, boys, on average, outperform girls in math and science, while girls outperform boys in reading. These gaps increase throughout high school (Dee 2006). Some of the reasons for these differences, as you'll see in Chapter 14, include parental socialization and teacher expectations. Also, many counselors still steer girls and boys into courses that focus on gender-stereotyped disciplines (such as social work for girls and engineering for boys).

In higher education, "women face more obstacles as faculty . . . than they do as managers and directors in corporate America" (West and Curtis 2006: 4). Even when women earn doctoral degrees in male-dominated fields, they are less likely to be hired than men. For example, women have received over 45 percent of all doctorates in biology for the past 15 years, but only 14 percent of full-time biology professors are women (Handelsman et al. 2005). Once hired, women faculty members are less likely than men to be promoted. For nearly 25 years, one-third of all recipients of doctoral degrees have been women, but men still dominate the rank of full professor (see *Table 9.1*).

GENDER AND THE WORKPLACE

There has been progress toward greater workplace equality, but we still have a long way to go. In the United States (as around the world), most occupations are gendered. Many jobs are segregated by sex, there are large wage gaps, and numerous women experience sexual harassment and pregnancy discrimination.

Gender-Segregated Work and Wage Gaps

American women still tend to be clustered in 21 of 500 occupational categories (WAGE 2006). Generally, women's jobs tend to be "feminine"—those that require nurturing, caregiving, serving others, and working with people. The occupations that men dominate tend to be "masculine"—manual, relatively autonomous, often containing an element of danger, and requiring technical skills or training.

A number of U.S. occupations are almost entirely filled by either women or men. For example, between 92 and 98 percent of all registered nurses, child care workers, secretaries, and preschool and kindergarten teachers are women. At least 98 percent of all steel workers, mechanics, plumbers, and loading machine operators are men. Women have made some progress in the higher-paying professional occupations, but 70 percent of physicians, 75 percent of architects, and 78 percent of dentists are men (U.S. Department of Labor 2007).

There is also considerable sex inequality *within* occupations. About 30 percent of all lawyers are women, up from 20 percent in 1992, but they have enjoyed only modest professional advances. Among other barriers, women attorneys are often criticized—regardless of how they act—for being too aggressive or not aggressive enough, too emotional or too unfeeling, and too easygoing or too demanding. And despite many law firms' official family-friendly policies, few women take maternity leave because they would be seen as "less seriously committed to the profession" and would jeopardize their chances for promotion. These are the same problems that women attorneys faced in the 1980s (American Bar Association 2006a, 2006b).

In 2007, women who worked full-time year-round had a median income of $35,102 compared with $45,113 for men. This means that women earned 77 cents for every dollar men earned. Between 1999 and 2005, the wage gap between women and men increased in 15 states. In effect, *the average woman would have to work almost 4 extra months every year to make the same wages as a man* (Hartmann et al. 2006; DeNavas-Walt et al. 2008).

gender pay gap the overall income difference between women and men in the workplace (also called the *wage gap*).

This income difference between women and men is the **gender pay gap** (also called the *wage gap*). Over a lifetime, the average woman who works full-time and year-round for 47 years is deprived of a significant amount of money because of the wage gap: $700,000 for high school graduates, $1.2 million for college graduates, and over $2 million for women with a professional degree, as in business, medicine, or law. Because raises are typically based on a percentage of one's annual income, the lower the wages, the lower the raises. Lower wages and salaries reduce women's savings as well as their purchasing power and quality of life and bring them lower Social Security payments after retirement (Murphy and Graff 2005; Horn 2006).

Not only do women earn less than men at all educational levels, but the gender pay gap increases as the level of educational attainment increases (see *Figure 9.1* on the next page). Female high school graduates earn 71 percent of what their male counterparts earn, and women's earnings drop to 69 percent of men's for women with at least a college degree. As you'll see in Chapter 12, men earn more than do women in *every* occupational category, and the earnings gap increases at the higher-paying managerial and professional levels.

Only 20 percent of the 3 million American teachers are men. In elementary schools, only 9 percent are males, down from 18 percent in 1981. A major reason for the decline is low salaries compared with other occupations, but gender stereotypes are an important factor. Especially in the lower grades, men who show nurturing skills or affection for children may be accused of being pedophiles (Provasnik and Dorfman 2005; Bonner 2007; Scelfo 2007).

© Creatas Images/Jupiterimages

TABLE 9.1
Proportion of Women Faculty by Rank

Proportion of full-time faculty members, by rank, that are female	
Professor	24%
Associate Professor	38%
Assistant Professor	45%
Instructor	52%

Note: Of the almost 632,000 full-time faculty in 2005, 49 percent were women.

Source: Based on U.S. Department of Education 2006, Table 227.

sexual harassment any unwanted sexual advance, request for sexual favors, or other conduct of a sexual nature that makes a person uncomfortable and interferes with her or his work.

FIGURE 9.1
The Gender Pay Gap, by Education

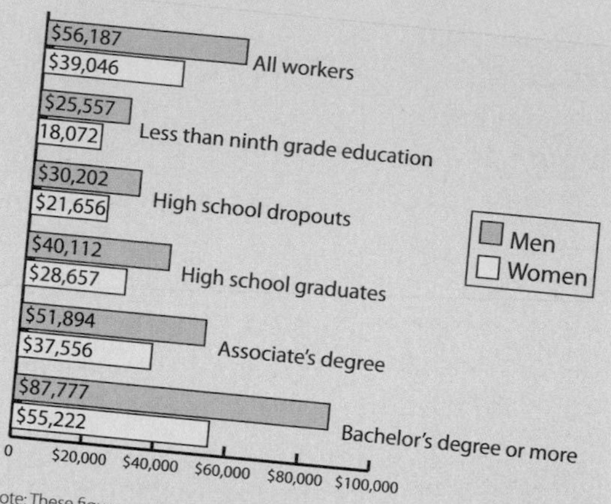

$56,187
$39,046
All workers
$25,557
18,072
Less than ninth grade education
$30,202
$21,656
High school dropouts
$40,112
$28,657
High school graduates
$51,894
$37,556
Associate's degree
$87,777
$55,222
Bachelor's degree or more
0
$20,000
$40,000
$60,000
$80,000
$100,000
Men
Women

Note: These figures represent the average earnings of year-round full-time workers 18 years and older in 2005.

Source: Based on U.S. Census Bureau 2008: Table 681.

The wage gap can be partially explained by differences in education, experience, and time in the labor force, but about 41 percent of the gap is the result of sex discrimination in hiring, promotion and pay, bias against mothers, and occupational segregation. That is, a wage gap remains even when women and men have the same education, occupation, number of years in a job, seniority, marital status, number of children, and are similar on numerous other factors (Boraas and Rodgers 2003; Blau and Kahn 2006).

Some of the wage gap can be attributed to the *glass ceiling,* an invisible barrier that keeps women from being more successful in the workplace. Among the *Fortune* 500 companies, for example, 91 percent of the corporate officers are men and the handful of women who get ahead do so by consistently exceeding, rather than just meeting, performance expectations (Catalyst 2006).

Sexual Harassment and Pregnancy Discrimination

Sexual harassment became an illegal form of sex discrimination in Title VII of the Civil Rights Act of 1964 and again in the 1980 Equal Employment Opportunity Commission (EEOC) guidelines. Sexual harassment complaints represent the fastest-growing type of employment discrimination complaints, with over 220,000 filed with the EEOC between 1992 and 2007 (U.S. Equal Employment Opportunity Commission 2008b).

Sexual harassment is any unwanted sexual advance, request for sexual favors, or other conduct of a sexual nature that makes a person uncomfortable and interferes with her or his work. It includes *verbal behavior* (such as pressures for dates or demands for sexual favors in return for hiring, promotion, or tenure as well as the threat of rape), *nonverbal behavior* (such as indecent gestures and displaying posters, photos, or drawings of a sexual nature), and *physical contact* (such as pinching, touching, and rape).

Sexual harassment cuts across many types of jobs. Dov Charney, the founder and CEO of clothing manufacturer American Apparel Inc., holds meetings in his underwear and regularly refers to women as sluts, whores, and other demeaning epithets. He is facing sexual harassment charges, but his lawyer defends Charney's behavior as "creative expression" (Bronstad 2008).

Sexual harassment is especially common in male-dominated occupations where female newcomers are unwelcome. Among firefighters, for example, 97 percent of whom are men, many women have complained about offensive behavior such as finding feces in the women's shower stalls, unwanted touching, being referred to as "bitches," and attempted rape (Banks 2006).

The federal Pregnancy Discrimination Act of 1978 makes it illegal for employers with more than 15 workers to fire, demote, or penalize a pregnant employee. Some state laws extend this protection to companies with as few as 4 employees. Despite such laws, the EEOC reports that charges of pregnancy discrimination have risen 33 percent in recent years. In 2007 alone, nearly 5,600 women filed complaints that they had been fired or demoted or had some of their responsibilities taken away from them when their employers learned that they were pregnant (U.S. Equal Employment Opportunity Commission 2008a). These complaints represent only the tip of the iceberg because only a fraction of women who face pregnancy discrimination ever take action: Many aren't aware of their rights, and others don't have the resources to pursue lengthy lawsuits.

GENDER AND POLITICS

Recently, voters in Atlanta, Georgia, and Cleveland, Ohio, elected their first female mayors, one of whom is black, and Puerto Rico elected its first woman governor. In 1872, Victoria Chaflin Woodhull of the Equal Rights Party, was the first female presidential candidate. Since then, about 36 women have sought the nation's highest office. In 2007, Senator Hillary Rodham Clinton engaged in a "whisker-close but losing battle to become the first woman major party presidential nominee," a historic run that came close to breaking the glass ceiling of a male-dominated U.S. presidency (Halloran 2008: 34).

When we stop reading about such firsts, can we be more confident that there is greater political equality between men and women? Consider, for example, women's participation in national and state politics. Unlike a number of other countries (including Great Britain, Germany, India, Israel, Pakistan, Argentina, Chile, and Philippines), the United States has never had a woman serving as president or even vice-president. In the U.S. Congress, 84 percent of the members are men. In other important elective offices, only one of four important decision makers is a woman (see *Table 9.2*), and this number hasn't changed for 15 years (Center for American Women and Politics 2008). Even though some women serve in Congress, they are typically absent from the most powerful and prestigious committees (Swers 2002).

Worldwide, the United States lags far behind many other countries, including some developing nations, in women's political empowerment (measured by the number of women serving as head of state or in decision-making government positions). Of 128 countries, the United States ranks only 77th in women's political empowerment (Hausmann et al. 2007). According to the president of a U.S. organization working to advance women in political leadership, "It will take [women] till 2063 to reach parity [with men]" (Feldmann 2008: 2).

As you'll see in Chapter 11, women's voting rates in the United States have been higher than men's since 1984. Why, in contrast, are there so few women in political office? One reason may be that women, socialized to be nurturers and volunteers, see themselves as supporters rather than active doers. As a result, they may spend many hours organizing support for a candidate rather than running for a political office themselves.

Second, women are less likely than men to receive encouragement to run for office from a political source (such as a party leader). As a result, even successful women are twice as likely as men to rate themselves as not at all qualified to run for office. In contrast, men are two-thirds more likely than women with similar credentials to consider themselves qualified or very qualified to run for office (Lawless and Fox 2005).

Third, a lingering sexism is still present in the thinking among both men and women, that "from the pulpit to the presidency," men are better leaders (Tucker 2007: A9). The pervasive sexism in media coverage of political candidates was especially evident during the 2008 presidential campaigns. From the beginning, reporters and news commentators criticized Senator Clinton's appearance ("She looked haggard") and voice ("cackling laugh"), described her as a "nagging wife," and referred to her as Hillary but never called Senators Obama and McCain by their first names (Long 2008: 11A; Wakeman 2008: 61).

TABLE 9.2
U.S. Women in Elective Offices, 2008

	TOTAL NUMBER OF OFFICE HOLDERS	PERCENTAGE WHO ARE WOMEN
U.S. Congress	535	16
Senate	100	17
House of Representatives	435	17
State Elective Offices		
Governor	50	16
Lieutenant Governor	50	20
State Legislator	7,382	24
Attorney General	50	8
Secretary of State	50	24
State Treasurer	50	22
State Comptroller	50	8
Mayor (100 largest cities)	100	11

Source: Based on material at the Center for American Women and Politics 2008; Morales 2008.

3 Sexual Orientation

biological sex affects just about everything we do as well as the quality of our lives. Similarly, our sexual orientation also shapes our (and others') attitudes and behavior.

In the movie *Annie Hall*, a therapist asks two lovers how often they have sex. The character played by Woody Allen answers, "Hardly ever, maybe three times a week." The character played by Diane Keaton replies, "Constantly, three times a week." As this example illustrates, sex is more important for some people than others. Our sexuality is considerably more complex than just

sexual orientation a preference for sexual partners of the same sex, of the opposite sex, or of both sexes.

homosexuals those who are sexually attracted to people of the same sex.

heterosexuals those who are sexually attracted to people of the opposite sex.

bisexuals those who are sexually attracted to members of both sexes.

asexuals those who lack any interest in or desire for sex.

physical contact, however, because it is a product of sexual identity and sexual orientation.

WHAT IS A SEXUAL ORIENTATION?

Our sexual identity incorporates a **sexual orientation**, a preference for sexual partners of the same sex, of the opposite sex, of both sexes, or neither sex:

- **Homosexuals** (from the Greek root *homo*, meaning "same") are sexually attracted to people of the same sex. Male homosexuals prefer to be called *gay*, female homosexuals are called *lesbians*, and both gay men and lesbians are often referred to as *gays*.

Coming out is a person's public announcement of a gay or lesbian sexual orientation.

- **Heterosexuals**, often called *straight*, are attracted to partners of the opposite sex.
- **Bisexuals**, sometimes called *bis*, are attracted to members of both sexes.
- **Asexuals** lack any interest in or desire for sex.

Heterosexuality is the predominant sexual orientation worldwide, but homosexuality exists in all known cultures. When Iran's president declared that there are no homosexuals in Iran, some Iranians were surprised. One gay man from that country said that, in fact, "There are plenty of gay men and women in Iran," and "You can have a secret gay life as long as you don't . . . start demanding rights." Doing so would result in being rejected by relatives, imprisoned, or executed ("Iranian Gays . . ." 2007: 16A). Similarly, in Egypt and many African countries, gays can be stoned, imprisoned, or killed. Other nations, however, have accepted gay relationships, especially informally. In Afghanistan, for instance, homosexuality—includ-

Lynn and Meredith Bacon (above right) remain married despite Meredith's sex change in 2005. Meredith, who used to be Wally, is a popular faculty member at the University of Nebraska. Some people were shocked by the transformation, but most students, faculty, and administrators were supportive (Wilson 2005). Such acceptance isn't typical. In 2007, Steven B. Stanton, city manager for 14 years of Largo, Florida, was dismissed after his plans to have a sex change operation became public. Now Susan Stanton, above left, standing with attorney Shelbi Day, acknowledges Judge Cynthia Newton's declaration of her official name change.

ing relationships with young boys—has "long been a clandestine feature of life" even though not practiced openly (Smith 2002: A4).

Transgendered Sexual Orientations

Our cultural expectations dictate that we are female or male, but a number of people are transgendered and are "living on the boundaries of both sexes" (Lorber and Moore 2007: 141). **Transgendered people** encompass several groups:

- *Transsexuals* are people born with one biological sex but who choose to live their life as another sex—either by consistently cross-dressing or by having their sex surgically altered.
- *Intersexuals* are people whose medical classification at birth is not clearly either male or female (this term has replaced *hermaphrodites*).
- *Transvestites* are people who cross-dress at times but don't necessarily consider themselves a member of the opposite sex.

Transgendered people include gays, heterosexuals, bisexuals, and men and women who don't identify themselves with any specific sexual orientation.

Extent of Homosexuality

How many Americans are gay men, lesbians, and bisexuals? No one knows for sure, largely because it is difficult to define and measure sexual orientation. For example, are people homosexual if they have ever engaged in same-sex behavior? What about people who have other-sex partners and desire same-sex partners but are afraid to act on this wish?

Researchers generally measure the extent of homosexuality by simply asking people whether they identify themselves as heterosexual or gay, lesbian, bisexual, or transgendered (the acronym is GLBT). About 4 percent of Americans identify themselves as homosexual, and almost 5 percent describe themselves as bisexual. However, 18 percent have had same-sex sexual contact (Mosher et al. 2005).

WHAT DETERMINES SEXUAL ORIENTATION?

No one knows why individuals are heterosexual, homosexual, bisexual, or transgendered. *Biological theories* maintain that sexual orientation has a strong genetic basis because, at least for males, the more genes one shares with a homosexual relative, the more likely it is

that one will be homosexual. These studies suggest that there is a gene for male homosexuality. Some biologists theorize that brain structure may also be associated with sexual orientation because the size of the hypothalamus—an organ deep in the center of the brain that is believed to regulate the sex drive—differs between heterosexual and gay men. Other scientists have found no evidence of a biological influence on homosexuality. If there's a strong genetic predisposition, gay children would come from gay households and straight children from straight households. This isn't the case, however, because most gay men and lesbians are raised by heterosexuals (Burr 1996; Golombok and Tasker 1996; Rice et al. 1999).

Social constructionist theories hold that sexual behavior is largely the result of social pressure and that culture, not biology, plays a large role in forming sexual identity. For example, and despite their homosexual inclinations, many straight men who have sex with other men refuse to accept the possibility that they're gay or bisexual. In effect, then, and because of societal pressure to be straight, many gay men and lesbians are living heterosexual lives.

So far, no study has shown conclusively that there is a gay gene or that the environment causes sexual orientation (see Brookey 2002). At this point, many researchers speculate that a combination of genetic and cultural factors may influence sexual orientation.

SOCIETAL REACTIONS TO HOMOSEXUALITY

Americans are ambivalent about homosexuality. The number who believes that homosexuality is an acceptable way of life increased from 43 percent in 1977 to 59 percent in 2007, and 89 percent say that gay men and lesbians should have equal rights in the workplace (Saad 2007).

Although there is growing acceptance of gay rights (except for same-sex marriage, as you'll see shortly), why do millions of Americans oppose homosexuality? Two concepts—heterosexism and homophobia—provide some answers. **Heterosexism** is a belief that heterosexuality is superior to and more natural than homosexuality or bisexuality. Like sexism, heterosexism pervades societal customs, laws, and institutions.

transgendered people those who are transsexuals, intersexuals, or transvestites.

heterosexism the belief that heterosexuality is superior to and more natural than homosexuality or bisexuality.

homophobia the fear and hatred of homosexuality.

abortion the expulsion of an embryo or fetus from the uterus.

Examples of heterosexism in the United States include the continuing ban against lesbian and gay military personnel and widespread hostility toward gay couples who want to formalize their committed relationships through marriage. As a result, many gays and lesbians still hide, deny, or try to suppress their sexual orientation to avoid attacks by heterosexuals.

Homophobia, the fear and hatred of homosexuality, is less overt today than in the past, but is still widespread. Homophobia often manifests itself in *gay bashing*—threats, assaults, or acts of violence directed at homosexuals. In 2006, 15 percent of all hate crimes were against homosexuals, but much gay bashing is not reported. In one national study, for example, nearly 75 percent of lesbians, gays, and bisexuals reported being targets of verbal abuse, nearly 33 percent were assaulted physically, and 33 percent said that their families treated them like outcasts (*Inside*-OUT. . . 2001; Federal Bureau of Investigation 2007).

Some Protestant denominations have welcomed gays and lesbians as members and ordained them as ministers, and even bishops. Others—Episcopal, Lutheran, Methodist, Presbyterian, and United Church of Christ—find themselves polarized and divided over homosexuality. Among black congregations, similarly, welcoming homosexuals may help some gay men who have left their churches and are even suicidal because of the community's rejection. On the other hand, a number of African American churches "have lost many of [their] most loyal, generous parishioners, who could not accept a message that contradicted what they saw as the Bible's condemnation of same-sex relations" (Banerjee 2007: A12). The Catholic Church denounces homosexuality as immoral and contrary to God's law, but many American Catholics feel that the church's view is outdated, and some say they've known homosexual couples who are more committed to each other than are many heterosexuals (Janofsky 2003).

Companies that welcome homosexual workers often enjoy positive public relations and attract well-educated gay men and lesbians. Numerous municipal jurisdictions, corporations, and smaller companies now extend more health-care coverage and other benefits to their gay and lesbian employees and their partners than to unmarried heterosexuals who live together. In 2008, 125 of the *Fortune 500* companies included "gender identity" in their nondiscrimination policies, compared with almost none in 2002. And, among the largest corporations, 58 percent provided transgender benefits such as regular hormone treatment, psychological counseling, and paying for gender reassignment surgery (Belkin 2008). In effect, then, there is considerably less job discrimination against GLBTs than even five years ago.

4 Some Current Controversies about Sexuality

m ost Americans see sex as a private act, but many also feel that that there should be governmental control of some sexual expressions and decisions. Three of the most controversial issues today are abortion, same-sex marriage, and pornography.

ABORTION

Abortion is the expulsion of an embryo or fetus from the uterus. It can occur naturally—in *spontaneous abortion* (miscarriage)—or it can be induced medically. After the United States outlawed abortion in the nineteenth century, the procedure was illegal until 1973, when the U.S. Supreme Court legalized abortion (in its *Roe v. Wade* decision).

Trends

Half of all pregnancies in the United States are unintended: Many are unwanted at the time of conception or at any time, while others are mistimed because they have occurred sooner than the woman wanted. About half of unintended pregnancies end in abortion, and 35 percent of American women have an abortion by age 45 (Boonstra et al. 2006).

About 2 percent of American women, or 1.2 million individuals, terminate pregnancies by abortion each year. The *abortion rate*, or the number of abortions per 1,000 women aged 15 to 44, increased during the 1970s but has decreased steadily since 1980. At its peak, the abortion rate was 29 in 1981; it fell steadily and reached an all-time low of 19.4 in 2005. Abortion is most common among women who are young (in their 20s), white, and never married (see *Figure 9.2*). Proportionately, however, African American women are more than three times as likely as white women to have an abortion, and Latinas are 2.5 times as likely to do so.

FIGURE 9.2
Who Gets Abortions?

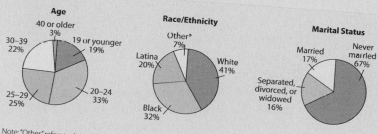

Age

40 or older 3%
19 or younger 19%
30–39 22%
20–24 33%
25–29 25%

Race/Ethnicity

Other* 7%
Latina 20%
White 41%
Black 32%

Marital Status

Married 17%
Never married 67%
Separated, divorced, or widowed 16%

Note: "Other" refers to Asian/Pacific Islanders, American Indians, and Alaska Natives.
Source: Based on Guttmacher Institute 2005, 2008.

Why Is Abortion Controversial?

Since its legalization, abortion has been one of the most persistently divisive issues in American politics and culture. Almost equal numbers hold opposing positions: Abortion should be legal (24 percent) or illegal (20 percent) in all circumstances. About 55 percent of Americans support abortion in cases where there is rape, incest, or a life-threatening condition for the mother. These divergent opinions have been fairly consistent since 1975 (Newport and Saad 2006).

Anti-abortion groups insist that the embryo or fetus is not just a mass of cells but a human being from the time of conception and, therefore, has a right to live. In contrast, many abortion rights advocates believe that, at the moment of conception, the organism lacks a brain and other specifically and uniquely human attributes, such as consciousness and reasoning. Abortion rights proponents also believe that a pregnant woman has a right to decide what will happen to her body (see Almond 2007).

Anti-abortion groups maintain that abortion is immoral and endangers a woman's physical and emotional health. Whether or not abortion is immoral is a religious and philosophical question. Safety, however, can be measured on two levels: physical and emotional. On the physical level, a legal abortion in the first trimester (up to 12 weeks) is safer than driving a car, using oral contraceptives, undergoing sterilization, or continuing a pregnancy. Abortions performed in the first trimester pose virtually no long-term risk of problems such as miscarriage, birth defects, or preterm or low-birth-weight delivery. There is also no evidence, despite the claims of abortion opponents, that having an abortion increases the risk of breast cancer, causes infertility, and leads to post-abortion stress disorders or suicide (Boonstra et al. 2006; Guttmacher Institute 2008).

What about emotional health? Most recently, a team of psychologists who evaluated 73 empirical studies in peer-reviewed journals published since 1989 concluded that, among adult women who had an unplanned pregnancy, the risk of mental health problems after an abortion was no greater than having a baby and experiencing normal *postpartum depression*—the "blues" that many women experience after the birth of a baby because of a sudden drop in some hormones. When women experience

Abortion opponents protesting

Abortion rights advocates protesting

The passion of abortion opponents is matched by that of abortion-rights activists. Both sides of this highly controversial issue often stage public demonstrations to support their views. Pictured here are the annual March for Life demonstration in Washington, D.C., and a rally in support of the Jackson Women's Health Organization clinic.

sadness, grief, and depression after an abortion, according to the researchers, such feelings are usually due to "co-occurring factors" including poverty (because low-income mothers worry about providing for a baby), a history of emotional problems and drug or alcohol abuse, or keeping the abortion secret from family and friends who stigmatize abortions. Also, these findings were consistent with previous national studies conducted in the early 1990s. The researchers noted, however, that they couldn't examine multiple elective abortions and their effect on women's emotional and mental health because of a scarcity of data (Major et al. 2008). Another unanswered question, so far, is whether teenage girls who have an abortion experience emotional problems that differ from those of adult women.

SAME-SEX MARRIAGE

Same-sex marriage (also referred to as *gay marriage*) is a legally or socially recognized marriage between two people of the same sex. Although still controversial, same-sex marriage is becoming more acceptable in the United States and several other countries.

Trends

In 1999, the Vermont Supreme Court ruled that same-sex partners should have the same legal rights as married couples through *civil unions*. Some of the rights include joint ownership of homes and other property, a share of the partner's medical or life insurance, and a right to inheritance and to survivors' benefits if a partner dies. Since then, several other states (like Connecticut and New Jersey) have passed similar legislation, but 41 states limit marriage to a man and a woman and 30 states have amended their constitutions to explicitly prohibit gay marriage.

Since 2001, six countries have legalized same-sex marriage (Belgium, Canada, Netherlands, Norway, South Africa, and Spain). In 2004, Massachusetts became the first state to recognize same-sex marriage. Four years later, the legislature defeated a proposed constitutional amendment to define marriage as legally existing only between a man and a woman, ensuring the legality of same-sex marriages in the state (Savage 2008).

TABLE 9.3

Why Are People For or Against Same-Sex Marriages?

Many states have passed laws that limit marriage to a man and a woman. How do you feel? What other reasons can you add for each side of the debate?

SAME-SEX MARRIAGE SHOULD BE LEGAL BECAUSE...	SAME-SEX MARRIAGE SHOULD NOT BE LEGAL BECAUSE...
• Attitudes and laws change. Until 1967, for example, interracial marriages were prohibited.	• Interracial marriages are between women and men, but gay marriages violate many people's notions about male-female unions.
• Gay marriages would strengthen families and long-term unions that already exist. Children would be better off with parents who are legally married.	• Gay households are not the best place to raise children. Children raised in gay households might imitate their parents' homosexual behavior.
• It would be easier for same-sex couples to adopt children, especially those with emotional and physical disabilities.	• All adopted children—those with and without disabilities—are better off with parents who can provide heterosexual gender role models.
• There are no scientific studies showing that children raised by gay and lesbian parents are worse off than those raised by heterosexual parents.	• There are no scientific studies showing that children raised by gay and lesbian parents are better off than those raised by heterosexual parents.
• Every person should be able to marry someone that she or he loves.	• People can love each other without getting married.
• Same-sex marriages would benefit religious organizations both spiritually and emotionally and bolster membership.	• Same-sex marriages would polarize church members who are opposed to gay unions and decrease the size of congregations.

Sources: Sullivan 2003; "The Merits of Gay Marriage" 2003; LaFraniere and Goodstein 2007.

In May 2008, California's Supreme Court ruled (four to three), that same-sex marriage is legal because "an individual's capacity to establish a loving and long-term committed relationship with another person and responsibly to care for and raise children does not depend upon the individual's sexual orientation" (Arnoldy 2008: 3). Less than six months later, however, California's voters passed Proposition 8, a constitutional amendment restricting the definition of marriage to a union between a man and a woman. As this book goes to press, GLBT organizations have filed petitions with California's Supreme Court to consider the constitutional legality of the proposition and, if necessary, will pursue this issue in the U.S Supreme Court. In November 2008, Connecticut became the second state to legalize same-sex marriage. The Connecticut Supreme Court ruling (again four to three) cannot be appealed.

Why Is Same-Sex Marriage Controversial?

A majority of Americans support civil unions, but 56 percent oppose same-sex marriages (down from 68 percent in 1997). Opposition to same-sex marriage is considerably higher among those who regularly attend religious services, those who live in the South, people aged 65 and older, and those who have conservative views on family issues (Pew Research Center 2006; Saad 2008).

Those who favor same-sex marriage argue that people should have the same legal rights regardless of sexual orientation and that marriage may increase the stability of same-sex couples and lead to better physical and mental health for gays and lesbians. Those who oppose same-sex marriage contend that such unions are immoral, weaken our traditional notions of marriage, and are contrary to religious beliefs (Sullivan 1997; King and Bartlett 2006). *Table 9.3* provides a summary of some of the major pro and con arguments in this debate.

PORNOGRAPHY: EROTIC OR OBSCENE?

Pornography is the graphic depiction of images that cause sexual arousal. The images include photographs, videos (including those on the Internet), and other visual materials. The pornography industry, a broader concept, includes massage parlors, prostitution rings, stripping, saunas, live sex shows, street prostitution, escort services, peep shows, phone sex, international and domestic trafficking of children but especially girls and women, mail-order bride services, and prostitution tourism (Whisnant and Stark 2004).

pornography the graphic depiction of images that cause sexual arousal.

Trends

In the United States alone, pornography is a $15 billion industry. According to some estimates, 25 percent of Americans look at Internet pornography every day: Over 90 percent of the consumers are men, and despite the costs, they spend at least 6 hours a day viewing these sites. Among American college students, 20 percent of men view online pornographic materials every day or nearly every day, compared with only 3 percent of women (Paul 2005; Carroll et al. 2008).

Why Is Pornography Controversial?

American humorist Mason Cooley once quipped that pornography supplies the only really safe sex. However, most Americans feel that children, especially, should be protected from pornography, both as viewers and victims. Some people see adults' viewing of pornography as a harmless erotic recreation, while others denounce pornography as debasing and obscene.

Many proponents argue that "because men like to look at naked women, they will inevitably look at pornography." Among other things, some maintain, pornography augments the U.S. economy, enhances some men's sexual lives, and provides a safe outlet for men's sexual fantasies about oppressing women rather than resorting to violence. In addition, pornography can be lucrative for women who strip at expensive casinos and who act in popular adult films. Thus, pornography meets many people's, especially men's, needs. It is also legal under the First Amendment (freedom of speech) (Paul 2005; Urbina 2007).

Others maintain that pornography should be banned because it devalues and exploits women, increases violence against girls and women, and makes people more accepting of rape. Many feminists, especially, argue that there's no scientific evidence that pornography enhances men's lives or improves their sexual relationships. Instead, they believe that women in pornography are debased, exploited, and abused physically and psychologically. Also, according to some researchers, pornography use can increase men's risky sexual behavior, including having multiple sexual partners, which raises the chances of contracting sexually transmitted diseases, including HIV and AIDS (Paul 2005; Carroll et al. 2008; Dines 2008).

Jenna Jamison left home at 16 to live with her boyfriend, a tattoo artist. At 19, she gave up posing nude for magazines and stripping in clubs and began to act in pornographic films. At 31, she published a memoir, *How to Make Love Like a Porn Star*, which was on the *New York Times* best-seller list for 6 weeks. Her porn site, adult films, sex phone messages, and sex toys have made her a billionaire (Miller 2005).

5 Sociological Explanations of Gender Inequality and Sexuality

Paama Island in the South Pacific has a population of 600 people. A few years ago, the male leaders decided to ban women's wearing trousers "to make sure that western influence does not erode our cultural values." The ban was carried out by the all-male police force, even though many women protested that it violated the national constitution and their democratic rights ("Island Women . . ." 2002). Why can a handful of men dictate how women should dress?

Looking at sex and gender, gender inequality, sexual orientation, and controversial issues about sexuality through the lenses of the four major sociological perspectives provides a broader understanding of what sociologists perceive is at stake in these issues (*Table 9.4* on page 174 summarizes these perspectives).

FUNCTIONALISM

Functionalism, you'll recall from Chapter 1, is based on the assumption that society is a complex system of interrelated parts that work together to ensure a society's survival. A division of gender roles, especially within the family, helps society operate smoothly, and sexuality is important for reproduction.

Division of Gender Roles

Some of the most influential functionalist theories, developed during the 1950s, proposed that gender roles differ because women and men have distinct roles and responsibilities. A man (typically a husband and father) plays an *instrumental role* of procreator, protector, and provider. He must produce children, especially boys, to carry on his family name and, as a protector, must ensure his family's physical safety. The provider is competitive and works hard even if he is overwhelmed by multiple roles, such as the responsible breadwinner, the devoted husband, and the dutiful son (Parsons and Bales 1955; Gaylin 1992; Betcher and Pollack 1993).

A woman (typically a wife and mother) plays an *expressive role* by providing the emotional support and nurturance that sustain the family unit and support the father/husband. She should be warm, sensitive, and sympathetic, for example, comforting a husband who has had a bad day at work. As expressive role players, women also often act as the family helper, problem solver, and mediator. For functionalists, the instrumental and expressive roles are complementary. The duties are specialized, but both roles are equally important in meeting a family's needs and ensuring a society's survival.

Why Is Sexuality Important?

For functionalists, sexuality is critical for reproduction, but it should be expressed in an orderly way. For example, many cultures rely on arranged marriages to select the best and most responsible partners for young unmarried people. Such marriages ensure that the woman's sexual behavior will be confined to her husband, avoiding any doubt about the offspring's parentage.

For functionalists, sex should be practiced within marriage, primarily to encourage the formation of families. Sex outside of marriage is seen as dysfunctional because most unmarried fathers don't support their children, and the offspring often experience instability and poverty as well as a variety of emotional, behavioral, and academic problems (Seltzer 2004; Avellar and Smock 2005).

Critical Evaluation

The characteristics of the instrumental and expressive roles are useful in understanding some of the differences between traditional and nontraditional gender roles. In traditional roles, because each person knows what is expected of him or her, rights and responsibilities are clear. Men and women do not have to argue over who does what: If the house is clean, she is a "good wife"; if the bills are paid, he is a "good husband."

One of the criticisms of the functionalist perspective is that even during the 1950s, white middle-class male views ignored almost a third of the labor force that was composed of working class, immigrant, and minority women who played both instrumental and expressive roles. A related criticism is that functionalists tend to devalue unpaid labor, such as child care and housework, and see the labor as less important than provider roles. And as you saw earlier, even when women work outside the home, they do much of the housework, while men help with it, suggesting that women should be responsible for domestic obligations. In effect, then, according to some critics, functionalist views discourage greater equality between the sexes (Beeghley 2000).

Functionalism emphasizes mate selection and reproduction as contributing to a society's organization and stability over time and across cultures, but it frowns on sexual relationships outside of marriage. For example, even though cohabitation (living together) fulfills many people's sexual and intimacy needs, many functionalists see cohabitation as deviant because it rarely leads to marriage (Popenoe and Whitehead 2006). As you'll see in Chapter 13, however, marriage doesn't guarantee long or happy relationships.

CONFLICT THEORY

Conflict theorists see gender inequality as built into the social structure, both within and outside the home. In many developing nations, women perform the vast majority of the agricultural work, but in Uganda and other African countries, for example, they have virtually no property rights because only 8 percent of the women own the land they work on (Bakyawa 2002). In industrialized countries, few women penetrate the top ranks in economic, political, military, and other institutions. In effect, then, most women are still second-class citizens because men control most of society's resources. Like functionalists, conflict theorists see sexuality as a key component of a society's organization, but they view sexuality as reflecting and perpetuating gender inequality.

Capitalism and Gender Inequality

Unlike functionalists, conflict theorists maintain that capitalism, not complementary roles, explain gender roles and men's social and economic advantages. A full-time stay-at-home American mother would earn about $135,000 a year if paid for all her work (Wulfhorst 2006). In effect, traditional gender roles are profitable for business: Companies can require their male employees to work long hours or make numerous overnight business trips and don't have to worry about workers' demanding child-care services.

Most conflict theorists agree that all men are not equally privileged and that women in upper classes have more economic power, status, and prestige than men in lower classes. However, within social classes, men typically enjoy more power and control than women.

Is Sexuality Linked to Gender Inequality?

According to conflict theorists, gender inequality has many effects on sexuality. Most of the victims of domestic violence and rape, in the United States and around the world, are women and girls. In some societies, especially in the Middle East, men dictate how women should dress (even at universities), dismiss women's charges of sexual assaults (including gang rapes), punish real or imagined instances of women's, but not men's, marital infidelity, and blame girls for child rape because they're "seducing" older men (Neelakantan 2006).

Conflict theorists see men's domination of women, especially in economic terms, as a result of gender power differences. In sexual harassment cases, the offender is typically a male supervisor. In prostitution and sex trafficking, 80 percent of the victims are women and girls who live in poverty (Farr 2005).

Critical Evaluation

Conflict theory has been useful in showing how social structures reinforce men's domination in income, the division of domestic work, and access to valued resources. Some social scientists have criticized conflict theorists, however, for emphasizing male-female competition rather than cooperation. At home, female and male spouses or partners may reach agreement on who should do what ("I'll dust and vacuum while you shop for the groceries"). Outside the home, women and men often barter with their employers to get what they want ("I'll work late all this week if I can have Friday off"). A related criticism is that conflict theory emphasizes the differences between women and men rather than their common goals and similar attitudes. Much of our everyday life, after all, reflects consensus rather than power plays and a struggle for economic dominance.

Conflict theory helps us understand how power over women's sexual lives benefits men personally and economically, but it usually ignores the ways women use sexuality to exploit other women. During the 1980s, for example, a woman known as the Mayflower Madam (because of her higher-class background) ran an elite prostitution ring in New York City that catered to wealthy and powerful men. In other cases, women provide men with sexual favors in exchange for resources for themselves or their children (Barrows 1989; Rhoads 2004). Thus, men aren't the only ones who profit economically by exploiting women's sexuality.

FEMINIST THEORIES

Like conflict theorists, feminists see gender stratification as benefiting men and the capitalist system. They emphasize, however, that women's subordination also includes their daily vulnerability to violence: "If they are not getting harassed on the street, living in an abusive relationship, recovering from a rape, or in therapy to deal with the sexual abuse they suffered as children, [women] are ordering their daily lives around the *threat of men's violence*" (Katz 2006: 5). Most feminist scholars agree with conflict theorists that sexuality is linked to gender inequality, but they go further in arguing that male dominance is especially harmful because it results in men controlling women's sexual behavior.

Living in a Gendered World

All feminists (female and male) agree on three general points: (1) men and women should be valued equally, (2) women should have more control over their lives, and (3) gender inequality can be remedied by changing political, economic, family, and other institutions as well as everyday interactions, attitudes, and behaviors. Feminism has a number of branches, but let's look briefly at three explanations of gender roles.

Liberal feminism maintains that gender equality can be achieved through equal civil rights and equal opportunities. Although the sexes are more similar than different in their abilities, liberal feminists maintain that gendered socialization creates inequality by teaching girls and women to be passive and maternal and by tracking women and men into different educational and employment fields (Jagger and Rothenberg 1984).

Radical feminism contends that patriarchy is the major reason for women's oppression, especially men's control over women's bodies and their sexuality. Men assert their power in the form of rape, battering, exploiting women through prostitution and pornography, and sexual harassment. For radical feminists, all women are potential victims of sexual violence and all men are capable of such violence. In this sense, patriarchy is more critical than social class in giving men power over women's lives (MacKinnon 1982; Bart and Moran 1993).

Multiracial feminism (sometimes referred to as *racial ethnic feminism* or *multicultural feminism*) maintains that gender, race, and social class intertwine to form a hierarchical stratification system that shapes women's and men's attitudes, experiences, and behavior. Privileged women have less status than privileged men, but upper-class white men *and* women subordinate lower-class women *and* minority men. In this sense, multiracial feminism addresses the social subordination of both sexes and the oppression that is due not only to race, class, and gender but also to sexual orientation and nationality (Collins 1990; Lorber 2005).

Is Sexuality Linked to Social Control?

Feminist theorists often point to sexuality as the root of inequality between women and men, both interpersonally and within economic and other institutions. Among other things, feminist scholars have demanded an end to sexual slavery and have called attention to the prevalence of marital rape. In effect, feminist thinkers have insisted that as long as men control women's bodies and behavior, women's sexual well-being will be inferior to men's (hooks 2000).

Feminist scholars emphasize that some sexual groups have power over others. Even though all states have laws against hate crimes, for example, police in many cities treat lesbians, gay men, bisexuals, and transgendered people more brutally than heterosexuals. Feminist theories have also gone further than conflict theory in documenting men's control of women's sexuality across cultures and over time. Examples include forcing women to marry against their will and *honor killings*—murders committed by male members of a family if they believe that a female relative has brought shame on the family by refusing an arranged marriage, being the victim of a sexual assault, seeking divorce, or committing adultery (Enloe 2004; Amnesty International 2006).

Critical Evaluation

Feminist perspectives have provided insightful analyses of gender roles and gender inequality, especially in challenging functionalist views of instrumental and expressive roles as outdated and as reinforcing sexism and patriarchy. Feminist perspectives are not without their weaknesses, however. *Liberal feminism* typically emphasizes women's but not men's oppression in capitalistic societies. *Radical feminism* has been criticized for being too narrow. For example, men are not universally violent or sexually exploitative; many, in fact, respect women and teach their sons to do the same. *Multiracial feminism* is very inclusive, but this strength can also be a weakness. When it comes to economic issues like the gender pay gap, for example, which variable is the most important—race, ethnicity, or social class? Because gender inequality is seen as having interlocking causes, it's difficult for multiracial feminists to agree on which factors carry more weight and whether some should be given a priority in implementing social change.

Feminist theorists, more than any other group, have challenged male sexual control and heterosexism, but the theories have several weaknesses. In focusing on men's sexual domination, many feminist researchers still neglect racial, ethnic, religious, age, and social class differences among women and men. Sometimes feminist scholars ignore women's sexual oppression by other women, especially when women promote and participate in global sex trafficking (Dietrich 2007). Despite their insistence on mutual respect, some feminists see themselves as having sexual value only when they are objects of male desire (Lorber 2005; Calasanti

For feminist scholars, some popular TV shows have sunk to an all-time low in their portrayal of women. In *America's Next Top Model*, for example, the contestants pretended to be dead by drowning, shooting, electrocution, and even decapitation. In a similarly objectionable episode, models were shown in coffins, as pictured here. If such shows demean females, why are they so popular among women?

and Slevin 2006). Also, feminist theories tend to minimize the importance of sexuality that reflects love and affection rather than only power.

SYMBOLIC INTERACTIONISM

While functionalist, conflict, and some feminist theories are macro level, symbolic interactionists focus on the everyday processes that produce and reinforce gender roles. Most of us accommodate our behavior to gender role expectations by "doing gender" (West and Zimmerman 1987). We do gender, sometimes consciously and sometimes unconsciously, by adjusting our behavior and our perceptions depending on the sex of the person with whom we're interacting, such as voting for a man to chair a committee but a woman to be the secretary. Our expression of our sexuality, similarly, is not inborn but reflects socialization processes and what families and other societal groups construct as appropriate and inappropriate behavior (Hubbard 1990).

Gender Inequality as a Social Construction

For symbolic interactionists, society is *socially constructed* (see Chapter 1). Because our view of reality is what people agree it is, gender is a social creation, and gender inequality is not shaped by social structures as much as by learning gender roles through daily social interaction. That is, "people act on the basis of their perceptions of equality if and when equality is a relevant concern for them" (Harris 2006: 8). For example, religious couples might try to be equal only in ways that matter to them—such as equality of love or respect—while also interpreting literally the Bible's injunction that "wives should submit to their husbands."

In 2007, the Vienna City Council in Austria launched a campaign to change people's perceptions of gender roles, with one feature being public restroom signs that showed both sexes for a baby changing station. The Council received hundreds of supportive e-mails and calls, but well over 3,000 complaints—about 80 percent of them from men who said the campaign was a waste of money (Brickner 2007). For symbolic interactionists, symbols (such as signs) can stimulate change or evoke hostility.

How Do Societies Construct Sexuality?

For symbolic interactionists, sexuality, like gender, is socially constructed through human interpretation. As with gender roles, we *learn* to be sexual and to express our sexuality differently depending on the culture in which we live. There is no natural human sexuality. Instead, the society in which we live channels, guides, and limits our sexual behavior. For example, out-of-wedlock births are common in most Western countries not because Americans and Europeans are innately more sexual but because these societies don't stigmatize out-of-wedlock births. By contrast, in many nations a man would not marry a woman with an out-of-wedlock child because of the belief that only promiscuous women have sexual relations outside of marriage (Hubbard 1990; Ore 2006).

Sexual behavior, because it's constructed socially, can and does change over time. Many Americans have become more comfortable with categories such as bisexual and even with men who sometimes have sex with other men but see themselves as straight. A number of symbols have emerged over the years that denote sexual orientation and signal the growing recognition of sexual relationships that are not exclusively heterosexual (see *Figure 9.3*).

Critical Evaluation

Symbolic interactionist theories have been valuable in explaining how gender and gender roles shape our everyday lives. Through both verbal and nonverbal communication, we determine what behavior is appropriate or not. We can also change expected gender roles, as when women enter the work force and men take more responsibility for domestic tasks and caring for children. Despite their contributions, symbolic interactionists have been criticized by some sociologists

FIGURE 9.3
Sex and Sexual Orientation Symbols

People attach meaning to symbols and use them to communicate with others. Here are some common symbols that denote sex and sexual orientation. From left to right, the symbols designate female, male, lesbian, bisexual, gay male, and transgendered.

for ignoring the social structures that create and maintain gender inequality such as military policies that dictate whether or not women can participate in combat, an important criterion in military promotions.

Symbolic interactionists have enhanced our understanding of sexuality by examining the ways in which behavior is socially constructed and why, consequently, there are cross-cultural variations in sexual attitudes and practices. However, symbolic interactionism doesn't explain why siblings, even identical twins, have different sexual orientations although they were socialized similarly. A related weakness is that symbolic interactionists don't explain why, historically and currently, women around the world are considerably more likely than men to be subjected to sexual control and exploitation. Such analyses require macro-level analyses that examine family, political, and economic institutions.

TABLE 9.4
Sociological Explanations of Gender Inequality and Sexuality

THEORETICAL PERSPECTIVE	LEVEL OF ANALYSIS	KEY POINTS
Functionalist	Macro	• Gender roles are complementary and equally important. • Agreed-upon sexual norms contribute to a society's order and stability.
Conflict	Macro	• Gender roles give men power to control women's lives instead of allowing the sexes to be complementary and equally important. • Most societies regulate women's, not men's, sexual behavior.
Feminist	Macro and micro	• Women's inequality reflects their historical and current domination by men, especially in the workplace. • Many men use violence—including sexual harassment, rape, and global sex trafficking—to control women's sexuality.
Symbolic Interactionist	Micro	• Gender inequality is a social construction that emerges through day-to-day interactions and reflects people's gender role expectations. • The social construction of sexuality varies across cultures because of societal norms and values.

SOC was built on a simple principle: to create a new teaching and learning solution that reflects the way today's faculty teach and the way you learn.

Through conversations, focus groups, surveys, and interviews, we collected data that drove the creation of the current version of SOC that you are using today. But it doesn't stop there—in order to make SOC an even better learning experience, we'd like you to SPEAK UP and tell us how SOC worked for you. What did you like about it? What would you change? Are there additional ideas you have that would help us build a better product for next semester's sociology students?

At **4ltrpress.cengage.com/soc** you'll find all of the resources you need to succeed in principles of sociology—
Printable and Interactive Flash Cards, Interactive Quizzing, Study Worksheets, Note-Taking Outlines, Games, Videos & Review Questions, and more!

Speak Up! Go to **4ltrpress.cengage.com/soc**.

America's multicultural

umbrella includes about 150 distinct ethnic or racial groups.

what do you think?

I consciously think of myself as belonging to a particular race.

1 2 3 **4** 5 6 **7**

strongly agree   strongly disagree

10

Race and Ethnicity

Two researchers sent almost 5,000 fictitious résumés in response to advertisements for a variety of jobs in Chicago and Boston. They randomly assigned either a stereotypically African-American name (Lakisha Washington or Jamal Jones) or a name that sounded typically white (Emily Walsh or Greg Baker) to both high-quality and low-quality résumés. The results were startling:

- Applicants with white-sounding names received 50 percent more calls for interviews. In fact, a white-sounding name yielded as many more calls as an additional 8 years of work experience for an applicant with a black-sounding name.

- Applicants with black-sounding names received fewer calls across all occupations (from managers to clerical workers) and in all industries (communications, manufacturing, finance, and social services).

- Employers who posted "Equal Employment Opportunity Employer" in their ad discriminated as much as other employers (Bertrand and Mullainathan 2003).

As this example shows, some of us enjoy more opportunities than others simply because of the color of our skin. The situation has improved during the last 50 years or so, but not as much as most people think. This chapter examines the significance of race and ethnicity, their impact on Americans' lives, and why racial-ethnic inequality is still widespread.

Key Topics

In this chapter, we'll explore the following topics:

1 Racial and Ethnic Diversity in America

What do you call a person who speaks three languages? Multilingual.
What do you call a person who speaks two languages? Bilingual.
What do you call a person who speaks one language? American.

As this joke suggests, many people stereotype the United States as a single-language and single-culture society. In fact, it's the most multicultural country in the world, a magnet that draws people from hundreds of nations and that is home to millions of Americans who are bilingual or multilingual.

One in five Americans is either foreign-born or a first-generation resident. America's multicultural umbrella includes about 150 distinct ethnic or racial groups among more than 305 million inhabitants (U.S. Census Bureau 2008). By 2025, only 58 percent of the U.S. population is projected to be white—down from 86 percent in 1950 (see *Figure 10.1*). By 2050—just a few generations away—whites may make up only half of the total population because Latino and Asian populations are expected to triple in size (Population Division, U.S. Census Bureau 2008).

Multiracial groups are also growing. The number of Americans who identify themselves as being two or more races is projected to more than triple—from 5.1 million in 2008 to 16.2 million in 2050 (2 percent and almost 4 percent of the total population, respectively (U.S. Census Bureau News 2008).

2 The Significance of Race and Ethnicity

Oprah Winfrey once commented that no matter how successful and wealthy she is, her experiences differ from those of her white male counterparts because she's still "only" a black woman. All of us identify with some groups and not others in terms of sex, age, social class, and other factors. Two of the most common and influential sources of self-identification as well as labeling by others are race and ethnicity.

RACE

Race refers to a group of people who share physical characteristics, such as skin color and facial features, that are passed on through reproduction. Race is a *social construct,* a societal invention, that labels people based on physical appearance. Possibly only 6 out of the human body's estimated 35,000

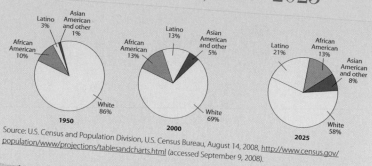

FIGURE 10.1
Racial and Ethnic Composition of the U.S. Population, 1950–2025

1950
- Latino 3%
- Asian American and other 1%
- African American 10%
- White 86%

2000
- Latino 13%
- Asian American and other 5%
- African American 13%
- White 69%

2025
- Latino 21%
- African American 13%
- Asian American and other 8%
- White 58%

Source: U.S. Census and Population Division, U.S. Census Bureau, August 14, 2008, http://www.census.gov/population/www/projections/tablesandcharts.html (accessed September 9, 2008).

genes determine the color of a person's skin. Because all human beings carry 99.9 percent of the same genetic material (DNA), the "racial" genes that makes us look different are miniscule compared with the genes that make us similar (Graves 2001; Pittz 2005).

If our DNA is practically identical, why are we so obsessed with race? People react to the physical characteristics of others, and those reactions have consequences. Skin color, hair texture, and eye shape, for example, are easily observed and mark groups for unequal treatment. As long as we sort others into racial categories and act on the basis of these characteristics, our life experiences will differ in terms of access to jobs and other resources, how we treat people, and how they treat us (Duster 2005).

There's much variation in skin color across and within groups. People of African descent have at least 35 different shades of skin tone (Taylor 2003).

© Comstock Images/Jupiterimages

ETHNICITY

An **ethnic group** (from the Greek word *ethnos*, meaning "nation") is a set of people who identify with a common national origin or cultural heritage that includes language, geographic roots, food, customs, traditions, and/or religion. Ethnic groups in the United States, for example, are Puerto Ricans, Chinese, Serbs, Arabs, Swedes, Hungarians, Jews, and many others. Like race, ethnicity can be a basis for unequal treatment, as you'll see shortly.

RACIAL-ETHNIC GROUP

A group of people who have both distinctive physical and cultural characteristics is a **racial-ethnic group.** Some people use the terms *racial* and *ethnic* interchangeably, but remember that *race* refers to physical characteristics with which we are born, whereas *ethnicity* describes cultural characteristics that we learn. The term *racial-ethnic* incorporates both physical and cultural traits.

Describing racial-ethnic groups has become more complex because the U.S. government allows people to identify themselves in terms of both race and ethnicity. In the 2000 Census, for example, a Latino/a could check off "Black" for race and "Cuban" for ethnic origin. Such choices generate dozens of racial-ethnic categories.

Most people maintain that they are color-blind, but we learn at an early age that some physical attributes are more valued than others. A notable example is light-colored skin, which confers *white privilege,* the advantages that white people enjoy simply because they happen to belong to a particular category. Most white people don't feel privileged because they are not wealthy. Nonetheless, they have everyday benefits that they take for granted, such as not being followed by store detectives when they shop and being able to buy products, like greeting cards, dolls, toys, and children's magazines, that represent their own race (McIntosh 1995; Johnson 2008; Rothenberg 2008).

3 Our Immigration Mosaic

the current proportion of foreign-born U.S. residents is 11 percent, compared with 15 percent in 1900. Since the turn of the twentieth century, however, there has been a significant shift in immigrants' country of origin. In 1900, almost 85 percent of immigrants came from Europe; now immigrants come primarily from Asia (mainly China and the Philippines) and Latin America (mainly Mexico) (Martin and Midgley 2006).

Another major change has been the rise of unauthorized or undocumented (illegal) immigrants—from 180,000 in the early 1980s to 12 million in 2006, representing one-third of all foreign-born U.S. residents. About 56 percent of undocumented immigrants are Mexican, 22 percent are from elsewhere in Latin America, and 22 percent are from other countries (Capps et al. 2007; Hoefer et al. 2007).

> **ethnic group** a set of people who identify with a common national origin or cultural heritage that includes language, geographic roots, food, customs, traditions, and/or religion.
>
> **racial-ethnic group** a group of people who have both distinctive physical and cultural characteristics.

ATTITUDES ABOUT IMMIGRATION

Most Americans are ambivalent about immigration: 40 percent say that immigrants have made the country better in terms of food, music, and the arts, but a majority feels that immigrants have increased taxes and crime. African Americans, especially, contend that immigration has decreased job opportunities for themselves or their family (Kohut et al. 2007). Generally, however, attitudes about legal immigration vary depending on the social context. Shortly after the 9/11 terrorist attacks, for example, 58 percent of Americans favored cutbacks in immigration compared with 39 percent in 2008 (Jones 2008).

Illegal immigration is more controversial, but attitudes vary, as in the case of legal immigration, depending on the importance of other issues. In 2006, for instance, 53 percent of Americans contended that illegal immigrants should be required to go home. In 2008—as many Americans focused on other issues like a struggling economy and record-high gas and housing prices—illegal immigration issues faded from public consciousness (Barabak 2006; Jones 2008).

IS IMMIGRATION HARMFUL OR BENEFICIAL?

José Tenas came to the United States from Guatemala in 1987. He spent 16 days crossing through Mexico on foot and rode illegally into California in the sweltering back of a truck. Tenas washed dishes in kitchens, became a U.S. citizen, brought over his wife and children, saved his money, got a loan, and opened his own restaurant in Virginia, which is now popular with local native-born residents and Latino immigrants (Williams 2005). Tenas represents one of many illegal immigrants who have worked hard to become successful. Still, the general public and some scholars feel that all immigration has costs.

Some immigration critics allege that, regardless of legal status, low-skilled immigrants reduce the standard of living and overload schools and welfare systems. Others note that because immigrants are younger, poorer, and less well educated than the native population, they use more government services and pay less in taxes. As a result, many Americans feel that immigrants don't share equally in the expenses they incur, especially at the local level (Bean and Stevens 2003).

Proponents argue that many immigrants, like José Tenas, provide numerous economic benefits for their host countries. They clean homes and business offices, toil as nannies and busboys, serve as nurses' aides, and pick fruit—all at low wages and in jobs that most American-born workers don't want. Accounting for 5 percent of the 148 million U.S. workers, unauthorized immigrants comprise 24 percent of all farm workers, 17 percent of the work force in cleaning occupations, and almost 14 percent of those in the construction and food preparation industries (Passel 2006; Reynolds 2007).

Many scholars argue that, on balance and in the long run, immigrants bring more benefits than costs. For example, they constitute an important labor force for an aging (and primarily white) American population that will require many workers to support Social Security and Medicare payments for the elderly. With-out large numbers of new workers, many funding programs for the elderly will be cut or decreased, requiring millions of older Americans to work well into their 70s (Bean and Stevens 2003; Myers 2007).

4 Dominant and Minority Groups

t he president of the Boston City Council tried, unsuccessfully, to strike the word *minority* from the official city documents because, he argued, the term is insulting and inaccurate in a city whose population is 51 percent people of color (Wiltenburg 2002). For sociologists, *minority group* and *dominant group* are descriptive terms that have little to do with a group's size.

WHAT IS A DOMINANT GROUP?

A **dominant group** is any physically or culturally distinctive group that has the most economic and political power, the greatest privileges, and the highest social status in a society. As a result, it can treat other groups as subordinate. For example, in most societies, men are a dominant group because they have more status, resources, and power than women (see Chapter 9).

Dominant groups aren't necessarily the largest groups in society. From the seventeenth century until 1994, about 10 percent of the population in South Africa was white and had almost complete control of the black population. Because of **apartheid**, a formal system of racial segregation, the black inhabitants couldn't vote, lost their property, and had minimal access to education and politics. Apartheid ended in 1994, but most black

Guest workers, who are permitted to work in the United States on a temporary basis because of labor shortages, especially in agriculture, often live in little more than shacks (1). Arizona lettuce farmers, some of whom pay their guest workers from Mexico up to $8.50 an hour, don't have enough field hands to harvest the crop (2). Undocumented immigrants, most of them Latinos, comprised a quarter of the construction workers who helped rebuild parts of New Orleans after Hurricane Katrina (3). Many of these construction workers made an average of $6.50 an hour less than legal workers and had more trouble collecting their wages (Fletcher et al. 2006).

South Africans are still in a minority group because whites "hold the best jobs, live in the most expensive homes, and control the bulk of the country's capital" (Murphy 2004: A4).

WHAT IS A MINORITY GROUP?

Sociologists describe Latinos, African Americans, Asian Americans, Middle Eastern Americans, and American Indians as minority groups. A **minority group** is a group of people who may be subject to differential and unequal treatment because of their physical, cultural, or other characteristics, such as gender, sexual orientation, religion, ethnicity, or skin color. Minority groups may be larger than a dominant group, but they have less power, privilege, and social status. For example, American minorities have fewer choices than dominant group members in finding homes and apartments because they are less likely to get help—from either a real estate agent or a bank—with the intricacies of mortgage financing that most people need, or to get the same information about housing possibilities (Massey 2007).

PATTERNS OF DOMINANT-MINORITY GROUP RELATIONS

To understand some of the complexity of dominant-minority group relations, think of a continuum. At one end of the continuum is genocide; at the other end of the continuum is pluralism (see *Figure 10.2*).

Genocide is the systematic effort to kill all members of a particular ethnic, religious, political, racial, or national group. By 1710, for example, the colonists in America had killed thousands of Indians in skirmishes, poisoned others, and promoted scalp bounties. And, in 1851, the governor of California officially called for the extermination of all Indians in the state (de las Casas 1992; Churchill 1997).

One of the most widely cited examples of genocide is the massacre of 6 million Jews in Nazi Germany during World War II. Although horrendous, the Holocaust is only one case of genocide. As *Table 10.1* on page 182 shows, well over 63 million people have been victims of genocide during the twentieth century alone.

Internal colonialism refers to the unequal treatment and subordinate status of groups within a nation. Sociologist Robert Blauner (1969, 1972) described the position of African Americans as an "internal colony" because blacks entered the United States involuntarily as slaves, were controlled by the dominant group, and were exploited economically and sexually (rapes of slave women, for instance).

Internal colonialism often manifests itself in **segregation**, the physical and social separation of dominant and minority groups. In 1954, the Supreme Court's ruling in the *Brown vs. the Board of Education* case declared *de jure*, or legal, segregation unconstitutional

> **minority group** a group of people who may be subject to differential and unequal treatment because of their physical, cultural, or other characteristics, such as gender, sexual orientation, religion, ethnicity, or skin color.
>
> **genocide** the systematic effort to kill all members of a particular ethnic, religious, political, racial, or national group.
>
> **internal colonialism** the unequal treatment and subordinate status of groups within a nation.
>
> **segregation** the physical and social separation of dominant and minority groups.

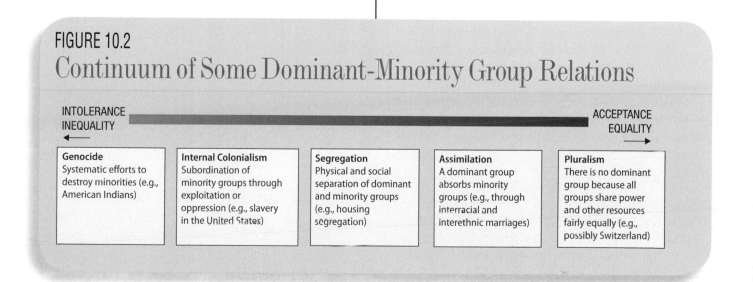

FIGURE 10.2
Continuum of Some Dominant-Minority Group Relations

INTOLERANCE INEQUALITY ← → ACCEPTANCE EQUALITY

Genocide
Systematic efforts to destroy minorities (e.g., American Indians)

Internal Colonialism
Subordination of minority groups through exploitation or oppression (e.g., slavery in the United States)

Segregation
Physical and social separation of dominant and minority groups (e.g., housing segregation)

Assimilation
A dominant group absorbs minority groups (e.g., through interracial and interethnic marriages)

Pluralism
There is no dominant group because all groups share power and other resources fairly equally (e.g., possibly Switzerland)

and was followed by the passage of a variety of federal laws that prohibited racial segregation in public schools as well as discrimination in employment, voting, and housing.

De facto, or informal, segregation has replaced *de jure* segregation in the United States. Some *de facto* segregation may be voluntary, as when members of racial or ethnic groups prefer to live among their own group. In most cases, however, *de facto* segregation is due to discrimination, as when realtors steer minorities away from white neighborhoods. Such residential segregation deprives minorities of access to quality schools, retail stores, leisure activities, and jobs that are burgeoning in the suburbs (Farley and Squires 2005).

Many minority-group members blend into U.S. society through **assimilation**, the process of conforming to the culture of the dominant group (by adopting its language and values) and intermarrying with that group. Some feel that many newcomers are assimilating less than in the past. For example, many Mexican immigrants aren't learning English, and Cubans have trans-

formed much of Miami, Florida, into an ethnic enclave that tends to exclude whites and African Americans (Huntington 2004).

Others maintain that immigrants who arrived after 1995 are assimilating more rapidly than their predecessors, as witnessed by higher rates of citizenship, service in the military, home ownership, ability to speak English, and intermarriage. However, the degree of assimilation varies by country of origin, educational level, and the legal status of the immigrant. Mexicans, for example, are considerably less likely to be assimilated (than immigrants from the Philippines, Vietnam, or South Korea) because they are more likely to have entered the country illegally, which cuts off the possibility of getting a good job, an education (especially for migrant workers), and becoming a citizen. Still, although Latino immigrants speak little English in the first generation, English dominates the second generation, and Spanish fades in the third generation (Hakimzadeh and Cohn 2007; Vigdor 2008).

Assimilationists want to eliminate racial and ethnic boundaries, while pluralists seek to maintain them. **Pluralism**, sometimes called *multiculturalism*, is a situation in which minority groups retain their culture but have equal social standing in a society. One historian describes the United States as a pluralistic country because it is multicultural, multicolored, and multilin-

TABLE 10.1
Twentieth-Century Genocide

EVENT	YEARS	ESTIMATED NUMBER OF DEATHS
Turkish government's massacre of Armenians living in Turkey	1915–1918	1.5 million
Joseph Stalin's (Russian dictator) massacre of nearly 25 percent of the population of Ukraine and of many other Eastern European countries (including Latvia, Lithuania, and Estonia)	1932–1953	50+ million
Japanese Imperial Army's murder of inhabitants of China's capital city, Nanking	1937–1938	300,000
Nazi Holocaust in Germany	1938–1945	6 million
Khmer Rouge leader Pol Pot's slaughter of Cambodians	1975–1979	2+ million
Murders of Muslims by Serb majority in former Yugoslavia	1992–1995	200,000+
Massacres of minority Tutsis by dominant Hutu group in Rwanda, Africa	1994 (nine months)	800,000

The total number of people killed in these countries equals about 22 percent of the current U.S. population.
Source: United Human Rights Council 2004, http://www.unitedhumanrights.org (accessed April 29, 2004).

gual. Thus, an American might have diverse neighbors like a German architect and his Iranian wife, a Palestinian contractor, a Korean scientist, and a car salesman from Madagascar (Karnow 2004).

5 Sources of Racial-Ethnic Friction

California Governor Arnold Schwarzenegger remarked that the lone Latina Republican lawmaker in California had a very fiery personality because of her ethnicity, and on the television program *The View*, Rosie O'Donnell mocked how Chinese-Americans speak English (Bonisteel 2006; Salladay 2006). Such public comments, especially by very visible people, suggest that racism is still common in U.S. culture.

RACISM

Racism is a set of beliefs that one's own racial group is naturally superior to other groups. It is a way of thinking about racial and ethnic differences that justifies and preserves the social, economic, and political interests of powerful groups (Essed and Goldberg 2002).

Some social scientists have suggested that African Americans are born with lower average intelligence than other groups, especially whites (Herrnstein and Murray 1994). Human intelligence is a product of both genetic and environmental influences, and there is no evidence that one cultural group is more intelligent than another (Montagu 1999). Nonetheless, such racist beliefs fuel prejudice and discrimination.

PREJUDICE

Prejudice is an attitude, positive or negative, toward people because of their group membership. We often prejudge those who are different from us in race, ethnicity, or religion ("Asians are really hard workers" or "White people can't be trusted"). Prejudice is not one-sided because *anyone* can be prejudiced, use stereotypes, and exhibit ethnocentric behavior.

A **stereotype** is an oversimplified or exaggerated generalization about a category of people (see, for example, stuffwhitepeoplelike.com). Stereotypes can be positive ("All African Americans are athletic") or negative ("All African Americans are lazy"). Whether positive or negative, stereotypes distort reality. Some blacks are great athletes and some are lazy, just like people in other groups. Once established, however, stereotypes are difficult to eliminate because people often dismiss any evidence to the contrary as an exception ("For an Asian, she's really athletic").

Ethnocentrism is the belief that one's own culture, society, or group is inherently superior to others. If we are ethnocentric, we reject those not from our group as strange, deviant, and inferior: "*Our* customs, *our* laws, *our* food, *our* traditions, *our* music, *our* religion, *our* beliefs and values, and so forth, are somehow better than those of other societies" (Smedley 2007: 32).

Stereotypes and ethnocentrism can result in a displacement of anger and aggression on **scapegoats**, individuals or groups whom people blame for their own problems or shortcomings ("They didn't hire me because the company wants blacks" or "I didn't get into that college because Asians Americans are at the top of the list"). Minorities are easy targets for scapegoating because they typically differ in physical appearance and are usually too powerless to strike back (Allport 1954; Feagin and Feagin 2008). At one time or another, almost all newcomers to the United States have been scapegoats. Especially in times of economic hardship, the most recent immigrants often become scapegoats ("Latinos are replacing Americans in construction jobs"). Stereotypes, ethnocentrism, and scapegoating may be limited to attitudes, but they often lead to discrimination.

DISCRIMINATION

Discrimination is any act that treats people unequally or unfairly because of their group membership. Discrimination encompasses all sorts of actions, ranging from social slights (such as not inviting minority co-workers to lunch) to rejection of job applications and hate crimes. Discrimination can be subtle (such as not sitting next to someone) or blatant (such as making racial slurs), and it occurs at individual and institutional levels.

racism a set of beliefs that one's own racial group is naturally superior to other groups.

prejudice an attitude, positive or negative, toward people because of their group membership.

stereotype an oversimplified or exaggerated generalization about a category of people.

ethnocentrism the belief that one's own culture, society, or group is inherently superior to others.

scapegoats individuals or groups whom people blame for their own problems or shortcomings.

discrimination any act that treats people unequally or unfairly because of their group membership.

Individual discrimination is harmful action directed intentionally on a one-to-one basis by a member of a dominant group against a member of a minority group. In a recent survey, for example, more than half of blacks said that they face everyday discrimination when eating in restaurants, shopping, renting an apartment, buying a house, or applying for a job (Kohut et al. 2007).

Individual discrimination also occurs *within* racial-ethnic groups. In one national poll, an overwhelming 83 percent of Latinos said that they had experienced discrimination from other Latinos. The most recent immigrants, especially Colombians and Dominicans, said that they had encountered employment discrimination by U.S.-born Latinos (Brodie et al. 2002).

In **institutional discrimination**, minority-group members experience unequal treatment and opportunities as a result of the everyday operations of a society's laws, rules, policies, practices, and customs. Institutional discrimination is widespread. In health care, for instance, minorities are less likely than whites to receive appropriate heart medicine, undergo bypass surgery, and receive transplants and cancer treatment. Minorities also tend to receive lower-quality care than whites, even when they have private health insurance and are treated by the same doctors (*National Healthcare Disparities Report* 2003; Sequist et al. 2008).

RELATIONSHIP BETWEEN PREJUDICE AND DISCRIMINATION

Nearly 60 years ago, sociologist Robert Merton (1949) described the relationship between prejudice and discrimination that is still useful today. Merton's model includes four types of people and their possible response patterns (see *Figure 10.3*).

Unprejudiced nondiscriminators, or "all-weather liberals," as Merton called them, are neither prejudiced nor do they discriminate. They sincerely believe in the American creed of freedom and equality for all and cherish egalitarian values, but don't do much, individually or collectively, to change discrimination. In contrast, *prejudiced discriminators* are "active bigots" who are consistent in attitude and action: They are prejudiced and discriminate and are willing to break laws to express their beliefs and protect their vested interests.

Sociologists consider unprejudiced discriminators and prejudiced nondiscriminators more interesting because they are more common, but their inconsistent attitudes and actions are more elusive to study. *Unprejudiced discriminators* are "fair-weather liberals" who aren't prejudiced but discriminate because it's expedient or in their own self-interest to do so. If, for example, an insurance company charges higher automobile insurance premiums to people with low occupational and educational levels (as is often the

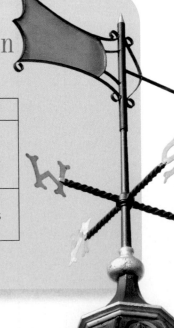

© Junghee Choi/iStockphoto.com

FIGURE 10.3
Relationships between Prejudice and Discrimination

IS THE PERSON PREJUDICED?	DOES THE PERSON DISCRIMINATE?	
	Yes	**No**
Yes	Prejudiced discriminator (e.g., a prejudiced person who attacks minority-group members verbally or physically)	Prejudiced nondiscriminator (e.g., a prejudiced person who goes along with equal employment opportunity policies)
No	Unprejudiced discriminator (e.g., an unprejudiced person who joins a club that excludes minorities)	Unprejudiced nondiscriminator (e.g., an unprejudiced employer who hires minorities)

Source: Based on Merton 1949.

case for many minorities), agents will implement these policies even though they themselves aren't prejudiced (Florida Office of Insurance Regulation 2007). *Prejudiced nondiscriminators* are "timid bigots" who are prejudiced but don't discriminate. Despite their negative attitudes, they hire minorities and are civil in everyday interactions because they feel that they must conform to anti-discrimination laws or situational norms. If, for example, most of their neighbors or co-workers don't discriminate, prejudiced nondiscriminators will go along with them.

6 Sociological Explanations of Racial-Ethnic Inequality

as with other topics, functionalism, conflict theory, feminist theories, and symbolic interactionism help us understand racial-ethnic relations (*Table 10.2* on p. 186 summarizes the key points of each perspective). Each has particular strengths and weaknesses.

FUNCTIONALISM

Those who criticize immigrants for not becoming Americanized quickly enough reflect a functionalist view of racial-ethnic relations. That is, if a society is to work harmoniously, newcomers must assimilate by adopting the dominant group's values, goals, and, especially, language. Otherwise, the society will experience discord and conflict.

Stability and Cohesion

Functionalists note that racial-ethnic inequality sustains a pool of cheap labor for jobs that require little or no training. Many farmers rely on immigrants to work in fields, orchards, and vineyards at low wages; otherwise, many crops, including lettuce, apples, and grapes would rot in the fields, leading to farmers' economic ruin and increased cost of food (Bustillo 2006). Thus, dominant group members benefit from inequality by avoiding undesirable jobs while providing employment to people with low educational or skill levels.

Racial-ethnic inequality also maintains or increases many dominant group members' current status, power, and profits. Keeping minorities from acquiring decent jobs as well as education and housing makes more of these scarce resources available to the white population. In inner-city neighborhoods, fringe bankers (such as check cashers and pawnshops) profit by offering financial services to low-income residents who can't get bank loans or credit cards (Gans 2005; Squires and Kubrin 2006). From a functionalist perspective, fringe bankers enjoy huge profits (because they charge about 33 percent a *month* in interest versus about 6 percent a *year* in traditional banks), but they also provide financial stability in poor neighborhoods where people have few borrowing alternatives (see Mahon 2004).

Functionalists acknowledge that discrimination can be dysfunctional. For example, racism prevents a society from recognizing or rewarding talented people who could make important contributions. And, recently, anti-Arab visa policies have cost the United States billions in lost tourism and student tuitions to colleges and universities (Smith and Hsu 2007).

Critical Evaluation

Functionalism is useful in understanding how inequality benefits some groups over others and why, as a result, it persists. Functionalists acknowledge that inequality can be dysfunctional, but this is not their major focus. By emphasizing that racial-ethnic inequality maintains a society's stability, functionalists seem to accept the inequality as inevitable. Another weakness is that functionalists tend to gloss over the fact that racial-ethnic inequality can lead to tension and violence (Chasin 2004).

CONFLICT THEORY

Conflict theorists see ongoing strife between dominant and minority groups. Dominant groups try to protect their power and privilege, while subordinate groups struggle to gain a larger share of societal resources. Once a system of racial oppression is in place (for example, through segregation), racial hierarchies are supported and perpetuated through economic inequality and the resulting social stratification.

Economic and Social Class Inequality

For many conflict theorists, economic inequality generates racial-ethnic inequality. For example, jobs in the *primary labor market,* held primarily by white workers, provide better wages, health and pension benefits, and some measure of job security. In contrast, workers in the *secondary labor market,* who are largely minorities (such as fast food workers), are easily replaced.

Their wages are low, there are few fringe benefits, and working conditions are generally poor (Doeringer and Piore 1971; Bonacich 1972).

Such economic stratification pits minorities against each other and low-income whites. Because these groups compete with each other instead of uniting against exploitation, capitalists don't have to worry about increasing wages or providing safer work environments. Other conflict theorists maintain that race is a more important factor than social class because even middle-class African Americans (and those in other racial-ethnic groups) experience severe discrimination on a daily basis that reminds them of their subordinate place in U.S. society (Feagin and Sikes 1994).

Watch or rewatch the Oscar-winning movie **Crash.** *How does the movie portray some of the benefits and costs of racism, prejudice, and discrimination for both dominant and minority groups?*

Critical Evaluation

Conflict theories are valuable in explaining why discrimination occurs and persists as when members of privileged groups benefit by subordinating minorities economically. However, conflict theorists often assume that inequality is typically conscious, deliberate, widespread, and inescapable. In contrast, a majority of African Americans attribute the differences between poor and middle classes to a "values gap": 53 percent of African Americans say that black people who don't get ahead are mainly "responsible for their own condition," while only 30 percent blame racial discrimination (Carroll 2007; Kohut et al. 2007). Such data suggest a substantial consensus among blacks that racial inequality is not due entirely to racism. Another weakness is that many conflict theorists tend to focus primarily on men, a limitation that feminist theories address.

FEMINIST THEORIES

Walk through almost any hotel, large discount store, nursing home, or fast-food restaurant in the United States. You'll notice two things: Most of the low-paid

TABLE 10.2
Sociological Explanations of Racial-Ethnic Inequality

THEORETICAL PERSPECTIVE	LEVEL OF ANALYSIS	KEY POINTS
Functionalist	Macro	• Prejudice and discrimination can be dysfunctional, but they provide benefits for dominant groups and stabilize society.
Conflict	Macro	• Powerful groups maintain their advantages and perpetuate racial-ethnic inequality primarily through economic exploitation.
Feminist	Macro and micro	• Minority women suffer from the combined effects of racism and sexism.
Symbolic Interactionist	Micro	• Hostile attitudes toward minorities, which are learned, can be reduced through cooperative interracial and interethnic contacts.

employees are women, and they are predominantly minority women. For feminist scholars, such segregation of minority women reflects gendered racism.

Gendered Racism

Gendered racism refers to the combined and cumulative effects of inequality due to racism *and* sexism. Many white women encounter discrimination on a daily basis (see Chapter 9). Minority women, however, are also members of a racial-ethnic group, bringing them a double dose of inequality. If social class is also considered, some minority women experience *triple oppression*. Affluent females, especially, often have no qualms about exploiting recent immigrants, especially Latinas who perform demanding housework at very low wages (Segura 1994; Hondagneu-Sotelo 2001).

For feminist scholars, men, especially those in dominant groups, exploit many minority women sexually. For example, the family of Senator Strom Thurmond of South Carolina (who died at age 100 in 2003) acknowledged that, at age 22, he had fathered a daughter with a black woman, aged 16, who had worked for the family as a maid. Throughout his 48 years in the Senate, Thurmond was a staunch segregationist: "He lived 100 years and never acknowledged his daughter. He never let her eat at his table. He fought for laws that kept his daughter segregated and in an inferior position" (Janofsky 2003: A28).

Critical Evaluation

Feminist perspectives have sharpened our understanding of the effects of gendered racism and minority women's subordinate status in U.S. society (Newman 2005).

Like conflict theorists, however, feminist scholars often assume that gendered inequality is deliberate when, in fact, it can be unintentional. Because all of us have internalized institutional discrimination, minority-group members may also be guilty of reinforcing inequality in schools, workplaces, and other situations. A related weakness is that feminist explanations seldom explore deliberate oppression of minorities by other minorities (such as affluent African American women and men who exploit domestic workers).

SYMBOLIC INTERACTIONISM

According to symbolic interactionists, we learn attitudes, norms, and values throughout the life cycle (see Chapters 1 and 4). Because, as you saw earlier, race and ethnicity are constructed socially, labeling, selective perception, and social contact can have powerful effects on everyday intergroup relations.

Labeling, Selective Perception, and the Contact Hypothesis

We learn attitudes toward dominant and minority groups through labeling and selective perception, both of which can increase prejudice and discrimination. For example, a comprehensive study of major U.S. news magazines (such as *Time* and *U.S. News & World Report*) concluded that labeling immigrants as a "problem" and a "menace" ignored "the broader array of Latino roles and contributions to American communities" (Gavrilos 2006: 4).

There are ways to decrease labeling and selective perception. For example, the **contact hypothesis** states that the more people get to know members of a minority group personally, the less likely they are to be prejudiced against that group. Such contacts are most effective when dominant and minority group members have approximately

> **gendered racism** the combined and cumulative effects of inequality due to racism and sexism.
>
> **contact hypothesis** the idea that the more people get to know members of a minority group personally, the less likely they are to be prejudiced against that group.

For symbolic interactionists, images shape our perceptions of racial and ethnic groups.

Looting

After Hurricane Katrina, the caption on a photo of a young black man wading through water described him as "looting".

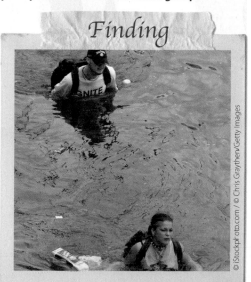

Finding

A caption described a white man and a light-skinned woman as "finding" goods at a flooded local grocery store.

the same status (such as co-workers or bosses), share common goals (such as working on a project), and cooperate rather than compete and when an authority figure supports such interaction (such as when an employer requires mentoring for minority workers) (Allport 1954; Kalev et al. 2006).

Critical Evaluation

Symbolic interactionism is valuable in helping us understand how race and ethnicity shape our everyday lives. If we are aware of the negative effects of labeling and selective perception, for example, we can change our attitudes and behavior. A major weakness, however, is that symbolic interactionism tells us little about the social structures that create and maintain racial-ethnic inequality. For instance, people who aren't prejudiced can foster discrimination by simply going along with the inequitable policies that have been institutionalized in education, the workplace, and other settings.

7 Major Racial and Ethnic Groups in the United States

a mericans absorb many aspects of immigrants' cultures: "Our everyday lexicon is sprinkled with Spanish words. We are now just as likely to grab a burrito as a burger. Hip-hop is tinged with South Asian rhythms. And Chinese New Year and Cinco de Mayo are taking their places alongside St. Patrick's Day as widely celebrated American ethnic holidays" (Jiménez 2007: M1). Of the major U.S. racial-ethnic groups, some experience more constraints than others, but all have numerous strengths that enhance U.S. society.

Many white ethnic immigrants worked hard to achieve the American dream despite low educational levels and numerous hardships. During high school, Estée Lauder, the daughter of Hungarian Jewish immigrants, started selling her uncle's face cream. In 1948, she convinced Saks Fifth Avenue to sell the product. When Lauder died in 2004, at age 97, her beauty empire was valued at $10 billion.

EUROPEAN AMERICANS: A DECLINING MAJORITY

During the seventeenth century, English immigrants settled the first colonies in Massachusetts and Virginia. Other white Anglo-Saxon Protestants (WASPs), who included people from Wales and Scotland, quickly followed. Most of these groups spoke English. Some of the immigrants were affluent, but many were poor or had criminal backgrounds.

Diversity

About 58 percent of the U.S. population has a European background. Many have ancestors from Germany, Ireland, England, Italy, Poland, and France, but others have Scandinavian roots (see *Table 10.3*).

TABLE 10.3
Americans of European Descent, 2006

Of the many European ancestries that Americans report, the following comprise at least 1 percent of the U.S. population.

ANCESTRY	NUMBER (IN MILLIONS)	PERCENTAGE OF TOTAL U.S. POPULATION
German	35.5	11.9
Irish	22.3	7.4
English	19.1	6.4
Italian	13.9	4.6
Polish	6.8	2.3
French	5.6	1.9
Scottish	3.8	1.3
Norwegian	3.1	1.0

Note: This survey was based on a total estimated population of almost 300 million Americans.
Source: Based on U.S. Census Bureau 2006 Community Survey.

Constraints and Strengths

WASPs generally looked down on later waves of immigrants from southern and eastern Europe. They viewed the newcomers as inferior, dirty, lazy, and uncivilized because they differed in language, religion, and customs. New England, which was 90 percent Protestant, was particularly hostile to Irish Catholics, characterizing them as irresponsible and shiftless (Feagin and Feagin 2008).

All the later waves of European immigrants faced varying degrees of hardship in adjusting to the new land because the first English settlers had a great deal of power in shaping economic and educational institutions. In response to prejudice and discrimination, many of the immigrants founded churches, schools, and recreational activities that maintained their language and traditions (Myers 2007).

Despite stereotypes, prejudice, and discrimination, European immigrants began to prosper within a few generations. They surmounted many barriers and became influential in all sectors of American life, including education, law, politics, social work, and the military. Overall, they now fare much better financially than most other groups in the United States. This doesn't mean that all are rich. In fact, in absolute numbers, poor whites outnumber those of other racial-ethnic groups (see Chapter 12).

LATINOS: A GROWING MINORITY

About one in three Americans is a member of a racial or ethnic minority group, but one of every two newcomers is Latino, making Latinos the fastest-growing minority group. Worldwide, only Mexico and Colombia have larger Latino populations than the United States ("Hispanic Heritage Month . . ." 2007).

Diversity

Some Latinos trace their roots to the Spanish and Mexican settlers who established homes and founded cities in the Southwest before the arrival of the first English immigrants on the East Coast. Other Latinos are recent immigrants or children of the immigrants who arrived in large numbers at the beginning of the twentieth century. Of the almost 41 million Latinos living in the United States, 64 percent were born in Mexico or are of Mexican heritage (U.S. Census Bureau 2008). Spanish-speaking people from Mexico, Puerto Rico, Ecuador, the Dominican Republic, and Spain differ widely in their customs, cuisine, and cultural practices.

Dominican-born Alfredo Rodriguez is one of numerous successful Latino businessmen who are rebuilding neglected inner-city neighborhoods. In 1985, Rodriguez bought his first grocery store in Queens, New York, with the $25 a week his mother had been setting aside for him for a decade. In 2002, he purchased a 53,000-square-foot supermarket in Newark, New Jersey, to meet the needs of local Latino shoppers. His Xtra Supermarket has annual sales of $9 million (Rayasam 2007).

Constraints and Strengths

Most Latinos are taking their place in mainstream America, but many still encounter obstacles. The median household income of Latinos is 72 percent of that of white Americans (see *Figure 10.4*). Almost 37 percent of Latino families earn $50,000 a year or more, up considerably from only 7 percent in 1972, but almost 24 percent of Mexican Americans and Puerto Ricans live below the poverty line, compared with 11 percent of Cuban Americans (U.S. Census Bureau 2008).

As with other groups, the socioeconomic status of Latinos reflects a number of interrelated factors, especially education level, being able to speak English, recency of immigration, and occupation. Many Latinos who were professionals in their native land find only low-paying jobs. They often don't have time to both work and learn English, which would help them gain the accreditation they need to practice as doctors, lawyers, and accountants. About 40 percent of Latinos don't have a high school degree, but there are subgroup variations. For example, 75 percent of Cuban Americans have at least a high school education, compared with only 47 percent of Mexican Americans ("Living Humbled . . ." 1996; Thompson 2007; U.S. Census Bureau 2008).

Despite their generally lower economic and educational attainment, many Latinos are successful. They

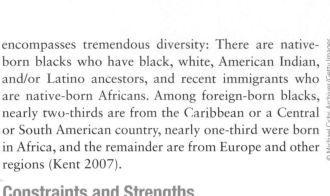

Among African Americans who have made major contributions, some are better known than others. For example, few people know that Madame C. J. Walker (1857–1919), above left, was a manufacturer of hair care products for African American women and one of the first American women millionaires. Or that Dr. Charles R. Drew (1904–1950) was a renowned surgeon, teacher, and researcher. He was responsible for founding two of the world's largest blood banks, which saved untold lives during and since World War II.

own almost 1.6 million businesses, or 15 percent of all U.S. businesses. And, between 1995 and 2005, many foreign-born Latino immigrants earned better hourly wages because they tended to be older, better educated, and more likely to be employed in construction than in agriculture (*Hispanic-Owned Firms . . .* 2006; Kochhar 2007). In addition, many Latino family networks have been resilient and adaptive in protecting their members' health and emotional well-being, coping with economic hardship, and adjusting to a new environment.

AFRICAN AMERICANS: A MAJOR SOURCE OF DIVERSITY

The 40 million African Americans are the second largest minority group in the United States, making up over 13 percent of the population (U.S. Census Bureau 2008). This percentage includes those of more than one race, like President Barack Obama, whose father was Kenyan and whose mother is white.

Diversity

Most African Americans share a common characteristic: They are members of the only group ever brought to the United States involuntarily and legally enslaved. Despite a common history of oppression, the term *African American*

encompasses tremendous diversity: There are native-born blacks who have black, white, American Indian, and/or Latino ancestors, and recent immigrants who are native-born Africans. Among foreign-born blacks, nearly two-thirds are from the Caribbean or a Central or South American country, nearly one-third were born in Africa, and the remainder are from Europe and other regions (Kent 2007).

Constraints and Strengths

The effects of 350 years of slavery and legal segregation are still evident. Compared with 12 percent of the U.S. population, 24 percent of African Americans live in poverty. The median family income of African Americans is the lowest of all racial-ethnic groups (see *Figure 10.4*). Especially in the country's inner cities, many young black men are high school dropouts, jobless, or incarcerated (Mincy 2006; U.S. Census Bureau 2008).

The gap in educational achievement between whites and African Americans has decreased in the last four decades. For example, the proportion of blacks who earned at least a high school diploma increased from 26 percent in 1964 to 80 percent in 2005. Similarly, during this same period, those with a college degree increased from 4 percent to 18 percent (U.S. Census Bureau 2008). Still, and as you'll see in Chapter 14, many black children receive a lower quality of education through high school, which leaves them unprepared for college and the job market.

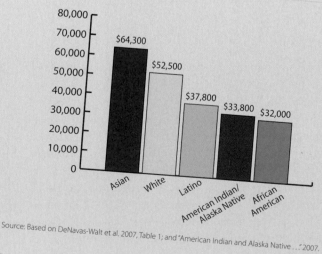

FIGURE 10.4
U.S. Median Household Income, by Race and Ethnicity, 2006

Race/Ethnicity	Median Household Income
Asian	$64,300
White	$52,500
Latino	$37,800
American Indian/Alaska Native	$33,800
African American	$32,000

Source: Based on DeNavas-Walt et al. 2007, Table 1; and "American Indian and Alaska Native . . ." 2007.

Despite centuries of oppression, many African Americans are successful. They own almost 6 percent of U.S. businesses, generating almost $90 billion in annual sales. African Americans are still rare at the largest *Fortune 500* corporations, but 84 percent of the top 200 companies have at least one black member on the board of directors, usually a male, a number that has doubled since 1987 (*Black-Owned Firms* . . . 2006; Crockett 2006).

African Americans have numerous strengths—strong kinship bonds, an ability to adapt family roles to outside pressures, a strong work ethic despite recessions and unemployment, a determination to succeed in education, and an unwavering spirituality that helps people cope with adversity. Single-parent families headed by mothers, especially, show exceptional fortitude and coping skills (Edin and Lein 1997; McAdoo 2002).

ASIAN AMERICANS: A MODEL MINORITY?

The almost 15 million Asian Americans comprise about 5 percent of the U.S. population. Of this group's most recent population increase, 43 percent is due to births and 57 percent to immigration ("Nation's Population . . ." 2006).

Diversity

Asian Americans encompass a broad swath of cultures and traditions. They come from at least 26 countries in East and Southeast Asia (including China, Taiwan, Korea, Japan, Vietnam, Laos, Cambodia, and the Philippines) and South Asia (especially India, Pakistan, and Sri Lanka). These diverse origins mean that there are drastic differences in languages and dialects (and even alphabets), religion, cuisines, and customs. Asian Americans also include Native Hawaiian and other Pacific Islanders from Guam and Samoa. Chinese form the largest Asian American group, followed by Filipinos and Asian Indians (see *Figure 10.5*). Combined, these three groups account for 58 percent of all Asian Americans.

Constraints and Strengths

The most successful Asian Americans are those who speak English relatively well *and* have high educational levels. Asian Americans have the highest median income of all U.S. racial-ethnic groups (see *Figure 10.4*). Asian Indians—two-thirds of whom have advanced degrees—have the highest annual median household income (almost

$74,000) of Asian American subgroups, compared with less than $40,000 for Cambodians, many of whom are less educated, experience language barriers, and have few marketable skills. Like Latinos, many recent Asian American immigrants who have top-notch credentials from their homeland experience underemployment. For example, some Korean doctors work as hospital orderlies and nurses' assistants because they need income to support a family while they prepare for English-language tests and the medical exam in their field of specialization (Jo 1999; U.S. Census Bureau 2007).

Overall, Asian Americans have higher educational levels than any other U.S. racial-ethnic group. Almost half have at least a college degree compared with 31 percent of the white population. There is considerable variation across groups, however. For example, 36 percent of Asian Indians have a graduate or professional degree compared with 7 percent of Vietnamese and 2 percent of Cambodians (U.S. Census Bureau 2007; "Asian/Pacific American . . ." 2008).

Because of their educational and economic success, Asian Americans are often hailed as a "model minority." Such labels are misleading, however, because many households are larger than average, including more workers, and because children tend to live with their parents longer, contributing to the household income (Adler 2003; Aguirre and Turner 2004). And, as you'll see in Chapter 14, there is considerable variation in educational attainment across Asian American and Pacific Islander subgroups.

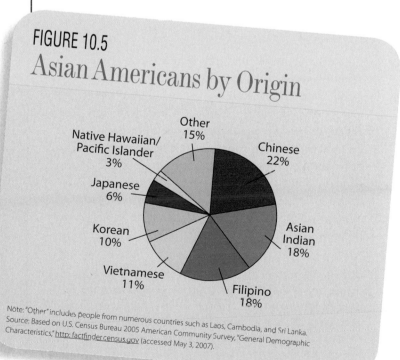

FIGURE 10.5
Asian Americans by Origin

- Other 15%
- Native Hawaiian/Pacific Islander 3%
- Japanese 6%
- Korean 10%
- Vietnamese 11%
- Chinese 22%
- Asian Indian 18%
- Filipino 18%

Note: "Other" includes people from numerous countries such as Laos, Cambodia, and Sri Lanka. Source: Based on U.S. Census Bureau 2005 American Community Survey, "General Demographic Characteristics," http: factfinder.census.gov (accessed May 3, 2007).

Asian Americans own 5 percent of all U.S. businesses, including many restaurants, dry cleaning stores, and landscaping services, and more than 3 in 10 Asian-American–owned firms provide professional, scientific, and technical services. Asian Indians own about 43 percent of the 47,000 U.S. hotels and motels. In many cases, these owners bought run-down lodgings and converted them to upscale Sheraton and Hilton hotels (Park 2002; *Asian-Owned Firms* . . . 2006; Yu 2007).

Asian Americans have also been more successful than other minority groups in penetrating corporate suites. In 1995, for example, all of the *Fortune 500* CEOs were white. By 2008, four were black males, five were Latino males, and seven were Asian Americans, two of them women (DiversityInc 2008).

AMERICAN INDIANS: A GROWING NATION

American Indians used to be called the "vanishing Americans," but have "staged a surprising comeback" due to higher birth rates, a longer life expectancy, and better health services (Snipp 1996: 4). The almost 5 million American Indians and Alaska Natives make up 1.5 percent of the U.S. population. Because the median age of this group is only 31, some scholars predict that the future population growth of American Indians may outpace that of Latinos (Ogunwole 2006; "American Indian and Alaska Native . . ." 2007).

Diversity

Like Asian Americans, American Indians are a heterogeneous group. Of the more than 560 federally recognized tribes, those numbering over 100,000 members include the Cherokee, Navajo, Chippewa, and Sioux. Tribes speak 150 native languages (although many are quickly vanishing) and vary widely in their religious beliefs and cultural practices. Thus, a Comanche-Kiowa educator cautions, "lumping all Indians together is a mistake. Tribes . . . are sovereign nations and are as different from another tribe as Italians are from Swedes" (Pewewardy 1998: 71; Ashburn 2007).

Mohegan Sun in southern Connecticut is one of the largest casinos in the United States. It has spent some of its profits on college tuitions, a $15 million senior center, other social programs, and health insurance for tribal members. In contrast, some of the poorest tribes, such as the Navajo and Hopi, who have rejected gambling for religious reasons, have many members who live in poverty.

Constraints and Strengths

American Indians are a unique minority group because they are not immigrants and have been in what is now the United States longer than any other group. Nevertheless, they have experienced centuries of subjugation, exploitation, and political exclusion (such as not having the right to vote until 1924) and have endured a legacy of broken treaties, stolen lands, and tribal extinction (Wilkinson 2006).

American Indians are better off today than they were a decade ago, but long-term institutional discrimination has been difficult to shake. Consider education. In 1824, the federal government created the Bureau of Indian Affairs (BIA). In the 1870s, the government began operating and financing boarding schools, run by Christian missionaries. By 1931, nearly one-third of American Indian children were in such boarding schools where the goal was obliterating all that was Indian by immersing Indian children in white society (Kelley 1999). American Indian communities got substantial control over school policies and programs only with the passage of the Educational Amendment Act of 1978. Since then, American Indian educators have implemented curricula that include knowledge of and pride in American Indian culture and heritage.

Almost 27 percent of American Indians live below the poverty line, compared with 12 percent of the general population. The median household income of American Indians is slightly higher than that of African American households, but lower than other racial-ethnic groups (recall *Figure 10.4* on p. 190). On many reservations, where 36 percent of American Indians live, unemployment rates run about 50 percent and

some are as high as 90 percent. Substandard housing is common: Many homes on reservations are overcrowded and lack kitchen facilities (including stoves and refrigerators) and indoor plumbing (Vanderpool 2002; Taylor and Kalt 2005; Ogunwole 2006).

Despite numerous obstacles, American Indians have made considerable economic progress by insisting on self-determination and the rights of tribes to run their own affairs. Since passage of the Indian Gaming Regulatory Act in 1988, the number of casinos owned by American Indian tribes has grown to more than 400, and they account for 37 percent of the U.S. gambling industry. Few tribes have benefited, however, because many casinos are in remote areas that don't attract tourists. In other cases, tribal leaders who control official membership sometimes refuse to recognize people as tribal members to increase their own share of the profits. Outside of gambling, the number of American Indian–owned businesses grew from fewer than 5 in 1969 to nearly 202,000 in 2002, most of them in construction, retail trade, and health care (Taylor and Kalt 2005; *American Indian-and Alaska Native-Owned Firms . . .* 2006; Kestin and Franceschina 2007).

Regardless of their socioeconomic status, strengths of American Indians include *relational bonding,* a core behavior that is built on values such as respect, generosity, and sharing across the tribe and kin group. Harmony and balance include putting community and family needs above indi-

vidual achievements. Another strength is a spirituality that sustains a person's identity and place in the world (Stauss 1995; Cross 1998).

MIDDLE EASTERN AMERICANS: AN EMERGING GROUP

The Middle East is "one of the most diverse and complex combinations of geographic, historical, religious, linguistic, and even racial places on Earth" (Sharifzadeh 1997: 442). The Middle East encompasses about 30 countries that include Armenia, Turkey, Israel, Iran, Afghanistan, Pakistan, and 22 Arab nations (such as Algeria, Iraq, Kuwait, Saudi Arabia, and the United Arab Emirates).

Diversity

Those from the Middle East comprise a heterogeneous population that is a "multicultural, multiracial, and multiethnic mosaic" (Abudabbeh 1996: 333). Most are Muslims, but many are Christians or Jews. Arabic is the most common language, but people from the Middle East also speak Turkish, Farsi, Kurdish, and other languages. There is also a multitude of ethnic and linguistic groups with very different customs and cultural practices.

Most Middle Eastern Americans are of Arab ancestry. About 1.2 million Americans, less than 0.5 percent of the total population, report that their ancestry is solely or partly Arab. Almost half were born in the United States, but nearly half arrived during the 1990s (de la Cruz and Brittingham 2003; Brittingham and de la Cruz 2005). Those who classify themselves as Arab Americans come from many Middle Eastern countries (see *Figure 10.6*).

Constraints and Strengths

Middle Eastern Americans tend to be better educated and wealthier than other Americans. Arab Americans are nearly twice as likely as the average U.S. resident to have a college degree—41 percent compared with 24 percent. The median income for an Arab American family, $52,300, is about $2,300 more than the median income for all U.S. families. As in other groups, there are wide variations. For example, Lebanese have higher median family incomes (almost $61,000) than Moroccans ($41,000). Because of their generally higher educational levels, the proportion of Arab Americans working in management jobs is higher than the proportion of all Americans, 42 percent versus 34 percent (Brittingham and de la Cruz 2005).

Not all Middle Eastern Americans are successful, of course. The proportion living in poverty (17 percent) is higher than that of the general population (12 percent). Lebanese and Syrians have the lowest poverty rates, 11 percent, compared with over 26 percent of those from Iraq and many other Middle East countries (Brittingham and de la Cruz 2005).

Generally, Middle Eastern Americans are well integrated into American life. Besides their high educational levels, three out of four speak only English at home or speak English very well, more than half are homeowners, 75 percent of the men are in the labor force, and more than half of those born in another country are U.S. citizens (Brittingham and de la Cruz 2005).

For decades, many Middle Eastern Americans have encountered prejudice and discrimination in the workplace, school, and public places. Years after the 9/11 terrorist attacks, many Middle Eastern Americans still follow self-imposed restrictions to prevent threats of violence: They avoid flying and crossing the border into Canada, change their names (Osama Nimer, electrician, is now Samuel Nimer), trim their beards, don't wear distinctive head scarves, speak English instead of Arabic in public, display the flag, and watch what they say (Marvasti and McKinney 2004; Hampson 2006).

Middle Eastern Americans cope with prejudice and discrimination because they have a strong ethnic identity, close family ties, and religious beliefs that secure children to their communities. Most importantly, perhaps, many families have extended kin networks and relatives on whom they can count for help during hard times (Ajrouch 1999; Hayani 1999).

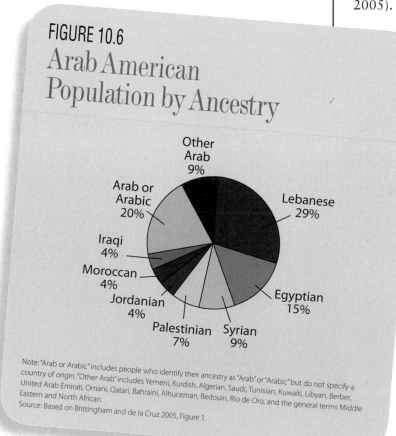

FIGURE 10.6
Arab American Population by Ancestry

- Lebanese 29%
- Egyptian 15%
- Syrian 9%
- Palestinian 7%
- Jordanian 4%
- Moroccan 4%
- Iraqi 4%
- Arab or Arabic 20%
- Other Arab 9%

Note: "Arab or Arabic" includes people who identify their ancestry as "Arab" or "Arabic" but do not specify a country of origin. "Other Arab" includes Yemeni, Kurdish, Algerian, Saudi, Tunisian, Kuwaiti, Libyan, Berber, United Arab Emirati, Omani, Qatari, Bahraini, Alhuceman, Bedouin, Rio de Oro, and the general terms Middle Eastern and North African.

Source: Based on Brittingham and de la Cruz 2005, Figure 1.

8 Interracial and Interethnic Relationships

n 1997, professional golfer Tiger Woods said he was "Cablinasian," a word he'd made up as a boy, because he was one-eighth Caucasian, one-fourth black, one-eighth American Indian, one-fourth Thai, and one-fourth Chinese. Many blacks were upset that Woods seemed to downplay his African American roots, but he said that he was embracing all parts of his multicultural heritage.

Interracial relationships are now more acceptable than in the past, but more in principle than as a lived reality. That is, Americans are more likely to approve of interracial dating and intermarriage than to engage in them.

GROWING MULTIRACIAL DIVERSITY

The 2000 Census allowed people to mark more than one race for the first time, because of increasing rates of interracial dating and marriage, and growing numbers of biracial children. Almost 99 percent of Americans reported only one race, but 1 in 40 is the product of two or more racial groups. Hawaii has the largest proportion of people—21 percent—who identify themselves as two or more races, followed by Alaska and California (5 percent each) (Jones and Smith 2001; Population Division, U.S. Census Bureau 2008).

INTERRACIAL DATING AND MARRIAGE

It's now fairly common for people to date someone from a different racial or ethnic group—48 percent of Americans overall say they have done so, including 69 percent of Latinos, 52 percent of blacks, and 45 percent of whites (Jones 2005). Laws against **miscegenation**, marriage or sexual relations between a man and a woman of different races, existed in America as early as 1661. It wasn't until 1967, in the U.S. Supreme Court's *Lov-ing v. Virginia* decision, that antimiscegenation laws were overturned nationally.

> **miscegenation** marriage or sexual relations between a man and a woman of different races.

In 1958, only 4 percent of Americans approved of marriage between blacks and whites. By 2007, 77 percent approved of such unions, but whites were more likely to disapprove (19 percent) than were blacks (10 percent) and Latinos (10 percent). Attitudes about interracial marriages have changed, but what about behavior? Racial-ethnic intermarriages have increased slowly—from 0.7 percent of all marriages in 1970 to 5.4 percent in 2000 (Fields and Casper 2001; Carroll 2007). Thus, about 95 percent of all Americans marry someone of the same race.

The increase of intermarriage reflects many inter-related factors—both micro and macro level—that include everyday contact and changing attitudes. For example, we tend to date and marry people we see on a regular basis. The higher the educational level, the greater the potential for intermarriage because highly educated minority-group members often attend integrated colleges and their workplaces and neighborhoods are more integrated than in the past (Kalmijn 1998; Qian 2005).

People often marry outside of their racial-ethnic groups because of a shortage of potential spouses within their own group. Because the Arab American population is so small, for example, 80 percent of U.S.-born Arabs have non-Arab spouses. In contrast, intermarriage rates for Latinos and Asian Americans have decreased since 1990—for both women and men—because the influx of new immigrants has provided a larger pool of eligible mates (Kulczycki and Lobo 2002; Qian and Lichter 2007).

Also, racial-ethnic groups that are the most assimilated are the most likely to intermarry. Asian Americans who are least likely to out-marry are those who live in ethnic enclaves, do not speak fluent English, and have lived in the United States a short time. Among Japanese Americans, in contrast, intermarriage rates are high because many families have been in the country for four or five generations and their members are generally more accepting of intermarriage than are more recent immigrants (Shinagawa and Pang 1996; Rosenfeld 2002).

Many people

feel strongly about politics.

what do you think?

The wealthy have disproportionate control over the government.

1 2 3 4 5 6 7

strongly agree   strongly disagree

11 Politics

government a formal organization that has the authority to make and enforce laws.

In 2008, the election of Barack Obama to the office of President of the United States affirmed many Americans' belief that the United States is a land of political opportunity.

Such optimism isn't entirely warranted because politics is highly complex. As you'll see in this chapter, there are many global political systems, special interest groups play a major role in elections, many Americans don't participate in the government, and democracies aren't always as open as they seem.

1 Government

government is a formal organization that has the authority to make and enforce laws. Governments can maintain order, provide social services, regulate the economy, and establish educational systems. Besides maintaining armed forces to discourage (real or imagined) attacks by other countries, governments also try to protect their citizens from internal assaults that range from individual crimes to organized paramilitary groups.

Most governments are huge bureaucracies (see Chapter 6). The U.S. government, for example, consists of executive, legislative, and judicial branches, with numerous departments in the executive branch alone (see *Figure 11.1*). Among other powers, the President proposes laws, the Congress writes laws, and the judicial system interprets laws and legitimates their enforcement.

Government is often affected by a *civic society,* the nongovernment group of citizens that includes community-based organizations, the mass media, lobbyists, and voters. If a government seems unjust, a country's civic society can exert considerable power by replacing elected officials or rebelling against appointed leaders.

2 Politics, Power, and Authority

harles de Gaulle, one of France's past presidents, once remarked, "I have come to the conclusion that politics are too serious a matter to be left to the politicians." All of us should view politics as a serious matter, however, because it can decrease or improve the quality of our everyday lives.

POLITICS AND POWER

Many of us can't discuss politics with relatives or friends without getting into heated disagreements. Most people feel strongly about **politics**, a social process through which individuals and groups acquire and exercise power and authority, two important concepts (Lasswell 1936).

A neighborhood bully picks on little kids. A bank forecloses on a home mortgage. A government quashes a protest. These are all examples of **power**, the ability of a person or group to affect the behavior of others despite resistance and opposition. There are elements of power in almost all social relationships, such as those between parents and children, students and faculty, officers and soldiers, and employers and employees.

One of the best examples of political power is *earmarks* (known commonly as "pork")—government funds, appointments, or benefits that officials distribute to gain favor with their constituents. Earmarks are measures tacked on to legislation outside normal budgetary processes.

Despite numerous promises by Congress to curb pork barrel spending, the funding has grown considerably—from $10 billion in 1995 to $18 billion in 2008 (Williams et al. 2008). In 2008, the national pork average was $34 per person. Alaska led the nation with $556 per person ($380 million), followed by Hawaii with $220 per person ($284 million). Some examples of recent pork-barrel projects include the following:

- $50 million for a 5-acre indoor rainforest in Iowa
- $13.5 million for a waterway in Louisiana that is rarely used
- $8 million for grape and wine research
- $3 million each for golf programs in Florida and South Carolina to develop young people's character
- $2.2 million to improve the recreational facilities for a population of 1,570 people in North Pole, Alaska
- $500,000 to improve the sidewalk and landscaping between two streets in Montezuma, Georgia
- $188,000 for the Lobster Institute in Maine, one of whose major accomplishments has been launching lobster dog biscuits
- $98,000 to develop a walking tour of Boydton, Virginia, population under 500 people (Javers 2007; Williams et al. 2008).

FIGURE 11.1
The Government of the United States

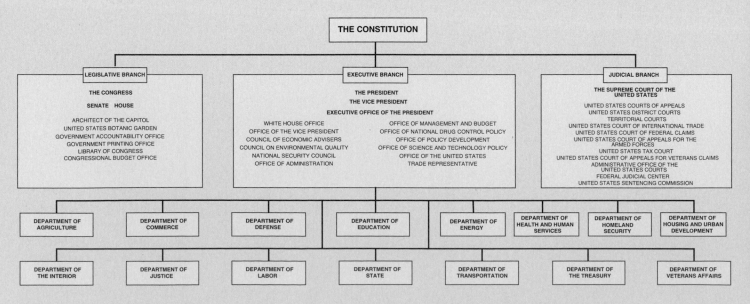

Source: *The United States Government Manual, 2005/2006*, p. 21.

TABLE 11.1
Weber's Three Types of Authority

TYPE OF AUTHORITY	DESCRIPTION	SOURCE OF POWER	EXAMPLES
Traditional	Power is based on customs, traditions, and/or religious beliefs.	Personal	Medieval kings and queens, emperors, tribal chiefs
Charismatic	Power is based on exceptional personal abilities or a calling.	Personal	Adolf Hitler, Gandhi, Martin Luther King, Jr.
Rational-legal	Power is based on the rules and laws that are inherent in an elected or appointed office.	Formal	U.S. presidents, Congressional members, state officials, police, judges

Both Democrats and Republicans fight for pork because, they maintain, the money fulfills pressing local needs, like helping ex-convicts reenter society. Critics contend that the earmarked monies decrease funding for important federal programs, influence reelections, go to groups that are not at the forefront of research or do not have a good record of completing projects, are typically awarded to high-ranking members of the Appropriations Committee, and are given to companies that in turn make generous campaign contributions to the politicians (Brainard and Hermes 2008; Klein 2008; O'Harrow 2008).

Whether based on persuasion or coercion, power—especially political power—is about controlling others. Because people may revolt against sheer force, many governments depend on authority to establish order, shape people's attitudes, and control their behavior.

AUTHORITY

Authority, the legitimate use of power, has three important characteristics. First, people *consent* to authority because they believe that their obedience is for the greater good (for example, in following traffic rules). Second, people see the authority as *legitimate*—valid, justifiable, and necessary (so they pay taxes that support public education, increase national security, and remove trash and snow). Third, people accept authority because it is *institutionalized* in organizations (such as police departments and government agencies).

Max Weber (1925/1978) described three ideal types of legitimate authority: traditional, charismatic, and rational-legal (see *Table 11.1*). Remember that ideal types are models that describe the basic characteristics of any phenomenon (see Chapters 1 and 5). In reality, the types of authority often overlap.

Traditional Authority

Traditional authority is power based on customs that justify the position of the ruler. The source of power is personal because the ruler inherits authority due to long-standing customs, traditions, or religious beliefs.

Traditional authority is most common in nonindustrialized societies where power is based on kinship, tribes, and clans. In the past, kings and emperors ruled because of heredity and the belief that they had a divine right to power, regardless of intelligence or ability. In many African, Middle Eastern, and Asian countries today, power is passed down among men within a family line (Tétreault 2001; see, also, Chapter 8).

Charismatic Authority

Charismatic authority is power based on exceptional individual abilities and characteristics that inspire devotion, trust, and obedience. Like traditional authority, charismatic authority is personal and based on extraordinary deeds or even a belief that a leader has been chosen by God, but the leaders don't pass their power down to their offspring.

Charismatic leaders can inspire loyalty and passion whether they are heroes or tyrants. Examples of the latter include historical figures such as Adolf Hitler, Napoleon, Ayatollah Khomeini, and Fidel Castro. These and other dictators have been spellbinding orators who radiated magnetism, dynamism, and tremendous self-confidence and promised to improve a nation's future (Taylor 1993).

authority the legitimate use of power.

traditional authority authority based on customs that justify the position of the ruler.

charismatic authority authority based on exceptional individual abilities and characteristics that inspire devotion, trust, and obedience.

Rational-Legal Authority

Rational-legal authority is power based on the belief that laws and appointed or elected political leaders are legitimate. Unlike traditional and charismatic authority, rational-legal authority comes from rules and regulations that pertain to an office rather than to a person. For example, anyone running for mayor must have specific qualifications (like U.S. citizenship). When a new mayor (or governor or other politician) is elected, the rules don't change because power is vested in the office rather than the person currently holding the office.

Mixed Authority Forms

Weber's ideal types are useful in identifying the major characteristics of the three types of authority. In reality, people may experience more than one form of authority. For example, some historians believe that presidents Abraham Lincoln, Theodore Roosevelt, John F. Ken-nedy, and Ronald Reagan enjoyed charismatic appeal beyond their rational-legal authority.

In some countries, such as Japan and England, emperors, kings, and queens have traditional authority and perform symbolic state functions such as attending religious ceremonies and granting titles (like that of Sir Elton John, a singer). Both of these countries also have parliaments that exercise legal-rational authority in determining laws and policies.

We see, then, that political leaders can exercise power and authority in many ways. They exert control differently, however, depending on a state's political system.

3 Types of Political Systems

in 1997, Britain returned Hong Kong to China as part of its territory. Hong Kong had experienced democracy under British rule, but China endorses *communism,* a political-economic system in which wealth and power are to be shared by the community (see, also, Chapter 12). Almost half a million people in Hong Kong demonstrated and chanted slogans such as "Fight for democracy" and "Return power to the people," but the protests were ignored by the newly instated Chinese government (Marshall 2004).

This example illustrates a population's reaction to two polar types of political system: democracy (under British rule) and totalitarianism (under the Chinese Communist Party). Both of these types of political system exist in many countries, but there are also totalitarian governments and monarchies.

Joan of Arc (1412–1431) is a well-known example of a charismatic woman who defied the status quo. She was convinced that she was destined by God to lead the French to victory over English invaders. She was burned at the stake as a heretic and witch for claiming that she communicated with God. During the 1960s, California's César Chávez (1927–1993) successfully organized a number of strikes to call attention to grape-pickers' and lettuce workers' low wages. Chávez's boycotts and strikes, supported by many consumers nationally, generally improved farm workers' wages and working conditions.

DEMOCRACY

A **democracy** is a political system in which, ideally, citizens have control over the state and its actions. Democracies are based on several principles:

- Individuals are the best judges of their own interests and participate in governmental decisions.

- Citizens select leaders who are responsive to the wishes of the majority of the people.

- Suffrage (the right to vote) is universal and elections are free, fair, secret, and occur frequently.

- The government recognizes individual rights such as freedom of speech (including dissent), press, and assembly and the right to organize political parties whose members compete for public office.

In reality, democracy doesn't always guarantee equality and respect for human rights. For example, the U.S. Constitution initially limited voting to white male landowners; African American men got the right to vote in 1870, and women of all racial-ethnic groups gained the vote only in 1920. Today, according to some critics (and as you'll see shortly), U.S. democratic values are endangered because special-interest groups have unprecedented influence on the political process. A minority of hardliners often imposes its views on the majority regarding war, abortion, homosexuality, gun laws, and other social issues, and there is an increased blurring of lines between politics and rigid religious fundamentalism (Carter 2005).

Some analysts have described the twentieth century as the "democratic century" because there was a "dramatic expansion of democratic governance" (Fukuyama et al. 2002: 2). In 2007, 46 percent of the world's population lived in free societies, and another 18 percent lived in partly free societies. This means, however, that 36 percent of the world's population (almost 2.4 billion people) still live in authoritarian countries and suffer "intense repression," including "brutal violations of human dignity" and no political freedom (Freedom House 2008b). People in sub-Saharan and North Africa, the Middle East, East Asia, and Russia are the least likely to have political freedom (see *Figure 11.2*).

TOTALITARIANISM AND DICTATORSHIPS

At the opposite end of the continuum from democracy is **totalitarianism**, a political system in which the government controls every aspect of people's lives. Totalitarianism has several distinctive characteristics:

- A pervasive ideology that legitimizes state control and instructs people how to act in their public and private lives.

- A single political party controlled by one person, a *dictator*—a supreme, sometimes idolized leader—who stays in office indefinitely.

- A system of terror that relies on secret police and the military to intimidate people into conformity and to punish dissenters.

- Total control by the government over other institutions, including the military, education, family, religion, economy, media, and all cultural activities, including the arts and sports (Taylor 1993; Tormey 1995; Arendt 2004).

Contemporary examples of repressive totalitarian governments include modern China, Burma (Myanmar), Cuba, Libya, North Korea, Somalia, Sudan, Turkmenistan, Uzbekistan, and Chechnya. Within these countries and territories, "state control over daily life is pervasive and wide-ranging, independent organizations and political opposition are banned or suppressed, and fear of retribution for independent thought and action is a part of daily life" (Freedom House 2007b: 5).

Whereas democracy assumes that people are rational beings who can manage their own affairs, totalitarian philosophy holds that "humans are by nature either too irrational or too ignorant to be entrusted with self-government" (Pauley 1997: 3). Most totalitarian countries have become less successful in controlling people's actions because their inhabitants are becoming part of the global economy and the government is less able to control the media.

Consider China. Since 1949, when the Chinese Communist Party (CCP) took power, analysts have consistently described this country as totalitarian and its government as among the most repressive in the world. President Hu Jintao, in power since 2002, is typically on "The World's 10 Worst Dictators" list (Wallechinsky 2008). Citizens cannot vote against the top leaders or

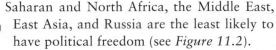

authoritarianism a political system in which the state controls the lives of citizens but generally permits some degree of individual freedom.

monarchy a political system in which power is allocated solely on the basis of heredity and passes from generation to generation.

express any opposition to government policy; journalists who criticize leaders or their policies are routinely harassed, jailed, or tortured. Freedom of assembly is severely restricted, colleges and universities must support official CCP ideology, and the death penalty can be given for nonviolent crimes like burglary, counterfeiting, and bribery. In addition, although antidiscrimination laws exist, religious groups, minorities, the disabled, and people with HIV/AIDS suffer severe discrimination (Fowler 2006; Montlake 2006; Freedom House 2008b).

In 2007, pro-democracy activists led the initial demonstrations in Rangoon, Burma, when the military-run government increased the price of fuel, which hiked the costs of staples such as rice and cooking oil. Tens of thousands of Buddhist monks joined the protests, withdrew their religious services from the military and their families, and issued a statement denouncing the government as "the enemy of the people." The monks' participation was significant because they are highly revered and influential.

© Reuters/Landov / © Michael Krinke/iStockphoto.com

AUTHORITARIANISM AND MONARCHIES

Most nations have some version of democracy or totalitarianism. There are also a number of countries that are characterized by **authoritarianism**, a political system in which the state controls the lives of citizens but generally permits some degree of individual freedom. In the Middle East, for example, the ruler of Qatar has absolute power and discourages public criticism of his policies, but he has implemented a constitution that specifies that two-thirds of the officeholders must be elected (rather than appointed) and has supported numerous progressive reforms, such as women's active participation in politics (Harman 2007).

A **monarchy**, the oldest type of authoritarian regime, is a political system in which power is allocated solely on the basis of heredity and passes from generation to generation. In this form of government, a member of a royal family, usually a king or queen, reigns over a kingdom. A monarch's power and authority are legitimized by religion (the right to govern is bestowed on the monarch by God) and tradition.

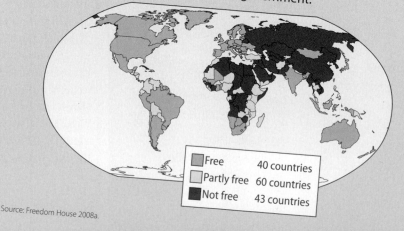

FIGURE 11.2
Political Freedom around the World 2008

This map shows the degree of political freedom, measured by political rights and civil liberties, in countries around the world. *Political rights* include free and fair elections, competitive parties, and no discrimination against minorities. *Civil liberties* include freedom of speech and the press; freedom to practice one's religion; and the freedom of administrators, teachers, professors, and students to discuss political issues without fear of physical violence or intimidation by the government.

Free	40 countries
Partly free	60 countries
Not free	43 countries

Source: Freedom House 2008a.

There are about 29 monarchies around the world. In some countries—especially in the Middle East and parts of Africa—monarchs have absolute control. Those in Norway, Denmark, Belgium, Britain, Japan, and many other countries have little political power because they are limited by democratic constitutions. Instead, many modern monarchs serve primarily ceremonial roles, by acting as goodwill ambassadors to other countries, for example.

So far, we've examined some of the worldwide variations of power and authority in political systems. We now turn to another important topic—politics in U.S. society today.

4 Power and Politics in U.S. Society

president Ronald Reagan once joked, "Politics is not a bad profession. If you succeed, there are many rewards. If you disgrace yourself, well you can always write a book." Despite Reagan's levity, politics plays a critical role in our everyday lives. In the United States, you can contact your government officials, be active in a political party, join a special-interest group, and, of course, vote. Do Americans take advantage of these democratic processes often enough? Let's begin with political parties.

POLITICAL PARTIES

A **political party** is an organization that tries to influence and control government by recruiting, nominating, and electing its members to public offices. A political party is a

In 1995, at age 3, Oyo Nyimba Kabamba-Iguru became the world's youngest monarch when he was crowned as king of the Ugandan Kingdom of Toro. When he reaches 18, he will officially take over full control of the kingdom.

© Peter Busomoke/AFP/Getty Images

legitimate way for citizens to shape public policy at the local, state, and national levels.

Functions of U.S. Political Parties

Humorist Will Rogers quipped, "I belong to no organized party. I am a Democrat." In fact, political parties are highly organized. They engage in a wide variety of activities—everything from stuffing envelopes and calling voters to drafting laws if a candidate is elected. Through their activities, parties perform a number of vital political functions, especially recruiting candidates for public office, organizing and running elections, and, if elected, running the government (Schmidt et al. 2001).

Ideally, that's how political parties *should* work. In reality, parties perform some of these functions more effectively than others. For example, parties typically concentrate on winning elections rather than making changes once candidates are in office. In addition, because the United States has a two-party system, many people who don't identify with either group may feel alienated and excluded.

The Two-Party System

Many democracies around the world have a number of major political parties: Canada has 3, Germany 5, Italy 9, and Israel more than 20. The two-party system of Democrats and Republicans in the United States is almost unique. In the past, the percentage of people who identified with either party was about equally divided. Since the early 2000s, however, a large number of voters (34 percent) have identified themselves as independent (Welch et al. 2004; "Trends in Political Values . . ." 2007). Because there is no national Independent Party, independents usually vote for Democrats or Republicans or not at all.

Those who identify with the Democratic Party typically believe that the government should provide social programs, especially for the poor; those in the Republican Party usually believe that the federal government should be involved in few social programs. *Table 11.2* provides some examples of other major ideological differences between Democrats and Republicans.

Other parties have emerged in the United States, but they usually have little weight. Most third-party candidates have little chance of being elected because the media don't take them seriously and give them little coverage. Sometimes, though, third parties meet with surprising success.

© AP Images

In 1998, former wrestler Jesse "The Body" Ventura, a Reform Party candidate, surprised everyone when he was elected governor of Minnesota. Many analysts felt that Ventura was successful because he professed to be anti-party and anti-establishment. His posture struck a chord among many Minnesotans who were disenchanted with both Democrats and Republicans. More often, however, campaign contributors and voters pay little attention to third-party candidates because they don't want their donations and ballots to have no effect.

SPECIAL-INTEREST GROUPS

Some people are active in political parties. Others join a **special-interest group** (sometimes called an *interest group*), a voluntary and organized association of people that attempts to influence public policy and policymakers on a particular issue.

Political parties often include diverse individuals, while special-interest groups are

TABLE 11.2
How Do Democrats and Republicans Differ?

ISSUE	MANY DEMOCRATS BELIEVE THAT . . .	MANY REPUBLICANS BELIEVE THAT . . .
Family	Government programs should implement universal health insurance for all families, especially those with low incomes.	Government should cut all welfare benefits to unwed mothers, stigmatize divorce, and emphasize "traditional" family values.
Abortion	Women should have the right to choose abortion in practically all cases; the government should pay for abortions for low-income women.	The unborn should be protected in all cases; there should be no public money for abortions.
Gay Rights	Gay men and lesbians should have the right to marry and to adopt children. They should be fully accepted in the military and other institutions.	Marriage and adoption should be limited to heterosexuals. Inclusion of gay men and lesbians in the armed forces creates interpersonal problems.
Education	The government should strengthen public schools by raising salaries for teachers and decreasing classroom sizes. It should not use tax dollars to help students attend private schools.	The government should increase state and local control of schools. It should use tax dollars to fund students to attend private schools of their choice if their local school is underperforming.
Environment	The government should provide incentives to promote a clean environment, penalize corporations that pollute, and seek environmental protections in trade agreements.	The government should balance businesses' rights and economic development with environmental protection. Insisting on environmental protection will discourage global trade agreements.

Sources: Based on Benokraitis 2000; Welch et al. 2004.

usually made up of people who are very similar in social class and political objectives.

There are thousands of special-interest groups, large and small, in the United States. A few examples include associations of teachers, farmers, firefighters, religious congregations, and a variety of environmental groups. If you've ever sent a donation to such a group or signed a petition, you're a member of a special-interest group. Special-interest groups use many tactics to influence the government. Three of the most effective—although their aims are rarely representative of the general population—are lobbying, campaign contributions, and political action committees.

Lobbyists

A **lobbyist** is a representative of a special-interest group who tries to influence political decisions on the group's behalf. At the federal level alone, over 64,000 firms, organizations, and individuals lobby about a variety of subjects such as bankruptcy, casinos, immigration, railroads, alcohol, and veterans. At the state level, lobbyists and the companies that hire them spend more than $1 billion every year to influence lawmakers. In 2006, nearly 40,000 registered lobbyists worked for 50,000 companies and organizations. Nationwide, that averages out to about five lobbyists per lawmaker (Political Money Line 2004; Laskow 2006).

Many lobbyists use their expensive townhouses close to the Capitol to host fund-raising events for at least 40 percent of the Congress members. Such activities are legal, but the lawmakers then usually support legislation promoted by the lobbyists who hosted the events (Dilanian 2008).

Some of the best-paid lobbyists are former members of Congress who have retired or have lost a re-election. They know how the federal government works and often maintain close ties with incumbents (Kelley 2007). Also, a growing number of lobbyists are relatives of members of Congress. For example,

> Senator Orrin Hatch's son works for a firm that has made more than a million dollars lobbying for the diet supplements industry, which the [Utah Republican] senator has long championed. . . . Senator Harry Reid, a Nevada Democrat, has had three sons and a son-in-law involved in firms that lobby the government or litigate for industries seeking government benefits, paying hundreds of thousands of dollars for their services. ("Kith & Kin Inc." 2004: A18)

Between 2000 and 2005, lobbyists paid almost $50 million to send members of Congress, their staffs, and their family members on at least 23,000 "seminars" (actually more like expensive vacations) overseas and across the United States to curry favor with the legislators. Such expense is miniscule compared with the benefits. On average, companies generate about $28 in earmarked revenue for every $1 they spend lobbying. In some cases, companies pull in well above $2,000 in earmarks for every lobbying dollar (Morris 2006; Kirkpatrick 2007).

lobbyist a representative of a special-interest group who tries to influence political decisions on the group's behalf.

Campaign Contributions

Money is the most important factor in winning an election. Having the most money doesn't guarantee a victory, but candidates with little financing usually have little public visibility. Among all presidential candidates, the spending soared from $67 million in 1976 to $1.32 billion by the third quarter of 2008. In 2004, George Bush and John Kerry each spent almost $75 million in the presidential campaign. By late October, 2008, Democratic presidential candidate Barack Obama had spent almost $574 million, and Republican nominee John McCain had spent almost $296 million. In state elections, some candidates have spent well over $15 million to win a seat in the Senate. Michael R. Bloomberg, a billionaire, spent $74 million of his own money in 2001 to be elected mayor of New York City. He was re-elected in 2005 after spending $85 million, seven times the amount that his Democratic opponent raised (Birnbaum 2004; Rutenberg and Healy 2005; Center for Responsive Politics 2008a, 2008b).

How do most political candidates raise money for their campaigns? There are two types of political contributions: hard money and soft money. *Hard money* refers to political donations that the Federal Election Commission regulates. It's illegal for corporations and labor unions to contribute to campaigns for federal elections, but individuals can contribute up to $25,000 per year to any national party and $10,000 per year to any state or local party.

Soft money is any contribution that is not regulated by federal election laws. Companies, unions, and individuals may give donations in any amount not to support a particular candidate, but to a political party for the purpose of *party building*. Party building includes increasing voter registration and running ads that educate voters about issues, as long as the ads don't mention which candidates to vote for. Soft money actually pays for much more: office overhead, the purchase of expensive computer equipment, and workers' meals and travel expenses

political action committee (PAC) a special-interest group set up to raise money to elect a candidate to public office.

during a candidate's campaign (Samples 2005).

Political Action Committees (PACs)

A legal but controversial type of special-interest group is a **political action committee** (PAC), a group set up to raise money to elect one or more candidates to public office. It's illegal for an individual entity—such as a corporation, labor union, or trade association—to donate money *directly* to a candidate, but it's legal for groups that represent business, unions, or other interests to contribute to a political action committee. Thus, for example, a state can have several hundred PACs supported by surgeons that have officially different names (for example, Baltimore County Surgeons, Howard County Surgeons, and so on). At the federal level, a PAC can contribute no more than $15,000 to a national party, and an individual's contribution to a PAC is limited to $5,000. Some states have similar limits but others don't.

PACs have mushroomed—from 608 in 1974 to almost 4,200 in 2006, 38 percent of them representing corporate interests (U.S. Census Bureau 2008). The top PAC contributors are members of corporations, but professional groups and labor unions are also generous donors (see *Table 11.3*). Notice, also, that some of the PACs give to both parties to cover their bases regardless of who's elected.

Why are PACs controversial? They provide access to politicians and create a sense of obligation, a need to reciprocate. And, the larger an organization, the greater its clout. In contrast to small business firms with few resources, national and global corporations can contribute the maximum amount to a similar PAC in every state, increasing their influence. In contrast, low-income and part-time workers aren't likely to have a PAC that represents their interests (Clawson et al. 1998).

Besides through soft money, hard money, and funds from PACs, interest groups and wealthy individuals have other legal ways to buy a lawmaker's goodwill. Businesses and affluent individuals can give generous gifts (like luxurious vacations and tickets to major sports events) to lawmakers' spouses and other family members, donate to the lawmakers' favorite charities, and hire the lawmakers or their family members as consultants. Many members of Congress routinely use campaign contributions and money from PACs to pay their relatives for fundraising and other campaign work. These payments sometimes

TABLE 11.3
Top 12 PAC Contributors to Political Parties, 2007–2008

ORGANIZATION SUPPORTING PACS	TOTAL AMOUNT (IN MILLIONS)	PERCENT FOR DEMOCRATS	PERCENT FOR REPUBLICANS
National Association of Realtors	$3.7	57%	43%
AT&T Inc.	$2.8	45%	55%
International Brotherhood of Electrical Workers	$2.7	98%	2%
Operating Engineers Union	$2.7	86%	14%
American Bankers Association	$2.6	40%	60%
National Beer Wholesalers Association	$2.5	52%	48%
Airline Pilots Association	$2.3	85%	15%
National Auto Dealers Association	$2.3	34%	66%
American Association for Justice	$2.3	95%	5%
International Association of Fire Fighters	$2.2	75%	25%
Laborers Union	$2.0	92%	8%
Machinists/Aerospace Workers Union	$2.0	97%	3%

Source: Based on material in Center for Responsive Politics 2008c.

amount to hundreds of thousands of dollars per family member (Vogel 2007; Citizens for Responsibility and Ethics in Washington 2007, 2008). Most Americans don't contribute to political campaigns, but many exercise their right to vote.

5 Who Votes, Who Doesn't, and Why

The ballot is stronger than the bullet. (President Abraham Lincoln)

Several Democratic and Republican primaries were just held all across the country. It was evenly split between those who forgot to vote and those who chose not to vote. (comedian Conan O'Brien)

Abraham Lincoln was right about the power of the ballot, but as Conan O'Brien's quote suggests, many Americans are disillusioned with politics. Even before the financial turbulence began in October 2008, 74 percent of Americans said that they were dissatisfied with the way the nation was being governed, up from 49 percent in 2004 (Jones 2008; Newport 2008).

One might expect that this dissatisfaction would result in higher voter turnout, but this isn't the case. Of 163 countries around the world that hold elections, the United States ranks only 140th in voter turnout. In 34 nations, at least 80 percent of the eligible population votes (and over 90 percent in Italy and Iceland) compared, on average, with about half of U.S. citizens. In local elections (as for mayor and city council), only 25 percent or less vote. This means that a relatively small number of people determines who will be in office and which laws will be passed.

Some of the high turnout rates in other countries may be partially due to compulsory voting, in which the government requires explanations or imposes fines for not voting, may disenfranchise nonvoters (which makes it difficult to get benefits like public assistance), or requires evidence of voting to get a passport or a driver's license. Generally, these kinds of sanctions send the message that voting is not a privilege but a civic responsibility (Holder 2006; International Institute for Democracy and Electoral Assistance 2007, 2008; U.S. Census Bureau 2008). In 2008, 66 percent of Americans voted, the most since 1908 (Marks 2008). It's not clear, however, whether voter turnout rates will drop in the future if the economy is more stable and if the candidates don't include women or minority members.

In the United States, who votes, who doesn't, and why reflect demographic characteristics, attitudes about politics, and situational and structural factors. Because presidential elections have the highest turnouts, let's look at some voting patterns in these elections.

DEMOGRAPHIC FACTORS

Many demographic factors influence registration and voting, but the most important are age, marital status, social class, race and ethnicity, and religion. Sex isn't a major variable because women are only slightly more likely to vote than men (52 percent versus 48 percent) (Holder 2006.) (As this book goes to press, there are no comprehensive national data on demographic variables for the 2008 election. Whenever possible, however, I've included some preliminary figures.)

Age

The voting rate is much higher among older than younger people (72 percent of those 55 and older versus 47 percent of those aged 18–24) and increases with age (see *Figure 11.3a*). Young people (ages 18-29) represent about 21 percent of the voting-eligible population; 18 percent voted in the 2008 presidential election, up from 17 percent who voted in the 1996, 2000, and 2004

When minorities vote, they can make a huge difference in the results. Democratic governor Janet Napolitano, pictured left, defeated her Republican opponent by about 20,000 votes, most of them from the Navajo reservations.

FIGURE 11.3

Selected Characteristics of Voters in the November 2004 Election

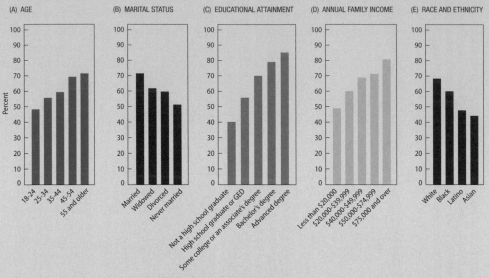

(A) AGE • (B) MARITAL STATUS • (C) EDUCATIONAL ATTAINMENT • (D) ANNUAL FAMILY INCOME • (E) RACE AND ETHNICITY

Note: For information on voting and registration in the November 2006 congressional election, see File 2008.

Source: Based on Holder 2006, Table B.

presidential elections. Obama captured 66 percent of the youth vote, but some political analysts suggest that the higher voting participation may reflect a generally greater interest in civic engagement (Education Commission of the States 2008; Keeter et al. 2008).

Why are young adults less likely to vote than their older counterparts? A key factor is registration. Young people have low registration rates because they are more mobile than older people and are less likely to re-register after a move. They may also be preoccupied with major life events such as going to college, finding jobs, getting married, and starting a family. Other young adults don't register or vote because they grew up in homes where their parents paid little attention to politics (60 percent) or feel that political officials don't care about their attitudes (38 percent) ("A Generational Look . . ." 2002; Schachter 2004). In effect, then, the low participation rate of young voters means that they have little impact on political processes.

Marital Status

Married people are more likely to vote than those who are widowed, divorced, or have never married (see *Figure 11.3b*). People who have not been married tend to be younger, which influences their voting rate, as you've just seen. Married couples are more likely to be registered voters, homeowners, to live in established neigh-

borhoods, to be parents of young children in school, and to be employed (Day and Holder 2004). This combination of characteristics suggests that married people, compared with never married singles, have a bigger stake in society and as a result are more likely to vote.

Social Class

Social class has a significant impact on voting behavior. At each successive level of educational attainment, the voting rate increases. For example, the voting rate of those with a college degree (78 percent) is almost twice as high as that for people who have not completed high school, and those with advanced degrees are the most likely to vote (see *Figure 11.3c*). People with the highest educational levels are usually more informed about and interested in the political process and feel that their vote counts ("Who Votes . . ." 2006).

Voting rates also increase with income levels. The voting rate of people with annual incomes of $75,000 or more is 81 percent, compared with 48 percent of those with annual incomes under $20,000 (see *Figure 11.3d*). People with higher incomes are more likely to be employed, to have assets (like houses and stock), and, therefore, to have a stake in who is elected. In contrast, people who are unemployed usually have few assets and may be too depressed or disillusioned with the political system to vote ("Who Votes . . ." 2006).

Race and Ethnicity

Whites are typically the most likely to vote (see *Figure 11.3e*). Here again, a key to voter turnout is registration. Across all racial-ethnic groups—including both native-born and naturalized individuals—the higher the voter registration, the higher the voting rate.

Immigrant status is a key variable in voting. Latinos and Asians, especially, who tend to be recent immigrants, are often under age 18 and are less likely to be citizens, both factors that make them unable to vote (Frey 2008). As the numbers of eligible voters in these groups increase in the next few years, they're expected to have considerable political clout. We saw some of the influence of minority groups during the 2008 primary election, when Senators Clinton, Obama, and McCain actively courted both African Americans and Latinos.

Low voter registration rates among racial-ethnic groups are associated with other demographic and social characteristics. For example, compared with white citizens, minorities tend to be younger, poorer, unemployed, and have lower educational levels. Because they are less likely to be homeowners or married, they may be less invested in election outcomes. In addition, minorities are more likely than whites to be pessimistic about government and politics (Frey 2008).

Religion

Only 14 percent of Americans say that their religious beliefs influence their political thinking, but there are strong links between political participation and religious beliefs. Among Jews and mainline Protestants (like Lutherans and Methodists), over 80 percent are registered to vote, compared with 69 percent of Catholics, 48 percent of Muslims, 42 percent of Hindus, and only 13 percent of Jehovah's Witnesses. The low voter registration rates of Muslims and Hindus are primarily due to the fact that many of them are recent immigrants who aren't eligible to vote (Pew Forum on Religion & Public Life 2008).

Religion also shapes political affiliation. Mormons, mainline Protestants, and members of evangelical churches (including born-again Christians) tend to describe themselves as conservative and support the Republican Party. Jews, Buddhists, Hindus, and those not affiliated with a religious group describe their political beliefs as liberal and favor the Democratic Party. The connection between religious affiliation and political opinion appears to be especially strong for certain issues such as abortion and homosexuality (Pew Forum on Religion & Public Life 2008; see, also, *Table 11.2* on p. 204).

ATTITUDES

Of the almost 14 million Americans who did not vote in 2004, 25 percent said they were "too busy" or had "conflicting schedules." Another 33 percent said they weren't interested in the election, didn't like the candidates, forgot to vote, or that the polling place was inconvenient. All of these responses suggest that voting is a low priority, whether because of apathy or because of other factors (Holder 2006).

In October 2008, only 25 percent of Americans said that the President was doing a good job, and only 12 percent felt this way about Congress. Many Americans simply don't trust politicians: "They all lie," "I've never met an honest politician," "They're all corrupt," and so on. Others feel that their lives won't improve regardless of who's elected (Jones 2008; Newport 2008).

Is such cynicism warranted? Recently, the lengthy Iraq war, the bungled response to Hurricane Katrina, the subprime mortgage and credit crises, and the plunging stock market have given many voters reason to doubt the ability and integrity of America's leaders. In early 2008, when gasoline prices averaged $3.20 a gallon (and some Californians were already paying $4.29 a gallon), a reporter asked President Bush about the possibility of the price going to $4.00 a gallon nationally. President Bush was surprised: "What, what did you say? You're predicting $4-a-gallon gasoline? That's interesting. I hadn't heard that . . ." Many Americans were dismayed that the president did not seem to be aware of what they had to face every day (Reynolds et al. 2008).

SITUATIONAL AND STRUCTURAL FACTORS

There are also a number of situational and structural factors that discourage voting. For example, nearly one in three Americans didn't vote in the 2004 presidential election because of illness, disability, or out-of-town travel. Another 10 percent said that they had experienced transportation or registration problems. Because a key to voter turnout is registration, some states have removed structural barriers to registering. North Dakota, for example, has no voter registration. In other states—Idaho, Maine, Minnesota, New Hampshire, Wisconsin, and Wyoming—people can register the same day they vote. In Oregon, all ballots have been mailed in since 2000. In all of these states, the voting rates are much higher than the national average, and sometimes well above 72 percent (Welch et al. 2004).

pluralism a political system in which power is distributed among a variety of competing groups in a society.

In many European and other countries, where voter turnout is typically high, voters are registered automatically when they pay taxes or receive public services, and elections are held on Saturdays. In the United States, in contrast, the government does little to help people register to vote, and elections are held on a weekday (usually Tuesday). Most of our polls are open from 7:00 A.M. to 8:00 P.M., but voting is still difficult for some commuters who spend up to 4 hours traveling to and from work every day (Welch et al. 2004).

We've seen that officeholders, political parties, and voters participate in a democracy. But who has the most power? And do government leaders represent the average citizen? Such questions have generated considerable debate among sociologists and other social scientists.

Taxpayers pay all costs for the cars that members of the House of Representatives lease for themselves and some of their top aides. There are few restrictions on what kind of car the members can choose, and no limit on what they spend. Some New York representatives lease modest vehicles (like a Chevrolet Impala for $219 a month), but Democrat Charles B. Rangel (pictured here) doesn't worry about price or gas mileage. He leases a Cadillac Deville for almost $800 a month to project an image of success and says, cheerfully, "I've got a desk in it. It's like an airplane . . . I want [my constituents] to feel that they are somebody and their congressman is somebody" (Hernandez 2008).

6 Who Rules America?

Who has power in the United States? According to symbolic interactionism, whenever there are social interactions, people can assume power and influence social life. And, as you saw earlier, people learn to be loyal to a political system—whether it's a democracy or a monarchy—and to show respect for its symbols and leaders. For the most part, however, sociologists rely on macro-level theories to analyze the power balance in the United States. Functionalists argue that the people hold most of the power ("government of the people, by the people, for the people"), conflict theorists maintain that power is concentrated in the hands of a few people at the top, and feminist theorists assert that men have most of the decision-making power (*Table 11.4* summarizes these perspectives).

FUNCTIONALISM: A PLURALIST MODEL

The pluralist model originates from functionalists' theories. For functionalists, the people rule through **pluralism**, a political system in which power is distributed among a variety of competing groups in a society (Riesman 1953; Polsby 1959; Dahl 1961).

Key Characteristics

According to the pluralist model, individuals have little direct power over political decision making but can influence government policies through special-interest groups—trade unions, professional organizations, and so on (see *Figure 11.4* on page 212). Because there is so much heterogeneity between and within groups, no single group dominates or controls political life. The numerous groups rarely join ranks because they focus on single issues like abortion, pollution, or animal rights. Their focus on separate single issues fragments groups, but the competition among groups, according to pluralists, also results in a broad representation of a variety of interests and a distribution of power.

Because there are a number of single-issue groups, functionalists maintain, there are multiple leaderships: "Those who exercise power in one kind of decision do not necessarily exercise power in others" (Dye and Ziegler 2003: 12). As a result, many leaders, not just a few, can shape decisions that represent numerous groups and issues. Pluralists note that people also have power outside of interest groups: They can vote, run for office, contact officeholders, and collect signatures to place specific issues on a ballot. Therefore, there are continuous checks and balances as individuals and groups vie for power and try to influence laws and policies.

Critical Evaluation

Does pluralism work as democratically as functionalists maintain? Critics argue that interest groups have unequal resources. The poor and disadvantaged rarely have the skills and educational backgrounds to organize or to promote their interests. In contrast, wealthy individuals and organizations can influence government through political contributions, and large corporations can mobilize like-minded executives and pressure employees to contribute to particular PACs.

Critics also maintain that pluralists aren't realistic about the power of groups to change the status quo. Only 19 percent of Americans register their political opinions with officeholders or government agencies because they don't trust politicians or they feel that political leaders pay little attention to people's complaints or input (Horrigan 2004).

Reflecting a functionalist view, historian Rich Shenkman (2008) contends that if Americans don't have the government they want, they themselves are to blame because they know little about politics (only two out of every five voters can name the three branches of government, for example), don't question government officials, are easily manipulated because they're uninformed, and don't vote. Conflict theorists argue, however, that many voters aren't ignorant; rather, they feel powerless or apathetic because most political decisions are secretive and it's very difficult to get accurate information about political corruption (see, for instance, Citizens for Responsibility and Ethics in Washington 2008).

> **power elite** a small group of influential people who make a nation's major political decisions.

CONFLICT THEORY: A POWER ELITE MODEL

In contrast to pluralists, conflict theorists contend that the United States is ruled by a **power elite**, a small group of influential people who make the nation's major political decisions. Sociologist C. Wright Mills (1956) coined the term *power elite* to describe a pyramid of power that he believed characterized American democracy (see *Figure 11.4*).

TABLE 11.4
Sociological Explanations of Political Power

	FUNCTIONALISM: A PLURALIST MODEL	CONFLICT THEORY: A POWER ELITE MODEL	FEMINIST THEORIES: A PATRIARCHAL MODEL
Who has political power?	The people	Rich upper-class people—especially those at top levels in business, government, and the military	White men in Western countries; most men in traditional societies
How is power distributed?	Very broadly	Very narrowly	Very narrowly
What is the source of political power?	Citizens' participation	Wealthy people in government, business corporations, the military, and the media	Being white, male, and very rich
Does one group dominate politics?	No	Yes	Yes
Do political leaders represent the average person?	Yes, the leaders speak for a majority of the people.	No, the leaders are most concerned with keeping or increasing their personal wealth and power.	No, the leaders—who are typically white, elite men—are most concerned with protecting or increasing their personal wealth and power.

Key Characteristics

According to Mills, the power elite is made up of three small but influential groups of people at the top level who run the country: political leaders (specifically chief executives, who include the president and his top aides), business heads (the corporate rich, who have enormous wealth), and military chiefs (who govern the Pentagon). Practically all of the members of these groups are white, Anglo-Saxon, Protestant males who form an inner circle of power.

For Mills, there are also a number of groups at the middle level of power: Congressional members, lobbyists, influential media commentators, active professionals, leaders of labor unions and other interest groups, and the heads of local and state governments. The bottom level, and the largest and least powerful group, is comprised of the masses, consisting of everyone else (see *Figure 11.4*). The power elite tolerates the masses—including their elections and laws—but, in the end, simply does what it wants, such as declaring wars, decreasing taxes for the rich, and cutting off benefits for the poor (see, for example, Kivel 2004, and Domhoff 2006).

Many contemporary power-elite theorists maintain, like Mills, that the United States is not a democracy because only about 6,000 individuals have formal authority over institutions that control about half of the nation's resources in industry, finance, insurance, mass media, education, law, and civil and cultural affairs. According to a political scientist, this group constitutes less than three-thousandths of 1 percent of the total population (Dye 2002).

Other conflict theorists emphasize that the corporate rich is a more powerful group than Mills originally thought. The owners and top-level managers of corporations, banks, and agribusinesses shape government policies for their own benefit and have "a major impact on the income, job security, and well-being of most other Americans" (Domhoff 2006: xi; see, also, Zweigenhaft and Domhoff 2006).

According to conflict theorists, the upper classes have enormous power over political institutions. Those with great wealth have a tremendous advantage in financing political campaigns and getting elected to offices. Well over half of the members of Congress are multimillionaires. Michelle and Barack Obama are millionaires, but many others—including the Clintons, the Bush family, John McCain (whose wife is an heiress to a beer distributing company), Mitt Romney, and all but one of the U.S. Supreme Court Justices (Anthony M. Kennedy)—are multimillionaires. Because many law-

FIGURE 11.4
Pluralist and Power Elite Perspectives on Political Power

PLURALIST MODEL

Power is dispersed among multiple groups that influence the government. For example,

- government employees
- victims' rights groups
- labor unions
- banks and other financial institutions
- realtors and home builders
- teachers
- environmental groups
- women's rights groups

POWER ELITE MODEL

Power is concentrated in a very small group of people who make all the key decisions. For example,

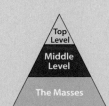

Top level (1 percent)—CEOs of large corporations, high-ranking political leaders in the executive branch, and top military leaders

Middle level (8 percent)—most members of Congress, lobbyists, entrepreneurs and owners of small businesses, leaders of labor unions and other interest groups (in law, education, medicine, etc.), and influential media commentators

The masses (91 percent)—people who are unorganized and exploited and either don't know or don't care about what's going on in government

makers are wealthy, they have the power to vote for tax laws and other legislation that promote their own interests (Becker 2001; Becker 2008; Feldmann 2008).

Critical Evaluation

Functionalists maintain that the conflict theorists' claims about the power elite are exaggerated. They point out, for example, that "the masses" aren't just puppets but are very aware of the influence of the power elite. As a result, they vote, mobilize, support particular PACs, and protest current administrative and corporate policies (such as high prescription prices, especially for many of the low-income elderly).

Also, according to critics, the power elite perspective assumes, incorrectly, that the wealthy people at the top are unified in their interests and goals. For example, Vice President Dick Cheney broke away from President Bush's strong endorsement of a constitutional amendment to ban same-sex marriage because Cheney's daughter, Mary, is a lesbian who has a life partner and a child. In addition, critics maintain, because the Democratic and Republican political parties and their top officials endorse very different agendas, any power elite that exists is rarely unified.

FEMINIST THEORIES: A PATRIARCHAL MODEL

Feminist theories emphasize that women are generally excluded from the most important political positions. Women's status in the Western world has improved considerably during the last 100 years, but as you saw in Chapter 9, relatively few women, especially minority women, serve in Congress and at the highest levels of government.

Key Characteristics

Of 187 countries worldwide, women occupy only 18 percent of positions in the decision-making bodies that are comparable to our Congress. In Argentina, Cuba, Finland, Sweden, and Rwanda, women make up at least 40 percent of these governing bodies. The United States ranks 81st, well below many African, European, and Asian countries, and even below most of the Arab countries that many Westerners view as repressive towards women (Inter-Parliamentary Union 2008). Thus, U.S. democracy is not as egalitarian as most of us assume.

Feminist theorists maintain that most American women have been shut out of the political process because the United States is still a patriarchal society. For example, three-quarters of President George W. Bush's top-level appointments were men; in comparison, dur-

ing the Clinton administration, 37 percent of important political appointments went to women. And, during the Bush administration, of the 17 White House staffers who earned $157,000—the top of the federal pay scale—12 were men (Tessier 2001; Milbank 2004).

American women are successful fund-raisers, vote in higher numbers than men, and often run for office. Compared with men, however, women get considerably less media coverage of their political campaigns or receive coverage that is more disparaging. Recently, for example, many commentators, especially feminist scholars, have accused the media of "outrageous and deplorable" sexist attacks that hurt the public's image of Senator Hillary Clinton as an experienced and competent lawmaker in her bid for a presidential nomination (Kahn 2003; Reed 2008; see, also, Chapter 9).

Feminist theorists also note that the contributions of women in rank-and-file positions in political organizations and campaigns are often undervalued and unrewarded (Kramer 2005). For example, few women are asked to be campaign managers, positions that often lead to political visibility and appointments.

Critical Evaluation

Conflict theories maintain that those among the powerful elite—who want to perpetuate and even strengthen their rule—make decisions about war, the economy, taxes, education, and other important issues that affect all Americans. Feminist theorists go much further by noting that these elites are typically men.

Critics reject several aspects of feminist explanations of political power. They claim, for example, that women aren't elected because they don't venture into political realms. Often, critics say, women work as volunteers rather than seek office. Some critics also argue that most women's organizations typically focus on single issues such as abortion, child care, domestic violence, and others. Such splintering dilutes women's mobilization within a political party and decreases their chances of winning office.

One might also question many feminists' contention that patriarchy is the root of political power differences between women and men. Some of the most patriarchal societies—those in much of Africa, Latin America, and many Arab countries, for example—have many more women in high-ranking positions in political bodies than does the United States, which professes equality between the sexes. Thus, patriarchy may not be as critical in explaining political power differences between the sexes as many feminist scholars claim.

Money is a major motivator

to work, but work provides other benefits.

what do you think?

If people enjoy their work, there's no difference between "good" and "bad" jobs.

1 2 3 4 5 6 7

strongly agree   strongly disagree

12

Work and the Economy

Anita Epolito was fired after refusing to stop smoking when not on the job. She complained that employers are "trying to change behavior after 5 o'clock," and asked, "What's next? No McDonald's? No caffeine? No Krispy Kreme?" (Jones 2007: A4).

Maybe so, because 21 states so far have passed laws that permit employers to fire or not hire people because they smoke, are obese, or have high cholesterol or blood pressure levels that increase the chances of diabetes, heart disease, and strokes. Employers, whose health-care costs are spiraling, believe that such policies are legitimate because rising health costs cut into wages, and people make choices about smoking, eating, and other lifestyle factors. Many workers like Anita Epolito, however, contend that state laws and employers' practices are intrusive because they dictate how people must behave when they're not working.

> **economy** a social institution that determines how a society produces, distributes, and consumes goods and services.
>
> **work** physical or mental activity that accomplishes or produces something, either goods or services.

This example illustrates the close linkage between an individual's personal life and the **economy**, a social institution that determines how a society produces, distributes, and consumes goods and services. Let's begin by considering why work is important in our lives.

Key Topics

In this chapter, we'll explore the following topics:

1 The Social Significance of Work

Work is physical or mental activity that accomplishes or produces something, either goods or services. Work has many forms: It can be legal or illegal, paid or unpaid (such as raising children and volunteering), and essential for a society's survival (such as producing food) or peripheral (such as creating entertainment). Some work is routine and mechanical, but some involves considerable stress and anxiety.

For most of us, money is a major motivator to work, but work provides other benefits. Generally, employment leads to better health and a sense of accomplishment and usefulness, and is a major source of social identity. Almost 80 percent of Americans say that having a good relationship with their co-workers is more important than earning a high salary or having a prestigious job because the social contacts often provide friendships and opportunities to participate in mutually satisfying activities. In addition, work provides a sense of stability and order and a daily rhythm that we don't get from many other activities. Some retirees continue to work part-time or do volunteer work because they enjoy the opportunities for learning and growing (Katzenbach 2003; Mirowsky and Ross 2007).

As with politics (see Chapter 11), societies differ in the kinds of economic systems they develop. Societies worldwide are also experiencing profound changes, as some nations gain unprecedented economic power because of globalization, political transitions, and a revolution in communications technology.

2 Global Economic Systems

t he two major economic systems around the world are capitalism and socialism. In actual operation, economies are usually some mixture of these two types.

CAPITALISM

Capitalism is an economic system in which wealth is in private hands and is invested and reinvested to produce profits. Ideally, capitalism has four essential characteristics (Smith 1937; Heilbroner and Thurow 1998):

- *Private ownership of property.* Property—such as real estate, banks, and utilities—belongs to individuals or organizations rather than the state or the community.

© iStockphoto.com / © Walt Disney/The Kobal Collection / © ABC-TV/The Kobal Collection / © AP Images / © AP Images / © AP Images / © PRNews Foto/Walt Disney Studios / © Todd Williamson/WireImages/Getty Images / ©Frazer Harrison/Getty Images

Pirates of the Caribbean: At World's End, featuring Johnny Depp (top left), was one of the Walt Disney Corporation's hit movies in 2007. The corporation is part of an oligopoly. Among other holdings, it owns 4 publishing companies, 11 major television stations (including ABC), 8 major cable stations (including ESPN and the Disney Channel), at least 60 radio stations in major cities, numerous magazines, and 6 theme parks and resorts (in the United States, Tokyo, Paris, and Hong Kong). All of these outlets promote and sell Disney products, increasing the corporation's profits even further.

- *Competition.* Capitalists compete in producing goods and services that offer consumers the greatest value in terms of price and quality.
- *Profit.* Selling something for more than it costs to produce generates profits and an accumulation of wealth for individuals and companies. Profits can invigorate competition because producers create goods and services that consumers want.
- *Investment.* By investing profits, capitalists can increase their own wealth. Workers, also, can accumulate savings and make investments that augment their income.

In reality, capitalism rarely functions in the ideal way because of abuses, greed, and worker exploitation, and it usually reflects monopoly and oligopoly rather than a free market that encourages competition. A **monopoly** is domination of a particular market or industry by one person or company. With little or no competition, a company can raise prices and reduce production.

An **oligopoly** is a market dominated by a few large producers or suppliers. For example, six companies—including CBS, Time Warner, and the Walt Disney Corporation—now own 90 percent of the U.S. media market. And, until fairly recently, a few large American companies dominated the automobile and steel industries.

The companies in an oligopoly often agree to set prices to reduce competition. Such agreements are illegal, but violations are rarely prosecuted. New entrepreneurs have little chance of breaking into an industry dominated by an oligopoly, and consumers have fewer choices in buying products or services (such as fuel and its delivery) because a small number of firms controls most of the market. Often, companies involved in an oligopolistic market—such as Barnes and Noble or Microsoft

Corporation—continue to grow through mergers and acquisitions of smaller companies.

SOCIALISM

Socialism is an economic and political system based on the principle of the public ownership of the production of goods and services. Ideally, socialism is the opposite of capitalism; in a socialistic system, there is

- *Collective ownership of property.* The community, rather than the individual, owns property. The government owns utilities, factories, land, and equipment but distributes them equally among all members of a society.
- *Cooperation.* Working together and providing social services to all people are more important than competition. People get what they need instead of being influenced by advertising.
- *No profit motive.* Goods and services are distributed equally, and private profits that are fueled by greed and exploitation of workers are forbidden.
- *Collective goals rather than individual investments.* The state is responsible for all economic planning and programs. People are discouraged from accumulating individual profits and investments and are expected to work for the greater good.

There have been many socialist governments during the last 150 years, including those in China, Cuba, and the Soviet Union (before its collapse in the early 1990s). None, however, has reflected pure socialism. Competition and individual profits were officially frowned upon, but there were always considerable differences in social equality. For example, government officials, athletes, and high-ranking party members enjoyed more freedom, larger apartments, higher incomes, and greater access to education and other resources than the rank and file.

MIXED ECONOMIES

After World War II, socialist parties came to power in many nations throughout the world, and much private industry was nationalized. Great Britain, Germany, Sweden, Belgium, the Netherlands, and some countries in Latin America, Asia, and Africa established socialist programs that included national health care and government-

monopoly domination of a particular market or industry by one person or company.

oligopoly a market dominated by a few large producers or suppliers.

socialism an economic and political system based on the principle of the public ownership of the production of goods and services.

communism a political and economic system in which all members of a society are equal.

globalization the growth and spread of investment, trade, production, communication, and new technology around the world.

owned enterprises (such as education and child-care programs). All of these countries, however, incorporated capitalistic features such as competition, private ownership of property, and free trade in the market.

Karl Marx predicted that **communism**, a political and economic system in which all members of a society are equal, would become dominant around the world because exploited workers would revolt against the yoke of capitalism. His prediction hasn't come true. China is a good example of a developing country that espoused communism and practiced socialism, but is now endorsing many aspects of capitalism (Becker 2002). In China, many state-owned enterprises are becoming private. Numerous leaders, who see capitalism as a source of economic development, approve foreign-owned factories that produce electronic goods, clothing, toys, and other products for export. As Communist-run industries and unions have dissolved, workers no longer have the security they enjoyed in the past. The income gap between the rich and the poor has increased dramatically, and there has been a surge in corruption, land seizures, arbitrary taxes, and social unrest. Learning about communism and socialism is mandatory from kindergarten through college, but college students, especially, often don't take the courses seriously because they see little if any connection between such ideologies and a "capitalist panorama" of wealthy executives, luxury apartments, shopping malls, and fast-food restaurants (Landsberg 2007; Lee 2007).

GLOBAL ECONOMIES

One of the most significant economic changes of the late twentieth century was **globalization**, the growth and spread of investment, trade, production, communication, and new technology around the world. Globalization also changes political systems, culture, language, migration, the environment, and many other aspects of life.

Proponents argue that everyone benefits from globalization because it increases economic freedom and democracy, creates millions of jobs, and brings affordable goods and services (like cell phones) to millions of households around the world. Most people among those polled in 47 industrialized and developing nations were enthusiastic about globalization but also worried that foreign influences and immigration may threaten their traditional cultures and national identities (The Pew Global Attitudes Project 2007).

Critics contend that globalization exploits poor workers, spreads pollution, and destroys indigenous cultures, natural habitats, and animal species (see Chapter 16). Critics also maintain that many people around the world don't need or benefit from globalization: They have little money to save or invest in global financial industries, consume food they grow, and use products they make rather than acquiring them on global markets, and don't rely on technologies that produce computers, fax machines, or even televisions and refrigerators. Thus, according to opponents, globalization benefits only the world's most powerful companies by expanding their worldwide base of consumers to reap more profits (Hunter and Yates 2002; Schaeffer 2003).

Large companies drive both capitalism and globalization. Major corporations, especially, wield enormous influence both at home and abroad.

3 Corporations and Capitalism

in 1999, Bob Thompson sold his construction company for $442 million. Because he and his wife believed this was far more than they needed for retirement, Thompson distributed $128 million to his 550 workers as well as

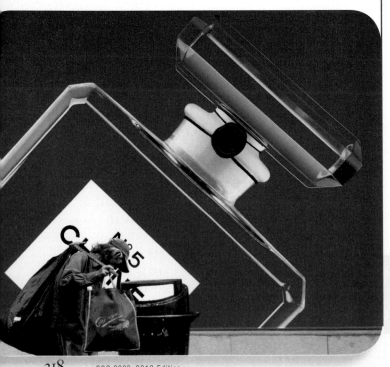

© AP Images

China's economic boom has not benefited all of its citizens. Many urban centers are thriving—offering those in upper and middle classes high-rise apartments, stores, and restaurants. In contrast, millions of Chinese, including many elderly people, like the one pictured left, survive by scouring trash bins for plastic bottles to recycle.

some retirees and widows. About 80 became instant millionaires ("Boss Sells His Company . . ." 1999). Unlike Thompson, most corporate heads have luxurious lifestyles, cut their employees' pensions, retire as billionaires, and pass on their massive wealth to their heirs (see Chapter 8). You'll see shortly that corporate and political power are interwoven, but let's begin by looking at some of the characteristics of corporations.

SOME CHARACTERISTICS OF A CORPORATION

A **corporation** is a social entity that has legal rights, privileges, and liabilities apart from those of its members. Until the 1890s, there were only a few U.S. corporations—in textiles, railroads, and the oil and steel industries. Today, there are more than 5 million, most created for profit. Corporations make up only 20 percent of all U.S. businesses but account for 85 percent of all annual business income, and a mere 0.5 percent of corporations bring in 90 percent of all corporate income (U.S. Census Bureau 2008).

Joel Bakan, a legal scholar, describes corporations as governing our lives: "They determine what we eat, what we watch, what we wear, where we work, and what we do." In amassing wealth, corporations exercise more power than do governments: "No internal limits, whether moral, ethical, or legal, limit what or whom corporations can exploit to create wealth for themselves and their owners" (Bakan 2004: 5, 111–112).

Even when stockholders suffer a financial loss, a corporation's chief executive officer (CEO) may still benefit greatly. In 2007, for example, Stan O'Neal, the CEO of Merrill Lynch, a financial corporation, retired after the bank reported its largest-ever quarterly loss of almost $2.3 billion. He walked away with almost $162 million in stock, options, and retirement benefits. If the company's stock rose under a new CEO, a $10 per share jump in stock could mean another $30 million for O'Neal ("Ex-Merrill Lynch CEO . . ." 2007).

CONGLOMERATES

When federal antitrust laws reined in monopolies during the twentieth century, many companies formed conglomerates. A **conglomerate** is a giant corporation that owns a collection of companies in different industries. Conglomerates emerged during the 1960s and were popular with investors because they protected against placing all of one's financial eggs into one basket.

One example of a conglomerate is Kraft Foods, which owns companies that produce snacks, beverages, pet foods, a variety of groceries, and convenience foods. The home page of Kraft Foods describes the company as one "that touches more than a billion people in 151 countries. Every day . . . We're there at breakfast, lunch and dinner, and anytime in between." Besides owning Maxwell House, Oscar Meyer, Life Savers, Ritz Crackers, and hundreds of other brand names, Kraft produces 23 cereals under the Post label and has ties with other corporations such as Starbucks (see www. kraft.com/brands). Conglomerates grow by acquiring companies through mergers. Shareholders like mergers because their stock usually rises in value, but few investors realize that mergers make chief executives "truly, titanically, stupefyingly rich" (Morgenson 2004: C1).

> **corporation** a social entity that has legal rights, privileges, and liabilities apart from those of its members.
>
> **conglomerate** a giant corporation that owns a collection of companies in different industries.
>
> **interlocking directorate** a situation in which the same people serve on the boards of directors of several companies or corporations.

INTERLOCKING DIRECTORATES

Conglomerates aren't the only locus of corporate power and wealth. In an **interlocking directorate**, the same people serve on the boards of directors of several companies or corporations. For instance, members of the board of directors of General Electric Company are also board members for at least 25 other companies, such as Johnson & Johnson, Home Depot, Kellogg Company, FedEx, Motorola, AT&T, ChevronTexaco, and several powerful brokerage and investment firms.

Some interlocking directorates are especially powerful because they include past U.S. presidents and members of Congress. For example, Sam Nunn was a partner at a prestigious law firm in Atlanta, Georgia, that represented numerous corporations. He was a Democratic senator from Georgia for almost 25 years, chaired several influential Senate committees, and then became a board member at ChevronTexaco Corporation, Coca-Cola, Dell Inc., and Internet Security Systems, Inc., among others. Because they sit on each other's boards, the members of interlocking directorates can set the prices for food, gasoline, automobiles, and even going to the movies (Krugman 2002; Draffan 2003).

TRANSNATIONAL CORPORATIONS AND CONGLOMERATES

Interlocking directorates have become more influential than ever because of the proliferation of transnational corporations. A **transnational corporation** (sometimes called a *multinational corporation* or an *international corporation*) is a large company that is based in one country but operates across international boundaries. By moving production plants abroad, large U.S. corporations can avoid trade tariffs, bypass environmental regulations, and pay low wages. The political leaders of many countries welcome transnational corporations to stimulate their economies, create jobs, and possibly enrich their personal bank accounts (Caston 1998; see also Chapter 16).

The most powerful corporations are **transnational conglomerates**, corporations that own a collection of different companies involved in various industries in a number of countries. General Electric is a good example of a transnational conglomerate. It owns hundreds of companies in the United States that are involved in, among other activities, manufacturing aircraft engines and parts, engineering consulting, insurance, and financial services. It also owns hundreds of subsidiaries in at least 27 countries and on every continent. These companies provide a wide range of products and services such as manufacturing light bulbs (in Hungary), ultrasound systems (in Germany and Thailand), and chemicals and medical technology systems (in Denmark, France, and Canada) (*LexisNexis Corporate Affiliations* 2003).

Transnational corporations and transnational conglomerates produce goods and services that they can sell to Americans at lower prices, largely because they develop local talent in emerging markets (like India and China) where the pay is 70 to 80 percent lower than in the United States. A small number of U.S. transnational corporations dominate world trade. Of the 25 most profitable companies worldwide, 17 are based in the United States. In effect, then, a very small group of corporations and interlocking directorates holds considerable global economic power (McGregor and Hamm 2008).

4 Work in U.S. Society Today

according to a recent Gallup poll, 55 percent of Americans said they were worse off in late 2008 than the year before—a record high in the poll's 39-year history. In addition, four in ten Americans felt that their standard of living was lower than 5 years earlier (Jones 2008; Newport 2008).

In mid-2008, 47 percent of Americans were very concerned about their financial situation. For example, 22 percent worried about being able to pay their rent or mortgage, 54 percent didn't have enough savings to handle a personal crisis like losing a job, and 58 percent of 18-to-29-year-olds had to borrow money from a friend or relative to meet expenses ("Dents in the Dream" 2008).

Scholars' views of such negative assessments are mixed. According to some economists, Americans are too gloomy because, since 2001, employers haven't laid off nearly as many people, the unemployment rate (which rose to 6.7 percent in early December 2008, and reached its highest level since 1993) is still relatively low compared to some European countries, and the domestic revenues from goods and services have steadily increased. Thus, "we have gotten so used to things being good, even when conditions become somewhat bad it feels terrible" (Irwin 2008: A1).

Others contend that many Americans aren't as pessimistic about the economic situation as perhaps they should be. Since 2006, for example, the share of the national income going to wages and salaries has been at its lowest level since 1929, whereas corporate profits have been at the highest level on record. White House aides maintained that the lower wages and salaries are due to employers' higher health care costs, but such arguments don't explain why corporate profits continue to rise, why low- and middle-income wages and salaries haven't kept up with inflation, and why only high-income households (that also receive health benefits) have reaped large gains, even during troubled economic times (Aron-Dine and Shapiro 2007; Bernstein 2008).

Millions of Americans are in debt because they purchase products they don't need or can't pay for, but many are using credit cards to stay afloat financially. During the housing boom between 2005 and 2007, many people took out mortgages they couldn't afford and now owe more than their houses are worth. When the housing bubble burst in 2007 and foreclosures increased, credit counselors were deluged by people, even some in the middle-income bracket, who had maxed out their credit cards purchasing everyday necessities like groceries and gas. In addition, food pantries and homeless shelters report that they've been swamped with unprecedented numbers of Americans, many of them middle-income wage earners, who can't keep up with mounting utility, food, and gas prices (Wolff 2007; Armour 2008).

You may not feel much sympathy toward people who live beyond their means, but even Americans who live modestly are losing economic ground because of deindustrialization and globalization.

DEINDUSTRIALIZATION AND GLOBALIZATION

Michael Wey, a 58-year-old father of three, expected to move into a secure retirement from his factory job as a toolmaker at Olofsson Corporation in Lansing, Michigan. Wey had worked there for 15 years and was earning $22.10 an hour. Instead, the company shut down the plant, taking Wey's retirement pension with it. In Michigan alone, 170,000 manufacturing jobs have been eliminated since 2000, many lost to overseas labor (Barnett 2004).

Michael Wey, like many others, is a casualty of **deindustrialization**, a process of social and economic change due to the reduction of industrial activity, especially manufacturing. Between 2000 and 2006, 20 percent of U.S. manufacturing jobs disappeared. Americans purchase more manufactured goods than in the past, but globalization has meant that these goods are increasingly produced in other countries. Because people in developing countries work for much lower wages, they are displacing workers in Western

economies, including the United States (Bivens 2008; Helper 2008). Among other consequences, deindustrialization has resulted in weaker labor unions and has been accelerated by offshoring.

Labor Unions

Some of your parents and grandparents probably reminisce about the days when labor unions provided job security, good benefits, pensions, and annual wage increases due to collective bargaining. This was especially true of some blue-collar jobs in manufacturing, transportation (such as truck drivers), and craft trades (such as plumbers and carpenters).

Times have changed. About 38 percent of government employees are unionized, but the rate of unionization among all other workers has been decreasing steadily—from a high of 27 percent in 1953 to 12 percent in 2007 (Eisenbrey et al. 2007). Labor union membership has declined for several reasons. First, some groups—such as retail salespeople and office workers, including secretaries—often see their jobs as short-term and hope to be upwardly mobile. Thus, there's little incentive to join a union. Second, many employees are afraid to unionize because employers often harass union supporters, fire them, and hold mandatory anti-union meetings with workers (Zellner 2002). Third, many people feel that unions don't have much clout because many jobs are offshored.

Offshoring

Offshoring refers to sending work or jobs to another country to cut a company's costs at home. Sometimes called *international outsourcing,* the

> **deindustrialization** a process of social and economic change due to the reduction of industrial activity, especially manufacturing.
>
> **offshoring** sending work or jobs to another country to cut a company's costs at home.

Strikes by unionized workers who can't be easily replaced through globalization (such as hotel workers, janitors, and teachers) are the most successful because the workers have considerable public backing (Dixon and Martin 2007). In contrast, airline pilots (pictured here in a strike against United Airlines in 2007 protesting management's large bonuses and labor shortages), who also can't be easily replaced, receive little public support. Why?

© AP Images

transfer of manufacturing jobs overseas has been going on since at least the 1970s. Labor unions can do little more than watch.

Most of the offshored jobs go to India and China, but large percentages are relocated to Canada, Hungary, the Philippines, Poland, Russia, Egypt, Venezuela, Vietnam, and South Africa. In India, for example, a computer programmer making $20,000 a year or less can replace an American programmer making $80,000 a year or more (Thottam 2004). Because wages in many countries are substantially lower than in the United States for comparable work, offshoring is profitable for both large and small companies.

Public debate about the practice became more heated when there was a jump in offshoring, especially in information technology (IT). Some economists predict that 14 million jobs (about 11 percent of all U.S. employment), many in IT, will be exported overseas by 2015. Often, those who are telling Americans to calm down have secure jobs with six-figure incomes or are executives of large corporations who derive a significant profit from offshoring (McGinn 2004; Bivens 2006).

Some economists maintained that the more creative aspects of IT (such as design and research and development) would never be offshored, but offshored jobs now include high-level and well-paid jobs in radiology, accounting, computer science, and engineering (Hira 2008). Thus, offshore workers are no longer simply telemarketers who work in back offices; they prepare tax forms and banking and medical analyses for large U.S. companies as well as for federal and state governments in the United States.

Advocates contend that offshoring by governments saves taxpayers money. Opponents say that there's something wrong when a government uses taxpayers' money to create jobs offshore, especially when experienced programmers and analysts are being forced out of the field. Critics also contend that offshoring increases risks of identity theft and privacy violations, especially when offshore workers have access to financial and medical records (Aspray et al. 2006; Bivens 2006; Whoriskey 2008).

HOW AMERICANS' WORK HAS CHANGED

Because deindustrialization and globalization have decreased job security, many Americans have had to adopt various coping strategies, including taking low-paying jobs and working shifts. These strategies, once characteristic of primarily working-class families, are now resorted to by increasing numbers of middle-class people. In addition, being laid off and able to find only part-time work is now common.

Low-Wage Jobs

One researcher has described the United States as "a nation of hamburger flippers" because of the growth of low-wage jobs (Levine 1994: 1E). These are jobs that are safe from offshoring because they must be done on site, such as child care, security, and hotel and restaurant work.

Measured in buying power, hourly wages have declined 15 percent since 1973 and are now at levels not seen since the mid-1960s. Because minimum wages don't keep up with inflation, they depress the living standards of low-income workers. Until the early 1960s, the minimum wage was around 50 percent of the average hourly wage. Today, it is only 33 percent of the average hourly wage, at its lowest level in more than 50 years (Chasanov 2004; Mishel et al. 2007).

By mid-2009, the minimum wage will rise to $7.25 an hour, but many families with a wage earner who already makes more than $7.25 an hour are still poor because of increases in rent, food prices, and energy bills. A mother of three said, for example, that her husband's weekly paycheck from his job at a grocery store lasted four days a year ago but lasts only two days now. Some economists predict that many employers will hire fewer people in low-wage jobs because of the higher labor costs of minimum wage earners (D'Innocenzio 2007; Trumbull 2007).

Unethical supervisors often find ways to decrease the earnings of low-wage workers even further. For example, Drew Pooters said he was stunned when he found his manager at the Toys "R" Us store in Albuquerque, New Mexico, altering workers' time cards. When Pooters objected, the manager threatened to fire him. Such illegal doctoring of hourly employees' records, called *shaving*, is not unusual but very hard to detect. Many managers shave time cards because they fear losing their jobs if they fail to keep costs down. Also, shaving helps to maximize profits and may thus increase managers' bonuses (Greenhouse 2004).

Shift Work

In many countries, workers are needed around the clock because business is being conducted almost every hour of every day, including weekends. In the United States, two in five people work weekends, evenings, or nights. Almost 15 percent of full-time employees work such nonstandard hours. Shift work is more common among men, African Americans, and people in service occupations. Nontraditional schedules are also becoming more common in professional and technical occupations because employees must often be on call during evenings and weekends in case international clients have questions or need assistance (U.S. Department of Labor 2005).

Shift work has both benefits and drawbacks. In terms of benefits, child care costs less because parents who work different hours can take turns at this activity. A related benefit is that fathers interact more with their children and get to know them better, and do more housework, including cooking, shopping for groceries, and sometimes doing the laundry. Such tag-team parenting also has drawbacks: Even if only one parent works on evenings or weekends, the children tend to do less well in school because a parent is not available to supervise homework. Preschool children in such families, especially, experience more emotional and behavioral problems because they receive less parenting; and parents report more depressive symptoms, such as feeling angry and sometimes withdrawing from family life. These and other strains, especially among low-wage workers, increase the likelihood of parental conflict and divorce (Swanberg 2005; Strazdins et al. 2006).

Downsizing, Contingent Workers, and Part-Time Work

Layoffs have long been fairly common in the workplace. One dramatic recent development, however, is the widespread occurrence of **downsizing**, a euphemism for firing large numbers of employees at once. When a factory closes or a company sends jobs offshore, thousands of people lose their jobs.

When Circuit City cut 17,000 jobs in 2007, many of the casualties were salespeople who had received laudatory reviews. They were earning only $15 an hour but were replaced by new hires who were paid $10 an hour (Gardner 2007).

Even when workers aren't downsized, millions are working fewer hours because many businesses are grappling with a lower demand for goods and services. In mid-2008, more than 5 million Americans who wanted full-time work held part-time jobs. While they are only about 14 percent of the labor force, Latinos represent 33 percent of people who shifted from full-time to involuntary part-time work, losing their benefits and health insurance. Because more than half of Latino workers tend to be young immigrants with fewer skills, they are most susceptible to economic downturns, especially in construction and service industries (like home and office cleaning chains) (Lazo 2008).

Large-scale layoffs have created a large pool of contingent and part-time workers. **Contingent workers** are employees who don't expect their jobs to last or who say that their jobs are temporary. Contingent workers make up 4 percent of the American work force. Across all age groups, contingent workers are more likely than permanent workers to be high school dropouts, but 52 percent of contingent workers want a permanent job (U.S. Department of Labor 2005). Some contingent workers may work full-time hours, but they usually have month-to-month contracts. Because the jobs are typically low-paid and rarely include benefits such as medical insurance and

downsizing a euphemism for firing large numbers of employees at once.

contingent workers people who don't expect their jobs to last or who say that their jobs are temporary.

Levi Strauss, a maker of blue jeans, cut its workforce from 37,000 employees in 1996 to 9,750 in 2003, sending most of the work to suppliers in 50 countries, from the Caribbean to Latin America and Asia ("Levi's Set to Close..." 2003).

© iStockphoto.com

Self-Employment and Underemployment

Of all U.S. workers, about 7 percent (almost 10 million) are *self-employed;* that is, they earn a living without being on an employer's payroll. Almost one of four self-employed individuals works in professional, scientific, and technical areas that require a high degree of expertise or training (such as accounting, engineering, photography, veterinary science, and law). Another large group provides products and services (these people include small retail and pet store owners, mechanics, drycleaners, realtors, and morticians). The third largest category is in construction (for example, painters, plumbers, electricians, and home maintenance contractors) (Bergman 2004; U.S. Bureau of Labor Statistics 2008).

About 38 percent of the self-employed are women, up from 27 percent in 1976 (U.S. Bureau of Labor Statistics 2007). Nearly a third have started their own businesses because they wanted a better work-life balance or feel that they have been pushed out of the

pension plans, contingent workers are likely to live from paycheck to paycheck instead of building up any savings.

work force by job discrimination, lack of child care, little sick time, and rigid work schedules (Ewers 2007; Johnson 2008).

Over 8 percent of Americans are **underemployed;** that is, they have part-time jobs but want full time work or their jobs are below their experience and education level (Mishel et al. 2007; U.S. Bureau of Labor Statistics 2008). Professional men in their 50s who are laid off (such as engineers, physicists, and computer scientists) are especially likely to start their own consulting companies as an alternative to unemployment or underemployment.

JOB SATISFACTION AND STRESS

A majority of Americans (53 percent) aren't satisfied with their jobs. Those who say they are very satisfied typically work in occupations that involve teaching, caring for and protecting others, and creativity. The least satisfying jobs are usually in low-skilled manual and service occupations, especially jobs involving customer service and food/beverage preparation and serving (see *Table 12.1*).

Job satisfaction is related to occupational prestige, but the association isn't perfect. For example, 58 percent

TABLE 12.1
Which Occupations Are the Most and Least Satisfying?

RANK	OCCUPATION	PERCENTAGE WHO ARE VERY SATISFIED	RANK	OCCUPATION	PERCENTAGE WHO ARE LEAST SATISFIED
1	Clergy	87.2	1	Laborers, except construction	21.4
2	Firefighters	80.1	2	Food preparers	23.6
3	Physical therapists	78.1	3	Packers and packagers	23.7
4	Authors	74.2	4	Clothing salespersons	23.9
5	Special education teachers	70.1	5	Cashiers	25.0
6	Teachers	69.2	6	Furniture salespersons	25.2
7	Education administrators	68.4	7	Roofers	25.3
8	Painters, sculptors, and other artists	67.3	8	Freight, stock, and material handlers	25.8
9	Psychologists	66.9	9	Bartenders	26.4
10	Security and financial services salespersons	65.4	10	Waiters/servers	27.0

Source: Based on Smith 2007, Table 2.

of doctors, a profession that has a high social standing, report lower job satisfaction than many people in lower prestige occupations (such as clergy and firefighters). It may be that doctors are less satisfied because their jobs involve a great deal of responsibility and stress (Smith 2007). They spend less time with patients and more time on paperwork than they did several decades ago because Health Maintenance Organizations (HMOs) dictate the length of a patient's stay in a hospital and often determine the treatment that a patient will receive, decreasing many physicians' autonomy.

People define happiness and satisfaction differently, but across all occupations, those who are most satisfied with their jobs are individuals who (in order of importance):

- feel appreciated (being praised or just thanked)
- get respect from workers and supervisors
- feel that they are trusted
- have opportunities to grow and learn on the job
- have a boss who's fair, honest, and listens to them
- feel that the job they're doing—however modest or low-paid—is important (Gardner 2008).

Such factors increase workers' satisfaction and commitment to an organization but don't cost the employer any money.

Work plays a central role in many people's lives, but 59 percent of Americans say that work is the leading source of stress in their lives ("Americans Engage . . ." 2006; Carroll 2007). You may have heard the quip "Europeans work to live; Americans live

to work," implying that Americans don't know how to relax and enjoy life. In France and Germany, the typical work week is 35 hours—comparable to the definition of part-time work in the United States. German and French government officials have tried to increase the number of hours in a typical work week but backed off because of worker protests.

The United States is the only industrialized nation in the world that doesn't legally guarantee workers a paid vacation (that is, there are no federal laws or regulations governing companies' vacation policies). Indeed, 25 percent of American workers in the private sector don't get any paid vacation time or paid holidays. In contrast, the typical worker, especially in Western Europe, has at least 6 weeks of paid vacation, regardless of job seniority or the number of years worked (Ray and Schmitt 2007). Furthermore, even though U.S. workers have the least vacation days in the industrialized world, a third of them don't take all the time they've earned—14 days on average. In contrast, Europeans use most of their vacation days (see *Table 12.2*). And, of the Americans who do take vacations, 24 percent check work e-mail or voice mail, up from 16 percent in 2005 (Expedia.com 2008).

Why do so many Americans forgo vacations? Many work longer (and often harder)

TABLE 12.2
Allowed and Used Vacation Days Vary by Country

COUNTRY	VACATION DAYS ALLOWED (AVERAGE)	VACATION DAYS NOT USED
United States	14	4
United Kingdom	20	3
Spain	22	2
Germany	24	3
France	30	2

Source: Data on vacation days are based on Ray and Schmitt 2007; data on vacation days not used are based on Expedia.com 2008.

© Pixland/Jupiter Images

than their European counterparts for a variety of reasons, such as a fear of downsizing, underemployment, a lack of medical insurance, and low wages. About 60 percent shrink their vacations to a few days or long weekends because they have too much to do on the job, fear being laid off, or worry about not being promoted because their company keeps pushing employees to put in more hours (Alesina et al. 2005; Egan 2006; Joyce 2006).

Some corporations, such as the professional services firm Pricewaterhouse-Coopers (PwC), encourage their employees to take all of their vacation days to avoid job burnout, reduce stress, and recharge. This isn't typical, however, because U.S. employers save about $65 billion a year when workers don't take the paid vacations they're entitled to (Expedia.com 2008).

In both high- and low-income jobs, stress results in a variety of health problems, including headaches, constant fatigue, sleep problems, difficulty in concentrating, upset stomachs, irritability, violence at home or the workplace, a weakened immune system, and heart attacks. Some people cope with work-related stress by changing jobs (if they can), exercising, enrolling in stress management programs, and/or eating healthier food. Many, however, cope with stress by missing work, using drugs, or overeating, all of which worsen employment and health problems. Job stress is costly—it carries an annual price tag for U.S. businesses of over $300 billion annually—because it results in increased absenteeism, employee turnover, and low productivity. Many of these costs are passed down to employees, who therefore pay higher health insurance premiums or receive fewer health benefits (Schwartz 2004; "Americans Engage . . ." 2006; American Institute of Stress 2007).

WOMEN AND MINORITIES IN THE WORKPLACE

One of the most dramatic changes in the United States during the twentieth century was the rise of women in the work force (see *Table 12.3*). Many factors contributed to the surge in women's employment, especially since the 1970s. These factors include increases in women's education and job opportunities, an increase in the number of single mothers, the fall in men's wages since 1980 (after accounting for inflation), and the increased costs of homeownership, requiring two incomes.

TABLE 12.3
Women and Men in the Labor Force Since the Nineteenth Century

	PERCENTAGE OF ALL MEN AND WOMEN IN THE LABOR FORCE		WOMEN AS A PERCENTAGE
YEAR	MEN	WOMEN	OF ALL WORKERS
1890	84	18	17
1900	86	20	18
1920	85	23	20
1940	83	28	25
1960	84	38	33
1980	78	52	42
1990	76	58	45
2006	74	60	44

Source: U.S. Census Bureau, 2008.

Largely because of higher educational attainment, 26 percent of women in two-income marriages bring home the bigger paycheck, up from 16 percent in 1981 (U.S. Bureau of Labor Statistics 2007). As a group, however, women have lower earnings than men in both the highest- and lowest-paying occupations, and the wage gaps are greater in high-income jobs (see *Figure 12.1*).

As you saw in Chapter 9, a number of factors help explain the gender pay gap. For example, women are concentrated in fields associated with lower earnings (such as health care and teaching in elementary and middle schools), whereas men tend to dominate the higher-paying fields (such as engineering, mathemat-

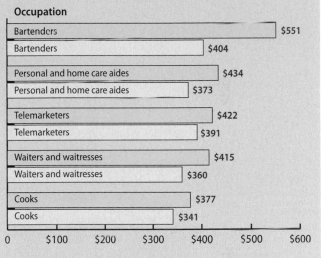

FIGURE 12.1

Women Earn Less Than Men Whether They're Chief Executives or Cooks

These were five of the highest- and lowest-paid occupations of full-time, year-round workers in the United States in 2007. How might you explain why the earnings differ by sex, especially in the highest-paid jobs?

☐ Median weekly earnings (Men)
☐ Median weekly earnings (Women)

Men Make More Than Women in Some of the Highest Paying Jobs

Occupation

Chief Executives	$1,918
Chief Executives	$1,536
Pharmacists	$1,887
Pharmacists	$1,603
Physicians and surgeons	$1,796
Physicians and surgeons	$1,062
Lawyers	$1,793
Lawyers	$1,381
Computer and information systems managers	$1,598
Computer and information systems managers	$1,363

0    $500 $1,000 $1,500 $2,000

. . . and in Some of the Lowest Paying Jobs

Occupation

Bartenders	$551
Bartenders	$404
Personal and home care aides	$434
Personal and home care aides	$373
Telemarketers	$422
Telemarketers	$391
Waiters and waitresses	$415
Waiters and waitresses	$360
Cooks	$377
Cooks	$341

0    $100 $200 $300 $400 $500 $600

Note: Some of the differences between men's and women's median weekly earnings may seem small, but multiply each figure by 52 weeks. For example, male physicians and surgeons earn over $90,000 a year compared with only about $55,000 for females.
Source: Based on material in U.S. Bureau of Labor Statistics 2008, Table 39.

---

ics, and the physical sciences). In addition, parenthood affects careers differently: Mothers are more likely than fathers (or women with no children) to work part time, take leave, or take a break from the work force to raise children—factors that reduce wages and salaries (Dey and Hill 2007). Still, a pay gap remains even when women and men have the same education, number of years in a job, seniority, marital status, and number of children and are similar on numerous other factors.

There are also big disparities in earnings across minority groups, but the differences are especially striking by sex. As you examine *Figure 12.2* on the next page, note two general characteristics. First, earnings increase—across all racial/ethnic groups and for both sexes—as people go up the occupational ladder. For example, managers and professionals make considerably more than car mechanics and truck drivers. But, across all occupations, men have higher earnings than their female counterparts of the same racial-ethnic group. At the bottom of the occupational ladder are African American women and Latinas, with the latter faring worse than any of the other groups. Thus, both sex *and* race or ethnicity affect the earnings of American workers.

5 Sociological Explanations of Work and the Economy

how do sociological theories help us understand work and the economy? During a recent conversation, I asked Matthew, one of my students, why, at age 51, he had decided to leave his

FIGURE 12.2
Median Weekly Earnings of Full-Time Workers by Occupation, Sex, and Ethnicity

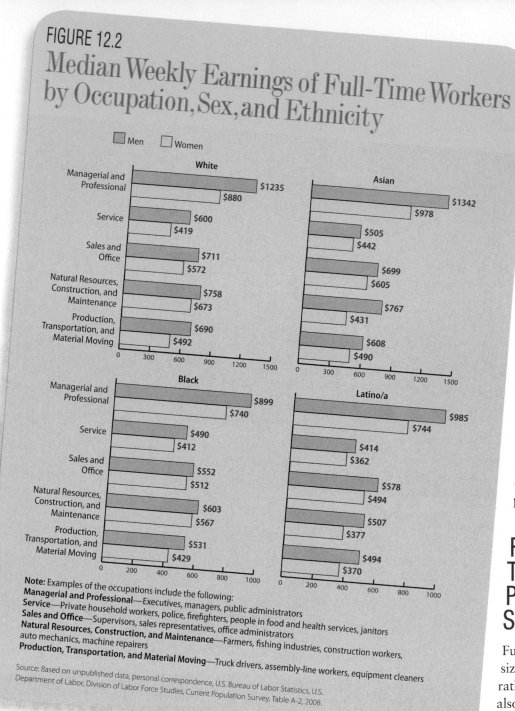

Note: Examples of the occupations include the following:
Managerial and Professional—Executives, managers, public administrators
Service—Private household workers, police, firefighters, people in food and health services, janitors
Sales and Office—Supervisors, sales representatives, office administrators
Natural Resources, Construction, and Maintenance—Farmers, fishing industries, construction workers, auto mechanics, machine repairers
Production, Transportation, and Material Moving—Truck drivers, assembly-line workers, equipment cleaners

Source: Based on unpublished data, personal correspondence, U.S. Bureau of Labor Statistics, U.S. Department of Labor, Division of Labor Force Studies, Current Population Survey, Table A-2, 2008.

ety. I know my salary will be at least 50 percent lower than it is now, but life is too short to worry about money all the time.

Matthew's comments—about job dissatisfaction and contributing to society—reflect functionalists' view that work provides numerous personal and societal benefits. In contrast, conflict theorists maintain that work is often dangerous to workers' health, feminist thinkers contend that work creates and reinforces sex inequality, and symbolic interactionists focus on how people learn work roles (*Table 12.4* on p. 231 summarizes each perspective's key points).

FUNCTIONALISTS' THEORIES: WORK PROVIDES MANY SOCIETAL BENEFITS

Functionalists typically emphasize the positive aspects of work rather than its constraints. They also see capitalism as bringing prosperity to society as a whole.

profitable job as a corporate accountant and return to college to get a degree in human services, which pays considerably less. According to Matthew,

> Accounting provided a steady salary and I made quite a bit of money preparing people's taxes on the side. After the kids graduated from college and we didn't have tuition payments, I became more fidgety. Work seemed more monotonous and I started dreading going to the office every morning. I figure that, after getting my [human services] degree, I can work with kids in low-income neighborhoods and contribute to soci-

Key Characteristics

For functionalists, work is important because it enhances a society's economy and defines many of its members' roles. Work is also functional because it bonds people. As jobs become more specialized, a small group of workers is responsible for getting the work done, and co-workers in such groups get to know one another well (Parsons 1954, 1960; Merton 1968). Such networks increase workplace solidarity and enhance a sense of belonging to a group where people listen to each other's ideas, converse, and "share a vision for the work [they] do together" (Gardner 2008: 16).

As you saw earlier, besides providing income, work has social meaning. When asked how much they like their jobs, many Americans see money as the third most important factor: less important than a sense of accomplishment, usefulness, and feeling valued; having a sense of stability, order, and a daily rhythm; and developing interesting social contacts (Katzenbach 2003; Brooks 2007).

Functionalists also maintain that wage inequities motivate people to persevere and to set higher goals. Low-paying jobs, for example, provide incentives to work harder or obtain further education to move up the economic ladder (see Chapter 8).

Critical Evaluation

Critics of the functionalists' perspective point out that work often leads to stress and myriad health problems. One author describes American jobs—at high and low income levels—as "digital assembly lines" with little room for creativity or independent thought and with supervisors who use software to micromanage and control employees' work (Head 2003). Many people have a McJob, an unstimulating and low-paid job that consists of performing repetitive and routine tasks, instead of feeling connected to a product or a group (see the material on McDonaldization on p. 109), and have little opportunity for advancement because of low educational attainment.

Functionalists can also be faulted for minimizing many U.S. corporations' disinterest in workers' well-being. About 135 developed countries have laws requiring private and public employers to provide paid sick leave to full-time workers. In contrast, as many as 43 percent of full-time nongovernment workers in the United States don't have paid sick days. Among workers who do have this benefit, many are finding their sick days reduced or counted as part of their vacation time. As a result, "many Americans simply tough it out when ill, going to work with pain, cramps, headaches, fevers, or worse" because they have no choice (Roan 2008).

CONFLICT THEORY: WORK CAN BE HAZARDOUS TO YOUR HEALTH

Whereas functionalists emphasize the benefits of work, conflict theorists argue that capitalism creates social problems. For example, they believe that globalization leads to job insecurity, a handful of transnational conglomerates have enormous power, jobs are offshored, and many workers have been laid off, especially because of offshoring.

Key Characteristics

Conflict theorists assert that low wages alienate employees rather than motivate them to work harder, because many workers realize that their labor benefits the wealthy and not themselves. Because much work is monotonous, employees arrive late, use company time for personal e-mail and online shopping, or quit. Matthew decided to change jobs because his work had become monotonous, but few people, especially those without college degrees and with heavy family responsibilities, have this option.

Functionalists say that if people have comparable skills, good attitudes, and perseverance, they'll have similar wages and chances for promotion. Conflict theorists challenge such claims because many women and minorities experience pay gaps and do not receive promotions regardless of their skills, attitude, or perseverance. Some contend, for example, that there is an "invisible hand" that excludes many African Americans from even blue-collar trades (Royster 2003).

Critical Evaluation

Are conflict theorists too quick to blame capitalism for economic problems? Middle-class Americans, especially, are borrowing more than they're earning and getting deeper into unmanageable debt. Having affordable health care, housing assistance, and low-cost student loans would decrease such debt, but 35 percent of Americans feel that people are simply living beyond their means—buying luxury products, taking expensive vacations, and often eating at restaurants (Center for American Progress 2006).

Critics also fault conflict theorists for emphasizing economic constraints rather than choices. Deindustrialization has lead to layoffs in assembly-line jobs, but many manufacturers complain that they are turning down business contracts because they are experiencing a shortage of skilled workers—such as welders, electricians, and machinists—despite offering decent pay

Are conflict theorists too quick to blame capitalism for **economic problems?**

(up to $25.00 an hour for a beginner), full health benefits, matching retirement funds, and annual bonuses (Hagenbaugh 2006).

FEMINISTS' THEORIES: WORK CREATES AND REINFORCES SEX INEQUALITY

Women make up 41 percent of the almost 3 billion workers around the world, but earn at least 20 percent less than men in every country (Hausmann et al. 2007). Feminist scholars agree with conflict theorists that there is widespread inequality in the workplace, but they see gender as a critical factor in explaining this inequality.

Key Characteristics

At all income levels and occupations, women—especially Latinas and black women—earn less than men (see *Figure 12.2* on p. 228). They rarely penetrate the top ranks, often regardless of ability, experience, or educational attainment. When a woman who worked at a large discount store complained to a department manager that she was earning less than men in the same position, he told her: "Women will never make as much as men [because] God made Adam first and so women will always be second to men" (Gibson 2004: A11). The message was clear—women are inferior to men.

Men tend to earn more and achieve higher positions in the workplace because family structure and the economy are intertwined. Because women are usually a source of cheap labor and few can support themselves and their children, they need a man's financial help. And because women provide most of the unpaid labor in the home, men are freed to devote most of their energy to their jobs and upward career mobility (England and Folbre 2005; Lorber 2005).

The effect of gender is especially evident during economic downturns. Men who work in manufacturing and construction do well in boom times and can build financial cushions, but those jobs are vulnerable during recessions, when business activities decline. However, some female-dominated occupations are also sensitive to economic fluctuations, as are the male dominated manufacturing and construction industries. During the recent real estate slump and subprime mortgage crisis, for example, many of those who lost their jobs worked at the lower levels of financial service organizations or as real estate agents, 77 percent of whom are women (Marks 2008).

Critical Evaluation

Critics of feminists' perspectives note that emphasizing men's domination of women in the workplace ignores many situations where high-status women control low-status men and women. Others criticize feminist scholars who accept the cultural expectation that mothers, not fathers, are responsible for hearth and home. For example, instead of pushing for paid leave for women to care for sick children or elderly family members, some critics argue that feminists should insist that men share such burdens and that the government implement family-friendly policies that reward men who share responsibilities at home (Kramer 2005; Graff 2007).

Some analysts, including women, also fault feminist groups for being too timid about shaming corporations that have few female officers at the top. Catalyst, a pioneering women's research organization, used to publish the names of companies that had few or no women at high-level positions. This practice motivated corporations such as General Electric to recruit women into the senior ranks. Recently, however, Catalyst has only praised corporations that have implemented greater diversity. One reason for the change may be that the

For symbolic interactionists, cultural expectations shape women's work and men's work, as well as determining whether women should even be employed. In Saudi Arabia, one of the world's most patriarchal societies, women are still not permitted to vote, travel abroad, work without the permission of a male relative, or drive, although this ban might soon be lifted ("Saudi Arabia . . ." 2008). Here, female employees work at the first car showroom in Saudi Arabia where women sell cars, but only to female buyers.

© AP Images

TABLE 12.4
Sociological Explanations of Work and the Economy

THEORETICAL PERSPECTIVE	LEVEL OF ANALYSIS	KEY POINTS
Functionalist	Macro	Capitalism benefits society; work provides an income, structures people's lives, and gives them a sense of accomplishment.
Conflict	Macro	Capitalism enables the rich to exploit other groups; most jobs pay little and are monotonous and alienating, creating anger and resentment.
Feminist	Macro and micro	Gender roles structure women's and men's work experiences differently and inequitably.
Symbolic Interactionist	Micro	How people define and experience work in their everyday lives affects their workplace behavior and relationships with co-workers and employers.

worst offenders (such as General Electric, Dell, and FedEx) are among the contributors to Catalyst's $11 million annual budget (Brady 2007).

SYMBOLIC INTERACTIONISTS' THEORIES: WE LEARN WORK ROLES

Unlike the other theorists, symbolic interactionists rely on micro-level approaches and try to explain the day-to-day meaning of work. They are especially interested in the formal and informal rules that develop in workplaces.

Key Characteristics

Interactionists have provided numerous insights on how people define and experience work. Many low-paid workers who feel disengaged from their jobs tolerate their situations because they have few options. They endure drug tests, constant surveillance, rat- and cockroach-infested buildings, physical pain, and work-related occupational injuries. In many of these jobs, "the trick lies in figuring out how to budget your energy so there'll be some left over for the next day" (Ehrenreich 2001: 195). Still, regardless of the occupation, co-workers may develop close relationships that become friendships outside of work.

Each of the four theoretical perspectives contributes to our understanding of work and the economy. They focus on different aspects but, taken together, provide us with a broad lens on the economic world.

Interactionists have also studied professionals to determine how they are socialized into their jobs. For example, sociology professors often encourage graduate students to attend and present papers at conferences to socialize them into becoming professionals. In medicine, surgeons teach their interns and residents that it is normal to make some mistakes, but that carelessness and continued errors are unprofessional (Bosk 1979). Assembly-line workers often control their co-workers who overproduce or underproduce by punishing or rewarding them, especially verbally (see Chapter 6).

Critical Evaluation

A common criticism is that interactionism, while providing in-depth analyses, sacrifices scope. For example, sociologist Deirdre Royster (2003) studied young men in Baltimore who had graduated from the same vocational high school at about the same time. She found that white male teachers tended to provide the white students, but not black students, with active assistance such as information about job vacancies and job references in trade occupations. We don't know, however, whether these findings are applicable to other cities or other types of jobs because the study was based on a small and nationally nonrepresentative sample of young men.

There is no

universal definition of family.

what do you think?

I would rather be married

than single.

1 2 3 4 5 6 7

strongly agree   strongly disagree

13

Families and Aging

In 2007, 95-year-old Nola Ochs received a college diploma, and was considering pursuing a master's degree. The mother of 4 and great-grandmother of 15 began taking classes after her husband died in 1972. Instead of viewing aging and her husband's death as setbacks, Ochs forged ahead: She had always yearned for a college education but was too busy raising the children on the family's farm. Ochs may not be typical, but her determination shows that, throughout the life course, we have choices and need not be held back by constraints.

In this chapter, we'll examine some of the most important aspects of families, how they've changed, their diversity, conflict and violence in families, and the aging of American society. Let's begin by considering what sociologists mean by *family*.

> **family** an intimate group consisting of two or more people who: (1) live together in a committed relationship, (2) care for one another and any children, and (3) share close emotional ties and functions.

1 What Is a Family?

sk five of your friends to define *family*. Their definitions will probably differ not only from each other but also from yours. For our purposes, a **family** is an intimate group consisting of two or more people who: (1) live together in a committed relationship, (2) care for one another and any children, and (3) share close emotional ties and functions. This definition includes households (such as those of foster families and same-sex couples) whose members aren't related by birth, marriage, or adoption.

There is no universal definition of *family* because contemporary household arrangements are complex. Family structures vary across cultures and have changed over time. In some societies, a family includes uncles, aunts, and other relatives. In other societies, only parents and their children are viewed as a family.

Key Topics

In this chapter, we'll explore the following topics:

HOW FAMILIES ARE SIMILAR

The institution of the family exists in some form in all societies. Worldwide, families are similar in fulfilling some functions, encouraging marriage, and trying to ensure that people select appropriate mates.

Family Functions

Families vary considerably in the United States and globally but must fulfill at least five important functions to ensure a society's survival (Parsons and Bales 1955):

- *Sexual regulation.* Every society has norms regarding who may engage in sexual relations, with whom, and under what circumstances. In the United States, having sexual intercourse with someone younger than 18 is a crime, but some societies permit marriage with girls as young as 8. One of the oldest rules that regulate sexual behavior is the **incest taboo**, a set of cultural norms and laws that forbid sexual intercourse between close blood relatives, such as brother and sister, father and daughter, or uncle and niece.

- *Reproduction and socialization.* Reproduction replenishes a country's population. Through socialization, children acquire language; absorb the accumulated knowledge, attitudes, beliefs, and values of their culture; and learn the social and interpersonal skills needed to function effectively in society (see Chapter 4).

- *Economic security.* Families provide food, shelter, clothing, and other material resources for their members. Increasingly in the United States, as you'll see shortly, both parents must work to purchase life's basic necessities such as housing and food.

- *Emotional support.* Families supply the nurturance, love, and emotional sustenance that people need to be happy, healthy, and secure. Our friends may come and go, but our family is usually our emotional anchor.

- *Social placement.* We inherit a social position based on our parents' social class. Family resources affect children's ability to pursue opportunities such as higher education, but we can move up or down the social hierarchy in adulthood (see Chapter 8).

Marriage

Marriage, a socially approved mating relationship that people expect to be stable and enduring, is also universal. Countries vary in their specific norms and laws dictating who can marry whom and at what age, but marriage everywhere is an important rite of passage that marks adulthood and its related responsibilities, especially providing for a family.

Endogamy and Exogamy

All societies have rules, formal or informal, defining an acceptable marriage partner. **Endogamy** (sometimes called *homogamy*) is the practice of selecting mates from within one's group. The group is similar in religion, race, ethnicity, social class, and/or age. Across the Arab world and in some African nations, about half of married couples consist of two first or second cousins because such unions reinforce kinship ties and increase a family's resources (Aizenman 2005; Bobroff-Hajal 2006).

Exogamy (sometimes called *heterogamy*) is the practice of selecting mates from outside one's group. In the United States, for example, 24 states prohibit marriage between first cousins, even though violations are rarely prosecuted. In some areas of India, where most people still follow strict caste rules, the government is encouraging exogamy by offering a $1,250 cash award to anyone who marries someone from a lower caste. This is a hefty sum in areas where the annual income is less than half that amount (Chu 2007; see also Chapter 8 on castes).

© Benjamin F. Fink, Jr./Brand X Pictures/Jupiterimages

HOW FAMILIES DIFFER

There are also considerable worldwide variations in many family characteristics. Some variations affect the structure of the family while others regulate household composition as well as other behaviors.

Nuclear and Extended Families

In Western societies, the typical family form is a **nuclear family** that is made up of married parents and their biological or adopted children. In much of the world, however, the most common family form is the **extended family**, a family consisting of parents and children as well as other kin, such as uncles and aunts, nieces and nephews, cousins, and grandparents.

As the number of single-parent families increases in industrialized countries, extended families are becoming more common. By helping out with household tasks and child care, adult members of extended families make it easier for a single parent to work outside the home. Because the rates of unmarried people who are living together are high, nuclear families now comprise only 23 percent of all U.S. families, down from 40 percent in 1970 (U.S. Census Bureau 2008).

Residence and Authority

Families also differ in their living arrangements, how they trace their descent, and who has the most power. In a **patrilocal residence pattern**, newly married couples live with the husband's family. In a **matrilocal residence pattern**, they live with the wife's family. In a **neolocal residence pattern**, the newly married couple sets up its own residence.

Around the world, the most common residence pattern is patrilocal. In industrialized societies, married couples are typically neolocal. Since the early 1990s, however, the tendency for young married adults to live with the parents of either the wife or the husband—or sometimes with the grandparents of one of the partners—has increased. At least half of all young American couples can't afford a medium-priced house; others have low-income jobs, are supporting children after a divorce, or just enjoy the comforts of a parental nest.

As a result, a recent phenomenon is the **boomerang generation**, young adults (and twice as many are men as women) who move back into their parents' home after living independently for a while or who never leave home in the first place. Parents try to launch their children into the adult world, but, like boomerangs, some keep coming back. Some journalists have called this group "adultolescents" because they're still "mooching off their parents" instead of living on their own. The English call such young adults "kippers" ("kids in parents' pockets eroding retirement savings"). In Germany, they are "nesthockers" (literally translated as "nest squatters") and in Japan "freeter" (young adults who job hop and live at home) (van Dyk 2005; Alini 2007).

Residence patterns often reflect who has authority within the family. In a **matriarchal family system**, the oldest

> **nuclear family** a form of family consisting of married parents and their biological or adopted children.
>
> **extended family** a family consisting of parents and children as well as other kin, such as uncles and aunts, nieces and nephews, cousins, and grandparents.
>
> **patrilocal residence pattern** newly married couples live with the husband's family.
>
> **matrilocal residence pattern** newly married couples live with the wife's family.
>
> **neolocal residence pattern** each newly married couple sets up its own residence.
>
> **boomerang generation** young adults who move back into their parents' home after living independently for a while or who never leave it in the first place.
>
> **matriarchal family system** the oldest females (usually grandmothers and mothers) control cultural, political, and economic resources and, consequently, have power over males.

In China's Himalayas, the Mosuo may be a matriarchal society. For the majority of Mosuo, a family is a household consisting of a woman, her children, and the daughters' offspring. An adult male will join a lover for the night and then return to his mother's or grandmother's house in the morning. Any children resulting from these unions belong to the female, and it is she and her relatives who raise them (Barnes 2006).

© Michael S. Yamashita/Corbis

patriarchal family system the oldest men (grandfathers, fathers, and uncles) control cultural, political, and economic resources and, consequently, have power over females.

egalitarian family system both partners share power and authority fairly equally.

marriage market a process in which prospective spouses compare the assets and liabilities of eligible partners and choose the best available mate.

monogamy one person is married exclusively to another person.

serial monogamy individuals marry several people, but one at a time.

polygamy a marriage in which a man or woman has two or more spouses.

females (usually grandmothers and mothers) control cultural, political, and economic resources and, consequently, have power over males. Some American Indian tribes were matriarchal, and in some African countries, the eldest females have considerable authority and influence. For the most part, however, matriarchal societies are rare.

A more widespread system is a **patriarchal family system**, in which the oldest men (grandfathers, fathers, and uncles) control cultural, political, and economic resources and, consequently, have power over females. In some patriarchal societies, women have few rights within the family and none outside the family, not being permitted to vote, drive, work outside the home, or attend college (see Chapter 9). In other patriarchal societies, women may have considerable decision-making power in the home, but few legal or political rights, such as the right to obtain a divorce or to run for a political office.

In an **egalitarian family system**, both partners share power and authority fairly equally. Many Americans think they have egalitarian families, but patriarchal families are actually more common. Employed women, especially, often complain that their husbands don't consult them before making important decisions such as buying a home or new car.

Love and Mate Selection

Sociologists often describe the dating process that occurs in the United States as a **marriage market**, a process in which prospective spouses compare the assets and liabilities of eligible partners and choose the best available mate. Marriage markets don't sound very romantic, but such open dating fulfills several important functions: fun, recreation, and companionship; a socially acceptable way of pursuing love and affection; opportunities for sexual intimacy and experimentation; and finding a spouse (Benokraitis 2007).

Many societies, in contrast, discourage dating because marriage is seen as a family decision rather than an individual one. Children may have veto power, but they are raised to accept arranged marriages. In much of India, marriages are usually carefully arranged to ensure a union that the family deems acceptable. In some of India's urban areas, however, arranged marriages also rely on nontraditional methods (such as online dating services) to find prospective spouses. (Pasupathi 2002).

Monogamy and Polygamy

Several types of marriages are common worldwide. In **monogamy**, one person is married exclusively to another person. Where divorce and remarriage rates are high, as in the United States, people are engaging in **serial monogamy**. That is, they marry several people but one at a time—they marry, divorce, remarry, divorce, and so on. One anthropologist concluded that only about 20 percent of societies are strictly monogamous; the others permit either polygamy or combinations of monogamy and polygamy (Murdock 1967).

Polygamy, a form of marriage in which a man or woman has two or more spouses, is subdivided into *polygyny* (one man married to two or more women) and *polyandry* (one woman married to two or more men). Polygyny is common in many societies, especially in Africa, South America, and the Mideast. In Saudi Arabia, for example, Osama bin Laden—who orchestrated the 9/11 terrorist attacks—has 4 wives and 10 children; his father had 11 wives and 54 children. No one knows the incidence of polygamy worldwide, but some observers feel that polygyny may be increasing (Nakashima 2003; Coll 2008).

Although extremely rare, polyandry may have existed in societies where it was difficult to amass resources. If there was a limited amount of available land, for example, the kinship group was more likely to survive if more than one husband contributed to the production of food (Cassidy and Lee 1989).

Western and industrialized societies forbid polygamy, but there are pockets of isolated polygynous groups in the United States and Canada. Most are fundamentalists who have broken away from the Church of Jesus Christ of Latter-Day Saints (Mormons); these groups perform marriages in secret ceremonies, and men—who are often in their 50s or 60s—marry girls as young as 10 years old. Law enforcement agencies rarely prosecute polygamists, but wives who have escaped from these groups have reported forced marriage, sexual abuse, pedophilia, and incest (Janofsky 2003; Madigan 2003).

How much do you know about contemporary U.S families and aging? Take this quiz to find out.

True	False	
☐	☐	1. **Out-of-wedlock births to teenagers have increased in the United States over the past 20 years.**
☐	☐	2. **Cohabitation (unmarried couples living together) decreases the likelihood of divorce.**
☐	☐	3. **Almost half of all U.S. children live in one-parent households.**
☐	☐	4. **Having children increases Americans' marital satisfaction.**
☐	☐	5. **About 20 percent of all Americans will never marry.**
☐	☐	6. **Women and men are equally likely to experience violence by an intimate partner.**
☐	☐	7. **Baby boomers (those born between 1946 and 1964) are the fastest-growing segment of the U.S. population.**
☐	☐	8. **Most people aged 65 and older are isolated and lonely.**

Turn the page for the answers.

©Ivan Kmit/iStockphoto.com

2 How U.S. Families Are Changing

the American family has changed dramatically over the past half century. Noteworthy changes involve divorce, singlehood, cohabitation, unmarried parents, and two-income families.

DIVORCE

Couples of all ages experience **divorce**, the legal dissolution of a marriage. Before the twentieth century, when life spans were much shorter, marriages typically ended with the death of one of the spouses. Now, over a lifetime, about 41 percent of Americans divorce. The U.S. divorce rate rose steadily during the twentieth century, plateaued, and then started dropping in 1995 (see *Figure 13.1*). Thus, divorce rates are *lower* today than they were between 1970 and 1990 (Kreider 2005).

Why do people divorce? There are many macro-level reasons. Divorce is easier than in the past because all states have enacted laws allowing **no-fault divorce**, meaning that neither partner need establish guilt or wrongdoing on the part of the other. Legally valid reasons for divorce range from infidelity and mental cruelty to irreconcilable differences and simple incompatibility. Technological advances, such as the Internet, have also made divorce more accessible than in the past. Some do-it-yourself divorce kits available online cost as little as $50 for all the necessary court forms and documents.

Changing gender roles, especially employed women's growing economic independence, have also contributed to higher divorce rates. Women with more resources—those who have college degrees and contribute at least half of the family income, for example—are filing for divorce instead of living with a husband who is emotionally distant or adulterous or who doesn't share domestic tasks (Rogers 2004).

Demographic variables also help explain divorce rates. Marrying at an early age—especially under

> **divorce** the legal dissolution of a marriage.
>
> **no-fault divorce** state laws that do not require either partner to establish guilt or wrongdoing on the part of the other to get a divorce.

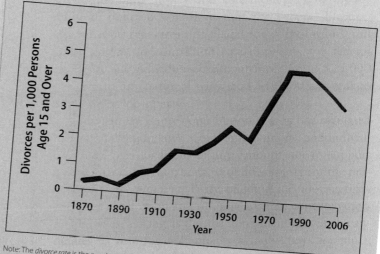

FIGURE 13.1
Divorce Rates in the United States, 1870–2006

Note: The *divorce rate* is the number of divorces that occur in a given year for every 1,000 marriages for women aged 15 and older.
Sources: Plateris 1973; U.S. Census Bureau 2007a; Eldridge and Sutton 2007.

stepfamily a household in which two adults are biological or adoptive parents, with a child from a prior relationship, who marry or cohabit.

18—increases the chance of divorce; in fact, it is one of the strongest predictors of divorce. For example, 48 percent of first marriages of women under 18 dissolve after 10 years, compared with only 24 percent of first marriages of women who are at least 25 at the time of the wedding (Kurdek 1993; Bramlett and Mosher 2002).

In most cases, young couples are not prepared to handle marital responsibilities. Among other things, they complain that their spouses become angry easily, are jealous or moody, spend money foolishly, or abuse alcohol and other drugs. People with a college degree are less likely to divorce than those with only a high school education, not because they're smarter but because going to college postpones marriage. As a result, these better educated couples are often more mature and capable of dealing with personal crises. They also have higher incomes and better health care benefits, which lessen marital stress over financial problems (Kreider and Fields 2002; Glenn 2005).

There are also interpersonal reasons for divorce. The most common causes are infidelity, conflict and communication problems, substance abuse, financial problems, and spousal abuse. For example, among those aged 40 and older, 23 percent of women, compared with only 8 percent of men, say that verbal, physical, or emotional abuse was "the most significant reason for the divorce" (Montenegro 2004; Glenn 2005).

According to a 2008 Gallup poll, 70 percent of Americans said that divorce is "morally acceptable" (up from 56 percent in 2001) but viewed 16 other activities—including gambling, premarital sex, abortion, homosexuality, and medical

If you answered true to any of the statements, you're wrong! Based on material in this chapter, all of the statements are **FALSE**:

1. Out-of-wedlock births to teenagers have decreased over the past 20 years, especially since the early 2000s.
2. Couples who cohabit before marriage have a higher divorce rate than those who don't.
3. A majority of U.S. children (68 percent) live in homes with married parents.
4. Generally, child-rearing lowers marital satisfaction for both partners.
5. At least 90 percent of Americans will marry at least once during their lifetime.
6. Women are almost three times more likely than men to experience violence by an intimate partner.
7. People aged 85 and over are the fastest-growing segment of the U.S. population.
8. Most people aged 65 and over visit friends and relatives daily, and two-thirds work or engage in numerous volunteer activities.

research on animals—as less acceptable (Saad 2008). Such tolerance may seem surprising because divorce causes stress, social isolation, ongoing parental conflict and hostility, and economic hardships, as well as behavioral problems among children (Harvey and Fine 2004; Amato and Afifi 2006).

The major positive outcome of divorce is that it provides an escape route for people in miserable marriages. Despite the high divorce rate, most people aren't disillusioned about marriage. Indeed, nearly 85 percent of Americans who divorce remarry, half of them within 3 years. Through remarriage, many couples form a **stepfamily**, a household in which two adults are biological or adoptive parents, with a child from a prior relationship, who marry or cohabit. More than 7 percent of all U.S. households have stepchildren. African American and American Indian children are most likely to live in stepfamilies, and Asian American children the least likely (Bumpass et al. 1995; Kreider 2003, 2005).

SINGLEHOOD AND POSTPONING MARRIAGE

The number of single Americans increased from 38 million in 1970 to 92 million in 2006, comprising 45

© Nicholas Belton/iStockphoto.com

© Mike Kemp/Rubberball/Jupiterimages

percent of those aged 15 and over. Singles include people who are divorced and widowed, but those who never married make up the largest and fastest-growing segment of the single population—29 percent in 2006, up from 22 percent in 1960 (U.S. Census Bureau 2007c).

Are Americans giving up on marriage? Not at all, because at least 90 percent will marry at least once during their lifetime. Instead, many people—including you, perhaps—are postponing marriage. In 1960, the median age at first marriage was 20 for women and 23 for men. By 2006, these ages had risen to 26 for women and 28 for men, the oldest average ages for first marriages ever recorded by the U.S. Census Bureau (U.S. Census Bureau 2007d).

There are numerous reasons for postponing marriage. On an individual level, most singles enjoy their independence and autonomy. Others stay single because they worry about divorce (Gibson-Davis et al. 2005). Moreover, if sex outside of marriage is available and socially acceptable, why marry?

Demographic factors also help explain the large number of single people. The longer one waits to marry, the smaller the pool of eligible partners because more of one's peers have paired off. The likelihood of singlehood also varies by education level. For example, 82 percent of unmarried people aged 25 and older are high school graduates compared with 23 percent who have at least a bachelor's degree (U.S. Census Bureau 2007a). More education often means more income, and more income reduces financial barriers to marriage.

Macro-level variables have also increased the number of single people. For example, because it's difficult to juggle a job and a family, many women have chosen to advance their professional careers before marrying and having children. For men, the well-paid blue-collar jobs that once enabled high school graduates to support families are mostly gone. Also, many young men don't want the responsibility of marriage and children, especially if they are in college or graduate school or seeking full-time work. Such factors delay marriage and encourage cohabitation.

COHABITATION

Another major change affecting U.S. families is an increase in **cohabitation**, an arrangement in which two unrelated people are not married but live together and have a sexual relationship (shacking up, in plain English). Because it is based on emotional rather than legal ties, "cohabitation is a distinct family form, neither singlehood nor marriage. We can no longer under-

stand American families if we ignore it" (Brown 2005: 33).

The number of U.S. households consisting of an unmarried couple has risen rapidly—from 50,000 in 1950 to over 5 million in 2006. If same-sex couples are included, the number increases by at least another 594,000 (Simmons and O'Connell 2003; U.S. Census Bureau 2007c).

> **cohabitation** an arrangement in which two unrelated people are not married but live together and have a sexual relationship.

Who cohabits? People who do so are a diverse group. Only 20 percent are 24 or younger, while over 60 percent are in their mid-30s to mid-40s (Martinez et al. 2006). Seniors sometimes cohabit rather than marry because of financial reasons. A 72-year-old woman who lives with her 78-year-old partner, for example, has no intention of getting married because she'd forfeit her late husband's pension: "My income would be cut by $500 a month if I got married, and we can't afford that" (Silverman 2003).

Does cohabitation lead to a better marriage? No. Most cohabiting relationships are relatively short-lived: About half end within a year, and over 90 percent end within 5 years. About 44 percent of these relationships result in marriage, but how stable are the unions? The probability of a first marriage ending in separation or divorce within 5 years is 20 percent for couples who did not cohabit before a marriage, but 49 percent for those who did. Thus, living together before marriage typically increases a couple's risk of divorce (Bramlett and Mosher 2002; Brown 2005; Lichter et al. 2006).

Why do cohabitors generally experience higher divorce rates? Some cohabitors are unsuccessful marriage partners because of drug problems, inability to handle money, trouble with the law, unemployment, sexual infidelity, or mental health problems. In addition, cohabitants are less likely than noncohabitants to put effort into the relationship, less likely to compromise, more likely to dissolve a relationship than work on it, and more likely to have poorer communication skills—characteristics that tend to lead to divorce (Dush et al. 2003; Stanley and Smalley 2005). Because of the higher divorce rate it leads to, some social scientists are adamantly opposed to cohabitation, but others see benefits (see *Table 13.1*).

UNMARRIED PARENTS

In 1950, only 3 percent of all U.S. births were to unmarried women. By 2006, there were almost 1.7 million such births, accounting for 39 percent of all U.S. births. Thus, nearly four in ten American babies are now born outside of marriage, a new record (Hamilton et al. 2007).

Births to unmarried women vary widely across racial-ethnic groups. White women have more out-of-wedlock babies than do other groups. Proportionately, however, nonmarried birth rates are highest for black women and lowest for Asian American women (see *Figure 13.2*). There is also considerable variation within groups. Among Asian/Pacific Islanders, for example, 6 percent of births to unmarried women are to Chinese women, compared with almost 20 percent to Filipinas and 51 percent to Hawaiian women. Among Latinas, 25 percent of births to unmarried women are to Cuban women and 60 percent to Puerto Rican women (Ventura et al. 2000).

The number of births to unmarried women aged 20 to 29 is three times higher than the number **to teenagers.**

TABLE 13.1
Benefits and Costs of Cohabitation

BENEFITS	COSTS
• Couples have the emotional security of an intimate relationship but can also maintain their independence by spending time with their friends separately and visiting family members alone (McRae 1999).	• Unlike married couples, cohabitants enjoy few legal rights. For example, there's no automatic inheritance if a partner dies without a will, and it's more difficult to collect child support from a cohabiting partner than a spouse (Silverman 2003).
• Couples can save money by sharing living expenses, dissolve the relationship without legal problems, and leave the relationship more easily if it becomes abusive (DeMaris 2001; Silverman 2003).	• Women in cohabiting relationships do more of the cooking and other household tasks than do wives (Coley 2002).
• Couples find out how much they really care about each other when they have to cope with unpleasant realities, such as a partner who doesn't pay bills or has low hygiene standards.	• People who cohabit before marriage tend to have fewer problem-solving skills (such as patience) than those who don't cohabit. In addition, cohabitors are more likely than biological parents to abuse their children sexually, physically, and emotionally (Cohan and Kleinbaum 2002; Popenoe and Whitehead 2002).
• Children in cohabiting households can reap some economic advantages from living with two adult earners instead of a single mother (Kalil 2002).	• Children who grow up in cohabiting households often lack role models for marital success because cohabiting adults don't always respect each other or communicate effectively (Martin et al. 2001).

FIGURE 13.2

Percentages of Births to Unmarried Women, by Race and Ethnicity, 2006

Legend:
- Black
- American Indian/Alaska Native
- Latinas
- White
- Asian American/Pacific Islander

Bar values:
- 71%
- 65%
- 50%
- 27%
- 16%

Source: Based on Hamilton et al. 2007, Table 1.

Every year, the media feature and applaud stay-at-home dads, but their numbers are negligible. An estimated 143,000 American men care for the family and do the housework while their wives are the wage earners. Among all married two-parent families, less than 1 percent of fathers are stay-at-home dads during a given year, and the role is usually temporary and undertaken because of unemployment or health problems.

© Comstock Images/Jupiterimages

dual-earner couples both partners are employed outside the home (also called *dual-income, two-income, two-earner,* or *dual-worker* couples).

The number of births to unmarried women aged 20 to 29 is three times higher than the number to teenagers. Still, 84 percent of teenage births are nonmarital, compared with 58 percent for women aged 20 to 24 and 31 percent for those aged 25 to 29 (Hamilton et al. 2007). Because teenage mothers are less likely than their older counterparts to have the resources, parenting skills, and maturity to raise healthy children, their offspring are especially vulnerable to poverty and neglect (Carlson et al. 2005).

Education is an important factor affecting out-of-wedlock births. For example, 63 percent of these births are to women who have not graduated from high school, compared with only 6 percent to women who have either a college or graduate/professional degree. Educated women usually have more job opportunities, more awareness of family planning, and more decision-making power in their relationships, such as the power to choose to use contraceptives (Downs 2003; Chandra et al. 2005).

TWO-INCOME FAMILIES

In the past 50 years, the proportion of married women in the labor force has almost tripled (see Chapter 12). In **dual-earner couples** (also called *dual-income, two-income, two-earner,* or *dual-worker couples*), both partners are employed outside the home. Employed married couples with children under 18 make up 65 percent of all married couples (U.S. Census Bureau 2008). When wives work full time outside the home, the median family income can be twice as high as when they do not, but the couples must also cope with conflicts between domestic and employment responsibilities.

For several decades, a number of sociologists described employed women, especially mothers, as enduring a *second shift*—having to perform housework and child-care tasks after coming home from a job. These days, however, *both* employed partners often experience a second shift. The total workload (in and out of the house) is high for both parents. Employed mothers shoulder twice as much child care and housework as their spouses, but employed fathers often work much longer hours than do mothers. Thus, the second shift can be equally stressful for both parents because mothers may feel overwhelmed and fathers may feel guilty for not taking on more domestic tasks and child-rearing responsibilities (Hochschild 1989; Robinson et al. 1999; Bianchi et al. 2006).

3 Diversity in American Families

If your ancestors were white ethnic immigrants (such as Irish, Italian, or Polish), they were probably viewed as deviant because of their language, customs, or religion. Despite discrimination, they contributed considerably to contemporary *family diversity*, "the variety of ways that families are structured and function to meet the needs of those defined as family members" (Stewart and Goldfarb 2007: 4). What about minority families? Let's begin with Latinos, the largest racial-ethnic minority in the United States.

LATINO FAMILIES

For many Latinos, *familism*—the belief that family relationships take precedence over individual decisions—and the strength of the extended family have traditionally provided emotional and economic support. In a national poll, for example, 82 percent of Latinos, compared with 67 percent of the general U.S. population, said that "relatives are more important than friends" ("The Ties That Bind" 2000).

About 66 percent of Latino children live in two-parent families, down from 78 percent in 1970 (Lugaila 1998; U.S. Census Bureau 2008). About 25 percent live in mother-only families (see *Figure 13.3*), and others live with relatives because recent immigrants often depend on family members until they can become self-sufficient (Garcia 2002; Sarmiento 2002).

Latinas, even when they're in the labor force, devote much of their lives to bearing and rearing children. Parental and marital conflicts can erupt, however, because mothers are often overloaded with

caring for children, husbands, and elderly relatives as well as working outside the home. Fathers don't do nearly as much parenting as mothers, but they're typically warm and loving with children. Compared with white fathers, Latino fathers are more likely to supervise and restrict their children's TV viewing and to ensure that they complete homework before they go out to play (DeBiaggi 2002; Toth and Xu 2002; Behnke 2004).

AFRICAN AMERICAN FAMILIES

Until 1980, married-couple families were the norm in African American families. Since then, black children have been more likely than children in other racial-ethnic groups to grow up with only one parent, usually a mother (see *Figure 13.3*). This shift reflects a number of social and economic factors: postponement of marriage, high divorce and separation rates, low remarriage rates, male unemployment, and out-of-wedlock births, especially among teenagers. Only 35 percent of all black children under 18 live in a two-parent home because "the formation of African-American households often originates not in marriage but in the birth of a child" (Barbarin and McCandies 2003: 52).

Some African American nuclear families expand to take in unemployed relatives and become extended families. Others welcome **fictive kin**, nonrelatives who are accepted as part of the family. The ties with fictive kin may be as strong as those established by blood or marriage because these members of the household provide support—such as caring for children—when parents are employed or negligent (Billingsley 1992; Dilworth-Anderson et al. 1993).

Affluent and middle-class black families, especially, emphasize success and the accumulation of material wealth through punctuality, hard work, diligence, and saving. Working-class black families also have high aspirations for their children, but they have few resources to achieve their goals. Low-income parents often have few contacts with any community institutions beyond church and school and depend on both of these to help their children succeed (Willie and Reddick 2003).

AMERICAN INDIAN FAMILIES

About 62 percent of the nation's American Indian/Alaska Native children live with two parents (see *Figure 13.3*).

Extended families are typical in these groups, especially among those living on reservations. Supportive family networks are especially important to American Indians who are isolated geographically because they migrate to urban areas to find jobs (MacPhee et al. 1996).

Children are important family members across all tribes. Parents and grandparents spend considerable time and effort making items for children to play with or use in activities and ceremonies (such as costumes for special dances and tools for gardening, hunting, and fishing). In many American Indian languages, there is no distinction between blood and marital relatives. Sometimes a father's brothers are called "father," uncles and aunts refer to nieces and nephews as "son" or "daughter," and a great-uncle may be referred to as "grandfather" (Sutton and Broken Nose 1996; Yellowbird and Snipp 2002).

American Indian families often emphasize values such as cooperation, sharing, personal integrity, generosity, harmony with nature, and spirituality—which differ from the individual achievement, competitiveness, and drive toward accumulation that many white parents emphasize. Families teach children that men and women may have different roles but that both should be respected for their contributions to the family (Kawamoto and Cheshire 1997).

ASIAN AMERICAN FAMILIES

Asian American family structures vary widely depending on the members' country of origin, time of arrival, status as either immigrants or refugees, and socioeconomic status. Asian American households are likely to be extended families, often including parents, children, unmarried siblings, and grandparents. Most children grow up in two-parent homes. Female-headed households—whether arising from divorce or out-of-wedlock birth—are much less common than in other racial-ethnic groups (see *Figure 13.3*).

Many Asian American parents are indulgent, tolerant, and permissive with infants and toddlers. As a child approaches school age, parents expect children to take on more responsibility—such as dressing, completing chores, and doing well in school. Most parents don't tolerate aggressive behavior or sibling rivalry and expect older children to serve as role models for their younger brothers and sisters. Parents also teach their children to conform to societal expectations because they are concerned about what other people—both within and outside of the Asian American community—think. Many Asian American parents not only pressure their children to excel in school but also endure hardships—even selling their house—to ensure the best college opportunities for their children (Chan 1997; Fong 2002).

MIDDLE EASTERN AMERICAN FAMILIES

"Wealth and children are the ornaments of this life," says the Qur'an, the sacred book of Islam. Despite the cultural emphasis on large families, U.S.-born Arab-American women average just under two children each (which is lower than the average among all U.S. women). Many postpone childbearing and have fewer children because they pursue college and

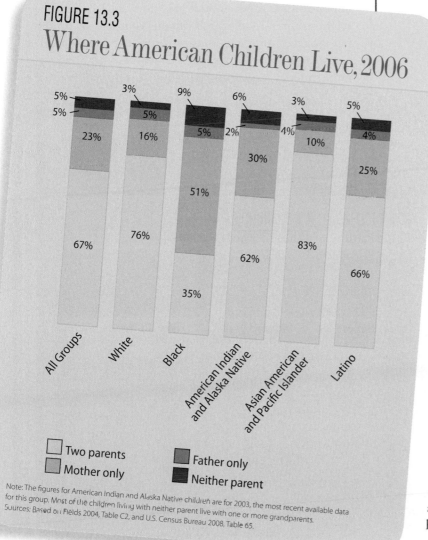

FIGURE 13.3
Where American Children Live, 2006

Two parents
Mother only
Father only
Neither parent

Note: The figures for American Indian and Alaska Native children are for 2003, the most recent available data for this group. Most of the children living with neither parent live with one or more grandparents.
Sources: Based on Fields 2004, Table C2, and U.S. Census Bureau 2008, Table 65.

professional degrees and careers (Kulczycki and Lobo 2001; Brittingham and de la Cruz 2005).

Most Middle Eastern American children (84 percent) live with both parents, compared with 67 percent of all U.S. children. Nuclear families are the norm, but extended family ties are important. Households composed of parents and children maintain close contact with their relatives and often provide financial, social, and emotional support. Parents also teach their children to feel a lifelong responsibility to their siblings and parents and to respect their aunts, uncles, cousins, and grandparents (Phinney et al. 2001).

Because many Middle Eastern American parents see education as critical for success, they expect their children to do well in school. Parents encourage their children to spend time on homework instead of watching television and hanging out with friends. The parents often serve as role models for acquiring education because 41 percent have at least a college degree, compared with 24 percent of the general U.S. adult population (Abu-Laban and Abu-Laban 1999; Brittingham and de la Cruz 2005).

GAY AND LESBIAN FAMILIES

In most respects, lesbian and gay families are like heterosexual families: The parents must make a living, may disagree about finances, and must develop problem-solving strategies. There are three major differences, however. One is that same-sex parents, compared with heterosexual parents, receive less social support from their relatives. Second, gay and lesbian parents face the added burden of raising children who may experience discrimination because of their parents' sexual orientation (Kurdek 2004). A third difference is that banning same-sex marriage denies gay and lesbian parents and their children numerous benefits that married couples and their children enjoy (see *Table 13.2*).

Lesbian and gay parents generally do a better job of disciplining their children than many heterosexual couples because they are more likely to use discussion instead of physical punishment. One of the reasons may be that lesbian and gay parents tend to be highly educated: Almost half have graduate degrees (Schorr 2001). Generally, high-socioeconomic-status parents are less likely than those at lower socioeconomic levels to hit or spank their children.

4 Family Conflict and Violence

Conflict is a normal part of family life, but violence is *not* normal. Families can be warm, loving, and nurturing, but they can also be cruel and abusive: "That violence and love can coexist in a household is perhaps the most insidious aspect of family violence, because we grow up learning that it is acceptable to hit the people we love" (Gelles 1997: 12).

Whether partners are married or not, tension, misunderstanding, and conflict are inevitable in intimate relationships. Couples fight about a variety of things, but the most common disagreements have four sources: gender role expectations (such as who does what housework), money (saving and spending), children (especially

TABLE 13.2
What Are the Legal Benefits of Marriage?

Some states, counties, and cities recognize civil unions and domestic partnerships (see Chapter 9). However, there are more than 1,138 federal benefits and protections tied to marriage that are denied to unmarried couples (U.S. General Accounting Office 2004). For example, partners who aren't married:

- aren't eligible for Social Security benefits if a partner dies;
- aren't entitled automatically to a share of the property if there is no will, and even if there is a will, a court can appoint a stranger to administer the estate;
- are unlikely to be required to pay child support or to have visitation rights;
- can't collect a one-time payment of $100,000 in line of duty benefits paid to a surviving spouse of a deceased firefighter, public prosecutor, police officer, or corrections officer;
- aren't eligible for veterans' compensation if a partner is killed or disabled in action;
- often pay higher federal and state income taxes because they can't file a joint return;
- don't have automatic privileges for hospital visits, access to intensive care, transferring a partner to a different health care facility, making decisions about organ donations, or deciding where the deceased will be buried.

© Larry Dale Gordon/Stone/Getty Images

discipline), and infidelity, both personally and online (Anderson and Sabatelli 2007; Benokraitis 2007). The conflict can escalate into abuse and violence.

INTIMATE PARTNER VIOLENCE

Nationally, almost 74 percent of intimate partner violence victims are women (Durose and Langan 2004; Catalano 2006). About 20 percent of women and 3 percent of men say that a current or former spouse, cohabiting partner, or girlfriend/boyfriend physically assaulted them at some time. These numbers are conservative because most people are often too ashamed or afraid to report the violence. Of all homicides, 33 percent of women and 3 percent of men are killed by their spouses or other intimate partners (Catalano 2006; Fox and Zawitz 2007).

Men are more likely than women to engage in repeated violence against their partners. Women are also more likely to sustain serious physical injuries because they are usually smaller than their partners, are attacked when they are pregnant (and especially vulnerable physically), and are more likely to use

fists rather than weapons (Weiss 2003; Koenig et al. 2006).

There is no "typical" batterer because abusive relationships reflect a combination of factors, some of which are macro-level variables. For example, although domestic violence cuts across all social classes and ethnic groups, women living in households with annual incomes less than $7,500 are nearly seven times more likely to be abused than those living in households with an annual income of $75,000 or more, because poverty and unemployment increase the likelihood of stress and violence. In addition, cultural factors, such as a strong orientation toward family and community, which is especially common among Latino and Asian American families, increase the likelihood of violence against women who don't "obey" their husbands. Women in Latino or Asian families, especially those who don't speak English well, may not report marital violence because they fear deportation or being ostracized by their community or they don't know about or trust social service organizations that provide help (National Latino Alliance . . . 2005; Macomber 2006; Thompson et al. 2006).

Micro-level variables also affect the likelihood of abuse. For example, violence escalates if one or both partners abuse drugs, if they have more children than they can afford (which intensifies financial problems), or if either partner has been raised in a violent household where abuse was a common way to resolve conflict (Benson and Fox 2004).

CHILD MALTREATMENT

Child maltreatment includes a broad range of behaviors that place a child at risk of serious harm or result in such harm or even death. Because only a fraction of the total number of child victimization cases is reported, federal agencies assume that millions of American children experience abuse and neglect on a daily basis (U.S. Department of Health and Human Services 2007).

Almost 80 percent of the perpetrators are parents, 10 percent are unrelated caregivers (such as foster parents and boyfriends), and 9 percent are relatives. Nearly 77 percent of the children who die due to abuse and neglect are younger than 4 years old. Victimization rates decline as children get older, presumably because older children can protect themselves or run away from home or because adults fear being reported by older children (U.S. Department of Health and Human Services 2007).

About 30 percent of children live in homes where parents or other adults engage in violence. Whether children are targets of abuse or see it, a growing body of research has linked violence with lifelong developmental problems, including depression, delinquency, suicide, alcoholism, low academic achievement, unemployment, and medical problems in adulthood (Currie and Tekin 2006; McDonald et al. 2006; Putnam 2006).

ELDER ABUSE AND NEGLECT

Elder abuse (sometimes called *elder mistreatment*) includes physical, psychological, and sexual abuse; neglect; financial exploitation; isolation from family and friends; deprivation of basic necessities, such as food and heat; and failure to administer needed medications. Family members and acquaintances mistreat an estimated 5 percent of people aged 65 and older every year. Some researchers call elder abuse "the hidden iceberg" because about 93 percent of cases are not reported to police or other protective agencies (National Center on Elder Abuse 2005).

Who are the victims of elder abuse? Over 77 percent are white, 43 percent are 80 years of age or older, and 66 percent are women; almost 90 percent of cases occur in domestic settings. Whites are more likely to engage in physical abuse of elders, while minorities are more likely to be guilty of neglect, emotional abuse, and financial exploitation (for example, keeping much of the income from the elderly person's Social Security checks) (Malley-Morrison and Hines 2004; Teaster et al. 2006).

Who are the perpetrators of elder abuse? Most are adult children (53 percent), spouses (19 percent), or other family members (18 percent). Only about 10 percent are friends, neighbors, or nonfamily service providers (Tatara 1998).

Family members neglect or abuse the elderly for a variety of reasons. On the micro level, for example, abuse of alcohol and other drugs is more than twice as likely among family caregivers who abuse elders as among those who do not. On the macro level, a shared residence is a major risk factor for elder mistreatment because the caregiver(s) may depend on the elderly person for housing while the elder is dependent on the caregiver(s) for physical help, increasing the likelihood of everyday tensions and conflict. Also, unlike low-income families, middle-class families are not eligible for admission to public nursing homes, and few can pay for the in-home nursing care and services that upper-class families can afford. As a result, cramped quarters and high expenses increase the caregivers' stress and may lead to elder abuse (Reay and Browne 2001; Bonnie and Wallace 2003).

The older a person is, the greater the likelihood of physical abuse. Elderly women are more likely than men to be victimized because they usually live longer than men, are more dependent on caregivers because they have few economic resources, and have little authority with family members. Pictured here, two adult children, whose mothers were victims of elder abuse inflicted by caregivers, testify before Congress on the subject of physical and sexual abuse in nursing homes.

5 Our Aging Society

at 97, Martin Miller of Indiana works full-time as a lobbyist on issues that concern older people. At 91, Hulda Crooks climbed Mount Whitney, the highest mountain in the continental United States. Julia Child was in her 50s when she became famous for her television programs on French cuisine. In her 80s, she embarked on a new show featuring world-renowned chefs.

Are these older people typical? Probably not, but many elders are vigorous and productive. The higher life expectancy of recent decades has forged many changes for older people and their families, some positive and some negative. (Many researchers use the terms *elderly, aged,* and *older people* interchangeably.) Let's begin by considering what is meant by "old."

WHEN IS "OLD"?

What images come to mind when you hear the word "old"? Someone who uses a cane or a walker? Or someone like Hulda Crooks, who climbed Mount Whitney? About a year before my mother died, she remarked, "It's very strange. When I look in the mirror, I see an old woman, but I don't recognize her. I know I'm 86, but I feel at least 30 years younger." She knew her health was failing, yet my mother's identity, like those of many older people, came from within, despite her physical age.

Age is largely a social construction. In societies where people rarely live past 50 (because of HIV/AIDS in many African nations, for example), 40 is old. In industrialized societies, where the average person lives to at least 75, 40 is considered young. Still, regardless of how we feel, society usually defines "old" in terms of chronological age. In the United States, for instance, people are typically deemed old at age 65, 66, or 67 because they can retire and become eligible for Medicare and Social Security benefits.

Gerontologists—scientists who study the biological, psychological, and social aspects of aging—emphasize that the aging population should not be lumped into one group. Instead, there are significant differences among the *young-old* (aged 65 to 74), the *old-old* (aged 75 to 84), and the *oldest-old* (aged 85 and older) in ability to live independently or to work and in health needs. Generally, for example, a 65-year-old woman is much less likely than an 85-year-old woman to need caregiving from family and friends.

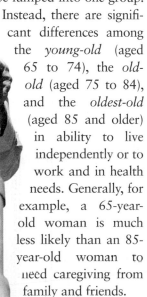

© BananaStock/Jupiterimages

LIFE EXPECTANCY AND MULTI-GENERATIONAL FAMILIES

In 1800, the average person's chance of living to the age of 100 was roughly 1 in 20 million; today, it's 1 in 50. More people are reaching age 65 than ever before, and American children born in 1990 have a **life expectancy,** the average length of time people of the same age will live, of 78 years (on average, 75 for men and almost 81 for women) (Jeune and Vaupel 1995; National Center for Health Statistics 2007). As a result, the number of Americans aged 65 and older is booming.

Almost 37 million Americans (about 12 percent of the total population) are 65 or older. This is a dramatic increase from just 4 percent in 1900. One of the fastest-growing groups is the oldest-old, whose number increased from 100,000 in 1900 to 4.2 million in 2000. By 2030, this group will constitute almost 3 percent of the U.S. population. By 2020, about 214,000 Americans will be *centenarians*, or people who are 100 years old or older (He et al. 2005; Federal Interagency Forum . . . 2006; National Center for Health Statistics 2007).

While the number of older Americans has increased, the proportion of young people has decreased. By 2030, there will be more elderly people than young people in the United States (see *Figure 13.4* on the next page). One result of longer life spans is the increase of multigenerational families. Many young children enjoy relationships not only with their grandparents, but also with their great-grandparents and even great-great-grandparents. On the other hand, millions of adults spend more years caring for frail and elderly parents, grandparents, and other relatives. These adults are often referred to as the **sandwich generation** because they are in the middle of two generations, caring for their own children as well as their aging parents.

GLOBAL GRAYING

In 2030, 12 percent of the planet's population (about 975 million people) will be 65 or older, up from 7 percent currently. Industrialized nations have the highest percentages of older people, but the less developed nations in many regions also have large numbers, and the proportions are increasing rapidly (see *Figure 13.5*).

This rapid aging suggests that many countries will face debates—which have already emerged in Europe, the United States, and Canada—over health care, social security, and other costs. Consider a few examples:

- Spain, which has a life expectancy of 80 years, has space in nursing homes for only one of every eight seniors who need such care.

- In China—which implemented a one-child policy in 1980 and whose proportion of people 65 and older is growing faster than that of any other large country—the pension system is already deeply in debt. One demographer predicts that China will soon become a 4-2-1 society in which one child will support two parents and four grandparents.

- In France, where the average retirement age is 59, four workers financed each pensioner in 1960, compared with two in 2000, and this ratio is expected to decrease to one worker per pensioner by 2020. Despite considerable protests, the government recently passed a law that requires public sector workers (more than 25 percent of the French work force) to work 40 years instead of 37.5 years before receiving a pension (Sciolino 2003; Longman 2004; French 2007).

Today, you have a 1 in 50 chance of living to be 100.

© David Freund/iStockphoto.com

6 Sociological Explanations of Family and Aging

the four sociological perspectives that we've considered in previous chapters are also useful in understanding families and aging. Functionalists focus on how families contribute to social stability, conflict theorists argue that the family reproduces social inequality, feminist scholars emphasize that gender roles burden women more than men, and symbolic interactionists consider how intimate relationships teach family roles throughout the life cycle (*Table 13.3* on p. 253 summarizes the key points of these theories).

The rapid global aging suggests that many countries **will face debates over health care, social security, and other costs.**

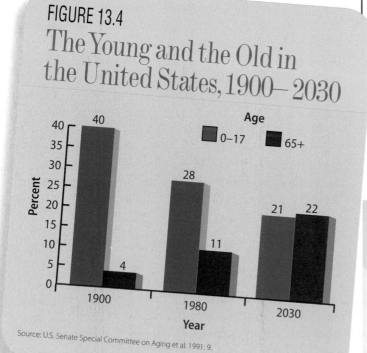

FIGURE 13.4
The Young and the Old in the United States, 1900—2030

Source: U.S. Senate Special Committee on Aging et al. 1991: 9.

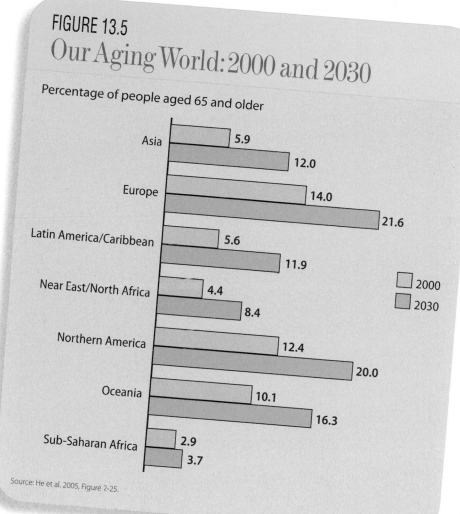
FIGURE 13.5
Our Aging World: 2000 and 2030

Percentage of people aged 65 and older

Asia — 5.9 / 12.0
Europe — 14.0 / 21.6
Latin America/Caribbean — 5.6 / 11.9
Near East/North Africa — 4.4 / 8.4
Northern America — 12.4 / 20.0
Oceania — 10.1 / 16.3
Sub-Saharan Africa — 2.9 / 3.7

2000
2030

Source: He et al. 2005, Figure 2-25.

FUNCTIONALISM

Functionalists spotlight the essential tasks that families carry out and that benefit society. The performance of these tasks, or *functions*, is critical for a society's survival. Even in our later years, society influences our roles and behavior.

Stability and Activity

We began this chapter by looking at the vital functions that families perform—such as procreation, socialization, and economic security—that promote societal stability and individual well-being. Functionalists recognize that families may differ in structure (nuclear versus extended, for example) but believe that the similarities ensure a society's continuity. Social problems arise, for instance, when parents can't or don't provide their children with the necessary financial and emotional support.

In the late 1950s, functionalists maintained that, as people age, they withdraw from their social roles and become isolated. This *disengagement theory* was abandoned when many researchers, including gerontologists, found that disengagement was due to age discrimination, not choice, and was limited to older people with disabilities and poor health. Functionalists replaced disengagement theory with **activity theory**, which proposes that many older people remain engaged in numerous roles and activities, including work, and that those who do so adjust better to aging and are more satisfied with their lives (see Atchley and Barusch 2004, for a summary of some of this research).

Critical Evaluation

For functionalists, the family is critical in promoting social order and stability. Critics maintain, however, that the functionalist approach reflects several weaknesses. First, the perspective seems to support traditional family arrangements, such as the nuclear family, but largely devalues other family forms, such as single-parent and same-sex families. Second, it is questionable whether some family functions are as universal or necessary as functionalists claim, because procreation often occurs outside of marriage and because the state has assumed some of the family's functions, such as caring for some children (as in foster homes) and the elderly (by providing Medicare payments, for example) (Lundberg and Pollak 2007).

Moreover, is activity theory as representative of older people as some functionalists claim? Many people continue to work, even in their 80s, not because they choose to do so but because they can't afford to retire, even though they are in poor health and unhappy in their low-income jobs (Kinsella and Phillips 2005).

Conflict theorists, especially, maintain that functionalists focus on family harmony, often ignoring the family's failings.

CONFLICT THEORY

Conflict theorists agree that families serve important functions. However, conflict theorists contend that some groups benefit more than others because families are sources of social inequality that mirror the larger society. The inequities are evident from birth to old age.

Inequality, Social Class, and Power

For conflict theorists, families perpetuate social stratification. Those in high-income brackets have the greatest share of capital, including wealth, that they can pass down to the next generation. Such inheritances reduce the likelihood that all families have equal opportunities or equal power to compete for resources such as education, decent housing, and health care (see Chapter 8). There's also a question of whether the government cherishes children as much as it professes. For example, 40 countries—including Cuba, Taiwan, and most of Europe—have lower infant mortality rates than the United States (Miniño et al. 2007).

Unequal power and access to resources continue as people age. Profit-seeking corporations lay off older workers because of their higher salaries and health insurance costs. Some employers say they value older workers' loyalty, work ethic, reliability, and experience, but many are less likely to hire or retain older workers because they view these individuals as less creative, less willing to learn new things, and less able to perform physically demanding jobs. Because many large companies have simply cut their pension plans, numerous older workers must work long after they expected to retire (Peterson 2006; Mermin 2007).

Critical Evaluation

Conflict theory is useful in highlighting family inequality across social classes, but it often glosses over the ability of low-income families to carry out important functions such as providing children with basic necessities and enough emotional support. Another limitation is that most of the elderly, especially those among the middle and upper classes, do not need to struggle for resources. Between 1960 and 2007, for example, the U.S. government's spending on children declined from 20 percent to 15 percent, but spending on people aged 65 and older grew from 22 percent to almost 46 percent because of programs like Social Security and Medicare. Some economists predict that even less of federal and state budgets will be allotted to children in the future because the American Association of Retired Persons (AARP) is the largest and most influential lobbying group in Congress (Chamberlain and Prante 2007; Carasso et al. 2008). A third weakness is that conflict theory links family inequality to capitalism and social class but often deemphasizes patriarchy and gender roles. Feminist theorists fill this gap.

FEMINIST THEORIES

Feminist scholars agree with conflict theorists that wealthy families differ significantly from families with few resources. However, feminist theorists go much further in emphasizing the inequality of gender roles in families, especially in patriarchal societies (including the United States).

Gender Roles and Patriarchy

Feminist scholars view the patriarchal family as a major reason for women's domination. Because employed mothers do twice as much housework and child care as men, many men benefit from women's unpaid labor at home. In most countries, males determine laws about property and inheritance rights, regulations on women's wages, and many other regulations that give men authority over women. Men hold most of the power, resources, and privilege, and many feel free to use women and children as sexual objects or targets of physical abuse, especially when legal systems, religious organi-

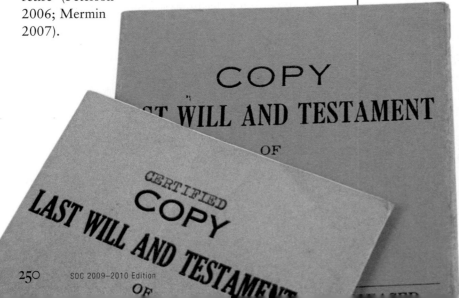

© Fred Dimmick/iStockphoto.com

zations, and medical institutions don't take violence against women and children as seriously as they should (Lindsey 2005).

Feminist sociologists have for some time challenged popular myths that children of single and employed mothers have emotional, cognitive, and behavioral problems. Feminist scholars have also been the first to document the ways in which sexism, racism, and classism intersect to oppress women and children. For example, the poorest older adults are most likely to be minority women, and caregivers of the old—who are predominantly women—must often leave their jobs or work only part-time to accommodate caregiving (Allen and Beitin 2007; Houser 2007).

Critical Evaluation

Feminist theories have called attention to the impact of gender and patriarchy in our everyday relationships. However, some critics question whether feminist perspectives overstate women's oppression. For example, patriarchal family structures limit women's rights, but they also provide important economic resources for women and children. Traditional gender roles discourage egalitarian relationships, but they provide stability and predictability (Lindsey 2005; Benokraitis 2007).

Feminist sociologists have deepened our understanding of family diversity, but some critics think feminist

© Image 99/Image 100/Jupiterimages

scholars have a tendency to view full-time homemakers as victims rather than as individuals who choose the role. Thus, some maintain, feminist scholars are "in danger of refusing to listen to a multiplicity of women's voices" (Johnson and Lloyd 2004: 160).

exchange theory contends that people seek through their interactions with others to maximize their rewards and to minimize their costs.

SYMBOLIC INTERACTIONISM

For symbolic interactionists, people create subjective meanings of what a family is and what its members' roles are. Thus, people learn, through interaction with others, how to act as a parent, a grandparent, a teenager, a stepchild, and so on throughout the life course.

Learning Family and Aging Roles

Interactionists often use exchange theory to explain mate selection and family roles. The fundamental premise of **exchange theory** is that people seek through their social interactions to maximize their rewards and to minimize their costs. In mate selection, people seek to trade their resources—such as wealth, intelligence, good looks, youth, and/or status—for more, better, or different assets.

Many people stay in unhappy marriages and other intimate relationships because the rewards seem equal to the costs ("It's better than being alone"). Similarly, many female victims tolerate an abusive relationship because they fear being alone or losing economic benefits a man provides. Rewards for perpetrators of family violence include the release of anger and frustration as well as the accumulation of power and control (Sherman 1992; Choice and Lamke 1997).

A well-known psychologist who interviewed more than 200 couples over a 20-year period found that the difference between lasting marriages and those that

Employed mothers do twice as much housework and child care as men, so many **men benefit from women's unpaid labor at home.**

split up was a "magic ratio" of 5 to 1—that is, five positive interactions between partners for every negative one: "As long as there was five times as much positive feeling and interaction between husband and wife as there was negative, the marriage was likely to be stable over time" (Gottman 1994: 41). The implication, often cited by sociologists, is that we can improve our family relationships by learning to interact in more positive ways.

Stereotypes about older people are rooted in U.S. society, but **continuity theory** posits that older adults can substitute satisfying new roles for those they've lost (Atchley and Barusch 2004). For example, a retired music teacher can offer private lessons.

Critical Evaluation

Symbolic interactionists offer valuable insights on social interaction as a dynamic process in which fam-

For symbolic interactionists, family traditions and rituals give people an identity and a sense of belonging to a group. Birthday parties (left), rituals that commemorate an important event, also unite family members and friends through positive social interaction. In most Latino communities, the quinceañera (pronounced "kin-say-NYE-ra") is a coming-of-age rite that celebrates a girl's entrance into adulthood on her fifteenth birthday. The quinceañera, an elaborate and dignified religious and social event (right), includes a traditional waltz with the father and the tossing of a bouquet to the boys to determine who will win the first dance with the young woman.

ily members—of all ages—continually modify their behavior throughout the life course. Nonetheless, a common criticism is that interactionism, a micro-level perspective, doesn't address macro-level constraints. For example, families living in poverty, and especially single mothers, are likely to be stigmatized and must often raise their children in unsafe neighborhoods. Such constraints increase stress, feelings of helplessness, and family conflict—all of which can derail positive everyday interactions (Seccombe 2007).

Exchange theory is useful in explaining why people stay in unhappy or abusive relationships, but individuals don't always calculate the potential costs and rewards of every decision. In the case of women who care for older family members, for example, genuine love and concern can override cost-benefit decisions, especially in many traditional Asian, Latino, and Middle Eastern families, where culturally defined kinship duties take precedence over individual rights (Hurh 1998; Do 1999; see also Chapter 10).

Another criticism is that continuity theory neglects societal obstacles that deter older people from engaging in many previous interactions. After retirement, for example, many people must get by on low fixed incomes and can't afford luxuries such as attending cultural or sports events that they might have enjoyed when they were employed. Also, older people who experience illnesses because of low-quality health care over a lifetime may become too sick to participate in family and community activities or pursue hobbies.

TABLE 13.3
Sociological Perspectives on Family and Aging

THEORETICAL PERSPECTIVE	LEVEL OF ANALYSIS	KEY POINTS
Functionalist	Macro	• Families are important in maintaining societal stability and meeting family members' needs. • Older people who are active and engaged are more satisfied with life.
Conflict	Macro	• Families promote social inequality because of social class differences. • Many corporations view older workers as disposable.
Feminist	Macro and Micro	• Families both mirror and perpetuate patriarchy and gender inequality. • Women have an unequal burden in caring for children as well as older family members and relatives.
Symbolic Interactionist	Micro	• Families construct their everyday lives through interaction and subjective interpretations of family roles. • Many older family members adapt to aging and often maintain previous activities.

Education is an

important source of formal knowledge
and socialization.

what do you think?

People don't need a college
degree to be successful.

1 2 3 4 5 6 7

strongly agree   strongly disagree

DICTIONNAIRE DE LA PEINTURE

L'Architecture
du XXᵉ Siècle

14
Education

Key Topics

In this chapter, we'll explore the following topics:

1 How Education in the United States Has Changed

education is a social institution that transmits attitudes, knowledge, beliefs, values, norms, and skills to its members through formal, systematic training. **Schooling**, a narrower term, refers to formal training and instruction provided in a classroom setting. U.S. education and schooling

> **education** a social institution that transmits attitudes, knowledge, beliefs, values, norms, and skills to its members through formal, systematic training.
>
> **schooling** formal training and instruction provided in a classroom setting.

have experienced four significant transitions during a relatively short period of time.

First, *universal education has expanded*. During colonial days, one of the family's many functions was teaching children to read. Because the teaching varied, in 1647, officials in Massachusetts ordered every town that had 50 families or more to provide schooling that would prepare all children for work and participation in the community. By the turn of the twentieth century, mass schooling was nearly universal in the United States.

Second, *community colleges have flourished*. In their early years, two-year colleges focused on vocational education in the fields of agriculture, home economics, and industrial arts. Besides preparing students to transfer to four-year institutions, the 1,600 community colleges in the United States today educate more than half of the nation's undergraduates, training 60 percent of nurses and 80 percent of firefighters, law enforcement officers, and other emergency workers (Boggs 2004).

Third, *public higher education has burgeoned*. Before World War II, most colleges were characteristically rural, private, small, and elitist, had students who were almost all white, male, and Protestant, and generally excluded married students. Passage of the GI Bill in 1944 changed the higher education landscape dramatically by providing to men who had served in the military during World War II funds for tuition, fees, books, and living expenses for up to 48 months. In 1947, veterans accounted for almost half of all college enrollments (Greenberg 2004).

Fourth, and especially during the last few decades, *student diversity has increased*. The proportion of all U.S. public school students who are members of racial-ethnic groups jumped from 31 percent in 1986 to 43 percent in 2006, largely due to the rise of Latino students. And, in higher education, members of minority groups and women have accounted for a large proportion of the increases in enrollment at colleges and universities since 1990 (Planty et al. 2008).

Because of these and other changes, 86 percent of Americans 25 or older have completed at least high school and 29 percent have attained at least a bachelor's degree. Both numbers are all-time highs. Young adults between the ages of 25 and 29, especially, are better educated than ever before (see *Figure 14.1*).

2 Sociological Perspectives on Education

Sociologists offer distinctive views of education. Functionalists emphasize its benefits, conflict theorists argue that it often perpetuates social inequality, feminist theorists point out that sexism prevents many women from gaining the full benefits they could get from education, and symbolic interactionists examine how the social context shapes teachers' and students' everyday experiences (*Table 14.1* on p. 265 summarizes these perspectives).

FIGURE 14.1
Educational Attainment of the U.S. Population, 1947–2007

High school graduate or more, 25 to 29 years

High school graduate or more, 25 years and over

Bachelor's degree or more, 25 to 29 years

Bachelor's degree or more, 25 years and over

Sources: Based on Stoops 2004, Figure 1; Planty et al. 2008, Tables 25-1 and 25-3; and Snyder et al. 2008, Table 8.

FUNCTIONALISM: WHAT ARE THE BENEFITS OF EDUCATION?

As in their analyses of other social institutions, functionalists maintain that education contributes to society's stability, solidarity, and well-being. Some of these contributions are *manifest functions* that are intended and visible. Others, *latent functions*, are unintended and less visible.

Manifest Functions of Education

For functionalists, education serves several important purposes in society, including the following:

- Schools are *socialization agencies* that teach children how to get along with others and prepare them for adult economic roles (Durkheim 1898/1956; Parsons 1959). Schools have even taken over some socialization functions traditionally handled by the family, such as teaching youngsters basic etiquette rules, self-discipline, anger management, and even acceptable sexual behavior (see Chapters 4 and 9).

- Education *transmits knowledge and culture*. Schools teach children skills such as reading, writing, and counting, but also convey cultural values that encourage competition, achievement, patriotism, and democracy (see Chapters 3 and 11).

© BananaStock/Jupiterimages

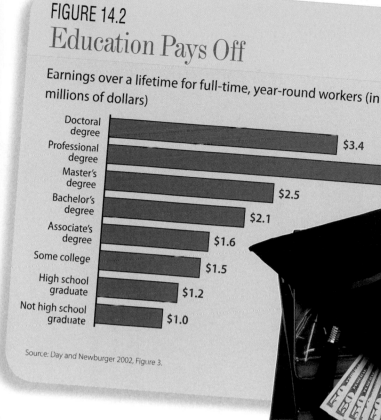

FIGURE 14.2
Education Pays Off

Earnings over a lifetime for full-time, year-round workers (in millions of dollars)

Doctoral degree	$3.4
Professional degree	$4.4
Master's degree	$2.5
Bachelor's degree	$2.1
Associate's degree	$1.6
Some college	$1.5
High school graduate	$1.2
Not high school graduate	$1.0

Source: Day and Newburger 2002, Figure 3.

© Comstock Images/Jupiterimages

- The transmission of cultural values increases *cultural integration*, the social bonds that people have with each other and with the community at large. One of the major reasons for the rise in public schools in the United States was the desire to Americanize the children of immigrants by teaching them the English language and the rules of responsible citizenship. Such schooling increases *societal cohesion* because it instills national values.

- Education encourages *cultural innovation*. For example, faculty at research universities, especially, receive billions of dollars every year to develop computer technology, treatments for diseases, and programs to address social problems such as crime and drug abuse.

- Education fosters a *meritocracy*, which allows people to move up the economic ladder because of their accomplishments rather than social background or connections (see Chapter 8). Indeed, 72 percent of first-year college students say that a major reason for attending college is job preparation (Eduventures 2006).

There's no doubt that education increases job prospects and upward mobility. A college graduate earns an average of $55,000 per year compared with only $29,000 for someone with a high school diploma (U.S.

Census Bureau 2008). By about age 55, college graduates bring home 90 percent more than people without a degree. Lifetime earnings are even higher for people with advanced degrees (see *Figure 14.2*). And, during economic downturns, people with the greatest amount of education are more likely to avoid long-term unemployment.

Many Americans seem to agree with functionalists that an education is important. In a national survey, 92 percent of recent graduates of two- and four-year institutions said that their education was worth the time and money, 84 percent stated that they were prepared with the skills and knowledge they needed in the workplace, and 50 percent felt that they used (in both their personal life and their work) the critical thinking, research, and communication skills they had learned in college. Furthermore, according to medical researchers, those who are better-educated—regardless of race or ethnicity— live longer because they are better informed about the benefits of healthful behaviors like keeping weight down, not smoking, and using condoms during sexual intercourse (American Council on Education 2008; Jemal et al. 2008).

Latent Functions of Education

Schools also fulfill many latent, or unintended, functions. For example,

- Schools *provide child care* for the growing number of single-parent and two-income families. Because many parents must work swing shifts or commute long distances to work, preschool, kindergarten, and after-school programs have mushroomed.

- High schools and colleges are *matchmaking institutions* that bring together unmarried individuals.

- Education *decreases job competition*; the more time that young adults spend in school, the longer the jobs of older workers they might replace are safe.

- Educational institutions *create social networks*. Especially in college, students can forge relationships with peers that can ultimately lead to jobs or business opportunities.

- Education is *good for business*. Thousands of companies (and jobs) have been created to measure students' intelligence and abilities and offer services (such as tutoring) to increase scores on standardized tests.

Critical Evaluation

Functionalist analyses are beneficial in highlighting the manifest and latent functions of education, but do functionalists exaggerate its benefits? Those who complete higher levels of education have, on average, higher vocabulary scores than those with lower educational levels. Over the last 70 years, however, the verbal abilities of graduates at all educational levels have slightly decreased (Nie and Golde 2008). Thus, higher educational attainment doesn't cause better communication skills. Instead, verbal ability reflects other variables such as social class.

Also, according to some critics, functionalists tend to gloss over education's dysfunctional aspects. For instance, schools aren't preparing young adults for the work force as well as they could; even many college graduates aren't punctual, dependable, or industrious; can't solve problems creatively; and don't write and speak well or evaluate information critically. Between 1992 and 2003, there was a 20 percent drop in college graduates' reading literacy, as in understanding news stories, job application forms, transportation schedules, and drug and food labels (EPE Research Center 2007; Kutner et al. 2007). Conflict theorists, especially, maintain that such problems reflect social stratification, an important macro-level factor that functionalists tend to deemphasize (see Chapter 8).

CONFLICT THEORY: DOES EDUCATION PERPETUATE SOCIAL INEQUALITY?

Whereas functionalists underscore the advantages of education, conflict theorists ask why education benefits some people more than others. From preschool to graduate school, conflict theorists maintain, education creates and perpetuates social inequality based on social class, race, and ethnicity and uses results of standardized tests and social control to maintain the status quo.

Social Class, Race, and Ethnicity

One of the best predictors of educational attainment is social class. Wealthy parents have the greatest amount of *economic capital* (such as income and other monetary assets), *cultural capital* (such as an elite education and positive attitudes toward education), and *social capital* (such as social networks that provide support and information about schooling). Differential access to all three types of capital reinforces and reproduces the

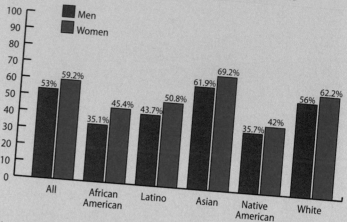

FIGURE 14.3

Racial/Ethnic and Gender Gaps in College Graduation Rates

Note: These are 6-year graduation rates for students who entered college in 2000.
Source: Kevin Carey, 2005, "One Step from the Finish Line: Higher College Graduation Rates Are Within Our Reach," p.2. Education Trust, Jan. 18, 2005. Reprinted by permission of Education Trust.

existing class structure from one generation to the next (Bourdieu 1984; Dickert-Conlin and Rubenstein 2007).

The children—whether intelligent or not—of influential, celebrity, and wealthy families attend elite elementary and high schools that guarantee them a place in the most prestigious colleges and universities. And affluent families not at that top level tend to live in school districts where public schools enjoy generous funding, up-to-date textbooks and laboratories, small classes, experienced teachers, and a high-quality curriculum ("Quality Counts . . ." 2007).

Minority students are enrolling in college in greater numbers than ever before, but graduation rates are higher for whites than for any minority group except Asian Americans (see *Figure 14.3*). Why are the attrition rates for some groups of students so high? There are numerous reasons, but most reflect social class. Students from lower socioeconomic backgrounds—many of whom are minorities—are less likely to attend high schools that offer high-level courses in mathematics and science that can mean the difference between passing or failing required courses during the first few years of college. Also, many college youth from low-income families face numerous obstacles: For example, they must work part time or full time to pay for tuition, books, and other costs, and therefore often have little time to interact with faculty who might motivate them to continue their studies or to seek financial aid (Chen 2005; Adelman 2006).

Standardized Testing and Gatekeeping

Gatekeeping refers to controlling the access to education or jobs of people from lower socioeconomic levels. Both IQ tests and standardized tests are examples of gatekeeping.

Researchers don't agree on a definition of intelligence but regularly measure an **intelligence quotient (IQ)**, an index of an individual's performance on a standardized test relative to the performance level of others of the same age. Schools typically use the IQ scores to place students into different ability groups.

Numerous scholars have challenged the validity of IQ scores. Among other things, they argue, social and environmental factors have at least as much influence on intelligence as inherited factors. For example, on the Wechsler Intelligence Test for children, the gap between the IQ scores of black and white students decreased from 17 to 11 points between 1978 and 2002. If schooling gains continue, the racial IQ gap could close in about 50 years. In addition, IQ scores can fluctuate because of major life stresses (Bowles et al. 2005; Dickens and Flynn 2006).

What about other standardized tests? The SAT Reasoning Test (called the Scholastic Aptitude Test until 1993, but now known simply as the SAT) includes sections titled "Critical Thinking," "Mathematics," and "Writing." According to the College Board, a nonprofit organization that owns and administers the test, the SAT "is a measure of the critical thinking skills you'll need for academic success in college. [It] assesses how well you analyze and solve problems—skills you learned in school that you'll need in college" (www.college board.com).

SAT defenders claim that scores on the test are accurate predictors of how well a high school graduate will perform in college. Thus, students with higher scores are better bets for institutions that are swamped with admissions applications and shrinking budgets. Critics contend that the SAT measures social class rather than ability because students at lower socioeconomic levels have less economic, cultural, and social capital. Commercial SAT preparation courses that cost $1,000 or more can sometimes increase an individual's total score by an average of 60 to 100 points. Such coaching gives an advantage to students from higher-income families. In addition, according to critics, the SAT and other standardized tests don't measure a variety of skills and abilities such as *creative intelligence* (using skills to create, invent, discover, and deal with novel situations) and *practical intelligence* (the ability to apply knowledge to everyday situations). Because of these and other criticisms, over 750 four-year colleges and universities in the United States don't use SAT scores in making admissions decisions (Lemann 2000; Sedlacek 2004; Yagelski et al. 2005; FairTest 2007; Soares 2007).

Advanced Placement (AP) courses and tests are also used in gatekeeping. Almost 25 percent of the 2.8 million students who graduated from U.S. public high schools in 2007 took at least one AP test. Students who do well on an AP test can skip one or more introductory courses in college in 37 subjects, ranging from Art History to World History. Many high school teachers encourage students to take AP courses and tests because they impress college admissions officers and can enable students to graduate a semester or two early, reducing the cost of a college education. As a result, the number of students taking AP courses has mushroomed, but only 7 percent of African American students, 10 percent of Asian American students, 14 percent of Latinos, and 0.6 percent of American Indian students take AP courses, compared with 62 percent of white students (College Board 2008a, 2008b).

Many minority, inner city, and rural students don't have access to AP courses. Even when such courses are available, students from low-income families may take

intelligence quotient (IQ) an index of an individual's performance on a standardized test relative to the performance level of others of the same age.

© Comstock Images/Jupiterimages

them less frequently than their peers from higher-income families. It may be difficult for low-income students to concentrate in school due to problems at home, or they may have teachers who discourage them because the courses are "too challenging" (Klopfenstein and Thomas 2005).

Education and Social Control

Functionalists see education as an avenue for upward mobility. For conflict theorists, such mobility is restricted by several factors: a hidden curriculum, credentialism, and privilege.

Every school has a formal curriculum that includes reading, writing, and learning other skills. Schools also have a **hidden curriculum** consisting of practices that transmit nonacademic knowledge, values, attitudes, norms, and beliefs that tend to legitimize "economic inequality and the staffing of unequal work roles" (Bowles and Gintis 1977: 108).

Schools in low-income and working-class neighborhoods stress obedience, following directions, and punctuality so that students can fill jobs (as in restaurants, retail stores, and hospitals) that require these characteristics. Instead of preparing students for college or careers, for example, some low-income urban high schools have curricula such as "medical careers and health professions" that focus on training nursing aides, health assistants, and other low-paid personnel for hospitals and nursing homes (Kozol 2005).

Schools in middle-class neighborhoods tend to emphasize proper behavior and appearance, cooperation, following rules, and decision making because many of these students will be working in middle-level bureaucracies that require such attributes. Routine is emphasized because it's good training for many bureaucratic jobs. In contrast, elite private schools encourage leadership, creativity, independence, and people skills—all prized characteristics in elite circles (see Chapter 8).

Teachers may tolerate rule breaking in these schools because their students will probably be making rather than following rules in the future.

Through the hidden curriculum, schools reproduce the existing class structure, ensure that social class position will be passed from one generation to the next, and provide workers to fill various jobs and occupations. According to conflict theorists, corporate values dominate much of higher education. College students are now "consumers," an education is a "product," canceling classes with low enrollments is "cost effective," colleges are "knowledge enterprises," nearly half of the faculty comprises a "contingent labor force" that works part time and for low salaries, and full-time faculty members are "shareholders" who are pressured to pursue "marketable" research that is supported financially by corporations and the government and leads to the development of products such as new drugs and defense technology.

From a conflict perspective, education—especially for lower and middle classes—has become a commodity, an object for sale, that is susceptible to the demands of the marketplace. One of these demands is for credentials.

Have you noticed that doctors', lawyers', and dentists' offices are usually wallpapered with framed degrees, certificates, and service awards? Faculty members, similarly, have letterheads and name plates that include their titles to signal their educational achieve-

© Rachel Epstein/PhotoEdit

ment. Whether the tangible symbols of people's achievements are framed, hung on office doors, or embellished on business cards, all of them reflect **credentialism**, an emphasis on certificates or degrees to show that people have certain skills, educational attainment levels, or job qualifications. Besides conferring social status, credentials connote knowledge or expertise in an area.

Functionalists maintain that credentialism rewards people for their accomplishments, sorts out those who are the most qualified for getting jobs, and fosters upward social mobility. Conflict theorists, however, contend that people don't need credentials for many jobs because most can gain skills on the job, during a few weeks of training, or will succeed because of ability or other factors. For example, many successful entrepreneurs like Ray Kroc, founder of the McDonald's restaurant chain, never attended college. Bill Gates of Microsoft, Steve Jobs of Apple Computers, and Michael Dell of Dell, Inc., were college dropouts. In addition, the brightest students aren't necessarily the only ones who go to college: 78 percent of students from low-income families who rank at the top of their high school class attend college, but so do 77 percent of students from high-income families who rank at the bottom of their class (Symonds 2003).

Because of a large supply of high school graduates, employers can demand higher levels of education even though some jobs (such as accounting and law enforcement) don't require a college degree for competent performance, a process called *credential inflation*. As more people acquire a college degree, its value diminishes, and students from low-income families, who are the least likely to have access to a college education, fall further behind (Collins 2002; Bollag 2007).

Many colleges say that they welcome low-income students. In fact, the proportion of financially needy undergraduates at the nation's wealthiest colleges and universities dropped between 2004 and 2007 (Fischer 2008). One reason is the schools' admission of *legacies*, the children of alumni who have "reserved seats" regardless of their accomplishments or ability. For example, President George W. Bush—who had ordinary high school grades and average standardized test scores—was admitted as a legacy at Yale University, where his father and grandfather attended (Golden 2006; Wickenden 2006; see also Chapter 8).

Another privileged group consists of *developmental cases*—students with inferior academic records who are children of celebrities, wealthy executives, or influential politicians who make million-dollar donations to the schools they want their children to attend. It is expected that the students will in turn also contribute after graduation. Because winning sports teams increase alumni donations, developmental cases also include outstanding athletes, even though athletes' dropout rates are higher than those of other students (Massey 2007).

> **credentialism** an emphasis on certificates or degrees to show that people have certain skills, educational attainment levels, or job qualifications.

Critical Evaluation

Conflict theory helps us understand the linkages between education and social inequality. Social class, standardized tests, the hidden curriculum, and credentialism enhance

Many sociologists note that graduating from a selective college or university has many benefits that include a network of useful contacts and better job offers. However, a student's motivation, ability, creativity, and ambition may be more important factors than his or her alma mater. Steven Spielberg, the well-known movie producer and director (pictured here while directing **Transformers**)*, graduated from California State University at Long Beach after being rejected by the more prestigious University of Southern California and University of California at Los Angeles film schools.*

privilege, whose effects continue well after college. Despite their contributions, however, conflict theorists have been criticized for ignoring the gains that many low-income and minority students have made, including graduating from college and graduate school (Marks 2007).

There is also a question of whether those in the social elite control education and job entry as much as conflict theorists claim they do. For example, several national surveys show that most CEOs of the biggest U.S. corporations didn't attend Ivy League or other elite private colleges. Instead, they went to state universities, big or small, to less well-known private colleges, and to community colleges (Jones 2005; Hymowitz 2006; Sowell 2008).

Feminist scholars agree with conflict theorists that education often perpetuates social inequality, but they attribute much of the inequity, both within and across racial/ethnic groups, to gender. And, unlike conflict theorists, feminist scholars supplement macro-level analysis with micro-level research.

© AP Images

The Taliban, a conservative religious group in Afghanistan, outlawed all education for girls in 1996, believing that educating girls violates Islam. The group was ousted from national power, but its influence is still strong in southern Afghanistan. In a country where 79 percent of women are illiterate and where coeducation is forbidden, gunmen have recently killed teachers and some schoolgirls. Other men have torched schools to stop girls from learning. Some elders are determined to rebuild the girls' schools, but many parents are extremely frightened and don't let their daughters attend school (Moreau and Yousafzai 2006).

FEMINIST THEORIES: HOW DOES GENDER AFFECT EDUCATION?

In 1967, the president of Columbia University said, "It would be preposterously naive to suggest that a B.A. can be made as attractive to girls as a marriage license" (Starr 1991: 198). About 35 years later, the president of Harvard University "explained" that fewer women than men have careers in math and science because of "innate ability differences between the sexes" (Margolis and Fisher 2002). Despite significant progress, women at home and around the world still experience striking inequality in education.

Worldwide Gender Gaps in Schooling and Literacy

Low levels of schooling, especially in developing countries, often reflect poverty: Poor parents simply can't afford to pay the fees for books (and often uniforms) and need their children's labor for economic survival. Girls, especially, encounter obstacles that limit their schooling, such as responsibility for household chores and caring for younger siblings, marriage at age 12 and sometimes younger, and parents' concerns that male teachers and classmates harass and even abuse female pupils (Sultana 2008).

Literacy is the ability to read and write in at least one language. In almost every country, women's literacy rates lag well behind those of men. Almost two-thirds of the world's 862 million illiterate people are women, and women's illiteracy rates are expected to increase in many regions, especially some African and Asian countries, by 2015 (UNESCO World Report 2005; World Bank 2007).

Both limited schooling and illiteracy diminish women's economic independence, increase their dependency on men, and decrease their ability to control their own property and understand their legal rights or vital information about AIDS, infant mortality, and birth control (see Chapter 9). Especially in developing countries, men often oppose educating women because doing so would empower women and encourage them to question current political and economic inequalities. In many cases, if parents can muster the resources to educate only one child, it will be a son whose adult responsibilities include providing for a family and elderly parents. In addition, if girls know they will be married at age 10 or so, they themselves may not see the value of literacy (Nussbaum 2003).

Gender Gaps in U.S. Education

Until the 1840s, U.S. higher education institutions banned women. How do girls and women fare now? Since 1971, national assessment tests have shown that there is little difference in average reading and mathematics scores for boys and girls aged 9 to 17. The biggest gaps are by ethnicity and social class: Students score lower if they are black or Latino, but especially if they come from low-income families that have few resources and lower parental involvement in the children's schooling. Also, in the highest- and lowest-scoring states, both girls and boys tend to do well or to have low scores, which suggest a connection between achievement scores and a state's population characteristics, such as the proportion of low-income students. Overall, however, girls and boys from similar socioeconomic backgrounds have similar scores (Perie et al. 2005; Lee et al. 2007; Hyde et al. 2008).

Since 1972, boys have consistently scored slightly higher than girls on the math and verbal portions of the SAT. In 2004, for example, the overall average for the verbal/critical reading score was 511 for males and 505 for females. The gender gap was slightly larger in mathematics—537 for males and 502 for females. As with the elementary to high school assessment tests, however, student performance across all racial-ethnic groups has been strongly linked to a family's social class, especially income (College Board 2006; Corbett et al. 2008).

In higher education, if we believe the media, college campuses are becoming all-female enclaves. Across all racial-ethnic groups, women have slightly higher college graduation rates than do men (see *Figure 14.3* on p. 258). As a result, many colleges have been giving men preferential treatment in admissions.

Slightly more women than men enroll in two-year and four-year U.S. colleges, but more men are earning college degrees today than at any time in history. The largest gap is not between women and men but between middle- and high-income white men and low-income Latino, African American, and white men. Overall, women have higher educational attainment rates because—in both high school and college—they earn better grades, spend more time studying, hold more leadership posts, and are more involved than men in student clubs and community volunteer work. These gender differences have existed since the early 1980s, probably because many young women realized that even if they married, they would be likely to work outside the home, and wanted higher-income jobs (Buchmann and DiPrete 2006; Goldin et al. 2006; King 2006; Pryor et al. 2007).

A comprehensive study of gender equity in U.S. education concluded that, from elementary to high school, both girls and boys are improving by most measures of educational achievement and that gender achievement gaps have narrowed. Despite the progress, however, if a crisis exists, it affects both sexes of African American and Latino students and those from lower-income families (Corbett et al. 2008).

Critical Evaluation

Feminist theorists extend conflict theorists' analyses by emphasizing how educational institutions produce gender-based inequality, but some critics fault feminist scholars for being too accepting of the inequality they document. For example, few women, including feminist faculty members, have challenged male privilege in schooling even though feminists have fought for educational equality since 1848 (Rhoads 2004).

Another criticism, although controversial because it seems to blame women, is that female faculty members in women's studies programs focus on topics such as sexual assault, eating disorders, and global women's issues rather than pay inequity in the United States. For example, one year after graduating from college, women earn 80 percent of what their male peers earn, but this percentage drops to 69 percent 10 years out of school, even in the same occupation (Dey and Hill 2007; Seligson 2007). It's not clear, however, whether courses don't address gender pay gaps or whether many female students aren't paying attention to the issue because they think, "It won't happen to me."

Like functionalist, conflict, and feminist theorists, symbolic interactionists examine educational outcomes. Their views supplement the other perspectives because

tracking (also called *streaming* or *ability grouping*) assigning students to specific educational programs and classes on the basis of test scores, previous grades, or perceived ability.

they seek to understand the schooling *process*, which requires asking different questions (see *Table 14.1*).

SYMBOLIC INTERACTIONISM: HOW DO SOCIAL CONTEXTS AFFECT EDUCATION?

None of us is born a student or a teacher. Instead, these roles, like others, are socially constructed (see Chapters 3 to 5). For symbolic interactionists, education is an active process in which students, teachers, peers, and parents participate and which often involves tracking, labeling, and student engagement.

Tracking

If a society believes that its education system is based on merit (which encompasses knowledge and effort), it justifies sorting students by aptitude. Such sorting results in **tracking** (also called *streaming* or *ability grouping*), assigning students to specific educational programs and classes on the basis of test scores, previous grades, or perceived ability.

Some educators believe that tracking is beneficial because students learn better in groups with others like themselves and because tracking allows teachers to meet the individual needs of students more effectively. Many symbolic interactionists maintain, however, that tracking creates and reinforces inequality. For example,

- High-track students take classes that involve critical thinking, problem solving, and creativity that high-status occupations require. Low-track students take classes that are limited to simple skills, like punctuality and conformity, that lower-status jobs usually require.

- High-track students have more homework, better-quality instruction, and more enthusiastic teachers. Moreover, high-track students see themselves as "bright," while low track students see themselves as "dumb" or "slow."

- The effects of tracking are usually cumulative and long lasting. Teachers tend to have low expectations for low-track students, who therefore fall further behind every year in reading, mathematics, and inter-action skills (Oakes 1985; Riordan 1997; Hanushek and Woessman 2005).

Tracking typically starts in kindergarten when teachers evaluate youngsters' ability to follow directions, to work

independently and in a group, and to communicate effectively—skills they'll need in the first grade. Tracking continues into and after middle school. High school becomes even more stratified as high-track students are sorted into gifted, honors, or advanced placement programs and courses. The reason for this tracking, typically, is to prepare the 90 percent of high school students who plan to attend college for the SAT and college admissions (Pianta 2002; Horn et al. 2003). In college, students continue to be tracked and sorted as some are accepted into honors' programs and accelerated undergraduate courses.

In 2002, the University of Connecticut received funding to identify and nurture math talent among elementary students, expecially those from low-income and minority backgrounds, even with poor language skills. When Project M³ (Mentoring Mathematical Minds) chose girls who spoke only Spanish at home, one teacher, who had planned to hold a girl back because she wasn't doing well in school, was surprised. After joining Project M³, the third grader "ended up being one of the top students when she left the fifth grade" (Teicher 2007: 16–17).

Labeling

Tracking often leads to labeling, a serious problem because "there is a widespread culture of disbelief in the learning capacities of many of our children, especially children of color and the economically dis-advantaged" (Howard 2003: 83). Labeling, in turn, can result in a *self-fulfilling prophecy* whereby students live up or down to teachers' expectations and evaluations that are affected by background, gender, skin color, hygiene, accent, and tests (see Chapter 4).

Teachers at all educational levels tend to praise and encourage students they consider bright and attentive, but are often disinterested in, critical of, or even impatient with students perceived as less bright, unmotivated, or troublemakers. Students with difficulties often meet their teacher's low expectations by not performing well. Teachers may try to avoid labeling, but they often convey their low expectations and negative attitudes through body language—for example, raising their eyebrows, frowning, or rolling their eyes—and by ignoring the student's accomplishments, however modest. Even very young children pick up on such cues (see Chapters 4 and 5).

Labeling and stereotypes can also affect a student's progress in higher education. For example, most Americans believe the myth that Asian American students are dominat-

TABLE 14.1
Major Sociological Perspectives on Education

THEORETICAL PERSPECTIVE	LEVEL OF ANALYSIS	VIEW OF EDUCATION	SOME MAJOR QUESTIONS
Functionalist	Macro	Contributes to society's stability, solidarity, and cohesion	What are the manifest and latent functions of education?
Conflict	Macro	Reproduces and reinforces inequality and maintains a rigid social class structure	How does education limit equal opportunity?
Feminist	Macro and micro	Produces inequality based on gender	How does gender inequality in education limit women's upward mobility?
Symbolic interactionist	Micro	Teaches roles and values through everyday face-to-face interaction and practices	How do tracking, labeling, self-fulfilling prophecies, and engagement affect students' educational experiences?

ing higher education institutions (especially in science, technology, engineering, and math departments), are enrolled in only the most elite colleges, and are all stellar students (see Chapter 10).

In fact, Asian Americans and Pacific Islanders are evenly distributed in two-year and four-year institutions; the majority attend public colleges and universities; they have a wide range of abilities and standardized test scores; and almost 70 percent earn bachelor's degrees in business, the social sciences, humanities, and education. Because of labeling and stereotypes, however, many Asian American college students are less likely to receive mentoring from instructors and are often held to a higher standard than white students (College Board 2008a).

Student Engagement

Many U.S. schools have been assessing performance not only through test results but also by *student engagement*, how involved students are in their own learning. A team of researchers who spent thousands of hours in more than 2,500 first-, third-, and fifth-grade classrooms concluded that the typical U.S. child has only a 1-in-14 chance of being in a school that encourages her or his engagement. Largely because of the federal law called No Child Left Behind (which we'll examine shortly), teachers in public elementary schools spend most of the school day on basic reading and math drills and little time on problem solving, reasoning, science, and social studies. As a result, the study found, classrooms are often "dull and bleak places" where kids don't get a lot of teacher feedback or face-to-face interaction with their peers (Pianta et al. 2007).

Especially before high school, parental involvement has a strong and positive effect on student achievement. Across public and private schools—and regardless of social class or race and ethnicity—parents can enhance their children's academic performance by frequently discussing what the children study in school, encouraging them to participate in school activities and events, selecting courses, attending school meetings, and volunteering at the school (Houtenville and Conway 2008).

Are most U.S. high school students engaged in their education? Not as much as they could be. For example, 26 percent admit that they usually don't do their homework, 18 percent come to school without their books or supplies, and 10 percent drop out. The students who are the most likely to be disengaged are from poor or low-income families, are African American or Latino, are not living with both biological parents, and attend financially strapped urban public high schools that are typically overcrowded and understaffed and have fewer resources such as computers and even textbooks (Finn 2006; Planty et al. 2007).

At the college level, the more engaged students are, the better their grades and their likelihood of graduating. The positive effects of engagement are just as great for lower-ability students and for those who attended disadvantaged high schools. But how engaged are most college students? The typical college student spends more time partying and socializing than studying. Some 65 percent of college students say that they study for 10 or fewer hours a week, well below the 24–30 hours faculty members say students should be spending on class preparation. Also, 32 percent of college students admit playing video games during classes. According to one college instructor, "Education is the only business in which the clients want the least for their money" (Perlmutter 2001: B1; Jones 2003; National Survey of Student Engagement 2006; Saenz and Barrera 2007).

Critical Evaluation

Symbolic interactionists shed considerable light on education and its everyday social context. Their theories show how, through everyday practices (such as tracking and labeling), educational institutions influence people's attitudes, values, and behavior and shape self-identity.

One weakness of this approach is that it deemphasizes individual power in changing the course of events, including overcoming the effects of labeling. For example, in a telephone survey, low-income Latino parents in Chicago, New York, and Los Angeles who were recent immigrants said that they contacted school personnel on a regular basis, despite the language barrier: "I had a constant call with the teachers. I would call them at least twice a month," said one mother. Other parents said that they went to college open houses and "pushed their way past the crowd" to speak to college representatives, with their children translating the questions and answers (Tornatzky et al. 2002).

Another limitation is that symbolic interactionists, because of their micro-level analysis, neglect the macro-level constraints that are built into society. As conflict and feminist theories show, education in the United States is affected by gatekeeping and control of resources by privileged and powerful groups, which often shape what teachers teach, what students learn, and who will and won't have access to higher education and to specific colleges (Sacks 2007). Such constraints have created numerous problems with U.S. education.

3 Some Problems with U.S. Education

Consider the following statistics:

- 50 percent of Americans aged 18–24 can't find the state of New York on a blank map of the United States (National Geographic Education Foundation 2006).

- 45 percent of employers say that public high school graduates don't have the basic skills to advance beyond entry-level jobs (*Crisis at the Core . . .* 2004).

- 40 percent of students at four-year colleges and 63 percent at two-year colleges require remedial coursework ("Rising to the Challenge. . ." 2005).

- 32 percent of Americans aged 18 and older don't know all the words to our national anthem (Corso 2008).

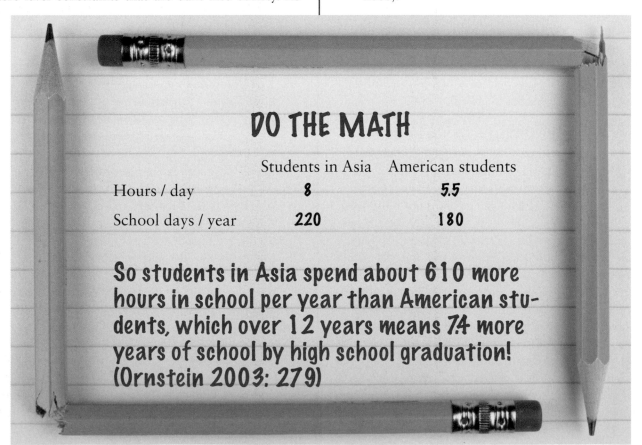

© Gary Woodard/iStockphoto.com

DO THE MATH

	Students in Asia	American students
Hours / day	8	5.5
School days / year	220	180

So students in Asia spend about 610 more hours in school per year than American students, which over 12 years means 7.4 more years of school by high school graduation! (Ornstein 2003: 279)

These statistics and others make the outlook seem bleak, even though the United States, compared with many other countries, has a high college graduation rate. Many high school and college students aren't as engaged in learning as they might be, but the U.S. educational system suffers from other problems, including inadequate schooling and funding, high dropout rates (in both high schools and colleges), and widespread grade inflation.

QUANTITY AND QUALITY OF SCHOOLING

When Gallup pollsters ask Americans about their dissatisfactions, public education is typically on the list of top problems every year (Saad 2007). Is such concern warranted? Yes. Fifteen-year-olds in 22 countries outperform their U.S. counterparts in science and those in 35 countries have much higher scores in math, surpassing even the students in the highest-achieving states in the Northeast (Baldi et al. 2007).

Why are many U.S. students doing so poorly? Some believe that the typical curriculum in the United States covers "too many topics too superficially," unlike the curricula in many other countries (Grant and Murray 1999: 25). One reason for the superficiality of coverage may be short school hours. For example, Asian students have longer and more school days than their American counterparts. From grades one through twelve, "this equates to 7,320 extra hours or 7.4 more years of schooling in Asia than in the United States (Ornstein 2003: 280).

Not all Americans endorse the idea of longer school days, even if the extra hours offer social studies, art, music, or creative projects like building model houses (to teach fractions). Instead, many parents feel that kids need some time to relax and play, and administrators worry about the additional costs of materials and teachers' salaries. Many Americans complain that the United States is falling behind other nations in academic achievement, but 44 percent feel that colleges and universities should not require students to take more math and science courses (The Winston Group 2006; Tyre 2007).

Unlike the United States, many countries identify young children with high innate ability in mathematical problem solving and then provide them with the resources (such as highly qualified teachers and specialized programs) to excel (Andreescu et al. 2008). A number of policy analysts maintain that many U.S. students

are performing poorly, in mathematics and other areas, mainly because of inadequate school funding.

SCHOOL FUNDING

The United States, the richest country in the world, devotes only 7 percent of its overall federal budget to public education, lagging behind eight other high-income countries. States and local school districts must provide the rest of the funding (*Public Education Finances* 2007). As a result, most public schools are struggling to survive. Because every state has experienced budget cuts since 2002, the rate of growth of funding for U.S. education has dropped to its lowest point since 1990. On average, states spend almost three times as much on each prisoner as they do on each public school student (Western et al. 2003; Goodman 2004; see also Chapter 7).

Despite the overall low federal funding, the "rich states" that have high median incomes get more than the "poor states," and school districts, similarly, spend less on high-poverty and high-minority schools than on more affluent white schools. In New York City, a high school in a poor district receives almost $3,500 a year less per student than one in a more affluent district (Liu 2006; Wiener and Pristoop 2006; Planty et al. 2007).

Such data suggest that American society doesn't provide a high-quality education for *all* students, especially neglecting those from low-income families who need the greatest support. However, some educators contend that high funding doesn't guarantee high-quality education, but that good results depend on teachers' effectiveness.

TEACHERS' EFFECTIVENESS

According to most college faculty, and in contrast to many high school teachers' belief, students are not prepared for college (see *Figure 14.4*). For example, 35 percent of college instructors but only 10 percent of high school teachers emphasize knowing the elements of basic grammar and usage, such as sentence structure and punctuation (ACT 2007).

Are many students unprepared for college because the United States does a poor job of educating teachers? Compared with teachers in countries in Europe and elsewhere, many U.S. teachers are out-of-field; that is, they have neither certification nor a major in the subject they teach. For example, only 8 percent of fourth-grade math teachers in the United States majored or minored in math, compared with 48 percent in Singapore.

Furthermore, classes in disadvantaged American schools are 77 percent more likely to be assigned to an out-of-field teacher than are classes in affluent schools (Holt et al. 2006; Levine 2006).

According to some policy analysts, raising wages would draw more talented individuals to teaching. In 2006, U.S. public school teachers earned 15 percent less on average (25 percent less in 15 states) than other employees (such as accountants, registered nurses, and personnel officers) with comparable credentials and experience. And the longer they teach, the more ground teachers lose financially. Teachers' economic survival is clearly threatened when the average starting salary is less than $32,000 a year, the average annual salary is about $47,000, and cost-of-living increases have not kept up with inflation (American Federation of Teachers 2007; Allegretto et al. 2008).

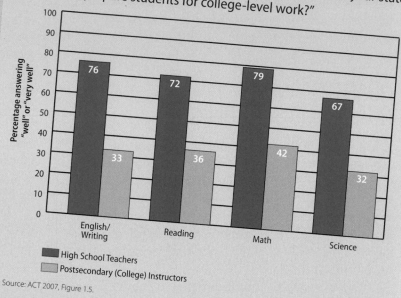

FIGURE 14.4
Are Students Prepared for College?

Responses to the survey question "How well do you think your state's standards prepare students for college-level work?"

- ■ High School Teachers
- ▨ Postsecondary (College) Instructors

Source: ACT 2007, Figure 1.5.

CONTROL OVER CURRICULA

Public school teachers in the United States have less control over their subject matter than ever before. In 2002, President George W. Bush signed, but didn't fund, the No Child Left Behind Act (NCLB) that requires all children to be proficient in reading and math by 2014. The major components of this act include the following:

- States must test public school students in reading and math every year from third through eighth grade, but are free to select the tests they use and to establish their own proficiency standards.

- If students at a particular school don't test well two years in a row, parents can transfer their children to another school. After three years of poor results, schools must provide tutoring services.

- After five years of failing to meet standards, the school must be restructured by replacing the staff and revamping the curriculum, being converted to a charter school (which we'll discuss shortly), or being taken over by a private management firm.

- States must set standards for teachers' qualifications.

Some have praised NCLB for making schools more accountable, for focusing attention on low-achieving students who had been overlooked (especially poor, minority, and disabled children in urban schools), and for giving principals greater leverage to dismiss teachers whose

In tests administered to students in 40 industrialized nations, Finnish 15-year-olds routinely score at or near the top in reading, science, and math, whereas their U.S. peers rank 18th, 22nd, and 28th, respectively, in those subjects. Finland attributes its success to several factors: Classes average 20 to 25 students (even in high school); local schools have enormous flexibility in choosing textbooks, designing curricula, and allocating funds; and teachers are well-trained. Teachers for all grades must obtain at least a master's degree, education programs at universities are highly competitive, and teachers enjoy high prestige in Finnish society—even those for elementary school-children, like the one pictured opposite. "The status of teachers is comparable to doctors and lawyers" (Moore 2007: 55).

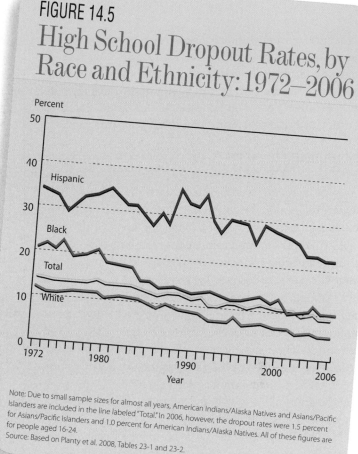

FIGURE 14.5
High School Dropout Rates, by Race and Ethnicity: 1972–2006

Note: Due to small sample sizes for almost all years, American Indians/Alaska Natives and Asians/Pacific Islanders are included in the line labeled "Total." In 2006, however, the dropout rates were 1.5 percent for Asians/Pacific Islanders and 1.0 percent for American Indians/Alaska Natives. All of these figures are for people aged 16-24.
Source: Based on Planty et al. 2008, Tables 23-1 and 23-2.

students have consistently low test scores. Critics, however, have denounced NCLB on numerous grounds. Most teachers teach to the test. That is, they cover only what will allow students to answer the items on the required yearly test, rather than trying to ignite creativity and keep youngsters interested in school. For example, the tests require memorization ("3 × 4 = 12") rather than the application of information ("If a car has four wheels, how many wheels do three cars have?"). Even worse, according to some critics, students who pass easy tests have an inflated sense of their knowledge and may study less (The Commission on No Child Left Behind 2007; Toch 2007).

Another complaint is that although NCLB and state accountability systems (that predate NCLB) have improved the reading and math scores of the bottom 10 percent of students, the brightest students are stagnating. Nationally, 81 percent of public school teachers admit that struggling students are a top priority and receive most of their attention; only 5 percent say that their most advanced students are likely to get the attention they need (Loveless et al. 2008).

Schools in European countries have uniform funding, curricula, and tests. In the United States, schools are largely funded and controlled at the state and local levels. NCLB is unfunded, but the federal government has threatened to reduce existing funding levels if students' test scores don't show progress each year. Thus, states that had set high standards in the past have lowered them because of constant pressure to raise test scores and to avoid financial penalties (Center on Education Policy 2006; National Center for Education Statistics 2007).

DROPPING OUT

Over 9 percent of Americans aged 16–24 are high school dropouts, down from 15 percent in 1972. Men are more likely to drop out than women. The dropout rate is highest among Latinos, but the gap between that group and others has decreased only slightly, especially since the early 1990s (see *Figure 14.5*).

Foreign-born youths, many of whom were already behind in schooling before arriving in the United States, make up 40 percent of the nation's teenage high school dropouts. They may attend school briefly but make little progress because of the language barrier, taking a job to help their families, teen pregnancy, and gang membership, all of which are related to higher dropout rates (Orrenius 2004; Fry 2005).

High school students who were born in the United States are more likely to drop out because they are unmotivated, are not challenged enough, or feel overwhelmed by family problems rather than because they are failing academically. Over time, many of these students become disengaged—arriving late, skipping classes, and leaving after lunch. In other cases, even stu-

dents with passing grades drop out in their junior or senior year, fearing that they won't pass the exit examinations that a growing number of states require before receiving a diploma (Bridgeland et al. 2006; Dee and Jacob 2006; Grodsky et al. 2008).

It's far easier to be accepted by a college than to graduate from one. After five years, for instance, only 53 percent of students who enrolled in a four-year college had completed a bachelor's degree and only 25 percent of students at two-year colleges had earned a certificate or associate's degree. The United States still ranks as one of the top five Western countries in the proportion of young people who attend college, but it ranks 16th among 27 industrialized and developing countries in the proportion who complete college. If current patterns persist, by 2025, the United States will have almost 16 million fewer people with a bachelor's or associate's degree than it needs to keep up with its top economic competitors worldwide (Wirt et al. 2004; Reindl 2007). The high dropout rates of U.S. students are somewhat surprising because there is considerable grade inflation at both high schools and colleges.

GRADE INFLATION

In 1973, only 20 percent of high school students earned an A average, compared with 47 percent in 2003. However, standardized tests show that, between 1992 and 2007, the percentage of twelfth-graders who performed at or above a basic level in reading decreased from 80 percent to 73 percent, while the percentage performing at or above a proficient level declined from 40 percent to 35 percent (Ewers 2004; Lee et al. 2007; Planty et al. 2008). Such statistics suggest that grades are rising but learning is lagging.

According to one critic, "The demanding professor is close to being extinct" because "professors are under pressure to accommodate students" (Murray 2008: 32). Grade inflation, especially in college, is due to a number of factors, both institutional and individual. Many faculty members give high grades because it decreases student complaints, involves less time and thought in grading exams and papers, and reduces the chances of students' challenging a grade. Some faculty also believe that they can get favorable course evaluations from students by handing out high grades, and others accept students' view of high grades as a reward for simply showing up in class. Moreover, administrators want to keep enrollment up. If students are unhappy with their grades, admission application rates may decrease. Thus, inflating grades satisfies

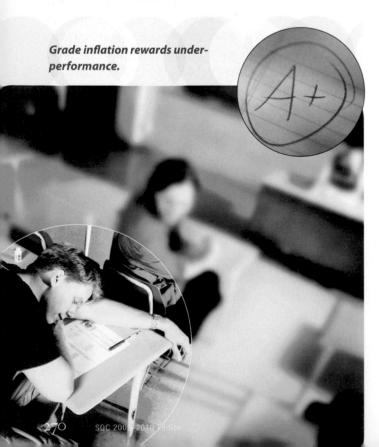

Grade inflation rewards under-performance.

Would *you* want a military officer, a financial planner, or a dentist who cheated her or his way through college or graduate school?

administrators and makes them look good, especially where state legislators base funding on graduation rates (Carroll 2002; Kamber and Biggs 2002; Halfond 2004; Bartlett and Wasley 2008).

Among other problems, grade inflation gives students an exaggerated and unrealistic sense of their ability and accomplishments. Sometimes, for example, students don't understand why their job searches are unsuccessful because "I'm an A student." In fact, many students' incorrect usage of grammar on job application forms and résumés, as well as during interviews, turns off prospective employers (Halfond 2004).

Despite grade inflation, cheating is common in high school, college, and graduate school. Some 60 percent of high school students admit to cheating on tests and 33 percent have plagiarized from the Internet. Between 50 and 75 percent of college students say that they have cheated at least once. Dozens of studies have uncovered cheating at the Air Force Academy and in numerous graduate programs, including those in business administration and dentistry (Josephson Institute of Ethics 2006; McCabe et al. 2006; Burrus et al. 2007; Frosch 2007; Wasley 2007a).

Grade inflation, dropping out, and low-quality instruction are serious problems in U.S. education. Because many traditional public schools are failing to educate students adequately, parents and legislators have turned to a variety of alternatives.

4 New Directions in U.S. Education

Several innovations, some more controversial than others, are transforming traditional education in the United States; these include vouchers, charter schools, and magnet schools. Home schooling has also been expanding.

SCHOOL VOUCHERS

Vouchers are publicly funded payments (up to $7,500 per child a year) that parents can apply toward tuition or fees at a public or private school of their choice. Supporters argue that vouchers provide parents with more options and thereby equalize the playing field, especially in low-income districts with failing schools. They also claim that the competition will improve public schools. Opponents maintain that directing public dollars to private schools diverts much-needed resources from public education and that taxpayers are taxed twice—once to support public schools and a second time to support vouchers for private schools. Also, some object to providing public funds to schools run by religious organizations on the grounds that this violates the constitutional principle of separation of church and state.

Are vouchers effective? The data are mixed, but there is no significant overall difference in achievement between the children participating in voucher programs and those who remain in public schools. Parents whose children have switched schools using vouchers are less likely to worry that school could be physically dangerous. However, vouchers don't affect the students who earned relatively high scores in standardized tests in their previous schools, possibly because they already had positive attitudes toward education, good teachers, and supportive parents (Heald et al. 2003; Glod and Turque 2008).

CHARTER SCHOOLS

Charter schools are self-governing public schools that have signed an agreement with their state government

vouchers publicly funded payments that parents can apply toward tuition or fees at a public or private school of their choice.

charter schools self-governing public schools that have signed an agreement with their state government to improve students' education.

Because many traditional public schools are failing to educate students adequately, parents and legislators have turned to a variety of alternatives.

to improve students' education. Some see charter schools, in contrast to regular public schools, as providing innovative teaching, offering smaller classrooms, reducing bureaucratic red tape, being friendlier toward parents and students, and having more dedicated teachers. Others maintain that charter schools sometimes avoid accountability by not reporting test scores, drain resources from traditional public schools, hire uncertified teachers, and are more racially segregated than other public schools. Also, a few have been forced to close because students' performance is poor or because of financial mismanagement (Teske et al. 2001; Renzulli and Roscigno 2007).

Do charter schools work? As with school vouchers, the data are mixed. Fourth- and eighth-graders in charter schools trail students in regular public schools in both reading and math, especially in central cities. Students perform about the same, however, if the teachers in the charter school are certified and have had at least five years' experience and if the charter school is not new (Hassel 2005; Braun et al. 2006).

MAGNET SCHOOLS

A **magnet school** is a public school that offers students a distinctive program and specialized curriculum. Magnet schools emphasize particular areas of study, such as business, science, the arts, or technology. Because magnet schools are typically small, there are close student-teacher relationships and a sense of community (Metz 2003). There are magnet schools for middle schoolers, but most are targeted at high school students.

A major advantage of magnet schools is that, because they focus on specific subject matter areas, students from diverse socioeconomic and ethnic backgrounds have similar opportunities in developing their shared interests. For example, actress Jada Pinkett Smith graduated from the Baltimore School for the Arts, a magnet school in music, theatre, dance, and visual arts; and Sarah Jessica Parker graduated from the School for the Creative and Performing Arts, a magnet school in

the Cincinnati Public School system. The major disadvantage is that few students can be accepted because enrollments are limited (Rickles et al. 2002).

HOME SCHOOLING

Home schooling refers to teaching children in the home as an alternative to enrolling them in a public or private elementary, middle, or high school. According to the most recent available data, the number of American children who were home-schooled grew from 850,000 in 1999 to more than 1.1 million in 2003 (about 2 percent of the school-age population aged 5 to 17). About 81 percent of home-schooled children live in a two-parent home where one parent (usually the mother) is not employed and both parents are likely to have a college education; 77 percent of these households are white, and the majority are in middle-income brackets. Some of the major reasons why parents opt for home schooling are their concerns about negative peer pressure, children's safety, and drugs; their desire to provide religious and moral instruction; and their feeling that schools don't challenge their children academically (Princiotta and Bielick 2006).

How effective is home schooling? No one knows for sure, because, unlike public school students, children who are home-schooled aren't required by law to undergo either state or national tests. Recently, a California appellate court ruled that only parents with state-recognized credentials can educate their children at home, but reversed this decision after a "nationwide uproar" from home-schoolers, religious activists, and others (Mehta 2008).

Opponents of home schooling argue that the children tend to be isolated and don't learn necessary social and academic skills. There are no national data on the effects of home schooling, but several small studies have found that some colleges see home schoolers as desirable applicants and that many of these students perform as well academically as their peers (Wasley 2007b; Coombs and Shaffer 2008).

As this chapter shows, the United States has high college graduation rates, but public schools, especially, struggle with serious problems in the quality of schooling. Schools will improve only to the extent that students, teachers, administrators, parents, local communities, and government officials make educational excellence a top priority, and regardless of children's social class, race/ethnicity, and gender.

More Bang for Your Buck

Religion is an important

and complex institution in the United States and
around the world.

what do you think?

It is necessary to believe in
God to be a moral person.

1 2 3 4 5 6 7

strongly agree   strongly disagree

15
Religion

Key Topics

In this chapter, we'll explore the following topics:

1 What Is Religion?

for sociologists, **religion** is a social institution that involves shared beliefs, values, and practices based on the supernatural, which unites believers into a community. The notion of community is important because different groups have different beliefs, values, and practices. For example, being Catholic involves confessing (telling one's sins to a priest), a practice that is not followed by Protestants, Jews, Muslims, and other religious groups.

> **religion** a social institution that involves shared beliefs, values, and practices based on the supernatural and unites believers into a community.
>
> **sacred** anything that people see as mysterious, awe-inspiring, extraordinary and powerful, holy, and not part of the natural world.
>
> **profane** anything that is not related to religion.
>
> **secular** the term sociologists use (instead of *profane*) to characterize worldly rather than spiritual things.
>
> **religiosity** the ways people demonstrate their religious beliefs.

THE SACRED AND THE SECULAR

Every society distinguishes between the sacred and the profane (Durkheim 1961). **Sacred** refers to anything that people see as mysterious, awe-inspiring, extraordinary and powerful, holy, and not part of the natural world. **Profane** refers to anything that is not related to religion. Sociologists typically use **secular** (rather than *profane*) to describe worldly rather than spiritual things. They also differentiate religion from religiosity and spirituality.

RELIGION, RELIGIOSITY, AND SPIRITUALITY

Religion refers to a community of people who have a shared faith, but the levels of individuals' commitment to their religion and the frequency and intensity of their religious expression can vary. When sociologists examine **religiosity**, the ways people demonstrate their religious beliefs, they often find that religion and religiosity differ. For example, 72 percent of Americans say that they are "deeply religious," but only 44 percent attend worship services once a week. And although 79 percent of college students say that they believe in God, only 40 percent

In 2005, the U.S. Supreme Court ruled, 5–4, that a 6-foot-high monument containing the Ten Commandments located outside the Texas state capitol (bottom) was constitutional and could stay, even though it's on government property. The same day, the Court barred, in a 5–4 vote, displaying the Ten Commandments amid other documents in two Kentucky courthouses (top) because the displays promoted religion. Do such rulings seem contradictory? Or not?

report that religion influences their everyday lives. Especially at schools that are not affiliated with any religion, a majority of students who describe themselves as religious also report engaging in casual sex, attending campus parties with pornographic themes, and cheating on exams (Astin and Astin 2005; Lyons 2005; Freitas 2008).

Spirituality is a personal quest to feel connected to a reality greater than oneself. About 25 percent of Americans who never go to church consider themselves spiritual: They say they feel united with the people around them, have made personal sacrifices "to make the world a better place" (such as working with others to decrease poverty), and often believe in miracles (Adler 2005; Pew Forum on Religion & Public Life 2008).

People who see themselves as spiritual may not be religious. For example, 77 percent of students at one college described themselves as "spiritual," but only 16 percent participated in religious activities on campus (Denton-Borhaug 2004; see also Stanczak 2006). Thus, belonging to a specific religious group, religiosity, and spirituality can involve different behaviors.

2 Types of Religious Organization

religion is important in all known societies, but there's considerable diversity in how people around the world reach out to something eternal. People express their religious beliefs most commonly through organized groups, including cults, sects, denominations, and churches.

CULTS (NEW RELIGIOUS MOVEMENTS)

A **cult** is a religious group that is devoted to beliefs and practices that are outside of those accepted in mainstream society. Some sociologists prefer to use **new religious movement (NRM)** rather than *cult* because the latter term has been used by the media in pejorative ways to describe any unfamiliar, new, or seemingly bizarre religious movement (Roberts 2004).

NRMs usually organize around a **charismatic leader** (like Jesus) whom followers see as having exceptional or superhuman powers and qualities. Some NRMs have become established religions. The early Christians were a renegade group that broke away from Judaism. Islam, the second-largest religion in the world today, began as a cult organized around Muhammad, and The Church of Jesus Christ of Latter-day Saints (the Mormon Church) began as a cult around Joseph Smith. Most contemporary cults are fragmentary, loosely organized, and temporary, but others have developed into groups that are more lasting, organized, and highly bureaucratic (such as the Unification Church, or Moonies, and the International Society for Krishna Consciousness, or Hare Krishnas).

SECTS

A **sect** is a religious group that has broken away from an established religion. Dissatisfied members who feel that the parent religion has become too secular and has abandoned key original doctrines usually begin sects. Like cults, some sects are small and disappear after a time, while others become established and persist. Examples of sects that have persisted include the Amish, the Jewish Hassidim, Jehovah's Witnesses, Quakers, Seventh Day Adventists, and the Fundamentalist Church of Jesus Christ of Latter Day Saints (FLDS) (Finke and Stark 1992; Bainbridge 1997). Sometimes sects develop into denominations.

DENOMINATIONS

A **denomination** is a subgroup within a religion that shares its name and traditions and is generally on good terms with the main group. Denominations can form slowly or develop rapidly, depending on factors such as geography, immigration, and culture. Some scholars describe a denomination as somewhere between a sect and a church. Like sects, denominations have a professional ministry. Unlike sects, however, denominations view other religious groups as valid and don't make claims that only they possess the truth (Niebuhr 1929; Hamilton 2001).

Denominations typically accommodate themselves to the larger society instead of trying to dominate or change it. As a result, people may belong to the same denomination as their grandparents or even great-grandparents did. Denominations exist in all religions, including Catholicism, Judaism, and Islam. In the United States, for example, the many Protestant denominations include Episcopalians, Baptists, Lutherans, Presbyterians, and Methodists.

CHURCHES

A **church** is a large established religious group that has strong ties to mainstream society. Because leadership of the group is attached to an office rather than a specific leader, new generations of believers replace older generations, and members follow tradition or authority

cult a religious group that is devoted to beliefs and practices that are outside of those accepted in mainstream society.

new religious movement (NRM) term used instead of *cult* by most sociologists.

charismatic leader a religious leader whom followers see as having exceptional or superhuman powers and qualities.

sect a religious group that has broken away from an established religion.

denomination a subgroup within a religion that shares its name and traditions and is generally on good terms with the main group.

church a large established religious group that has strong ties to mainstream society.

rather than a charismatic leader. As in a denomination, people are usually born into a church, even though they may decide to withdraw later.

Churches (such as the Roman Catholic Church and the Protestant church) are typically bureaucratically organized, have formal worship services and trained clergy, and often maintain some degree of control over political or educational institutions (see Chapters 11 and 14). Because churches are an integral part of the social order, they often become dependent on, rather than critical of, the ruling classes (Hamilton 1995).

3 Some Major World Religions

Worldwide, the largest religious group is Christians, followed by Muslims. If the world's population is represented as an imaginary village of 100 people, it has

- 33 Christians,
- 21 Muslims,
- 16 nonreligious individuals or atheists (people who don't believe in God),
- 14 Hindus,
- 6 Buddhists,
- 6 Chinese Universalists (followers of a complex set of beliefs and practices that combines ancestor cults, elements of Confucianism and Buddhism, Taoism, folk religion, and goddess worship), and
- 4 believers in another religion (including Judaism and Sikhism). (Based on "Major Religions of the World . . ." 2007).

It is noteworthy that there is no religious group that comes close to being a global majority, the third-largest group consists of nonbelievers, and non-Christians outnumber Christians two to one. Five religious groups, especially, have had a worldwide impact on economic, political, and social issues. *Table 15.1* provides a brief summary about these groups.

4 Religion in the United States

about 92 percent of Americans believe in God, but religion in the United States is complex and diverse because people may change their faith and those who describe themselves as religious identify with one of over 150 groups (Bader et al. 2006; Pew Forum on Religion & Public Life 2008). For sociologists, religiosity is a better measure of religious expression than simply asking people which religion they follow. When sociologists measure religiosity, they consider a number of variables, but especially religious belief, affiliation, and attendance at services.

RELIGIOUS BELIEF

Some 56 percent of Americans today say that religion is "very important in their lives," down from 75 percent in 1952. Not surprisingly, religion is not very important for the 4 percent of Americans who are agnostics (say that it's impossible to know whether there is a God) or atheists (believe that there is no God) or for others who are skeptics (Newport 2006b; Pew Forum on Religion & Public Life 2008).

© iStockphoto.com

The importance of religion in Americans' lives has declined.

GOD IS WITH YO

TABLE 15.1
Characteristics of Some Major World Religions

RELIGION	DATE OF ORIGIN	FOUNDER	PREVALENCE	NUMBER OF FOLLOWERS (IN MID-2007)	CORE BELIEFS
Christianity	0 C.E.	Jesus Christ	All continents, with largest numbers in Latin America and Europe	2.2 billion	Jesus, the son of God, sacrificed his life to redeem humankind. Those who follow Christ's teachings and live a moral life will enter the Kingdom of Heaven. Sinners who don't repent will burn in hell for eternity.
Islam	600 C.E.	Muhammad	Mainly Asia (including Indonesia), but also parts of Africa, China, India, and Malaysia	1.4 billion	God is creator of the universe, omnipotent, omniscient, just, forgiving, and merciful. Those who sincerely repent and submit (the literal meaning of *islam*) to God will attain salvation, while the wicked will burn in hell.
Hinduism	Between 4000 and 1500 B.C.E.	No specific founder	Mainly India, Nepal, Malaysia, and Sri Lanka, but also Africa, Europe, and North America	887 million	Life in all its forms is an aspect of the divine. The aim of every Hindu is to use pure acts, thoughts, and devotion to escape a cycle of birth and rebirth (*samsara*) determined by the purity or impurity of past deeds (*karma*).
Buddhism	525 B.C.E.	Siddhartha Gautama	Throughout Asia, from Sri Lanka to Japan	386 million	Life is misery and decay with no ultimate reality. Meditation and good deeds will end the cycle of endless birth and rebirth, and the person will achieve *nirvana*, a state of liberation and bliss.
Judaism	2000 B.C.E.	Abraham	Mainly Israel and the United States	15 million	God is the creator and the absolute ruler of the universe. God established a particular relationship with the Hebrew people. By obeying the divine law God gave them, Jews bear special witness to God's mercy and justice.

Note: C.E. (Common Era) is the nondenominational abbreviation for A.D. (*Anno Domini*, Latin for "In the year of Our Lord)" and B.C.E. (Before Common Era) is the nondenominational abbreviation for B.C. (Before Christ).
Source: Based on a number of sources including the Center for the Study of Global Christianity 2007 and "Religions of the World ..." 2007.

RELIGIOUS AFFILIATION

Americans' religious affiliations have been remarkably fluid. More than 40 percent of U.S. adults have changed their religion since childhood, many opting for no religion at all. A major change has been the decline of the so-called mainline Protestant groups (especially Methodists and Lutherans) that some sociologists now describe as a "vanishing majority" (Smith and Kim 2004; Pew Forum on Religion & Public Life 2008). There has also been a surge of evangelicals (also called born-again Christians and Pentecostals) who now comprise a significant proportion of Protestants (see *Table 15.2*). Evangelicals believe that the Bible is the literal word of God, that salvation can be attained only through a personal relationship with Jesus Christ, and that every Christian has a responsibility to spread his or her beliefs (to evangelize).

TABLE 15.2
Religious Affiliation in the United States

CHRISTIAN	**78.4%**
Protestant	51.3%
Evangelical churches	26.3%
Mainline churches	18.1%
Historically black churches	6.9%
Catholic	23.9%
Mormon	1.7%
Jehovah's Witnesses	0.7%
Other Christian	0.9%
OTHER RELIGIONS	**4.7%**
Jewish	1.7%
Buddhist	0.7%
Muslim	0.6%
Hindu	0.4%
Other	1.5%
UNAFFILIATED	**16.1%**
Atheist	1.6%
Agnostic	2.4%
Nothing in particular	12.1%

Note: Due to rounding and the fact that almost 1% of those polled refused to answer or said "don't know," the numbers may not add to 100%.
Source: Based on Pew Forum on Religion & Public Life 2008: 8.

RELIGIOUS PARTICIPATION

Turning to the third measure of religiosity, religious participation, about 4 in 10 Americans attend religious services at least once a week, while 27 percent seldom or never attend. Thus, many people are more likely to believe in a religion than to practice it by attending formal services regularly. Mormons (75 percent), evangelical Protestants (58 percent), and members of historically black churches (59 percent) have higher rates of weekly or almost weekly attendance at religious services than do Catholics (42 percent), Muslims (40 percent), or Jews (16 percent) (Pew Forum on Religion & Public Life 2008).

Megachurches—Protestant congregations that have a regular weekly attendance of more than 2,000 and tend to be evangelical—represent only 0.5 percent of all U.S. churches, but their number has more than doubled (to over 1,200) since 2000. In 2006, worshippers packed the grand opening of the Lakewood Church in Houston, Texas, which holds 16,000 and has a congregation of 30,000. Megachurches appeal to many people because they provide social services, have electronic altars and huge TV screens that encourage enthusiastic participation, and inspire a sense of a joyful spiritual community (Thumma et al. 2005).

SOME CHARACTERISTICS OF RELIGIOUS PARTICIPANTS

Americans differ in their beliefs and affiliations, and religious participation varies by one's sex, age, race and ethnicity, and social class.

Sex

Across all age groups, women tend to be more religious than men in believing in God, praying, attending services, and saying that religion is very important in their lives. This pattern is seen in many religious groups in the United States, though equal numbers of male and female Mormons (83 percent) say that religion is very important in their lives (Newport 2007; Pew Forum on Religion & Public Life 2008).

Age

Generally, Americans aged 65 and over are more likely than younger ones to describe themselves as religious, to say that religion is very important in their lives, and to attend services at least weekly. These age-related differences may reflect several factors: Older Americans

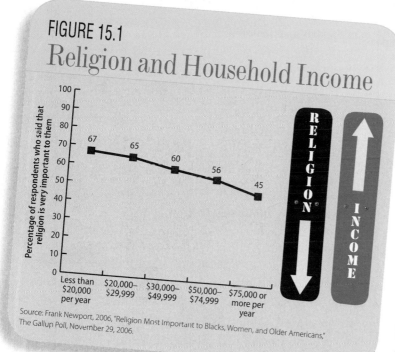

FIGURE 15.1
Religion and Household Income

Source: Frank Newport, 2006, "Religion Most Important to Blacks, Women, and Older Americans," The Gallup Poll, November 29, 2006.

grew up decades ago when church attendance was higher for all Americans, they seek spiritual comfort as elderly friends and relatives die, they want to lessen a sense of isolation or loneliness, and they are preparing for death. There is no similar age gap, however, among Mormons, Jews, and Muslims. Within these groups, those who are under 49 are about as likely as those who are older to say that religion is very important to them (Roof 1993; Newport 2006a; Pew Forum on Religion & Public Life 2008).

Race and Ethnicity

In the last decade or so, the group of Americans that is increasing the fastest consists of those who have no religious affiliation. This group now comprises over 16 percent of the population: 73 percent of the unaffiliated are white, 11 percent are Latino, 8 percent are black, 4 percent are Asian, and 4 percent are mixed race (Pew Forum on Religion & Public Life 2008).

One reason for whites' relative lack of religious affiliation may be that those in higher social classes (who are mostly white) tend to be less religious. And one reason for higher levels of religious affiliation among ethnic groups may be that religious groups provide important resources (such as housing and finding jobs) to newly arrived immigrants, many of whom are Latino or Asian (see Chapter 10). Some Catholic and evangelical Protestant churches in the United States have been especially successful in attracting recent Latino immigrants, because they are more likely to offer services in Spanish, their services are expressive

rather than formal, and the clergy are more responsive to their members' social and economic needs (Campo-Flores 2005; Wolfe 2008).

Asian Americans are the most diverse religious group in America: "Asian Buddhist and Hindu temples, Muslim mosques, and Catholic and Protestant churches have mushroomed in Los Angeles, New York, and other major cities over the past thirty years" (Min 2002: 5). Besides being places of worship, Chinese churches and temples provide weekend language instruction, religious study, and meditation. Similarly, Southeast Asian refugees (primarily Vietnamese, Cambodians, and Laotians) have established churches and temples that teach English, help newcomers find jobs and housing, and maintain their cultural identity and spirituality (Yang 2002; Zhou et al. 2002).

Social Class

People with lower levels of educational attainment are generally more religious than those with higher educational levels. For example, 60 percent of people with a high school degree or less say that religion is very important in their daily lives, compared with 50 percent of college graduates (Pew Forum on Religion & Public Life 2008). Also, because education and income are highly correlated, as income increases, the importance of religion generally decreases (see *Figure 15.1*).

About 44 percent of Americans have switched from the religion or denomination in which they were raised, but almost 95 percent identify with a religion. Thus, "it is clear that religion remains a powerful force in the private and public lives of most Americans" (Pew Forum on Religion & Public Life 2008: 19). Many Americans worry, however, that religion is waning in importance and that the United States is becoming a more secularized nation.

5 Secularization: Is Religion Declining?

In 2004, France banned religious clothing—including head scarves and veils that some Muslim women wear—in public schools. Four years later, France's highest administrative court upheld a decision to deny citizenship to a 32-year-old Moroccan-born woman who insisted on wearing a *burqa*, a long veil that covers her from head to toe. She speaks French fluently, has three children who were born in France, and a French husband. The court said, however, that her "radical practice" of Islam

secularization a process of removing institutions such as education and government from the dominance or influence of religion.

fundamentalism the belief in the literal meaning of a sacred text.

violated French values such as equality of the sexes and that she was unable to assimilate—a requirement for citizenship. Critics argued that the French Constitution guarantees freedom of religion, but others, including many women who are practicing Muslims, supported the ruling, maintaining that wearing a *burqa* or other religious clothing denotes women's submission to men and is incompatible with France's strict secularism (Bennhold 2008; Sokol 2008).

Other industrialized nations, besides France, are undergoing **secularization**, a process of removing institutions such as education and government from the dominance or influence of religion. Some sociologists believe that secularization is increasing rapidly in the United States, but others contend that this claim is greatly exaggerated.

IS SECULARIZATION INCREASING IN THE UNITED STATES?

There is evidence of increased secularization in the United States. For example, attendance at religious services has decreased; only 52 percent of Americans now say that they have a "great deal" of confidence in organized religion (down from 66 percent in 1973); 52 percent feel that religious groups should keep out of political matters; and fewer Americans say that religion is "very important" in their lives (Newport 2007; Pew Research Center 2008).

In addition, a majority of Americans complain that religious holy days are anything but religious. They say, for example, that secular cards wish "Happy Holidays" (instead of "Merry Christmas"), and radio stations stream songs like "Let It Snow" rather than Christmas carols. Thus, many grumble, Santa Claus and presents have replaced traditional symbols of Christmas, especially images of the birth of Christ.

IS THE EXTENT OF SECULARIZATION EXAGGERATED?

On the other hand, many sociologists argue that the extent of secularization is greatly overstated. Some point out that **fundamentalism**, the belief in the literal meaning of a sacred text (such as the Christian Bible, the Muslim Qur'an, or the Jewish Torah), has increased in the United States and worldwide (Woolf 2004). There's also evidence that religion has an enormous impact on many Americans' personal and political lives:

- Increasing numbers of pharmacists are refusing to fill prescriptions for oral contraceptives because doing so violates their religious beliefs (Bergquist 2006).

- 100 million Americans listen to Christian radio stations almost every day, 43 percent more than in 2000 (Piore 2005).

- 59 percent of Americans wouldn't vote for a well-qualified presidential candidate who is an atheist (Gibbs and Duffy 2007).

- Evangelicals, especially, have considerable influence on U.S. corporations, the military, and the White House (Lindsay 2007).

Over the years, one of the most divisive secularization issues has been the conflict between religion and science. Nearly 87 percent of Americans feel that scientific developments make society better (Masci 2007). However, large numbers don't want scientific theories about evolution taught in public school classrooms.

Sociologists (and other social scientists) who contend that secularization is not increasing in the United States and other Western countries also cite the prevalence of

A young woman wearing a tricolor headband takes part in a Muslim demonstration in Lille, France, to protest the government's plan to ban religious attire in public schools. The law ultimately passed, banning Muslim head scarves, Jewish skullcaps and large Christian crosses from public schools to keep them secular and avoid religious strife.

civil religion (sometimes called *secular religion*), practices in which citizenship takes on religious aspects. Examples of civil religion in the United States involve including the phrase "one nation under God" in the Pledge of Allegiance, inscribing coins with the phrase "In God We Trust," routinely asking God to bless the nation in presidential inaugural addresses and annual state-of-the-union messages, typically beginning a new session of Congress with a prayer, and the closing of the New York Stock Exchange (the country's most capitalistic and secular organization) on Good Friday, a major Christian holy day.

6 Sociological Perspectives on Religion

Y ou've seen that religion plays an important role in many people's lives. Why? How does society affect religion? And how does religion influence society? Functionalists view religion as benefiting society, conflict theorists see religion as promoting social inequality, feminist scholars emphasize how religion subordinates and excludes women, and symbolic interactionists focus on everyday religious experiences and processes (*Table 15.3* on p. 291 summarizes these perspectives).

civil religion (sometimes called *secular religion*) practices in which citizenship takes on religious aspects.

FUNCTIONALISM: WHAT ARE THE BENEFITS OF RELIGION?

Durkheim (1961: 13, 43) described religion as an "essential and permanent aspect of humanity" whose "essential task is to maintain, in a positive manner, the normal course of life." Religion can also be dysfunctional, but first let's consider its positive influence.

Religion as a Societal Glue

Religion fulfills a variety of functions on individual, community, and societal levels. All of the following products of religious practice, according to functionalists, contribute to a society's survival, stability, and solidarity:

- *Belonging and identity.* Communal worship increases social contacts and enhances a sense of acceptance

In 2006, a dairy truck driver burst into an Amish schoolhouse in Lancaster County, Pennsylvania. He killed five girls and wounded three others before shooting himself because, according to his suicide note, he was angry with life and angry at God. Pictured here, an Amish funeral procession escorts the victims to a burial ground. The Amish community, in which violence is unthinkable, accepted the tragedy as the will of God and started a charity fund to help not only the victims' families but also the gunman's widow and children. From a functionalist perspective, why does their religion help the Amish and other religious people cope with tragedy?

© Timothy A. Clary/AFP/Getty Images

and identity. Rituals for major life events (such as baptisms, weddings, and funerals) and every-day behaviors (such as prayers at meals and bed-time) reinforce the members' feeling of belonging to a group (Christiano 2000; Gruber 2005). Religion also bolsters social cohe-sion by bringing people together for worship services that reinforce their shared beliefs.

- *Meaning, purpose, and emotional comfort.* Religion provides a sense of meaning in life and offers hope for the future. Religious people are often less likely to despair when they experience injustice, suffering, and illness. Religion also helps people cope with anxiety-producing events. Consider, for example, how often you've prayed (or seen someone pray) for help in deal-ing with disappointment, stress (especially before taking an exam), frustration, or the death of a loved one.

- *Social service.* Religions provide numerous social services that benefit their members and others. For example, religious organizations raised millions of dollars for the victims of Hurricanes Katrina and Rita, distributed food and clothing, and found shelter for displaced families. Also, many people express their religious beliefs by volunteering at soup kitchens and AIDS clinics because they believe that they're deliv-ering God's love in a practical way (Bender 2003; Brown 2005). Many religious groups also help im-migrants obtain visas, locate housing and jobs, and enroll in schools and English classes, all of which help decrease culture shock by acclimating newcomers to a foreign environment (see Chapter 3 on culture shock). Ethnic churches, especially, offer immigrants services in their own language, which help newcomers main-tain continuity with their home culture (Greeley 1972; Ebaugh 2003; McRoberts 2003).

- *Social control.* Because people who are religious inter-nalize rules about right and wrong and fear damnation in the afterlife, they try to practice self-control, which encourages social conformity and discourages deviant behavior (Durkheim 1961). Religion also legitimizes political leaders and their authority. In some countries, state religions dictate how people should behave; in other countries, such as the United States, politi-cal leaders may use religion to justify their decisions.

President George W. Bush, for example, defended the 2003 U.S. invasion of Iraq in religious terms ("Liberty is God's gift to every human being in the world") (Milbank 2003; "Faith in the System" 2004).

Religion and Social Change

For functionalists, religion usually supports the status quo, but it can also spearhead social change. Mohandas Gandhi (1869–1948), a spir-itual leader in India, worked for his country's independence from Great Britain through non-violence and peaceful negotiations. In the United States, religious leaders, especially Reverend Martin Luther King, Jr., were at the forefront of the civil rights movement during the late 1960s. Recently, Buddhist religious leaders led hundreds of thousands of people in revolts against what they believed was an unjust authoritarian government in Myanmar (Burma).

Max Weber (1920) argued that religion sparks economic development. Weber's study of Calvinism, a Christian sect that arose during the sixteenth century, concluded that this belief system supported the domi-nance of capitalism in the Western world by the nine-teenth century. Calvinists believed that people were predestined for salvation or the fires of hell. A sign from God that someone was among the elect (those destined for heaven) was that person's material success on earth ("By their fruits shall you know them"). This view became a self-fulfilling prophecy that Weber called the **Protestant ethic**, a belief that hard work, diligence, self-denial, frugality, and economic success would lead to salvation in the afterlife.

Was Weber right about the relationship between the Protestant ethic and the rise of capitalism? The data are mixed. Some argue that people are diligent not because of religious beliefs but simply because amassing sav-ings provides resources when disaster strikes (as when crops are lost because of a drought). On the other hand, a study of 59 industrialized and developing countries found that religious beliefs that encourage hard work and thrift spur economic growth (Cohen 2002; Barro and McCleary 2003).

How Is Religion Dysfunctional?

Functionalists emphasize the benefits of religion, but they also recognize that religion can be dysfunctional when it harms individuals, communities, and societ-ies. For example, religious intolerance can lead to con-flict between groups (even to vandalizing churches, mosques, and synagogues) and be used to justify attacks on religious minorities (see Chapter 18 online).

Religion helps people deal with misfortune, but it can also heighten anxiety about death and the afterlife. As a result, many elderly and poor people contribute much of their meager income to religious organizations, believing that such donations will help them enter heaven. Also, some groups in the United States have gone to court over the teaching of evolution rather than creationism or intelligent design in science classes, creating considerable conflict in local school districts. Scientists claim that human beings (and other creatures) evolved from earlier animals though a process known as "natural selection" over several million years. In contrast, creationism argues that the earth and the universe were created less than 10,000 years ago by God, as described in the Bible. Intelligent design, sometimes described as a watered-down version of creationism, doesn't reject evolution outright but holds that the universe is so complex that it must have been created by a supreme being, not necessarily named God (National Academy of Sciences 2008; Roylance and Hill 2008).

Critical Evaluation

Functionalists' theories are useful in explaining why many people are religious, how religion affects other institutions (such as political and educational), and how religion benefits societies. However, critics accuse functionalists of emphasizing benefits and tending to gloss over the dysfunctional aspects of religion that create and maintain social cleavages. For example, disagreements over whose deity is the "real God" have led to wars, terrorism, and genocide (see Chapter 18 online).

Critics also note that functionalists, by emphasizing the critical needs that religion fulfills, imply that religion is indispensable to leading a good life. However, people who are not religious donate to many social service organizations, serve as volunteers when disaster strikes, and report having happy and fulfilling lives. In addition, some Catholic priests, presumably very religious people, have admitted to molesting thousands of young boys, and some Protestant pastors and church treasurers have embezzled church funds that have amounted to millions of dollars ("What's That Commandment . . ." 2006; Padget 2007; Spano 2007). Conflict theorists, especially, maintain that functionalist perspectives focus on social stability rather than addressing discord and social inequality.

CONFLICT THEORY: DOES RELIGION PROMOTE SOCIAL INEQUALITY?

During the 2004 presidential election, a number of U.S. Roman Catholic bishops, archbishops, and priests declared it would be a "grave sin" to vote for any Catholic politician who supports same-sex marriage or abortion. They said that such politicians and their supporters aren't fit to receive the sacrament of communion because their "intrinsically evil" views violate Catholic principles. Some Catholics were outraged by the church hierarchy's meddling in politics, but others felt comforted that their religious leaders were taking a political stand on controversial social issues (Sullivan 2004; Wakin 2004). Many functionalists would view the religious leaders' statements as enhancing solidarity among Catholics, but conflict theorists would see the remarks as examples of religious views that create strife and divisiveness.

"The Opium of the People"

Much conflict theory reflects the work of Karl Marx (1845/1972), who described religion as "the sigh of the oppressed creature" and "the opium of the people" because it encouraged passivity and acceptance of class inequality. For Marx, religion taught people to endure suffering and deprivation instead of revolting against

In 2003, the Episcopal Church, which is the largest Protestant denomination worldwide, ordained V. Gene Robinson, an openly gay man, as bishop of New Hampshire. The ordination created considerable divisiveness because many of the church's conservative members believe that homosexuality is incompatible with the teachings of the Bible and church doctrine. In the United States, dioceses in three states split off from the denomination as have its bishops and archbishops in Nigeria, Africa, and the West Indies.

injustice. In effect, Marx maintained, religion is like a drug that numbs the poor and downtrodden and encourages them to tolerate their misery on earth in hope of a better life after death.

Marx viewed religion as a form of **false consciousness**, an acceptance of a system that prevents people from protesting oppression. Those who own the means of production (such as factory owners) profit if the masses (the workers) console themselves through religion instead of rising up against exploitation. Contemporary conflict theorists don't view religion as an opiate, but they agree with Marx that religion legitimizes social inequality and sometimes leads to social disruption and violence.

A Source of Social Disruption and Violence

For conflict theorists, religion tends to promote strife because, typically, religious groups differentiate between "we" and "they" ("We're right and they're wrong"). Such distinctions spark numerous disputes within and across societies (Marty 2005). For example, in India, Hindus fight Muslims (and sometimes burn Christian homes and churches), the schism between Iraq's Sunnis and Shiites dates back more than 1,300 years, and there are longstanding conflicts between Muslims and Jews in Israel and Palestine (Ghosh 2007; Ridge 2008; see also Chapter 18 online).

For thousands of years, many governments and religious leaders have condoned or perpetrated widespread violence in the name of religion. Terrorists targeted the World Trade Center on 9/11 to show their disdain for all of the Western values that the structures symbolized, such as capitalism and profit making. Those who perform such acts of violence do not see themselves as "crazed terrorists" but as "true believers" who are carrying out "God's will" by imposing their religious beliefs on others or destroying their "religious enemies" (Juergensmeyer 2003).

Many conflict theorists see religion as a major source of hatred, prejudice, and discrimination. For example, although 60 percent of Americans say that they know little about Islam, almost half have a negative perception of that religion, 25 percent have "extreme" anti-Muslim views, and 20 percent believe that it's acceptable to limit the civil liberties of Muslims living in the United States. Religious stereotypes are remarkably similar across societies: Many non-Muslims in the West and many Muslims in the Middle East and Asia generally describe people on the other side as immoral, not religious, or fanatical in their religious views and practices (Council on American-Islamic Relations 2006; Pew Global Attitudes Project 2006).

A Legitimation of Social Inequality

Conflict theorists also see religion as a tool dominant groups use to control society and to protect their own interests. In the United States before the Civil War, many churches used a slave catechism "to justify domination by masters, to encourage work, and to attribute lack of work to personal laziness. White pastors told Blacks that God created the masters over them and that the Bible tells them that they must obey their White masters" (Hurst 2001: 310).

In thirteen European countries, Muslims experience inequality in employment, education, and housing, and are also the victims of negative stereotyping, especially by the media. Some of the inequality is due to *xenophobia* (the irrational fear and distrust of anything foreign) and factors such as immigrants' not being able to speak the adopted country's language and their lack of higher education, but religious discrimination is also prevalent because a majority of Europeans believe that Muslims isolate themselves from mainstream society and engage in terrorism (European Monitoring Centre on Racism and Xenophobia 2006; Sengupta 2006).

Critical Evaluation

Conflict theory offers valuable insights on how religion creates and maintains social inequality, but it is criticized on several grounds. Functionalists may overemphasize consensus and harmony, but conflict theorists often ignore the role that religion plays in creating social cohesion and cooperation. For example, many religious people are altruistic, volunteering their time in low-income neighborhoods rather than seeking economic rewards at work or enjoying recreational activities. Some scholars have also criticized Marx (and other conflict theorists) for ignoring the possibility that religious people feel more than a mere opiate-like resignation to difficult circumstances. Instead, many religious people, even among those who are relatively privileged, run charitable organizations and shape public policy to help the poor (Hamilton 2001; Wald 2003; Roberts 2004).

FEMINIST THEORIES: DOES RELIGION SUBORDINATE AND EXCLUDE WOMEN?

Despite criticism from conservatives, the International Bible Society has updated its best-selling modern Bible to include gender-neutral wording. Examples include changing "sons of God" to "children of God" and "a man is justified by faith" to "a person is justified by faith." Are such changes only cosmetic? Most feminist scholars maintain that gender-neutral language is important because it signals inclusion to female adherents. Feminist theorists agree with conflict theorists that religion can create violence and maintain inequality. They go further, however, by criticizing organized religions for being sexist and patriarchal and for shutting many women out of leadership positions.

Sexism and Patriarchy

From a feminist perspective, most religions are patriarchal because they emphasize men's experiences and a male point of view and see women as subordinate to men. According to the apostle Paul, for example, "Wives should submit to their husbands in everything" (*Ephesians* 5: 24). The idea that Eve was created out of Adam's rib is used to justify the domination of men over women. Beliefs such as those of fourth-century theologian Augustine, "We are men, you are women, we are the head, you are the members, we are masters,

you are slaves," are reflected today in "even supposedly holy and intelligent men's particularly low opinion of women" (Mananzan 2002: 207).

In Orthodox Judaism, a man's daily prayers include this line: "Blessed are thou, O Lord, our God, King of the Universe, that I was not born a woman." The Qur'an tells Muslims that men are in charge of women and that good women should obey men. Almost all contemporary religions worship a male God, and none of the major world religions—Judaism, Christianity, Islam, Buddhism, Hinduism, or the East Asian philosophical traditions of Confucianism and Taoism—treat women and men equally (Daly 1973; Gross 1996).

Since at least the 1970s, however, some feminist scholars have charged that the Bible has often been interpreted in a patriarchal manner. For example, Jesus encouraged women's intellectual pursuits when it was not the norm and appeared first to women after his resurrection and told them to report the news to male followers. There are also numerous passages in the Bible about women spreading Jesus's teachings. Muslim feminists, similarly, note that men have interpreted sacred Islamic texts in ways that serve to ensure male dominance and control. Women were among some of Muhammad's earliest converts, and the Qur'an has numerous passages that establish women's equal rights in inheritance and family roles (Menissi 1991, 1996; Gross 1996). Thus, women's subordination in many Islamic societies is not due to the tenets of the religion but to "a patriarchy that is reinforced and perpetuated through the fundamentalist brand of Islam" (Smith 1994: 306).

Exclusion of Women from Leadership Positions

Since the 1970s, the percentage of Americans who support having women as pastors, ministers, priests, and rabbis has increased from 40 percent to over 70 percent. Also, the percentage of theology degrees earned by women has risen considerably—from only 2 percent

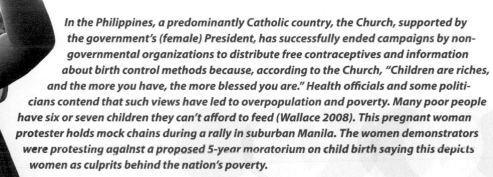

In the Philippines, a predominantly Catholic country, the Church, supported by the government's (female) President, has successfully ended campaigns by nongovernmental organizations to distribute free contraceptives and information about birth control methods because, according to the Church, "Children are riches, and the more you have, the more blessed you are." Health officials and some politicians contend that such views have led to overpopulation and poverty. Many poor people have six or seven children they can't afford to feed (Wallace 2008). This pregnant woman protester holds mock chains during a rally in suburban Manila. The women demonstrators were protesting against a proposed 5-year moratorium on child birth saying this depicts women as culprits behind the nation's poverty.

in 1970 to over 34 percent in 2007. Still, women make up only about 13 percent of the nation's clergy, and only 3 percent of female clergy lead large congregations (those with more than 350 people) (Winseman 2004; Carroll 2006; U.S. Census Bureau 2008).

Neither Roman Catholicism nor Orthodox Judaism allows women in the clergy because both groups believe that women should serve and not lead. In other religions, female members of the clergy report that they often encounter a stained-glass ceiling. The African Methodist Episcopal (AME) Church, the U.S. denomination with the largest number of African Americans, did not elect its first female bishop, Vashti McKenzie, until 2000. According to Bishop McKenzie, "For women, especially for African-American women, you always have to be better than men to get ahead" (Van Biema 2004: 60).

Protestant groups typically justify women's exclusion from leadership positions based on Biblical passages, such as "I permit no woman to teach or have authority over men; she is to keep silent" (I Timothy 2: 11–12). Many Protestant groups—including Southern Baptists and especially evangelical groups (such as born-again Christians)—interpret this passage and similar ones to mean that women should never, under any circumstances, instruct men, within or outside of religious institutions. Thus, Southern Baptist seminaries accept women for theological studies, but rarely hire them as faculty members (Bartlett 2007).

Even in liberal Protestant congregations, female clergy tend to be relegated to specialized ministries with responsibilities for music, youth, or Bible studies. The few women who become pastors frequently have small or financially struggling congregations instead of "tall-steeple churches" in affluent neighborhoods. And when women serve as associate pastors, some members of the congregation, especially men, are often unhappy because they prefer male preachers (Banerjee 2006).

Critical Evaluation

Feminist scholars have pressed for more rights for women—within Christianity, Islam, and Judaism—and for church leaders to take the religious aspirations and lives of women more seriously (Cooey et al. 1991;

Young 1999). However, feminist-oriented theology is still largely a Western movement, and some Muslim scholars have criticized feminists for misreading and misinterpreting Islamic and other sacred scriptures, such as reducing practically all discussions on gender to the *hijab* (a veil or scarf that Muslim women wear) instead of focusing on justice for both women and men in marriage, employment, and other areas (Ali 2003; Safi 2003; Choudhury et al. 2006; Esposito and Mogahed 2007).

Another criticism is that feminist scholars sometimes overlook how religious institutions—especially those that are conservative—encourage men to focus on their families. For example, churchgoing men in the United States are generally more involved and affectionate fathers and husbands and have lower rates of domestic violence than their peers who are not regular churchgoers (Bartkowski and Read 2003; Wilcox 2006).

© Asif Hassan/AFP/Getty Images

FIGURE 15.2
Religious Symbols

From left to right: Buddhist, Christian, Hindu, Indigenous and Ancient Religions, Islam, Judaism,

n mid-2005, a Danish newspaper published 12 cartoons lampooning Muslim terrorists, including one of the prophet Muhammad with a bomb in his turban. Many Muslims were furious because Muhammad is a revered symbol of Islamic religious beliefs and mocking him is considered blasphemous (Birch 2006). The cartoons sparked many demonstrations around Europe, including this one in London, England. Most European and American commentators agreed that the cartoons were in poor taste but argued that Muslims must learn to accept Western standards of free speech. Some of the commentators also accused the protestors of a double standard, citing Internet videos of radical Muslims slashing the throats of Western journalists and international humanitarian workers, Christians not being permitted to worship openly or to carry a Bible in public in Saudi Arabia, and cartoons by Muslims defaming Jews (Rubin 2006; Van Doorn-Harder 2006).

SYMBOLIC INTERACTIONISM: HOW DOES RELIGION PROVIDE MEANING IN EVERYDAY LIFE?

In 2007, a highly respected and experienced British teacher was teaching at an elementary school in Sudan, Africa. She began a project on animals and asked her 7-year-old students to name a teddy bear that the students had to photograph and write about. The class voted resoundingly for Muhammad, the name of Islam's founder and one of the most common names in the Muslim world. Within a few days, some of the children's parents objected because naming the teddy bear Muhammad violated Islamic beliefs that prohibit using the prophet's name for "idols." Shortly after that, there were numerous

Spirituality, Lutheran, Confucian, Baha'i, Scientology

protests by the Sudanese, many of whom demanded the teacher's execution. The teacher was arrested and went to trial, but the government decided to simply deport her (Crilly 2007; "Crowds Call for Teacher's Death" 2007). Thus, from a symbolic interactionist perspective, not understanding another culture's religion can lead to serious problems.

For symbolic interactionists, religion is a social phenomenon that is taught rather than being innate. Symbols, rituals, and beliefs are three of the most common vehicles for learning and internalizing religion.

> **ritual** (sometimes called a *rite*) a formal and repeated behavior in which the members of a group regularly engage.

Symbols

A *symbol* is anything that stands for or represents something else to which people attach meaning (see Chapter 3). Many religious symbols are objects (a cross, a steeple, a Bible), but they also include behaviors (kneeling and bowing one's head), words ("Holy Father," "Allah," "the Prophet"), and aspects of physical appearance (the wearing of head scarves, skull caps, turbans, clerical collars).

Religious symbols, like all symbols, are shorthand communication tools (see *Figure 15.2*). When people have the same symbols, they share the same definition of reality (including what is acceptable and what is not) and the same *worldview*, a concept of self, society, and the supernatural. Some symbolic interactionists define a religion as "a system of symbols" because it is a community that is unified by its symbols and shares a worldview (Berger and Luckmann 1966; Geertz 1966).

Rituals

A **ritual** (sometimes called a *rite*) is a formal and repeated behavior in which the members of a group regularly engage. There are many secular rituals, such as college graduations and the ceremonies marking the beginning and end of the Olympics.

Religious rituals, like secular ones, strengthen the self-identity of each participant (Reiss 2004). Religious rites of passage—such as the *bat mitzvah* for girls and the *bar mitzvah* for boys in the Jewish community and the first communion and confirmation for Catholic children—reinforce the individual's sense of belonging to a particular religious group. A group's rituals symbolize its spiritual beliefs and include a wide range of practices such as praying, chanting, fasting, singing, dancing, and offering sacrifices.

All religions have rituals that mark significant events in a person's life, such as birth, puberty, and marriage

Every year, over 2 million Muslims engage in an important ritual, the Hajj, by making a pilgrimage to Mecca (in Saudi Arabia). The Hajj is one of the five pillars of the Muslim faith that demonstrate the solidarity of Muslims and their submission to God. Every able-bodied follower who can afford it is expected to perform the pilgrimage at least once in a lifetime. The Hajj occurs from the 8th to the 12th days of the last month of the Islamic year (roughly in the November to January period of the Western calendar).

(see Chapters 3, 5, and 13). Death rituals are probably the most elaborate and sacred worldwide. These rituals vary across religious groups and societies, but all of them offer comfort to the living and show respect for the dead, for example:

- In *Buddhism*, many followers believe in reincarnation—that the spirit undergoes a series of rebirths. Because they believe that the emotions and thoughts of a dying person have a great deal to do with her or his experiences after death, Buddhists emphasize a peaceful, tranquil transition. They believe that a dying person should be surrounded by family members, close friends, and others who can provide spiritual comfort and guidance.

- In *Christianity*, a funeral service can take place in a church or at a funeral home and generally involves a minister or priest performing a service and someone who knew the person giving a eulogy. Following the service, the body is taken to the cemetery for burial, with prayers at the gravesite, or is cremated. Although all Christians believe in heaven and hell, Catholics also believe in purgatory, a place of purification where the soul goes before eventually entering heaven. Catholics believe that prayers and special masses dedicated to the soul of the dead can shorten a person's stay in purgatory.

- In *Hinduism*, the family often prepares the body at home, where family members say formal goodbyes and sing near the body. Some Hindus are buried, but most are cremated. A procession carries the body to the burning ground, usually on the bank of a river that Hindus consider sacred. The eldest son is typically the one who performs the rites, such as walking around the body, which symbolizes offering the dead to the next world.

- In *Islam*, followers believe the person will be resurrected and live after death in either paradise or hell. The body is usually cleansed with water and wrapped in a shroud. Embalming is not permitted because it is necessary that the earth reclaim the body as quickly as possible. A funeral service is held at a mosque, where the body may or may not be present. At the gravesite, a male family member or cleric recites verses from the Qur'an. The body is placed in the grave (a coffin isn't necessary), lying on its right side, facing Mecca.

- In *Judaism*, the corpse is washed, simply clothed, and placed in a plain coffin. During the funeral service, a rabbi or family members read specific Psalms and offer eulogies. The custom of sitting Shivah requires a period of 7 days devoted to mourning the deceased. Mourning ends after 30 days, except for a father or a mother, who is honored with daily prayers (the Kaddish) for 11 months and then with prayers on the anniversary of the death (Matthews 2004).

Beliefs

Rituals and symbols come from *beliefs*, convictions about what people think is true. Religious beliefs can be passive (believing in God but never attending formal services) or active (participating in rituals and ceremonies). Beliefs bind people together into a spiritual community.

One of the strongest beliefs around the world is that prayer is important. Islam requires prayer five times a day. In the United States, 58 percent of adult Americans say that they pray more than once a day and do so for a variety of reasons, such as feeling close to God as

TABLE 15.3
Sociological Perspectives on Religion

THEORETICAL PERSPECTIVE	LEVEL OF ANALYSIS	VIEW OF RELIGION	SOME MAJOR QUESTIONS
Functionalist	Macro	Religion benefits society by providing a sense of belonging, identity, meaning, emotional comfort, and social control over deviant behavior.	How does religion contribute to social cohesion?
Conflict	Macro	Religion promotes and legitimates social inequality, condones strife and violence between groups, and justifies oppression of poor people.	How does religion control and oppress people, especially those at lower socioeconomic levels?
Feminist	Macro and Micro	Religion subordinates women, excludes them from decision-making positions, and legitimizes patriarchal control of society.	How is religion patriarchal and sexist?
Symbolic Interactionist	Micro	Religion provides meaning and sustenance in everyday life through symbols, rituals, and beliefs and binds people together in a physical and spiritual community.	How does religion differ within and across societies?

well as requesting better health, more money, and cures for sick pets ("Polling Prayers" 2004; Pew Forum on Religion & Public Life 2008).

Is prayer effective? In a scientifically rigorous study, the researchers followed 1,800 patients who had received heart by-pass surgery and concluded that there were no differences in the number of post-operative complications or in the overall recovery rate of those who were and weren't prayed for. In fact, the patients who knew that they were being prayed for fared worse, presumably because they experienced anxiety that their recovery wouldn't live up to the expectations of the people who prayed for them or because the praying suggested that the patients were sicker than they thought (Krucoff et al. 2006).

Despite such empirical data, of the physicians who describe themselves as religious, 56 percent believe that religion, including prayer, affects a patient's health, and 33 percent say that religion and spirituality help prevent medical events like heart attacks, infections, and even death (Curlin et al. 2007). From a symbolic interactionist perspective, prayer provides psychological and spiritual benefits such as comfort and a sense of unity among those who pray together, but few Americans would feel comfortable with a physician who recommends prayer rather than aggressive treatment of a life-threatening illness.

Critical Evaluation

A major contribution of symbolic interactionists is that they help explain why religious behaviors vary within and across cultures and provide insights on the reasons for everyday religious practices. A common criticism, however, is that interactionists' focus on micro-level practices ignores that the ways that religion promotes social inequality at the macro level. Conflict theorists and feminist thinkers, especially, maintain that people often use religion to justify violence and women's subordination.

Some critics also wonder if symbolic interactionists are painting too rosy a picture of religion even on a micro level. After all, people's attachment to their religious symbols, rituals, and beliefs can create considerable conflict (as witnessed by the teddy bear incident). In many European and U.S. towns, for example, people often resist religious diversity by opposing proposals to build mosques in their communities, even though the plans include benefits such as housing for Islamic senior citizens and after-school activities for Muslim children (Landler 2006; "Plan to Build . . ." 2006).

Taken together, functionalist, conflict, feminist, and symbolic interactionist perspectives provide a multidimensional understanding of the beneficial and disruptive roles that religion plays on both individual and societal levels. That is, religion is an important institution that has different manifestations across societies.

©iStockphoto.com

Population growth
and urbanization are changing our lives.

what do you think?

Individuals can do little to prevent global warming.

1 2 3 4 5 6 7

strongly agree   strongly disagree

16

Population, Urbanization, and the Environment

On May 11, 2000, India gained its billionth citizen, one of an estimated 42,000 Indian babies born that day. In 2007, India had over 1.1 billion people, and it is expected to overtake China (with over 1.3 billion people) as the world's most populous nation by 2050 ("India Gives Mixed Welcome . . ." 2000; Haub 2004). This chapter examines population changes and urbanization and how both affect the physical environment in the United States and worldwide. Let's begin with population.

> **demography** the scientific study of human populations.
>
> **population** a collection of people who share a geographic territory.

1 Population Dynamics

Population growth was one of the most significant trends of the twentieth century. Since 1900, the world's population has more than tripled in size. On a typical day, in fact, the world gains almost 219,000 people (U.S. Census Bureau 2008). Even if most of the births are in other countries, population growth affects all Americans—now and in the future.

Information about population growth comes from **demography**, the scientific study of human populations. Demographers analyze populations in terms of size, composition, distribution, and why they change. A **population** is a collection of people who share a geographic territory. A population can inhabit a territory as small as a town or as vast as the planet, depending on a researcher's focus. Demographers also study personal data such as when and where you were born, your probability of getting married or divorced, the kind of job you'll probably have, how many times you'll move, and how long you'll probably live. Thus, according to one demographer, "If people are not interested in demographic phenomena, they are not interested in themselves" (McFalls 2007: 3).

WHY POPULATIONS CHANGE

Global population has grown rapidly since 1750 (see *Figure 16.1* on the next page). It reached 1 billion in 1804, 5 billion in 1987, 6.5 billion in 2005, and is expected to rise to 9.4 billion by 2050

Key Topics

In this chapter, we'll explore the following topics:

1 Population Dynamics

2 Urbanization

3 Environmental Issues

(U.S. Census Bureau 2008). When demographers examine population changes, they look at the interplay among three key factors: how many people are born (fertility), how many die (mortality), and how many move from one area to another (migration).

Fertility: Adding New People

The study of population changes begins with **fertility**, the number of babies born during a specified period in a particular society. There are several ways to measure fertility, but one of the most general and commonly used is the **crude birth rate**, also known as the *birth rate*, the number of live births for every 1,000 people in a population in a given year. "Crude" suggests that the rate is an imprecise measure of a society's childbearing pattern because it is based on the total population rather than more specific measures such as a woman's age or marital status. However, the crude birth rate allows comparisons for a given year across populations or countries. In 2008, for example, that rate was 21 worldwide, 37 for Africa, 14 for the United States, and 11 for Europe (Haub and Kent 2008).

The roads in many of India's largest cities, like this one in New Delhi, are chaotic: "Cars, trucks, buses, motorcycles, taxis, rickshaws, cows, donkeys, and dogs jostle for every inch of the roadway as horns blare and brakes squeal. Drivers run red lights and jam their vehicles into available space, ignoring pedestrians." India is booming economically but lacks a modern transportation system, reliable power and clean water are "in desperately short supply," and there is a growing chasm between the rich and the poor (Hamm 2007: 49–50).

© Manan Vatsyayana/AFP/Getty Images

Birth rates also vary within a country. In the United States, younger women—those between 20 and 34—have higher birth rates than those aged 35 to 44, and recent immigrants have higher birth rates than native-born women (Dye 2008). The more affluent, regardless of race and ethnicity, have fewer children than the poor; people with higher educational levels tend to postpone childbearing and are more likely to use contraceptives. Educated women are likely to delay childbirth until they have completed their education or started a career, which decreases the number of children they will have over a lifetime (Hamilton et al. 2003; see also Chapters 9 and 13).

Family Planning

UNITED STATES 8c

1972

© Sylvana Rega/iStockphoto.com

Mortality: Subtracting People

The second factor in population change is **mortality**, the number of deaths in a specified period in a population. Demographers typically measure mortality by the **crude death rate** (also called the *death rate*), the number of deaths per 1,000 people in a population

FIGURE 16.1
The World Population Explosion

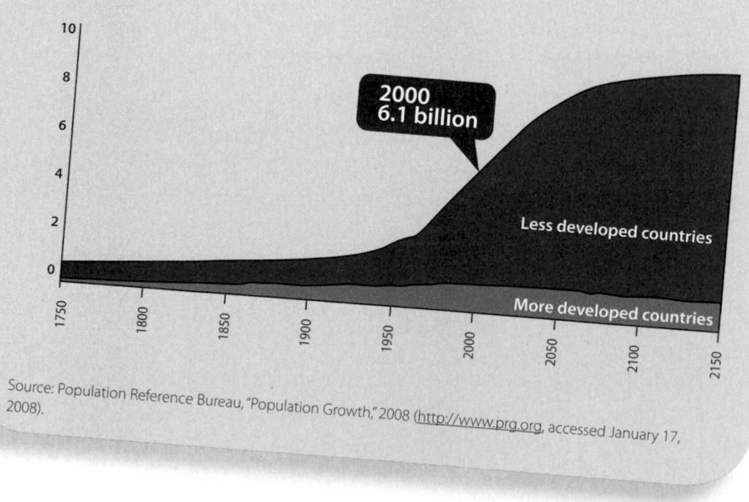

Population (in billions)

2000
6.1 billion

Less developed countries

More developed countries

in a given year. In 2008, for example, the crude death rate was 8 worldwide and for the United States, and went as high as 31 in Swaziland, Africa (Haub and Kent 2008).

A death rate isn't necessarily the best measure of a population's health, however. Death rates are high in developed countries—even though these nations have better medical services, better nutrition, and healthier environments than most developing countries—because industrialized nations also have large proportions of people who are 65 and older.

A better measure of a population's health is the **infant mortality rate**, the number of deaths of infants under 1 year of age per 1,000 live births. Generally, as the standard of living improves—meaning increased access to clean water, adequate sanitation, and medical care—the infant mortality rate decreases. The nations with the highest quality of life have the lowest infant mortality rates. In 2008, the infant mortality rate was under 3 in Singapore and Sweden, almost 7 in the United States, and over 157 in some African countries and Afghanistan (Haub and Kent 2008).

Lower infant mortality greatly raises *life expectancy*, the average number of years that people who were born at about the same time can expect to live. Worldwide, in 2008, the average life expectancy was 68 years—67 for men and 70 for women. Again, however, there are considerable variations across countries—from a high of 82 in Hong Kong and Japan to a low of 33 in Swaziland. The United States, with a life expectancy of 78, ranks below at least 25 other industrialized countries and only slightly higher than less developed countries such as Cuba, the Czech Republic, and Uruguay (Haub and Kent 2008).

In general, life expectancy has been increasing worldwide. Much of the increase is due to medical advances that have wiped out many diseases, and the availability of clean drinking water, sanitation, immunization, and antibiotics, all of which tend to prolong life. However, civil wars, genocide, and deaths due to AIDS have devastated many African countries (Ashford 2006).

U.S. life expectancy soared from 47 years in 1900 to 78 in 2008 (National Center for Health Statistics 2007; U.S. Census Bureau 2008). Still, mortality rates for Americans vary quite a bit by gender, race/ethnicity, and social class. Women have lower mortality rates than men at every age. Women probably live longer than men because they are less likely to work in physically dangerous jobs (such as construction and law enforcement), to engage in risky behaviors (such as fast driving), to serve in the military, and to commit suicide or to be homicide victims (Christenson et al. 2004). In virtually every society, people with a higher socioeconomic status live longer and healthier lives: They are more aware of the benefits of nutrition, work in jobs that are relatively safe, and have the resources to access medical services.

infant mortality rate the number of deaths of infants (under 1 year of age) per 1,000 live births in a population.

migration the movement of people into or out of a specific geographic area.

Migration: Adding and Subtracting People

The third demographic factor in population change is **migration**, the movement of people into or out of a specific geographic area. Migration can change the composition and size of a particular country's population but doesn't usually affect global population change because every person who leaves a country enters another country.

Migration is the product of two factors. *Push factors* encourage or force people to leave a residence; these factors include war, political or religious persecution, natural disasters, unemployment, high crime rates, and a high cost of living. *Pull factors* attract people to a new location; these factors include religious freedom, employment opportunities, better school systems, lower crime rates, and milder climates (Martin and Widgren 2002).

There are two types of migration: international and internal. *International migration* is movement to another country. Such migration includes *emigrants* (people who are moving out of a country) and *immigrants* (people who are moving into a country). You may be a product of international migration, for example, if your great-great-grandparents emigrated from Ireland and immigrated to the United States. International migration is often in the news, but only about 3 percent of the world's population migrates to a different country and ends up staying for a year or longer. Most emigrants move to a neighboring country (from Mexico to the United States, for example, rather than from Mexico to Canada). International migration is relatively uncommon both because most people have no desire to leave their family and friends and because governments try to regulate border crossings (Martin and Zürcher 2008).

The second type of migration is *internal migration*, movement within a country. About 40 million Americans move within the United States every year. Most are single and college-educated, and they are more likely to relocate to central cities than suburbs or rural areas because of job prospects (Franklin 2003; U.S. Census Bureau 2008).

POPULATION COMPOSITION AND STRUCTURE

Demographers examine age and sex to understand a population's composition and structure. Two of the most common measures are *sex ratios* and *population pyramids*.

The proportion of men to women in a population is a **sex ratio**. A sex ratio of 100 means that there are equal numbers of men and women, whereas a sex ratio of 95 means that there are 95 men for every 100 women (fewer males than females). Sex ratios are important because they affect the availability of marriageable partners, marriage rates, and childbearing (see Chapter 13).

Worldwide, on average, between 103 and 107 boys are born for every 100 girls. This newborn sex ratio decreases throughout life, however, as males experience higher rates of mortality at all stages of life. In fact, although it is not clear why, male fetuses even die in miscarriages at a higher rate than female fetuses (Christenson et al. 2004). However, sex ratios are skewed in favor of males in some countries, especially China and India, because of

the practice of *female infanticide*—the intentional killing of female infants due to a preference for male offspring (Haub and Sharma 2006).

Demographers also use population pyramids to illustrate the makeup of societies. A **population pyramid** is a visual representation of the makeup of a population in terms of the age and sex of its members at a given point in time. As *Figure 16.2* shows, Mexico is a young country because many of its people are 44 or younger

FIGURE 16.2
Population Pyramid Projections, 2025

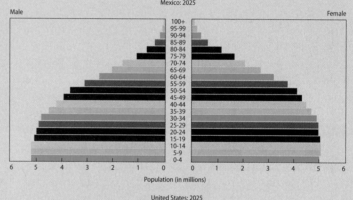

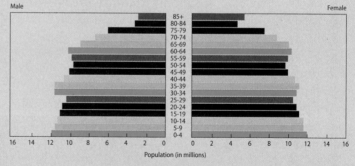

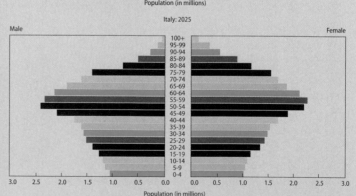

Source: U.S. Census Bureau, International Data Base, www.census.gov/ipc/www/idb/pyramids.html (accessed January 20, 2007).

(which also means that many women are in their child-bearing years) and there are relatively few people aged 65 and over. In contrast, Italy is an old country, and the United States is somewhere in the middle. The shape of the population pyramid (a triangle for Mexico, a rectangle for the United States, and a diamond for Italy) indicates future implications for young and old countries. In Italy, for example, the relatively small number of women aged 15 to 44 (in their reproductive years) and the bulge of people aged 45 to 79 suggests that there may not be enough workers to support an aging population in the future and that there will be a greater need for social services for the elderly than for children and adolescents.

POPULATION GROWTH: A TICKING BOMB?

Eight of the countries with the largest populations (many of them in the developing world) will increase even more by 2050 (see *Table 16.1*). So, has population growth gotten out of hand? There are many views on this issue, but two of the most influential have been Malthusian theory (which argues that the world can't sustain its unprecedented population surge) and demographic transition theory (which maintains that population growth is slowing).

Malthusian Theory

Malthusian theory the idea that population is growing faster than the food supply needed to sustain it.

For many demographers, population growth is a ticking bomb. They subscribe to **Malthusian theory**, the idea that the population is growing faster than the food supply needed to sustain it. This theory is named after Thomas Malthus (1766–1834)—an English economist, clergyman, and college professor—who maintained that humans are multiplying faster than the ability of the earth to produce sufficient food.

According to Malthus (1798/1965), population grows at a *geometric rate* (for example, from 2 to 4 to 8 and so on), whereas the food supply grows at an *arithmetic rate* (1, 2, 3, 4, and so on). That is, two parents can have four children and sixteen grandchildren within 50 years, whereas the available number of acres of land, farm animals, and other sources of food can increase in that time period, but certainly not quadruple. In effect, then, the food supply will not keep up with population growth. Because there are millions of parents, the results could be catastrophic, such as masses of people living in poverty or dying of starvation.

Malthus first posited that only war, famine, and disease act as *preventive checks* on population growth. In later essays, he also included "moral restraint" as a necessary preventive check, especially for lower classes. The lack of moral restraint he was referring to led people to marry at an early age and to not practice sexual abstinence prior to and outside of marriage, which often resulted in large families and out-of-wedlock children that the working men couldn't save from "rags and squalid poverty" (Malthus 1872/1991).

TABLE 16.1
The World's Largest Countries, 2008 and 2050

COUNTRY	MID-2008 POPULATION (IN MILLIONS)	ESTIMATED 2050 POPULATION (IN MILLIONS)
China	1,325	1,747
India	1,149	1,437
United States	305	420
Indonesia	240	297
Brazil	195	260
Pakistan	173	295
Nigeria	148	282
Bangladesh	147	231

Note: Remember to add six zeros to these figures. For example, China's population in mid-2008 was about 1,325,000,000, or 1.3 billion people.
Sources: Based on Haub 2007, and Haub and Kent 2008.

demographic transition theory the idea that population growth is kept in check and stabilizes as countries experience economic and technological development.

Except for the notions about moral restraint, Malthusian theory has had a lasting influence. *Neo-Malthusians* (or New Malthusians) agree that the world population is exploding beyond food supplies. For example, the world population reached its first billion in 1800. In the 200 years that followed, the world added 5 billion people (see *Figure 16.1* on p. 294). Because of this rate of growth, according to some influential neo-Malthusians, the earth has become a "dying planet"—a world with insufficient food and a rapidly expanding population that pollutes the environment (Ehrlich 1971; Ehrlich and Ehrlich 2008).

Demographic Transition Theory

Some demographers are more optimistic than neo-Malthusians. **Demographic transition theory** maintains that population growth is kept in check and stabilizes as countries experience economic and technological development, which in turn affects birth and death rates. According to this theory, population growth changes as societies undergo industrialization, modernization, technological progress, and urbanization. During these processes, a nation goes through four stages (see *Figure 16.3*), from high birth and death rates to low birth and death rates:

- *Stage 1: Preindustrial society.* In this initial stage, there is little population growth. The birth rate is high because people rarely use birth control and they want as many children as possible to provide unpaid agricultural labor and support their parents in old age. However, a high death rate offsets the high birth rate. Many children don't survive infancy, and mortality is high at all ages due to diseases and minimal access to health care.

- *Stage 2: Early industrial society.* There is a significant population growth because the birth rate is higher than the death rate. The birth rate may even increase over what it was in Stage 1 because mothers and their children enjoy improved health care. Couples may still have large numbers of children because they fear that many of them will die, but the death rate drops because of better sanitation, better nutrition, and modernization, especially medical advances (such as immuniza-

tions and antibiotics). It was during this stage of the demographic transition in Europe that Malthus formulated his ideas about population growth. Most of the world's poorest countries are currently in Stage 2.

- *Stage 3: Advanced industrial society.* As the infant mortality rate drops, parents have fewer children. Effective birth control reduces family size. This decrease in child-care responsibilities, in turn, provides women with time to work outside the home. China and many countries in Latin America are currently in Stage 3 (Gelbard et al. 1999; Brea 2003).

- *Stage 4: Postindustrial society.* In this stage, the demographic transition is complete; the society has low birth and death rates. Women tend to be well educated and to have full-time jobs or careers. If there is little immigration, the population may even decrease because the birth rate is low. This is the case today in Canada, Japan, Singapore, Hong Kong, Australia, New Zealand, the United States, and many European countries, including Italy and Scotland.

Critical Evaluation

The dire predictions of Malthus and his successors that global population growth would lead to worldwide famine, disease, and poverty have not come true. Still, today more than 20 percent of people live in abject pov-

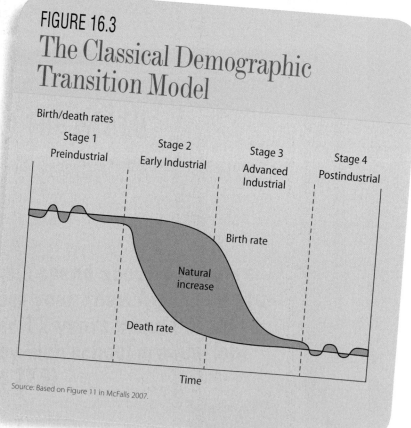

FIGURE 16.3
The Classical Demographic Transition Model

Birth/death rates

Stage 1 Preindustrial

Stage 2 Early Industrial

Stage 3 Advanced Industrial

Stage 4 Postindustrial

Birth rate

Natural increase

Death rate

Time

Source: Based on Figure 11 in McFalls 2007.

erty, subsisting on less than $1 a day (World Bank 2008; see also Chapter 8).

Despite the neo-Malthusians' fears, global fertility is half of what it was in 1972. The population of some industrialized countries is declining because people aren't having enough babies to replace themselves. These countries are experiencing **zero population growth (ZPG)**, a stable population level that occurs when each woman has no more than two children. In India, up to 83 percent of married couples—the world's highest rate—use effective contraceptives, such as condoms and birth control pills, that may result in a much lower fertility rate in the future (Larsen 2004).

Future population growth is difficult to predict because there are many unknowns. A number of countries with ZPG are now paying women to have more children because there won't be enough young workers to pay for social security systems and the rising cost of health care for aging populations. Some neo-Malthusians maintain, however, that it's irresponsible for *any* country to encourage higher fertility rates; these demographers worry about the consequences of adding 3 billion more inhabitants to the planet in less than 50 years, especially for many developing countries "with desperate economic outlooks" (Sachs 2005; Shorto 2008).

The birth rate in the United States has declined, but it's third in the world in population growth, behind India and China, largely because of a high rate of immigration, and many of the immigrants are young women with high fertility rates. Thus, U.S. annual population "is growing by more than all other developed countries *combined*" (Ryerson 2004: 21).

One of the results of population growth is urban growth. Cities attract both immigrants and native-born residents because of jobs and cultural activities, but the population growth of urban areas also creates numerous problems.

2 Urbanization

if you've flown over the United States, you've probably noticed that people tend to cluster in and around cities. After sunset, some areas glow with lights, while others are engulfed in darkness. The average person, in the United States and worldwide, is more likely to live in a city than a rural area, and this trend is rising. A **city** is a geographic area where a large number of people live relatively permanently and secure their livelihood pri-

marily through non-agricultural activities. **Urbanization**, which increases the size of cities, is the movement of people from rural to urban areas. Most of this discussion will focus on U.S. cities, but let's begin with a brief look at urbanization globally.

zero population growth (ZPG) a stable population level that occurs when each woman has no more than two children.

city a geographic area where a large number of people live relatively permanently and secure their livelihood primarily through nonagricultural activities.

urbanization population movement from rural to urban areas.

URBANIZATION: A GLOBAL VIEW

In 2008, for the first time in history, a majority of the world's population lived in urban areas. By 2030, urban dwellers will make up roughly 60 percent of the world's population (Population Reference Bureau 2007; United Nations Population Division 2008). Why is urbanization increasing? And where is most of it taking place?

The Origin and Growth of Cities

Cities are one of the most striking features of modern life, but they have existed for centuries. About 7,000 years ago, for example, people built small cities in the Middle East and Latin America to protect themselves from attackers and to increase trade. By 1800, 56 cities in Western Europe had a population of 40,000 or more (Chandler and Fox 1974; Flanagan 1990; De Long and Shleifer 1992).

Before the Industrial Revolution, which began in the late eighteenth century, urban settlements in Europe, India, and China developed largely because people figured out how to use natural resources (such as mining coal and transporting water efficiently for irrigation and consumption) that allowed many people to live in small areas. The Industrial Revolution spurred ever increasing numbers of people to move to cities in search of jobs, schooling, and improved living conditions. As a result, the urban population surged—from 3 percent of the world's population in 1800 to 14 percent in 1900 (Sjoberg 1960; Mumford 1961).

World Urbanization Trends

As industrialization advanced, urbanization increased. Between 1920 and 2007, the world's urban population increased from 270 million to 3.3 billion, and it

megacities metropolitan areas with at least 10 million inhabitants.

is expected to rise to 6.5 billion by 2050. The pace of urbanization is most rapid in the less developed regions of the world, especially Asia, Africa, and Latin America (see Table 16.2). By 2050, most of the world's urban population will be concentrated in Asia and Africa (United Nations Population Division 2008).

Because cities attract businesses, they become magnets for migrants who seek better education, employment, and health care. As a result, many of the world's largest cities are becoming **megacities**, metropolitan areas with at least 10 million inhabitants. In 1950, the three largest cities in the world were New York–Newark (12.3 million), Tokyo (11.3 million), and London (8.4 million). By 2025, there will be 27 megacities, but only two in North America (New York–Newark and Los Angeles) and one in Europe (Paris). Besides Tokyo—which will probably be the most populous city in the world, with nearly 36 million inhabitants in 2025—there will be 16 megacities in Asia, four in Latin America, and three in Africa, and most will be at least twice as large as those in North America and Europe (United Nations Population Division 2008).

Should the explosive growth of cities and megacities concern us? Generally, cities provide jobs, offer better health care, and stimulate technological innovation and cultural enrichment, but not everyone benefits from these and other advantages. The urban poor are often crowded into slums, where children are less likely to be enrolled in school and there is inadequate sanitation and a widening economic gap between the haves and the have-nots (Cohen 2005).

URBANIZATION IN THE UNITED STATES

Like many other countries, the United States is becoming more urban. During the Industrial Revolution, millions of Americans in agricultural areas migrated to cities to find jobs. Thus, between 1900 and 2000, the U.S. rural population shrank from 61 percent to 21 percent of the total population, while the urban segment increased from 39 percent to 79 percent (Riche 2000; U.S. Census Bureau 2008). This urbanization has been accompanied by greater racial-ethnic diversity in many cities and the booming of suburbs.

Shifts in Urban and Rural Populations

Despite rapid population growth in parts of the South and West, 45 percent of all U.S. counties have lost population since 2000. Of the 1,346 counties that shrank in population between 2000 and 2007, 85 percent were

With a population of almost 12 million in 2007, Cairo, Egypt, is one of the world's largest cities. Many see Cairo as the cultural center of the Arab world, but millions of Egyptians, including this fisherman, live in dire poverty and don't experience any of the city's cultural benefits. This man sleeps in his boat, makes tea from the water of the Nile River (which is infested with life-threatening parasites), often smiles and waves dutifully as tour boats motor up the river with tourists snapping his picture, and, on a good day, earns a few dollars (Slackman 2007).

© Shawn Baldwin/The New York Times/Redux

TABLE 16.2
Urbanization Around the World
Percentage of people living in urban areas

REGION	1950	2007	2025 (PROJECTED)
World	29	49	57
Africa	15	39	47
Asia	17	41	51
Latin America and the Caribbean	42	78	84
North America	64	81	86
Europe	51	72	76
Oceania	51	71	72

Source: Based on United Nations Population Division 2008, Table 1.

Between 1900 and 2000, the urban segment grew from 39 percent of the total U.S. population to 79 percent.

rural. Many rural communities, particularly in the Midwest, have been losing inhabitants for decades and "are on the brink of extinction" (Mather 2008).

The fastest-growing counties are located near large metropolitan areas, such as those around Atlanta, Chicago, Dallas, Houston, Los Angeles, Miami, New York City, and Washington, D.C. Some of this growth is due to migration from rural communities as blacks and whites search for jobs with decent wages. Cities also attract Latinos who seek jobs in rapidly growing economic sectors such as the construction and service industries (see Chapter 12). The Asian population continues to cluster in traditional immigrant magnet areas (Los Angeles, New York, and San Francisco), but large numbers are also moving to metropolitan areas in Illinois, Maryland, Virginia, and Wisconsin. Overall, a study of population makeup in metropolitan areas concluded that diversity is growing "at a pace that the nation has not seen for many decades" (Frey 2006: 21).

Suburbs and Exurbs

Another major factor in urban growth is **suburbanization**, population movement from cities to the areas surrounding them. When suburbs emerged during the 1930s, only the affluent could afford to commute between work in a congested and polluted city and a home in the tranquility and privacy of the countryside. During the 1950s, suburbs mush-roomed, attracting two-thirds of urban dwellers. The federal government, fearful of a return to the economic depression of the 1930s, underwrote the construction of much new housing in the suburbs. The general public obtained low-interest mortgages, veterans were offered the added incentive of being able to purchase a home with a $1 down payment, and massive highway construction programs enabled commuting by car (Rothman 1978). More than 60 percent of Americans reside in suburbs (Riche 2000), but, as you'll see shortly, there is still considerable racial and social class segregation in those areas.

Originally, most suburbs were bedroom communities, places from which commuters went daily to their jobs in the city. Over the last few decades, suburbanization has generated **edge cities**, business centers that are within or close to suburban residential areas and include offices, schools, shopping and entertainment, malls, hotels, and medical facilities. Examples of edge cities include Towson (Maryland), Framingham (Massachusetts), Cherry Hill (New Jersey), the Durham-Raleigh research triangle (North Carolina), Cool Springs (Tennessee), and Las Colinas (Texas).

In addition to edge cities, people have also created **exurbs**, areas of new development beyond the suburbs that are more rural but on the fringe of urbanized areas. Journalist A. C. Spectorsky introduced the concept in 1955, but it has become popular only recently because about 6 percent of Americans are exurbanites. The average exurbanite is white, middle-income, married with children, a "super commuter" (one who travels for two or more hours a day to get to work), and owns a large house outside of an expensive metropolitan suburb. Examples of exurbs include Geauga County, Ohio

urban sprawl the rapid, unplanned, and uncontrolled spread of development into regions adjacent to cities.

gentrification the process in which middle-class and affluent people buy and renovate houses and stores in downtown urban neighborhoods.

urban ecology the study of the relationships between people and urban environments.

(whose commuters go to Cleveland) and Yamhill County, Oregon (whose commuters go to Portland) (Berube et al. 2006; Lalasz 2006).

Some Consequences of Urbanization

Cities offer many benefits. Among other advantages, people can often walk, bicycle, or take a bus or subway to work; they are surrounded by a vast array of restaurants and shops; and they have easy access to numerous cultural activities (such as museums and theaters). Urbanization also creates problems, however, such as urban sprawl, increased traffic congestion, a scarcity of affordable housing, and racial segregation.

Urban sprawl—the rapid, unplanned, and uncontrolled spread of development into regions adjacent to cities—is widespread. Between 1995 and 2002, New Jersey, the nation's most densely populated state, lost 29 percent of its farmland, forests, wildlife habitats, and open recreational areas to urban sprawl (Lathrop and Hasse 2007).

In most cases, the only way to get around in areas of urban sprawl is by automobile. This means that most suburban households face the costs of buying, fueling, insuring, and maintaining several cars. The United States population increased by 23 percent over the last 25 years, but total highway miles have increased by only about 5 percent. One of the consequences has been greater traffic congestion within and outside cities. More than 3 million Americans (about 3 percent of workers) now travel 90 minutes or more to work every day, a proportion that has increased by 95 percent since 1990. Traffic snarls and long commutes increase air pollution and stress and decrease the time that people have for community involvement and leisure pursuits (Sullivan 2007; U.S. Census Bureau 2008).

Urbanization can also increase housing problems. Cities have high concentrations of poor and low-income people, many of them crowded into dilapidated neighborhoods. When city governments raze these areas, they often sell the property to developers who put in apartments or condominiums that the poor can't afford.

In other cases, the poor are pushed out through **gentrification**, the process in which middle-class

and affluent people buy and renovate houses and stores in downtown urban neighborhoods. Governments in many older cities (including Baltimore, Cincinnati, Detroit, Chicago, and New York City) encourage gentrification to increase dwindling populations, to revitalize urban areas, and to augment tax revenues. However, gentrification often adds to urban problems. For example, when the rent increases, it displaces low-income residents and small businesses.

Racial segregation is another urban problem. In addition to the displacement of numerous low-income white and African American families by gentrification, many middle-class families in both groups left the cities for the suburbs. Some highly educated African Americans migrated to mostly white neighborhoods, while others established their own high-income communities, including suburbs (Bayer et al. 2005; Booza et al. 2006). Many middle- and upper-class minorites have housing options, but millions of people with low incomes are stuck in inner cities that often have high rents but little employment that pays above a minimal wage (see Chapter 12).

SOCIOLOGICAL EXPLANATIONS OF URBANIZATION

How and why do cities change? And how do these changes affect their populations? In answering these and other questions, functionalists provide insights into urban development, conflict theorists emphasize the impact of capitalism and big business, feminist scholars focus on gender roles and space, and symbolic interactionists examine the quality of city life (*Table 16.3* on p. 305 summarizes these perspectives).

Functionalism: How and Why Cities Change

In the 1920s and 1930s, sociologists at the University of Chicago developed theories of **urban ecology**, the study of the relationships between people and urban environments. Initially, these sociologists based their theories on Chicago, the city in which their university was located, but social scientists later revised the descriptions (see *Figure 16.4*).

Sociologists Robert Park and Ernest Burgess (1921) proposed *concentric zone theory* to explain the distribution of social groups within urban areas. According to this model, a city grows outward

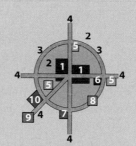

FIGURE 16.4
Four Models of City Growth and Change

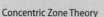

Concentric Zone Theory

1. Central business district
2. Zone in transition
3. Zone of workingmen's homes
4. Residential zone
5. Commuters' zone

Sector Theory

1. Central business district
2. Wholesale, light manufacturing
3. Lower-class residential
4. Middle-class residential
5. Upper-class residential

Multiple Nuclei Theory

1. Central business district
2. Wholesale, light manufacturing
3. Lower-class residential
4. Middle-class residential
5. Upper-class residential
6. Heavy manufacturing
7. Outlying business district
8. Residential suburb
9. Industrial suburb

Peripheral Theory

1. Central city
2. Suburban residential area
3. Circumferential highway
4. Radial highway
5. Shopping mall
6. Industrial district
7. Office park
8. Service center
9. Airport complex
10. Combined employment and shopping center

Sources: Based on Park and Burgess 1921, Hoyt 1939, Harris and Ullman 1945, and Harris 1997.

new urban sociology urban changes are largely the result of decisions made by powerful capitalists and other groups in the dominant social class.

and warehouses. Thus, heavy industry and high-income housing rarely exist in the same part of the city.

As cities grew after World War II, these models no longer fit the changes that occurred in urban spaces. Thus, Chauncey Harris (1997) proposed a *peripheral theory* of urban growth, which emphasized the development of suburbs around a city but away from its center. According to this model, as suburbs and edge cities burgeon, highways that link the city's central business district to outlying areas and beltways that loop around the city provide relatively easy access to airports and the downtown and surrounding areas.

from a central point in a series of rings. The innermost ring, the central business district, is surrounded by a zone of transition, which contains industry and poor-quality housing. The third and fourth rings have housing for the working and middle classes. The outermost ring is occupied by people who live in the suburbs and commute daily to work in the central business district.

In developing *sector theory*, economist Homer Hoyt (1939) refined concentric zone theory. He proposed that cities, including Chicago, develop in sectors instead of rings. Pie-shaped wedges radiate from the central business district, their orientation depending on transportation routes (such as rail lines and highways) and various economic and social activities. Thus, some sectors are predominantly industrial, some contain stores and offices, and others, generally further away from the central business district, are middle- and upper-class residential areas.

Geographers Chauncey Harris and Edward Ullman (1945) developed another influential model, *multiple-nuclei theory*, which proposed that a city contains more than one center around which activities revolve. For example, a "minicenter" often includes an outlying business district with stores and offices that are accessible to middle- and upper-class residential neighborhoods, whereas airports typically attract hotels

Conflict Theory: The Impact of Capitalism and Big Business

Functionalists see urban growth as a reflection of people's choices. In contrast, **new urban sociology**, a perspective, heavily influenced by conflict theory, that views urban changes as being largely the result of decisions made by powerful capitalists and other groups in the dominant social class. That is, economic and political factors and the rich, not ordinary citizens, determine urban growth or decline. For example, when a local government wants to rejuvenate parts of the inner city, it typically offers tax breaks, changes zoning laws, and allows real estate, construction, and banking interests to seek profits with little regard for the needs of low-income households or the homeless (Feagin and Parker 1990; Macionis and Parrillo 2007).

Conflict theorists see urban space as a commodity that is bought and sold for profit. It is not the average American, they argue, but bankers, corporate executives, developers, politicians, and influential businesspeople

© Fredrik Renander/Redux

Electronics City, India's version of California's Silicon Valley, is an industrial park that spans over 330 acres and houses more than a hundred businesses, such as Hewlett-Packard, Motorola, and Infosys, as well as a premier graduate school that focuses on information technology (Hamm 2007). If governments and corporations can find the capital to build such facilities, ask conflict theorists, why can't they provide affordable housing for people in surrounding neighborhoods?

who determine how urban space will be used. Increasing the value of some property has a higher priority than respecting community values, considering neighborhood needs, or maintaining a livable city (Logan and Molotch 1987; Gottdiener and Hutchinson 2000).

Feminist Theories: Gender Roles, Space, and Safety

Feminist scholars generally agree with conflict theorists that many of the problems associated with urbanization reflect macro-level factors—such as the provision of federal funds to developers to build houses regardless of people's needs. However, feminist theories emphasize gender-related constraints. Whether they live in cities or suburbs, women generally experience more problems than men because living spaces are usually designed by men who have tended to ignore women's changing roles.

In cities, both sexes tend to live in small apartments, but poor women and minorities still have the least access to decent housing. Especially for low-income single mothers, child rearing is difficult because of unsatisfactory schools, few after-school programs, high crime rates, limited recreational facilities, and lack of safe public places such as parks.

Feminist geographers Mona Domosh and Joni Seager (2001) have observed that many women fear the

city, especially urban public spaces such as streets, parks, and public transportation. They see these places as risky for their physical safety, despite the fact that most violence against women occurs at home. A few cities provide public transportation, such as minivans that operate seven nights a week, to prevent crimes against women, usually minorities, who must travel to work after 8:00 P.M. and return home before dawn. For the most part, however, such services are rare (Saegert and Winkel 1981; Hayden 2002).

In the suburbs, both women's and men's physical mobility is limited because of the scarcity of public transportation systems. Consequently, many suburban households have two or more cars, but women living in the suburbs usually experience greater problems than men. For example, there may be fewer job opportunities (especially if women have domestic responsibilities like needing to be at home when children return from school), and women often spend much of their time maintaining a single-family home, which decreases their time for educational or leisure activities (Cichocki 1981; Hayden 2002).

Symbolic Interactionism: How People Experience City Life

Symbolic interactionists are most interested in the impact of urban life on city residents. In a classic essay, sociologist Louis Wirth (1938: 14) described the city as a place where "our physical contacts are close, but our social contacts are distant."

Wirth defined the city as a large, dense, and socially and culturally diverse area. These characteristics produce urbanism, a way of life that differs from that of rural dwellers. Wirth saw urbanites as more tolerant of a variety of lifestyles, religious practices, and attitudes than residents of small towns or rural areas. However, he also emphasized urbanism's negative consequences, such as alienation, friction due to physical congestion, pursuit of self-interest, impersonal relationships, and a disintegration of kinship and friendship ties. Some studies have supported Wirth's theory of urbanism (see, for example, Guterman 1969), but others have challenged his views and have pointed out that many urbanites lead satisfying lives.

Critical Evaluation

Some functionalists use multiple nuclei theory and peripheral theory to describe some older industrial cities (such as Cleveland, New York, and Detroit), but patterns of urbanization have become more complex because of gentrification, edge cities, and peripheral expansion. Another limitation, as conflict theorists point out, is that functionalists tend to overlook the negative political and economic impact, especially when profit and greed guide urban planning.

Both functionalist and conflict theories rarely take into account the changing composition of urban populations, especially the increase in the numbers of single mothers who have the least access to affordable housing. Another limitation of conflict theory is the assumption that residents are helpless victims as developers and corporations raze low-income houses. In fact, environmental groups have had considerable success in pushing through legislation to maintain and even increase open public spaces and build energy-saving homes in low-income neighborhoods (see Moore 2008).

Recent feminist scholarship on urbanization has come primarily from historians, architects, and geographers, resulting in little sociological analysis. Feminist sociologists have made important contributions through studies of the everyday lives of low-income women, especially in central cities (see Chapter 12), but urbanization has received much less attention.

According to critics, urbanites are more diverse than some symbolic interactionists claim. People living in cities are not necessarily more self-centered or isolated than those in small towns or rural areas. Instead, many have close family bonds, friends, and satisfying relationships with co-workers (Gans 1962; Crothers 1979; Wilson 1993). Especially since the advent of the Internet and cell phones, people living in cities are interacting with family more than ever before (see Chapter 5). For symbolic interactionists, people interpret and actively shape their urban environment. However, this perspective does not show how social, political, cultural, educational, religious, and economic factors shape urban inhabitants' experiences of city life (Hutter 2007).

You've seen that the world's population is growing rapidly and becoming more urbanized. Many scholars worry that population increases are permanently damaging the earth because people are changing the environment through their high consumption of food, energy, water, and land. Others dismiss such concerns as alarmist; they maintain that natural resources are replaceable and technology will solve environmental problems. What do the data tell us?

3 Environmental Issues

When Americans are asked to name the country's top one or two problems, only 2 percent mention the environment or pollution, well behind the economy, health care, the Iraq war, crime, rising energy prices, and Social Security ("The Environment" 2008). Such concerns are understandable, but consider the following:

- Most of us have at least 116 toxic chemicals in our bodies that did not exist in the environment (much less in humans) just 75 years ago (*Second National Report . . .* 2003).

TABLE 16.3
Sociological Perspectives on Urbanization

PERSPECTIVE	LEVEL OF ANALYSIS	KEY POINTS
Functionalist	Macro	People create urban growth by moving to cities to find jobs and to suburbs to enhance their quality of life.
Conflict	Macro	Driven by greed and profit, large corporations, banks, developers, and other capitalistic groups determine the growth of cities and suburbs.
Feminist	Macro and micro	Whether they live in cities or suburbs, women generally experience fewer choices and more constraints than do men.
Symbolic interactionist	Micro	City people are more tolerant of different lifestyles, but they tend to interact superficially and are generally socially isolated.

ecosystem a system in which all forms of life live in relation to one another and a shared physical environment.

- Every American now produces, on average, 5 pounds of garbage a day compared with 2.7 pounds a day in 1960 ("Municipal Solid Waste" 2005).

- Commercial logging—spurred by high U.S. demand for hardwoods like teak, mahogany, and rosewood—results in the destruction of 50,000 species every year, including plants that produce life-saving Western medicines (Raintree Nutrition, Inc. 2008).

- Coastal counties comprise less than 25 percent of the land area of the United States, but they are home to more than 52 percent of the total population. Coastal development destroys wildlife habitat and degrades water quality (Bourne 2006).

- As many as 7 million Americans get sick every year from swimming in water contaminated with bacteria, viruses, or parasites that cause a wide range of diseases, including ear, nose, and eye infections; hepatitis; skin rashes; and respiratory illnesses (Natural Resources Defense Council 2007).

Such examples of environmental depletion are important because every person on the planet is part of an **ecosystem**, a system in which all forms of life live in relation to one another and a shared physical environment. This means that plants, animals, and humans depend on each other for survival. Because the ecosystem is interconnected worldwide, what happens in one country affects others. Let's look more closely at water and air pollution and global warming—two interrelated environmental problems that are threatening the ecosystem in the United States and globally.

WATER

"By means of water," says the Qur'an, "we give life to everything." And, in 1746, Benjamin Franklin noted, "When the well is dry, we learn the worth of water." Indeed, water has an enormous impact on human life. More than 1 billion people worldwide don't have clean water, almost 6 billion go without adequate sanitation, and almost 2 million children die every year because of contaminated water. Water-related diseases cause 50 percent of illnesses and deaths not due to accidents or violence worldwide every year (United Nations World Water . . . 2006).

The introduction of water filtration and chlorination in major U.S. cities between 1900 and 1940 was responsible for about 43 percent of the total decline in mortality over that period. Nearly half the world's population, however, still drinks contaminated water, and diarrheal diseases kill over 3 million children every year (Scommegna 2005; Water Quality & Health Council 2005). In some developing countries, families often spend up to 25 percent of their income to purchase water, and many women and children spend up to 6 hours per day carrying it home (United Nations World Water . . . 2006).

Consumption and Availability

A United Nations report predicts that water will become the dominant global problem in this century because the world's demand for water has tripled over the last half-century. Some people refer to water as "blue gold" because it is becoming one of the earth's most precious and scarce commodities. In fact, some current wars are being fought over water rather than oil or diamonds (see Chapter 18 online). By 2050, about 7 billion people in 60 countries may experience water scarcity, with the worst shortages occurring in poor countries (United Nations World Water . . . 2006).

Industrialized nations not only have greater access to clean water than the developing world, they use more and pay less for it. The average person in the United States uses about 151 gallons of water per day (for drinking, cooking, bathing, flushing toilets, and watering a yard), compared with 101 gallons in Italy, 23 in China, and less than 3 in Mozambique. Among people living in industrialized countries, Americans pay the lowest rate for water ($2.49 a gallon), while Danes and Germans pay the most (almost $9.00 a gallon). In the developing world, people typically pay five times as much as Europeans. Some of the poorest people living in urban slums and low-income areas often pay ten times as much or more for their water as high-income residents because they depend on vendors rather than a municipal provider (United Nations Development Programme 2006; Lavelle 2007). Thus, in many countries, clean water is a luxury rather than a basic human right.

Just 3 percent of the earth's water is fresh, and two-thirds of that water is locked up in the ground, glaciers, and ice caps. That leaves about 1 percent of freshwater for the world's almost 7 billion people (Schirber 2007).

Threats to Water Supplies

Precipitation (in the form of rain, snow, sleet, or hail) is the main source of water for the ecosystem. Clean water has been depleted for many reasons, including pollution, privatization, and mismanagement.

1. **Pollution.** When national polls ask Americans, specifically, about their top 10 environmental concerns, the top four involve water, especially the pollution of drinking water, rivers, and lakes (Carroll 2007). Are these concerns warranted? Yes, because toxins from cities, factories, and farms are spoiling U.S. freshwater supplies. Every year, more than 860 billion gallons of sewage, pesticides, fertilizers, automotive chemicals, and trash enter the country's rivers (Gurwitt 2005).

   In some places, water pollution has resulted in toxic mercury levels. Mercury occurs naturally in air, water, and soil. However, human activities—especially the operation of coal-fired power plants—account for about 40 percent of the highly toxic mercury emissions in the United States that contaminate the food we eat (Shore 2003). Mercury can affect the nervous systems of adults and children, especially when they eat polluted fish and shellfish. Because fetuses, infants, and children are still developing, their nervous systems are especially likely to be harmed by mothers who have mercury in their blood. As many as 600,000 babies are born in the United States every year with irreversible brain damage because the mother ate mercury-contaminated fish during her pregnancy ("Mercury Pollution . . ." 2005; Wheeler 2007).

Sugai, a farming village in China, was destroyed in a flood of sludge from two paper mills. Villagers tried to construct a makeshift dike, but all 57 homes sank into a black, polluted lake (Yardley 2006). In much of India (below), residents don't know when the next water delivery will arrive, so they must spend their days waiting for, and fighting over, the shipments.

Since 1990, the Environmental Protection Agency (EPA) has compelled municipal and medical waste incinerators to reduce their mercury emissions by over 90 percent. Power plants, however, are still largely exempt from EPA regulations and enforcement, primarily because the Bush administration has viewed these plants as an important source of energy (Clayton 2004; Janofsky 2005).

2. **Privatization.** Water is a big business because of *privatization*, transferring some or all of the assets or operations of public systems into private hands. Perrier, Evian, Coca-Cola, PepsiCo—and particularly the French giants Vivendi and Suez—have been buying the rights to extract water at will from aquifers (underground layers of rock that hold water), then bottling and selling it around the world. The bottling of water from aquifers is lucrative for corporations, but it contributes to global greenhouse gases and global warming (which we'll examine shortly), depletes local water sources, and produces—every year in the United States alone—2 million pounds of plastic bottles that clog landfills (Walsh 2007; Azios 2008).

3. **Mismanagement.** Most problems concerning water pollution and use are due not to nature but to human mismanagement, corruption, and bureaucratic bungling. In China and India, for example, most government officials support economic growth and rarely punish local or international polluters who dump chemicals and waste into rivers and lakes (United Nations World Water . . . 2006; Carmichael 2007; Ford 2007).

   The United States has the resources to decrease water pollution, but many government officials are apathetic about the problem or have close ties with industrialists. In 1972, Congress passed the Clean Water Act to modernize sewage treatment and to decrease water pollution. Local, state, and federal regulators often fail to enforce the Clean Water Act, however. They look the other way when farmers spread excess fertilizers on croplands that leach into aquifers or run off into rivers or when mining companies dump wastes into streams (Arax 2004; "Decapitating Appalachia" 2004).

   The EPA estimates that it would cost from $17 billion to $23 billion per year for the next 20 years to replace the country's substandard sewage pipes, especially in cities along the eastern seaboard. Most of the pipes in those cities are nearly 200 years old, and some are made of wood. A water pipe that leaks wastes large amounts of drinking water, and a broken water pipe can result in a large sinkhole in a street or highway that may require several weeks of repairs (American Rivers 2005; Lavelle 2007). Some environmentalists wonder why the federal government spends almost $611 billion a year for the war in Iraq (see Chapter 18 online) but less than $2 billion a year to replace aging sewage pipes that affect millions of Americans.

AIR POLLUTION AND GLOBAL WARMING

In 2007, the Intergovernmental Panel on Climate Change, an international panel representing more than a thousand scientists, shared the Nobel Peace Prize with former Vice-President Al Gore (who narrated the documentary *An Inconvenient Truth*), for disseminating information about global warming, a critical environmental problem. Let's begin by looking at air pollution, the major cause of global warming.

Air Pollution: Some Sources and Causes

Over 188 hazardous air pollutants can have negative effects on human health or the environment ("Hazardous Air Pollution . . ." 2005). There are many sources and causes of air pollution, but four are among the most common. First, a major source of air pollution is the burning of *fossil fuels*, substances obtained from the earth, including coal, petroleum, and natural gas. The exhaust gases of cars, trucks, and buses contain poisons—sulfur dioxide, nitrogen oxide, carbon dioxide (CO_2), and carbon monoxide. Power plants that produce electricity by burning coal or oil also spew pollutants.

Second, manufacturing plants that produce consumer goods pour pollutants into the air. Formaldehyde-based vapors that can lead to cancer and respiratory problems are emitted by many household and personal care products: pressed wood (often used in furniture), permanent-press clothes, grocery bags, waxed paper, latex paints, detergents, nail polish, cosmetics, shampoos, bubble baths, and hair conditioners ("Formaldehyde" 2004).

Third, winds blow contaminants in the air across borders and oceans. For example, air pollution from Asia, especially China, has affected the air quality in the Sequoia and Kings Canyon national parks in California. Air pollution originating in Europe has been tracked to Asia and the Arctic, and has polluted air from the United States that flows out over the Atlantic Ocean and across to Europe (Spotts 2004).

Finally, and most importantly, government policies can increase or lessen air pollution. In the United States, policies concerning air pollution vary from one administration to the next. Between 2002 and 2006,

for example, lawsuits by the Justice Department against polluters declined by 70 percent, and criminal and civil fines for polluting fell by more than half, compared with the period from 1996 to 2000. The Bush administration blocked the efforts of 18 states to cut emissions from cars and trucks because it judged current standards good enough and believed that tougher regulations would hurt the U.S. economy (Environmental Integrity Project 2007; Pelton 2007; "No Action on Greenhouse Gases" 2008).

One of the most dangerous effects of air pollution is climate change, especially global warming. As air pollution increases, it disrupts the earth's temperature and, as a result, the global ecosystem.

Global Warming and the Greenhouse Effect

If you've seen the movie *The Day After Tomorrow*, it probably left quite an impression. One scientist said that the film made him laugh because it's impossible for ground temperature to drop by 150 degrees, causing an ice age. Nonetheless, he warned that global warming is "the greatest challenge this planet has to face" (cited in Monastersky 2004: A8).

What causes global warming? The process begins with what scientists call the greenhouse effect. Every year, humans release at least 1 billion tons of CO_2 into the atmosphere, primarily from coal-fired power plants. At mid-2007, the United States, which has only 5 percent of the world's population, was responsible for 25 percent of all CO_2 emissions. Since then, China's CO_2 emissions have exceeded those of the United States

In 2007, the prestigious American Institute of Architects chose a Seattle branch of a public library as one of its "Top Ten Green Projects." The library has 4 inches of soil planted with native grasses and succulent ground covers designed to absorb and filter rainwater, remove CO_2 from the atmosphere, and insulate the building in both summer and winter. The roof also lasts longer than a traditional hard roof (Adler 2007).

Photo Courtesy of American Hydrotech, Inc.

by 8 percent (Netherlands Environmental Assessment Agency 2007).

Air pollutants, especially CO_2, can lead to the **greenhouse effect**, the heating of the earth's atmosphere due to the presence of certain atmospheric gases. Heat from the sun passes into the atmosphere: Some of it is absorbed by the Earth's surface, and some of it is reflected back to space. The presence of greenhouse gases in the atmosphere traps some of this heat. Heat is necessary to support life, but when greenhouse gases increase in the atmosphere, Earth becomes warmer than it would be otherwise (Fagan 2008).

Some Effects of Climate Change

Climate change is a change of overall temperatures and weather conditions over time. Since such record keeping began in 1850, after the invention of the thermometer, the years 1995–2006 have been the warmest. Climate change is reflected in **global warming**, an increase in the average temperature of the earth's atmosphere. The warming has resulted from numerous factors. Some are natural, such as changes in solar radiation, the Earth's orbit, and the frequency or intensity of volcanic activity. Most of the factors underlying global warming, however, particularly emissions of greenhouse gases, are due to human industrial activity (Intergovernmental Panel on Climate Change 2007).

Scientists predict that if CO_2 emissions continue at their current rate, the average temperature is likely to rise between 3 and 8 degrees Fahrenheit by 2100; sea levels could rise by almost 2 feet; there will be more intense droughts, floods, and storms; millions of people will be forced to move because of coastal erosion and flooding; and more than 40 percent of wildlife will face extinction. By 2050, the Southwest (one of the fastest-growing regions in the nation) may experience a permanent drought, and seven states—Colorado, Wyoming, Utah, Nevada, New Mexico, Arizona, and California—would have to battle each other over diminished Colorado River flows (Intergovernmental Panel on Climate Change 2007; Seager et al. 2007).

Some changes have already occurred because of global warming. As the sea level rises, there are more floods and more severe hurricanes, even outside of the normal hurricane season. On Alaska's coasts, ice shelves that acted as shields against storms and tidal forces are melting. As waters advance 80 feet a year because of coastal erosion, more than 180 Alaskan villages are in danger of being engulfed unless they relocate (Tizon 2008).

IS SUSTAINABLE DEVELOPMENT POSSIBLE?

Sustainable development refers to economic activities that meet the needs of the present without threatening the environmental legacy of future generations. Is there an inherent contradiction between "sustainable" and "development"? There are reasons to be both optimistic and pessimistic about achieving sustainable development.

> **greenhouse effect** the heating of the earth's atmosphere due to the presence of certain atmospheric gases.
>
> **climate change** a change of overall temperatures and weather conditions over time.
>
> **global warming** the increase in the average temperature of the earth's atmosphere.
>
> **sustainable development** economic activities that meet the needs of the present without threatening the environmental legacy of future generations.

Reasons to Be Pessimistic

Which country is the greenest? A recent study of environmental performance ranked 149 countries on factors including maintaining and improving air and water quality, cooperating with other countries on environmental problems, overfishing, exporting acid rain to other nations, and emitting greenhouse gases. Switzerland, Sweden, Norway, Finland, and Costa Rica had the top five spots, while the lowest-ranking countries were in Africa (Niger, Angola, and Sierra Leone). The United States ranked 39th, well below a number of developing countries, including Slovakia and Albania (Esty et al. 2008).

Why, compared with other industrialized and even many developing countries, does the United States rank so low in environmental performance? There are many reasons, but two may be especially important. First, politics often governs environmental issues. During 2006 and 2007, the oil and utilities industries were successful in pressuring Congress to drop laws that would have required them to provide renewable energy sources (such as wind and solar) and to pay $13 billion more in taxes for using only fossil fuels (Broder 2007; see also Chapter 11 on the close ties between government officials and corporations).

Second, environmental issues are a low priority for some administrations. Since 2000, for example, the EPA hasn't enforced many environmental laws because of funding cuts, which resulted in the agency's having fewer staff members to investigate polluters and to apply penalties (Stephenson 2007).

Some environmentalists are especially dismayed by *greenwashers*, companies and other organization, that pollute the planet while presenting an environmentally responsible public image. For example, Duke Energy—which supplies power to more than 4 million Americans and is the country's third-highest emitter of greenhouse gases—has publicly endorsed restricting greenhouse gas emissions but has also lobbied to be allowed to build new coal-burning power plants in several states (Elgin 2008).

Reasons to Be Optimistic

Is there hope for the environment? There has been some progress since 1970, when the United States implemented and celebrated its first Earth Day: Most cars no longer burn leaded gasoline; ozone-destroying chlorofluorocarbons (CFCs) have been generally phased out; and total emissions of the six major air pollutants dropped by 54 percent during the same period as the U.S. population increased by 47 percent (Hayward and Kaleita 2007; Sperry 2008). And, even just since 2005, there has been an explosion of interest in improving the environment.

Unlike the federal government, many companies are finding that being green is good for their profits and

Are Americans eco-friendly only when it's convenient?

image, as well as the environment. For example, Wal-Mart has slashed its electricity usage by 17 percent since 2002 by switching to more efficient lightbulbs and adding skylights to its stores for natural light (Carey 2007). And Johnson & Johnson, the world's largest manufacturer of health care products, now relies on renewable energy sources, such as wind and solar, for 30 percent of its electricity (Cohn 2007).

How much are Americans willing to sacrifice for a cleaner environment? Over 85 percent say that they take advantage of curbside recycling programs to recycle paper, glass, aluminum, or other items and that they have reduced their household energy use to save on bills, but only 9 percent have contacted a business to complain about products that harm the environment, a task that's more time-consuming. Americans also say that they want cleaner energy, but most oppose wind farms, for example, because the 40-foot windmills would "disturb our views of the landscape" (Dunlap 2007). Thus, are Americans eco-friendly only when it's convenient?

© Herman Wouters/*The New York Times*/Redux / © AP Images

Many groups are contributing to sustainable development. In the Netherlands, an oil refinery owned by Shell (one of Europe's largest corporations responsible for emitting tons of CO_2 into the air) pipes CO_2 gas into greenhouses. The CO_2 bolsters rose growers' crops and decreases the refinery's emission of greenhouse gases (left). In Vermont, a dairy farmer sells "poop power" by converting cow manure to electricity (right). Consumers pay about 4 percent more for the electricity, but the manure is abundant and endlessly renewable, and its conversion to power decreases air pollution and practically eliminates the smell of cow dung in rural areas (Moore 2006).

Learning Your Way

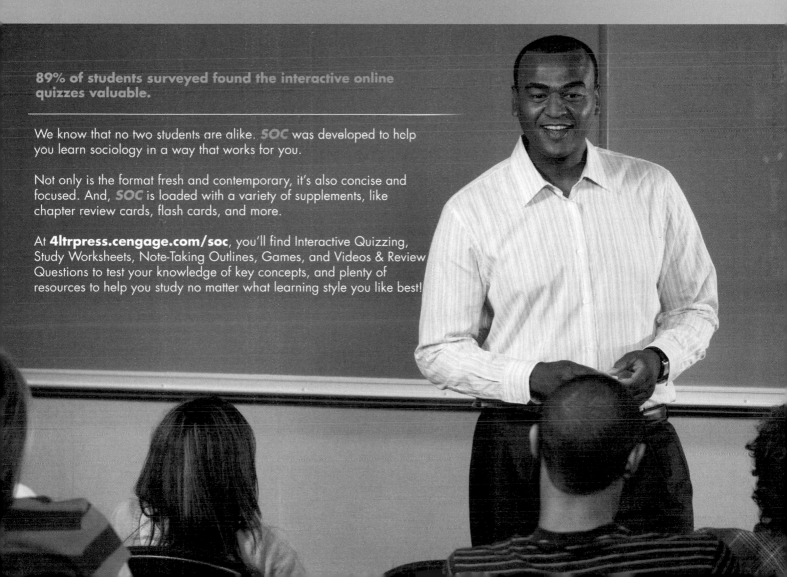

89% of students surveyed found the interactive online quizzes valuable.

We know that no two students are alike. *SOC* was developed to help you learn sociology in a way that works for you.

Not only is the format fresh and contemporary, it's also concise and focused. And, *SOC* is loaded with a variety of supplements, like chapter review cards, flash cards, and more.

At **4ltrpress.cengage.com/soc**, you'll find Interactive Quizzing, Study Worksheets, Note-Taking Outlines, Games, and Videos & Review Questions to test your knowledge of key concepts, and plenty of resources to help you study no matter what learning style you like best!

Every culture and society

experiences change.

what do you think?

Technological advancement

has more benefits than costs.

1 2 3 4 5 6 7

strongly agree   strongly disagree

17

Social Change: Collective Behavior, Social Movements, and Technology

Key Topics

In this chapter, we'll explore the following topics:

1 Collective Behavior

2 Social Movements

3 Technology and Social Change

For the most part, as you've seen throughout this textbook, sociologists examine behavior and social processes that are relatively institutionalized, routine, stable, and highly predictable. Often, however, change occurs in a society. Some forms of collective behavior, including fashions, are short-lived changes with few long-term consequences, while others, like social movements, can have lasting effects. Technology has also played a critical role in sparking social change. Let's begin with the concept of collective behavior.

> **collective behavior** the spontaneous and unstructured behavior of a large number of people.

1 Collective Behavior

do you have several boxes crammed with collectibles like Barbie dolls or baseball cards? A tattoo? Ever signed a petition? Joined a club? Created a site on Facebook or accessed one? If so, you've engaged in collective behavior.

WHAT IS COLLECTIVE BEHAVIOR?

Collective behavior is the spontaneous and unstructured behavior of a large number of people. Collective behavior encompasses a wide range of group phenomena, including riots, fads, fashion, panic, rumors, responses to disaster, and social movements.

Sociologists emphasize two important characteristics of collective behavior. First, it is an act rather than a state of mind. For example, you may *feel* panic when a hurricane threatens your town, but you don't engage in collective behavior until you actually *leave* your home and head for a safer location.

Second, collective behavior varies in its degree of spontaneity and structure. The residents of New Orleans who jammed highways to escape Hurricane Katrina demonstrated the least structured form of collective behavior, which came close to panic and was short-lived. Fads—like collecting baseball cards—are more structured. They may last several years, and many people carefully wrap and store their collectibles hoping to make money in the future. Other forms of collective behavior—such as pro- and anti-abortion groups—become highly institutionalized social movements that involve a staff, budget, and lobbying.

WHEN DOES COLLECTIVE BEHAVIOR OCCUR?

One of the biggest media stories in 2005 involved Cindy Sheehan. Her 24-year-old son and six other soldiers were killed in Sadr City in 2004, almost a year after President Bush had declared the end of major combat operations in Iraq. Sheehan pitched a tent near the president's ranch in Crawford, Texas, protesting the continuing war and demanding a face-to-face meeting with the president, but he refused to do so. Sheehan's stance elicited widespread national and international coverage. Gold Star Families for Peace, a coalition of military families who lost relatives in the war, aired a television ad in which Sheehan accused the president of dishonesty about the weapons of mass destruction: "You lied to us and because of your lies, my son died" (Madigan 2005: 1D). Some situations, like Sheehan's, are more likely to encourage collective behavior than others. Why?

Structural Strain Theory

According to sociologist Neil Smelser (1962), there are six macro-level conditions that encourage or discourage collective behavior. Smelser described these conditions as "value-added." That is, each condition leads to the next one, ending in an episode of collective behavior.

1. **Structural conduciveness.** Structural conduciveness consists of social conditions that allow a particular kind of collective behavior to occur. When channels for expressing a grievance either are not available or fail, for instance, like-minded people may resort to protests to voice their complaint. In Sheehan's case, the families who lost members in the Iraq war supported Sheehan because they felt that the Bush administration was insensitive to their personal losses.

2. **Structural strain.** Structural strain occurs when an important aspect of a social system is seen as discriminatory or unjust, creates problems, or interferes with people's everyday lives. Many of the families that sided with Sheehan, for example, did so because they felt that the administration, by not moving to end the war in Iraq, increased the number of U.S. casualties.

3. **Growth and spread of a generalized belief.** In the course of social interaction, people begin to see a situation as a widespread problem instead of just a personal experience. Along with this generalized belief that there is a problem is a general recognition that something should be done about it. In the Sheehan case, many mothers, especially, supported

Sheehan and were the most vocal leaders of a number of antiwar protests because they felt that their children were dying "for nothing" (Bumiller 2005).

4. **Precipitating factors.** Some incident or behavior triggers an event and inspires action. Sheehan's pitching a tent outside of President Bush's ranch spurred many other antiwar advocates to support her accusations on radio talk shows and in letters to the editor.

5. **Mobilizing people for action.** Mobilization often requires leaders who encourage agitation or changing the status quo. Sheehan was the leader in agitating for peace in Iraq, but antiwar groups supported her efforts through television ads.

6. **Social control.** In this stage, opposing groups may try to prevent, interrupt, or repress those advocating social change. Government officials, the police, community and business leaders, courts, the mass media, and other social control agents—all of whom benefit from the status quo—may quash, ridicule, or challenge the emerging collective behavior. In Sheehan's case, the administration and right-wing political commentators on television and radio dismissed Sheehan as a "crackpot" and described her protests as "treasonous" (Madigan 2005).

Critical Evaluation

Smelser's model made an important contribution to the analysis of the emergence and development of collective behavior as well as helping to predict when and where episodes of such behavior might break out. His model also provided insights on why, at every stage of development, the collective behavior may either fade or escalate (Locher 2002). If, for instance, the president had met with Sheehan to discuss the general anxiety surrounding the war (reducing structural strain), the initial criticisms of his administration would probably have died down (halting the spread of a generalized belief), and the antiwar groups would not have run the television ads (limiting the mobilization of participants).

Smelser's model doesn't explain all forms of collective behavior, however. With fads and rumors, as you'll see shortly, all six stages don't necessarily occur. A second criticism is that the sequence of the stages is not necessarily the same as Smelser outlined (Berk 1974). A third criticism is that the determinants don't always spark collective behavior. For example, many groups—such as college students who complain about the price of textbooks and tuition costs—experience structural strain and are free to protest and mobilize, but they rarely engage in collective behavior to change a situation that they complain about.

VARIETIES OF COLLECTIVE BEHAVIOR

There are many types of collective behavior, some more fleeting or harmful than others (Turner and Killian 1987). Let's begin by looking at some of the most common types beginning with rumors, gossip, and urban legends.

Rumors: Gossip and Urban Legends

There were widespread rumors that on January 1, 2000, a glitch (Y2K) in operating systems would cause computers around the world to crash, leading to global power outages, banks losing all of their customers' statements, and even airplanes falling from the skies. None of this occurred.

A **rumor** is unfounded information that people spread quickly. Through modern communication technology, a rumor can spread to millions of people over the Internet within seconds, especially when the subject line says something like "THIS IS REALLY TRUE!" Rumors can incite riots, panic, or widespread anxiety. Because of the Y2K rumor, thousands of people built underground shelters, and millions of others stocked up on bottled water, canned food, batteries, and medical supplies.

Most rumors (that rock stars Elvis Presley and John Lennon are alive, for example) are harmless. Others can wreak considerable damage. After a patron in northern California claimed that she found part of a human finger in her cup of Wendy's beef chili, the restaurant's business dropped by half nationally and rumors warning people to stop eating fast-food altogether spread over the Internet (Richtel and Barrionuevo 2005). Ultimately, the woman admitted that she had planted the finger to try to get a lucrative settlement. However, rumors can have a long life, as evidenced by customers who are still leery about eating at fast-food restaurants.

Rumors are typically false, so why do so many people believe them? First, rumors often deal with an important subject about which—especially during uncertain economic times or natural disasters—people are anxious, insecure, or stressed. This makes people especially suggestible. In Hurricane Katrina's aftermath, for example, the media reported numerous rumors of carjacking, murders, thefts, and rapes, the overwhelming majority of which subsequently proved to be false. Second, there is often little factual information to counter a rumor, or people distrust the sources of such information. During Y2K, the people who stockpiled groceries and so forth didn't believe computer scientists

or federal officials who said that there would not be a major calamity on January 1, 2000. Third, rumors offer entertainment, diversion, and drama in otherwise mundane daily lives (Turner and Killian 1987; Marx and McAdam 1994; Campion-Vincent 2005; Heath 2005).

There are two types of rumors: gossip and urban legends. **Gossip** is the act of spreading rumors, often negative, about other people's personal lives. Someone once said that "no one gossips about other people's virtues." Because of its tendency to be derogatory, gossip makes us feel superior ("Did you know that Margie just had breast implants? Isn't she pathetic?"). Sometimes, people gossip to control other people's behavior and to reinforce a community's moral standards. For example, comments about someone's drug abuse or marital infidelity reinforce notions of what's deviant or unacceptable (see Chapter 4). In other cases, people gossip because they resent or envy someone's success or accomplishments.

Another form of rumor is **urban legends** (also called *contemporary legends* and *modern legends*), stories—funny, horrifying, or just odd—that supposedly happened somewhere. Some of the most common and enduring urban legends, but with updated variations, deal with food contamination, like the finger at Wendy's. In 1991, Kentucky Fried Chicken shortened its name to KFC because management decided to offer a more varied menu than just chicken and to eliminate the word "fried" to please an increasingly health-conscious consumer market. The name change sparked numerous urban legends, such as the story that KFC was raising six-legged chickens (though the company doesn't even raise its own poultry). These and other urban legends about KFC have been circulating since 1970 when a couple reported that they found a fried rat in their Kentucky Fried Chicken bucket, a hoax (Brunvand 1999; Mikkelson and Mikkelson 2005).

Why do urban legends persist much longer than gossip? First, they reflect contemporary anxieties and fears—about contaminated food, unscrupulous companies, and corrupt and unresponsive governments. Second, urban legends are cautionary tales that warn us to watch out in a dangerous world. For example, there have been tales that sunscreens cause blindness, that

> **rumor** unfounded information that is spread quickly.
>
> **gossip** rumors, often negative, about other people's personal lives.
>
> **urban legends** (also called *contemporary legends* and *modern legends*) a type of rumor consisting of stories that supposedly happened.

Mountain Dew (a soft drink) decreases sperm count, and that women have died sniffing perfume samples sent to them in the mail. Third, we tend to believe urban legends because we hear them from people we trust—family members, co-workers, and friends. Finally, urban legends—such as the one about alligators living in New York City's sewage system—are fun to tell and "too beguiling to fade away" (Kapferer 1992; Brunvand 2001; Ellis 2005).

Panic and Mass Hysteria

In 2003, an indoor fireworks display meant to kick off a heavy metal concert in West Warwick, Rhode Island, set off a fire that killed 100 people and injured 200 others. Panic ensued as thick black smoke poured through the audience and hundreds of patrons stampeded for the front door (even though there were three other exits), trampling and crushing those who had fallen beneath them.

Most of the deaths in West Warwick were due not to the fire, but to **panic**, a collective flight, typically irrational, from a real or perceived danger. The danger seems so overwhelming that people desperately jam an escape route, jump from high buildings, leap from a sinking ship, or sell off their stock. Fear drives panic: "Each person's concern is with his [or her] own safety and personal security, whether the danger is physical, psychological, social, or financial" (Lang and Lang 1961: 83).

Panic can result in hundreds or thousands of casualties, but it is a relatively rare type of collective behavior (Smelser 1962). Most people try to rescue loved ones rather than fleeing a dangerous situation (such as a flood or hurricane), and most will leave a life-threatening situation in an orderly fashion (as in the World Trade Center on 9/11).

Panic is similar to **mass hysteria**, an intense, fearful, and anxious reaction to a real or imagined threat by large numbers of people. An example of mass hysteria is the scare over mad cow disease. In 1996, the government in Great Britain issued a warning that ten people had died from a brain-wasting syndrome called bovine spongiform encephalopathy (BSE) after consuming the brain, spinal cord, and parts of the intestine of an infected cow. BSE is extremely rare and smoking carries much higher risks, but the national media and many government officials described the cases of BSE as an epidemic, which led to mass hysteria. Many British cattle farmers and meat processors went bankrupt, and thousands of people lost their jobs because the countries of the European Union and others around the world banned British beef (Van Ginneken 2003).

Unlike panic, which usually subsides quickly, mass hysteria may last longer. Alarms similar to that in response to BSE will probably continue to occur in the future because urban legends—especially about health and food—reinforce our general fears and anxieties about life's dangers.

Fashions, Fads, and Crazes

Fashions, fads, and crazes are three more kinds of collective behavior that encompass broad geographical areas and involve large numbers of people. Of the three, fashion is the most structured and changes most slowly.

A **fashion** is a standard of appearance that enjoys widespread but temporary acceptance within a society. Whether they last for years or change after a few months, fashions are highly institutionalized products and styles that are popular among a large number of people (Smelser 1962; Blumer 1969). For example, African American women's hairstyles have changed—from Afros in the late 1960s, to straightened hair during the 1980s, to braids, cornrows, dreadlocks, and coloring more recently—to reflect gender politics, racial solidarity, generational differences, power, identity, and images of beauty (Banks 2000).

© iStockphoto.com / © Robert Blanchard/iStockphoto.com

BETTY CROCKER MAKEOVER

1936   *1955*   *1965*   *1968*

1972   *1980*   *1986*   *1996*

In 1921, General Mills created Betty Crocker, a fictitious woman, to answer thousands of questions about baking that came in from consumers every year. As fashions and hairstyles changed, so did Betty Crocker's image. The original image of a stern, gray-haired older woman has morphed over the years so that, by 1996, she had a darker complexion and was dressed in casual attire. Can you think of other brands that have changed their image over the years to keep up with changing trends?

that signal being an insider: "Others will see that I've made the right choice, and they will admire me for being the kind of person who makes stylish choices" (Best 2006: 85–86; for well-known analyses of fashion and collective behavior, see Veblen 1899/1953; Barber and Lobel 1952; Packard 1959; and Bourdieu 1984).

A **fad** is a form of collective behavior that spreads rapidly and enthusiastically but lasts only a short time (Turner and Killian 1987; Lofland 1993). Fads that have arisen and faded fairly quickly include *products* (such as the Rubik's cube and bean bag chairs), *activities* (such as disco dancing and a variety of diets, including Atkins and South Beach), and widespread enthusiasm for *popular personalities and television characters* (such as the Lone Ranger during the 1950s, and more recently, Britney Spears and the Backstreet Boys).

Fashion also involves periodic changes in the popularity of clothes, architecture, furniture, music, language usage, books, automobiles, sports, recreational activities, the names parents give their children, and even the dogs that people own. Teenagers who want to be fashionable buy clothes with prominent labels like Gap, Urban Outfitters, Sean John, Abercrombie and Fitch, Tommy Hilfiger, and Under Armour.

Why do fashions, especially in clothes, change fairly quickly? One reason is that designers, manufacturers, and retailers must continuously create demand for new fashions to maintain their profits. Second, many people keep up with fashion because they don't want to seem different or they fear being perceived as out-of-date or dowdy. Third, shopping for new clothes and other products decreases the boredom of everyday routine. Also, clothes and other products are status symbols

Why do fads occur? A major reason is profit. Because children and adolescents are especially likely to take up fads, manufacturers create numerous products and activities that they hope will catch on. Many of these products include toys, sportswear, and new cereals. The hottest fads are usually the must-have Christmas toys that children plead for (such as Cabbage Patch dolls, Elmo, and PlayStation). When such a fad arises, parents may stand in line for hours, drive to nearby cities, and scour the Internet to get the product. A few months later, the toy may be thrown away or stuffed in the back of a closet.

Some people may dismiss fads as "ridiculous" or "silly," but they serve several functions. In a mass society, where people often feel anonymous, participation in

© AP Images

craze a fad that becomes an all-consuming passion for many people for a short period of time.

disaster an unexpected event that causes widespread damage, destruction, distress, and loss.

public a collection of people, not necessarily in direct contact with each other, who are interested in a particular issue.

a fad can develop strong in-group feelings and a sense of belonging, especially among people who share similar interests. Fads can also be fun, promise to resolve a nagging problem (such as being overweight), celebrate progress, and help us keep up with technological changes (Marx and McAdam 1994; Best 2006).

Most fads are soon forgotten, but some become established. For example, Pez candy dispensers, which originated in 1952 and cost 49 cents, are still inexpensive (about $1.49), are sold in more than 60 countries, and have been continuously updated to include popular television characters such as the Simpsons (Paul 2002). Other fads, such as streaking (running around nude in public places), re-emerge from time to time. Streaking appeared on some college campuses during the early 1970s and may have been a way of releasing stress during the Vietnam War. After dying down for a time, it made a comeback during the late 1990s ("Streaking" 2005).

Some fads are **crazes**, forms of collective behavior that become all-consuming passions for a short period of time. The Beanie Babies fad of the late 1990s turned into a craze. After an entrepreneur published a highly successful magazine on these stuffed toys, there was a mad rush to buy them. The creator limited the sales in many stores to generate demand and retired earlier models to make them more sought after by collectors. The strategies worked because the entrepreneur became one of richest people on earth. What about the collectors? According to an owner of a large toy store, "[Beanie Babies] make great insulation if you stick them in the walls" (Mulligan 2004: 3C).

Crazes involving ways to make easy money are most common in capitalistic societies where many people want to acquire as much money as possible and as quickly as possible. Such crazes usually involve investments in high-risk ventures. The problem with economic crazes is that they may end quite suddenly and with very disappointing results. For example, most of the fast-growing dot-com businesses collapsed after 2000. Investors' portfolios fell by as much as 50 percent, thousands of workers were laid off, and almost all of the "hot" Internet sites disappeared nearly overnight (van Ginneken 2003; McCarthy 2004).

> Crazes involving ways to make easy money are most common in capitalistic societies.

© Rachel Weill/FoodPix/Jupiterimages

Disasters

Whereas people choose to participate in fashion, fads, and crazes, a **disaster** is an unexpected event that causes widespread damage, destruction, distress, and loss. Some disasters are due to *social causes,* such as war, genocide, terrorist attacks, and civil strife (see Chapter 18 online). Some are due to *technological causes,* including oil spills, nuclear accidents, burst dams, building collapses, and mine explosions. Others are the result of *natural causes,* such as fires, floods, landslides, earthquakes, hurricanes, tsunamis, and volcanic eruptions (Marx and McAdam 1994). Resulting in an estimated 1,800 deaths, Hurricane Katrina is one of the worst natural disasters to have occurred in the United States. The number of casualties was small, however, compared with many other natural disasters around the world. For example, floods in China killed an estimated 4 million people in 1931 and 2 million in 1959 (Crossley 2008).

Disasters often inspire organized behavior rather than chaos. Instead of panicking, most people are rational, cooperative, and altruistic. They often care for family members instead of fleeing, for example, and thousands of volunteers offer financial, medical, and other help. Days before the federal government responded to the Hurricane Katrina disaster, numerous individuals and organizations set up Web sites that offered the evacuees jobs and housing (Noguchi 2005).

Publics, Public Opinion, and Propaganda

A **public** is a collection of people, not necessarily in direct contact with each other, who are interested in a particular issue. A public is not the same as the general public, which consists of everyone in a society.

There are as many publics as there are issues—abortion, education, gun control, pollution, health care, and gay marriage, to name just a few. Even within one organization or institution, there may be several publics that are concerned about entirely different issues. At a college, for instance, students may be most concerned about the cost of tuition, faculty may spend much time discussing instructional technology, and maintenance employees may be most interested in wages.

In most cases, the interaction within a public is carried on indirectly through the mass media, newsletters, or professional journals. Because publics aren't organized groups with memberships, they are often transitory. Publics expand or contract as people lose or develop interest in an issue. For example, a public may surge during a highly publicized and controversial incident such as removing a patient's life support, but then evaporate almost overnight. In some cases, however, publics organize and become enduring social movements (a topic we'll examine shortly).

Some publics express themselves through **public opinion**, widespread attitudes on a particular issue. Public opinion has three components: It (1) is a verbalization rather than an action, (2) is about a matter that is of concern to many people, and (3) involves a controversial issue (Turner and Killian 1987). Like publics, public opinions wax and wane over time. People's interest in crime and education, for example, decreases when they are more concerned about pressing events such as an economic crisis.

Public opinion can be swayed through **propaganda**, the presentation of information to influence people's opinions or actions. Propaganda isn't a type of collective behavior, but it affects collective behavior in several important ways. First, it may create attitudes that will inspire collective outbursts such as strikes or riots. Second, propaganda may be used to try to prevent collective outbursts, as when corporations try to convince employees that the loss of jobs is due to the economy rather than to offshoring (see Chapter 12). Third, propaganda is often used to try to gain adherents for a cause, whatever it might be (Smelser 1962).

Propaganda is institutionalized in advertising, political campaign literature, and governmental pronouncements. It is conveyed to people in many ways: the mass media (newspapers, books, radio, television, and motion pictures), political speeches, religious groups, rumor, and symbols (such as flags). Much propaganda presents misinformation intended to sway an audience toward a particular viewpoint, but propaganda is not necessarily good or bad. Instead, it's a means of influencing people.

Crowds

Much collective behavior is scattered geographically, but crowds are concentrated in a limited physical space. After Pope John Paul II died in 2005, over 4 million people viewed his body at St. Peter's Basilica in Rome, one of the largest religious gatherings in the history of Christianity. This is an example of a **crowd**, a temporary gathering of people who share a common interest or participate in a particular event. Regardless of size—whether consisting of a few dozen people or millions—crowds come together for a specific reason, such as a religious leader's death, a concert, or a riot.

Crowds differ in their motives, interests, and emotional level:

- A *casual crowd* is a loose collection of people who have little in common except for being in the same place at the same time and participating in a common activity or event. There is little, if any, interaction, the gathering is temporary, and there is little emotion. Examples include people watching a street performer, spectators at the scene of a fire, and shoppers at a busy mall.

- A *conventional crowd* is a group of people that assembles for a specific purpose and follows established norms. Unlike casual crowds, conventional crowds are

public opinion widespread attitudes on a particular issue.

propaganda the presentation of information to influence people's opinions or actions.

crowd a temporary gathering of people who share a common interest or participate in a particular event.

© Yoshikazu Tsuno/AFP/Getty Images

structured; their members may interact, and they conform to rules that are appropriate for the situation. Examples include people attending religious services, funerals, graduation ceremonies, and parades.

- An *expressive crowd* is a group of people who exhibit strong emotions toward some object or event. The feelings—which can range from joy to grief—pour out freely as the crowd reacts to a stimulus. Examples include attendees at religious revivals, revelers during Mardi Gras, and enthusiastic fans at a football game.

- An *acting crowd* is a group of people who are motivated by intense, powerful emotions and have a single-minded purpose. The event may be planned, but acting crowds can also involve spontaneous demonstrations or other focused group behavior. Examples of acting crowds include people fleeing a burning building, soccer fans storming a field, and students engaged in water balloon fights.

- A *protest crowd* is a group of people who assemble in public to achieve a specific goal. Protest crowds demonstrate their support of or opposition to an idea or event. Most demonstrations—such as anti-war protests, civil rights marches, boycotts, and strikes—are usually peaceful. Peaceful protesters can become

aggressive, however, resulting in destruction and violence (Blumer 1946; McPhail and Wohlstein 1983).

One type of crowd can easily change into another. A conventional crowd at a nightclub can turn into an acting crowd if a fire causes people to panic and flee for safety. Indeed, any of the five types of crowds—from casual to protest—can become a mob or a riot.

A **mob** is a highly emotional and disorderly crowd that uses force or violence against a specific target. The target can be a person, a group, or a property. In the Oakland neighborhood of Chicago, a van veered off the street and struck a group of people sitting on a stoop of a home. Seven male spectators, ranging in age from 16 to 47, mobbed the van. They pulled the driver and passenger from the car, stomping and beating them to death with bricks and stones ("7 People Charged . . ." 2002).

After attacking, a mob tends to dissolve quickly. Mobs often arise in situations where people are demanding radical societal changes, like the removal of corrupt government officials. Compared to mobs, riots usually last longer. A **riot** is a violent crowd that directs its hostility at a wide and shifting range of targets. Unlike mobs, which usually have a specific target, rioters unpredictably attack whoever or whatever gets in their way during the rampage.

In 2005, two French teenagers hiding from the police were accidentally electrocuted. The incident sparked a week of rioting by angry youths in Paris's suburban housing projects; hundreds of cars and buses were torched, and several buildings were burned down. Similar riots occurred in 2007 after two teenagers' fatal collision with a police car. Policy analysts blamed the rioting on chronically high unemployment levels, overcrowded housing, and the failure of young men of African and Arab backgrounds to assimilate into French society.

Most riots, like those in Paris, arise out of long-standing anger, frustration, or dissatisfaction that may have smoldered for years or even decades. Some of the long-term tensions arise from discrimination, poverty, poor housing conditions, unemployment, economic deprivation, or other unaddressed grievances. There are numerous protests in the United States every year, but race riots have been the most violent and destructive. More than 150 U.S. cities experienced such riots after the assassination of Martin Luther King, Jr., in 1968. In 1992, riots broke out in 11 cities after the acquittal of police officers involved in the beating of Rodney King, a black motorist, in Los Angeles. That violence resulted in deaths, as well as considerable property damage due to fires and looting.

In 2006, hundreds of thousands of Latinos and their supporters participated in demonstrations and rallies in Washington, DC, and other cities, calling on Congress to offer citizenship to illegal immigrants. Is this kind of gathering an example of an expressive, acting, or protest crowd?

Riots are usually expressions of deep-seated hostility, but this isn't always the case. Sports celebration riots, such as the one that followed the Red Sox victory over the Yankees in the 2004 World Series, occur because of extreme enthusiasm and excitement rather than anger or frustration: "Participants smash, trample, and knock things down to express their ecstasy. Celebration riots are an orgy of gleeful destruction" (Locher 2002: 95).

Much collective behavior, such as a mob or riot, is spontaneous and short-lived. Bringing about long-term social changes, especially at a macro level, requires collective behavior that is structured and enduring. Social movements are important vehicles for creating or suppressing societal changes.

2 Social Movements

there are hundreds of social movements in the United States alone. Why are they so widespread? And how important are they in changing society?

WHAT IS A SOCIAL MOVEMENT?

A **social movement** is a large and organized activity to promote or resist some particular social change. Examples of social movements include groups that focus on civil rights, the rights of the disabled, crime victims, gun control, and drunk driving, to name just a few. "Social movements are as American as apple pie," notes sociologist Lynda Ann Ewen (1998: 81–82): "The abolition of slavery, women's right to vote, legal unions, open admissions to public colleges and student aid, and Head Start are all changes in our society that were won through social movements."

Unlike other forms of collective behavior, social movements are organized, goal-oriented, deliberate, and structured and can have a lasting impact on a society. And unlike many other forms of collective behavior (such as crowds, mobs, and riots), the people who make up a social movement are dispersed over time and space and usually have little face-to-face interaction (Turner and Killian 1987; Lofland 1996).

Some social movements in the United States, such as those focusing on white supremacy, are relatively small. Others are large and have subgroups that appeal to different segments of the population. For example, the U.S. environmental movement has at least 50 subgroups, including Earth First!, Greenpeace, the National Audubon Society, Union of Concerned Scientists, and the Wilderness Society.

social movement a large and organized activity to promote or resist some particular social change.

TYPES OF SOCIAL MOVEMENTS

Sociologists generally classify social movements according to their goals (changing some aspect of society or resisting such a change) and the amount of change that they seek (limited or widespread) (Aberle 1982). As *Table 17.1* suggests, some social movements may be perceived as more threatening than others because they challenge the existing social order.

Alternative social movements focus on changing some people's attitudes or behavior in a specific way. These movements typically emphasize spirituality,

© Feng Yu/ Stockphoto.com

TABLE 17.1
Five Types of Social Movements

MOVEMENT	GOAL	EXAMPLES
Alternative	Change some people in a specific way	Alcoholics Anonymous, transcendental meditation
Redemptive	Change some people, but completely	Jehovah's Witnesses, born-again Christians
Reformative	Change everyone, but in specific ways	gay rights advocates, Mothers Against Drunk Driving (MADD)
Resistance	Preserve status quo by blocking or undoing change	anti-abortion groups, white supremacists
Revolutionary	Change everyone completely	right-wing militia groups, Communism

self-improvement, or physical well-being. They are the least threatening to the status quo because they seek limited change and only for some people. For example, millions of non-Asian Americans, influenced by Asian religions, have embraced yoga, meditation, and healing practices such as acupuncture (Cadge and Bender 2004).

Redemptive social movements (also called *religious* or *expressive movements*) offer a dramatic change, but only in some peoples' lives. Redemptive movements are typically based on spiritual or supernatural beliefs, promising to renew people from within and to guarantee some form of salvation or rebirth. Examples include any religious movements that actively seek converts, such as the Jehovah's Witnesses and certain Christian evangelical groups (see Chapter 15).

Reformative social movements want to change everyone, but only with regard to a particular topic or issue. These movements, the most common type of social movement in U.S. society, do not want to remove or replace the existing economic, political, or social class arrangements but to change society in some specific way. Examples include civil rights groups, gay rights activists, labor unions, and animal rights groups.

Resistance social movements (also called *reactionary movements*) try to preserve the status quo by blocking change or undoing change that has already occurred. Resistance movements are often called *countermovements* because they usually form immediately after an earlier movement has succeeded in creating change within a society. For example, anti-abortion groups that arose in the United States shortly after the Supreme Court decision in *Roe v. Wade* (1973), which legalized abortion, seek to reverse that decision.

Revolutionary social movements want to completely destroy the existing social order and replace it with a new one. Their goal is the total transformation of society. Revolutionary movements range from utopian groups that withdraw from society and try to create their own to radical terrorists who use violence and intimidation. Examples of the latter include militia groups in the United States that believe the federal government is evil and want to overthrow it. Fidel Castro's socialist movement in Cuba, the French Revolution, and the Communist Revolution in China all succeeded in replacing the existing social order with a new one.

WHY SOCIAL MOVEMENTS EMERGE

A social movement is "an answer either to a threat or a hope" (Touraine 2002: 89). Yet not everyone who feels threatened or hopeful joins a social movement. Why not? Let's look at four explanations, beginning with the oldest.

Mass Society Theory

Early on, sociologists believed that the people who formed social movements felt powerless, insignificant, and isolated in modern mass societies, which are impersonal, industrialized, and highly bureaucratized. Thus, according to *mass society theory*, social movements offer a sense of belonging to people who feel alienated and disconnected from others (Kornhauser 1959).

Critical Evaluation

Mass society theory may explain why some people form extreme political movements, like Fascism and Nazism, but subsequent research has shown that movement organizers are typically not isolated but well-integrated into their families and communities. Also, historically, many political activists in the United States, such as those behind the civil rights and women's rights movements during the late 1960s, were not powerless but came from relatively privileged backgrounds (Davis 1991; McAdam and Paulsen 1994).

Relative Deprivation Theory

Relative deprivation theory is broader than mass society theory. **Relative deprivation** is a gap between what people have and what they think they should have based on what others in a society have. According to *relative deprivation theory*, what matters is not what people actually have—whether it's money, social status, power, or privilege—but what they *think* they should have compared with others.

Relative deprivation theorists note two other elements. First, people often feel that they *deserve* better than they have ("I've worked hard all my life"). Second, they believe that *they cannot attain their goals through conventional channels* ("I've written people in Congress, and they just ignore me"). Thus, shared beliefs combined with unfulfilled expectations can trigger change-oriented social movements, as witnessed by the gay rights and civil rights movements (Davies 1962, 1979; Morrison 1971).

Critical Evaluation

Because relative deprivation theory is more general than mass society theory, it provides a better explanation of why some social movements emerge. Critics point out, however, that there is a certain amount of relative deprivation in all societies, but that people don't always form social movements in reaction to it. In addition, relative deprivation theory doesn't explain why some people join movements even though they don't see themselves as deprived and don't expect to gain anything personally if the movement is successful (Gurney and Tierney 1982; Johnson and Klandermans 1995; Orum 2001).

Resource Mobilization Theory

It takes more than feeling alienated (mass society theory) or disadvantaged (relative deprivation theory) to sustain a social movement. Instead, according to *resource mobilization theory*, a social movement will succeed if it can put together (or mobilize) an organization and leadership dedicated to advancing its cause (Oberschall 1973, 1993; McCarthy and Zald 1977; Gamson 1990).

Organization and leadership are two critical types of resources for a social movement. Other resources include money, dedicated volunteers, paid staff, access to the mass media, effective communication systems, contacts, special technical or legal knowledge and skills, equipment, physical space, alliances with like-minded groups, and a positive public image. Despite many failing labor unions, those of teachers and police officers remain strong. This is because such unions have millions of users who pay dues, much of which goes to lobbyists who represent their interests in Congress (see Chapter 11). Further, these unions are well organized at state and national levels, and enjoy widespread support from the general public.

Critical Evaluation

One of the major contributions of resource mobilization theory is its emphasis on structural factors (such as organization and leadership) in explaining why some social movements thrive and others disappear. However, a major criticism is that resource mobilization theory largely ignores the role of relative deprivation in the formation of a social movement. If there aren't large numbers of dissatisfied people to initiate a movement, even plentiful resources will not be able to sustain it (Jenkins 1983; Klandermans 1984; Scott 1995; Buechler 2000). At many colleges, for instance, student government associations are well-funded, but attempts by some students to mobilize their classmates to protest tuition hikes or the war in Iraq have fizzled because many are focused on other priorities, including jobs.

New Social Movements Theory

New social movements theory, which became prominent during the 1970s, emphasizes the linkages between culture, politics, and ideology. Unlike the earlier perspectives, new social movements theory proposes that many recent movements (such as those that work for peace and environmental protection) promote the rights and welfare of *all* people rather than specific groups in particular countries (Touraine 1981; Laraña et al. 1994; Melucci 1995). Thus, new social movements theory is especially interested in "the struggle to liberate the voices of the dispossessed" (Schehr 1997: 6).

For these theorists, recent social movements differ from older ones in two ways. First, they attract a disproportionate number of members who are well-educated and relatively affluent, represent a wide variety of professions (such as educators, scientists, actors, businesspeople, and political leaders), and share a broad goal—improving the quality of life for all people around the world. Second, recent social movements pursue goals or advance values that may have no immediate personal benefit to the participants, such as eradicating measles or tuberculosis in developing countries (Obach 2004).

Critical Evaluation

Unlike earlier perspectives, new social movements theory contributes to our understanding of collective behavior that crosses international boundaries. According to some critics, however, neither this perspective nor the groups that it examines are novel. For example, some social movements (like feminism

"If there aren't large numbers of dissatisfied people to initiate a movement, even plentiful resources will not be able to sustain it."

and environmentalism) have been around for a long time and still focus on the same basic issues, such as women's second-class citizenship and population growth. In addition, some scholars point out that educated, middle-class or wealthy activists were as common in the old social movements as in more recent ones (Rose 1997; Buechler 2000; Sutton 2000). Critics also contend that new social movements theory often overstates people's altruistic motivations for forming or participating in social movements. For example, many people who join environmental groups do so for reasons referred to as NIMBY (not in my back yard). That is, they are concerned about some undesirable environmental condition in their own community but show little interest in environmental threats to people elsewhere (Obach 2004).

These four theories, despite their limitations, help us understand social movements because "no single theory is sufficient to explain the complexities of any social movement" (Blanchard 1994: 8). The next question is why some social movements thrive and others fail.

THE STAGES OF SOCIAL MOVEMENTS

Most social movements are short-lived. Some never really get off the ground; others meet their goals and disband. Social movements generally go through four stages—emergence, organization, institutionalization, and decline (see *Figure 17.1*) (King 1956; Mauss 1975; Spector and Kitsuse 1977; Tilly 1978).

Emergence

During *emergence*, the first stage of a social movement, a number of people are upset about some condition and want to change it. One or more individuals, serving as agitators or prophets, emerge as leaders. They verbalize the feelings of the discontented, crystallize the issues, and push for taking action. If leaders don't get support after the initial interest, the movement may die. If, on the other hand, the discontent resonates among growing numbers of people, public awareness increases, and the movement attracts like-minded people. For example, many scholars attribute the rise of the women's movement during the 1960s to the popularity of *The Feminine Mystique*, a book in which Betty Friedan (1963), then a full-time housewife, questioned whether educated women should give up careers to be full-time homemakers.

Organization

Once people's consciousness has been raised, the second stage of a social movement is *organization*. The most active members form alliances, seek media coverage, develop strategies and tactics, recruit members, and acquire the necessary resources. A division of labor is established in which the leaders make policy decisions and the followers perform necessary tasks such as preparing mass mailings, developing Web sites, and responding to phone calls or e-mail messages. At this stage, also, the movement may develop chapters at local, regional, national, and international levels.

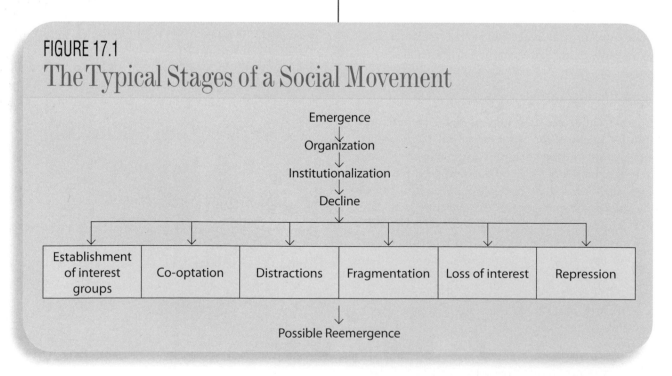

FIGURE 17.1
The Typical Stages of a Social Movement

Emergence
↓
Organization
↓
Institutionalization
↓
Decline

| Establishment of interest groups | Co-optation | Distractions | Fragmentation | Loss of interest | Repression |

Possible Reemergence

Smoking Deaths This Year 3077 And Counting

American Heart Association   AMERICAN CANCER SOCIETY   American Lung Association

© Bill Aron/PhotoEdit

Institutionalization

As a movement grows, it becomes *institutionalized* and more bureaucratic: The number of staff positions increases, members draw up by-laws governing the organization's activities, the organization may hire outsiders (such as writers, attorneys, and lobbyists) to handle some of the necessary tasks, and the charismatic leaders may spend more of their time on speaking tours, in media interviews, and at national or international meetings. As the social movement grows and becomes more bureaucratic and self-sufficient, the original leaders may move on to better-paying and more influential positions in government or the private sector.

Decline

All social movements end sooner or later. Their *decline*, the last stage, reflects a number of factors:

- If a social movement is successful, it can become an *interest group* and a part of society's fabric. For example, a small anti-smoking movement that began in the mid-1970s now has the enthusiastic support of numerous prestigious organizations, such as the American Cancer Society, the American Heart Association, the American Medical Association, the World Health Organization, and governing bodies within and outside the United States (Wolfson 2001).

- Members, and especially leaders, of social movements may be *co-opted* by governments or other groups. For example, a black leader (like Condoleezza Rice) may be appointed to a White House Cabinet position that requires focusing on international issues (such as the wars in Afghanistan and Iraq) rather than on the domestic issues of racial and ethnic discrimination (Kilson 2005).

- Those involved in a social movement may *become distracted* because the group loses sight of its original goals and/or their enthusiasm diminishes. For example, when Ralph Nader first criticized the automobile industry for car safety defects, he gained a large following. As Nader and his consumer rights groups expanded their focus to include environmental issues and corporate crimes, many of the initial followers lost interest.

- A social movement may experience *fragmentation* because the participants disagree about goals, strategies, or tactics. For example, the environmental movement has numerous groups that focus on different issues—air quality, marine life, land use, and global warming (see Chapter 16). Participants may also drift away from a movement because of time constraints, health problems, or other reasons.

- Social movements may also decline because of *repression*. Many autocratic governments quash dissent. A government can crush an emerging social movement by arresting protestors and imprisoning or even executing leaders (see Chapter 11).

A social movement that declines can sometimes experience a resurgence later on. For example, there have been several waves of the women's rights movement in the United States since the mid-nineteenth century. The first wave ensured women's right to vote in 1920; the second wave expanded employment and educational rights during the 1960s and 1970s; and the third wave, during the 1990s, focused on economic and other inequalities experienced by women of different social classes, sexual orientations, and nationalities (Kramer 2005).

Many people report that they avoid getting involved in social movements because "they don't make any difference." Are these people right?

WHY SOCIAL MOVEMENTS MATTER

As you've seen, social movements can either create or resist change. On an *individual level*, many of us enjoy a variety of rights as workers, consumers, voters, and even victims—rights that we owe to highly dedicated people who were determined to change inequitable laws and practices (see, for example, Jenness and Grattet 2001 and Switzer 2003).

technology the application of scientific knowledge for practical purposes.

On an *institutional level*, social movements can change general practices. For example, shopping for healthy food is much easier today than 20 years ago, when veggie burgers, tofu, and organic produce were rare. Now, mainstream grocery stores have large natural food sections, a much greater variety of fruits and vegetables, and breads and cereals that are made with whole grains, nuts, and less salt, sugar, and chemical additives. In effect, then, vegetarian and consumer groups have had a major impact on industries that control the choice of food products.

On a *societal level*, social movements have had a major effect in the United States and around the world. Most of the world's great religions began as protest movements (see Chapter 15). Also, democratic forms of government in the United States and other countries grew out of the activities of revolutionary groups that sought greater political and economic freedom (Giugni et al. 1999; della Porta and Diani 1999).

Social movements may seek change or seek to prevent change. However, "what all of these social movements have in common is the desire by ordinary citizens to have a say in the operation of their society" (Locher 2002: 246). Another important source of societal change is technology.

3 Technology and Social Change

t his brief history of how to cure an earache has circulated on the Internet:

- 2000 B.C. — Here, eat this root.
- 1000 A.D. — That root is heathen, say this prayer.
- 1850 A.D. — That prayer is superstition, drink this potion.
- 1940 A.D. — That potion is snake oil, swallow this pill.
- 1985 A.D. — That pill is ineffective, take this antibiotic.
- 2000 A.D. — That antibiotic is artificial. Here, eat this root.

As this joke suggests, despite technological progress over the centuries, some of the old remedies are enjoying renewed popularity (see, for example, The People's Pharmacy at www.peoplespharmacy.com, which offers home remedies for treating everything from arthritis to toenail fungus). Still, technological advances have brought enormous social changes. What are some of the major recent technological advances, their benefits and costs, and the ethical issues that they have generated?

SOME RECENT TECHNOLOGICAL ADVANCES

Technology, the application of scientific knowledge for practical purposes, is a vitally important aspect of human life: "For good or ill, [technologies] are woven inextricably into the fabric of our lives, from birth to death, at home, in school, in paid work" (MacKenzie and Wajcman 1999: 3).

In the next 75 years, technology is likely to change our lives more dramatically than ever before. Several companies are working on "smart" pills that will presumably improve mental ability and restore brains that are impaired by disease or injury. And some scientists predict that we won't have to vacuum our carpets because they will ingest dirt, spills, and stains. Our houses may be built with sensors that will automatically test for carbon dioxide, anthrax, environmental contaminants, allergens, and radioactivity, and with devices that will defend against the release of chemical and biological agents (Rubin 2004; Murphy 2005). Whether such predictions will become reality is anyone's guess. In the meantime, technological advances continue to affect our lives. Let's look briefly at a few of the most influential.

Computer Technology

In 1887, English mathematician Charles Babbage designed the first programmable computer. Since then, computers have gone through seven generations of evolution. One of the most practical aspects of computer technology is the development of *robots*, machines that are programmed to perform human-like functions. Robots are typically used for tasks that are too dull, dirty, or dangerous for humans—such as auto assembly, toxic waste cleanup, mining, minefield sweeping, and underwater and space exploration. Since 2001, American doctors have used a combination of computers, telecommunications, videoconferencing, and advanced robots to guide surgery that is carried out thousands of miles away (Rosen and Hannaford 2006).

Japan leads the world in robotics, especially in developing intelligent robots that are useful in the real world. Wakamaru (pictured right), which is 3 feet tall and costs about $14,000, wakes up and follows its owner around, keeps her/him on a schedule, recites the day's news headlines, and alerts friends or relatives if the owner doesn't respond to a message. AIBO (pictured left), which sells for about $2,000, seems to be almost as effective as a live dog in relieving the loneliness of nursing home residents. When petted or talked to, AIBO responds by wagging its tail, barking, and blinking its lights (Banks et al. 2008).

In time, humanoid robots may be able to discuss stock market investment strategies, give you advice about a personal problem, read a book to you in any desired language or in a voice of either gender, and be emotionally savvy companions who cheer you up when you're unhappy (Breazeal 2002; Bar-Cohen and Breazeal 2003).

Biotechnology

Biotechnology is a broad term that applies to all practical uses of living organisms. It covers anything from the use of micro-organisms (yeast) to ferment beer to *genetic engineering*, sophisticated techniques that can change the makeup of cells and move genes across species boundaries to produce novel organisms. One form of genetic engineering focuses on producing genetically modified (GM) crops to increase production and to make agriculture less costly, especially by developing new varieties of plants that can tolerate herbicides and/or resist pests (like the cotton bollworm). The three main GM crops in the United States are corn (45 percent of all corn acreage is planted with GM varieties), soybeans (85 percent of acreage), and cotton (76 percent of acreage) (Pew Initiative on Food and Biotechnology 2004).

Another application of biotechnology that is more controversial than GM crops is stem cell research. A *stem*

cell is a building block of the human body. It can replicate indefinitely and thus serves as a continuous source of new cells. These self–regenerating cells are found in embryos and umbilical cords, but also in parts of adult bodies, including the brain, blood, heart, skin, bone marrow, intestines, and other organs. Embryonic stem cells are more valuable in research because they can produce any cell type of the body, whereas adult stem cells are generally limited to the cell types of their tissue of origin (National Institutes of Health 2006).

Opponents of stem cell research argue that all embryos deserve protection. Proponents maintain that the tissue that yields stem cells doesn't constitute an embryo because it is not yet implanted in a woman's uterus. These scientists also argue that hundreds of thousands of embryos that fertility clinics dispose of every year are a wasted resource that could be used to treat heart disease, leukemia and other cancers, diabetes, Parkinson's disease, and numerous other health problems (Slevin 2005; Weiss 2005).

Nanotechnology

Another promising recent technology is *nanotechnology*, the ability to build objects one atom or molecule at a time. The key characteristic of these objects is tiny size. Nanotechnology is based on structures measured in nanometers, a unit of measurement equal to 1 billionth of a meter, or 1/80,000th the width of a human hair.

Nanotechnology will probably result in computer chips that are barely visible to the human eye. In medicine, researchers foresee the day when nanobots, or miniscule robots, will cruise the bloodstream and other tissues looking for signs of disease. Or, thousands of nanotubes could potentially be packed into a hair-like capsule the size of a splinter, which could be painlessly implanted under the skin to continuously monitor blood sugar, cholesterol, and hormone levels. Some ongoing research also suggests that nanotechnology might be able to destroy tumors in the body without also frying the adjacent healthy cells, one of the negative effects of current chemotherapy and radiation treatments for cancer (Weiss 2004, 2005).

SOME BENEFITS AND COSTS OF TECHNOLOGY

Technology provides numerous benefits. In 2004, for example, a paralyzed young Boston man became the first person to send an e-mail using only his thoughts. The military hopes that this technology will one day allow pilots to fly jets using mind control. In other research, scientists are working on vaccines that will block intense pain in less than 10 seconds and that will regrow body parts such as a hand lost during a war or a breast removed in a mastectomy (Garreau 2005).

Despite such benefits, many Americans are ambivalent about some aspects of technology. Let's consider DNA testing, privacy, and the digital divide.

DNA Testing

Technology has played a major role in detecting, apprehending, and prosecuting criminals. As DNA technology becomes more sophisticated, criminal investigation labs can match semen and other body fluids that are 30 years old with the people from whom the fluids came. Some of this technology has been presented in popular television programs such as *CSI: Crime Scene Investigation, Cold Case,* and *Forensic Files.*

DNA testing also provides people with information about their genetic predispositions for diseases such as emphysema, cancer, and Huntington's disease (an incurable neurological disorder). Having such information could help doctors and patients make more informed health care decisions, but many people avoid DNA testing or pay for it privately because they fear job discrimination, denial of health insurance, or much higher insurance rates. As a result, according to a research director at the National Institutes of Health, "It's pretty clear that the public is afraid of taking advantage of genetic testing" (Harmon 2008: 1).

Privacy

Cell phones, the Internet, and other forms of telecommunications technology bring people together and provide quick access to a wealth of information (see Chapter 5). Some people complain, however, that the Internet, especially, has made us more sedentary and lazy. A bigger concern is that virtually every technological advance in telecommunications brings a greater loss of privacy.

To find out how thoroughly people remove confidential information from a computer's hard drive before selling or discarding the hardware, two graduate students at the Massachusetts Institute of Technology bought 158 used hard drives—mostly off eBay but also from small businesses and used-computer stores. Of the 129 drives that worked, 49 had recoverable information that included, among other material, corporate personnel memos, love letters, pornography, credit-card numbers, and correspondence between physicians and patients. One drive from an automated teller machine (ATM) contained customers' Social Security and account numbers, names, addresses, transaction dates, and balances (Garfinkcl and Shelat 2003).

Almost 4 million U.S. households (more than 3 percent) have experienced identity theft.

© Timothy Goodwin/iStockphoto.com

Many companies routinely collect information about people as they click from site to site on the Internet. Much of this Web tracking is done anonymously, and marketing firms are typically aware only of the sites that a user has visited, not the person's name or address. As Web tracking technology becomes more sophisticated, however, experts on digital privacy say that marketers will soon know not only which sites a user has visited, but also who is surfing the Web for almost any topic, product, or service (Story 2008).

Almost 4 million U.S. households (more than 3 percent) have experienced *identity theft*, a term that describes a variety of illegal acts, including misuse of personal information to obtain loans or to commit other crimes (see Chapter 7). Contrary to popular belief, the majority of identity theft cases do not involve hacking into someone's computer but are carried out through dumpster diving (going through trash cans for discarded bank statements, preapproved credit card solicitations, old bills, computer printouts, and other financial documents) and phishing (pretending to represent financial institutions or companies and sending spam to get people to reveal personal information). Also, legitimate information brokers sell hundreds of thousands of sets of names, addresses, and Social Security numbers (Lewis 2005; Federal Trade Commission 2008).

Another increasingly common intrusion on privacy is instigated by health and life insurance companies that pay only about $15 per search to other companies that prepare health "credit reports." They have accessed at least 200 million Americans' five-year history of purchases of prescription drugs, the dosages and refills, and possible medical conditions. Originally, doctors used such data to treat patients, especially in emergencies when a patient was unconscious. Now, this kind of report also provides a numerical score predicting what a person may cost an insurer in the future, which has resulted in insurance companies' charging some customers higher premiums or excluding some medical conditions from policies. Even worse, according to some health experts, some insurance companies have denied coverage because of misinterpreting the prescription drug information—for example, by assuming that Prozac is always used for depression when

it can also treat other problems, such as menopausal hot flashes (Nakashima 2008; Terhune 2008).

The Digital Divide

The number of Internet users worldwide will be over 2 billion by about 2011, after a steady increase from only 2 million in 1990 (Computer Industry Almanac 2007). However, Internet usage is unequal across and within societies, creating a digital divide. An estimated 20 percent of the world's population now uses the Internet, but residents of wealthy nations are much more likely to do so than those living in developing countries. For example, 70 percent of Americans use the Internet, compared with only 5 percent of Africans (Internet World Stats 2008).

Technology also separates the haves and the have nots *within* a country. Many Americans take the Internet for granted, but social class is an important factor in its usage. In the United States, about 61 percent of lower-income adults use the Internet, whether at home or elsewhere, compared with 93 percent of adults with higher incomes. As income and educational levels increase, so does Internet usage (see *Figure 17.2*). In effect, then, technological development can reinforce social inequality.

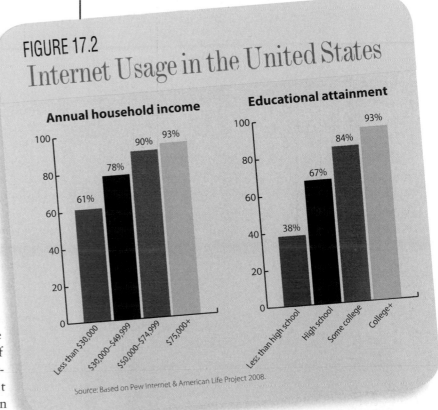

FIGURE 17.2
Internet Usage in the United States

Source: Based on Pew Internet & American Life Project 2008.

In 2005, after being mauled by her Labrador retriever, this French woman (who underwent several operations) became the world's first face transplant patient, receiving lips, a chin, and a nose from a brain-dead donor. Some doctors were appalled when they learned that she signed a film deal 3 months before the operation. However, an American bioethicist defended the woman. Even if her medical expenses were fully covered by the French national health system, he said, she might not find employment in the future (Victory 2005). Was signing the film deal ethical? Also, do you think that the success of such surgery will encourage wealthy people to trade their faces for those that are younger or prettier?

SOME ETHICAL ISSUES

You'll recall that the concept of *cultural lag* describes the gap when nonmaterial culture (including attitudes) changes more slowly than material culture (including technology) (see Chapter 3). The greater the technological advances, the larger the cultural lag.

A cultural lag can produce ethical dilemmas when government regulations do not keep up with technology advances and public opinion. For example, a federal judge in California initially ordered the shutdown of Wikileaks.org, a site that allows anonymous posting by "people of all countries to reveal unethical behavior in their governments and corporations." The case was brought by a Swiss banking company and its Cayman Islands subsidiary, which claimed that Wikileaks had posted confidential, personally identifiable information about some of its customers. The ruling drew considerable criticism from numerous organizations that maintained that the shutdown violated the First Amendment's protection of free speech. The judge reversed his original ruling, lamenting that constitutional law is not keeping up with technological change (Glater 2008).

Biotechnology sparks some of the most intense ethical controversies. One ethical dilemma, for example, is that biotechnological advances, like other technological developments, are most readily available to the wealthy. Whether in the United States or elsewhere, there's still a technological divide between the poor and the rich (see *Figure 17.2*). The inequality is especially evident in the use of biomedical technology to prolong life. For example, nearly 90,000 Americans are on lists for organ transplants, mostly kidneys and livers. Because of the scarcity of organs, many patients have turned to the Internet in search of donors. Proponents argue that the Web sites used by these people save lives because they motivate others to donate organs when they are touched by heart-wrenching personal stories. Opponents maintain that such pleas give an edge to those who are affluent, educated, and computer-literate (Stearns 2006; Morais 2007). Thus, technological advances are fraught with both promise and ethical dilemmas.

"7 People Charged in Fatal Mob Attack." 2002. *Baltimore Sun*, August 4, 2A.

Aberle, David. 1982. *The Peyote Religion among the Navaho*, 2nd ed. Chicago, IL: Aldine.

Abudabbeh, Nuha. 1996. "Arab Families." Pp. 333-346 in *Ethnicity and Family Therapy*, 2nd edition, edited by Monica McGoldrick, Joe Giordano, and John K. Pearce. New York: The Guilford Press.

Abu-Laban, Sharon M., and Baha Abu-Laban. 1999. "Teens Between: The Public and Private Spheres of Arab-Canadian Adolescents." Pp. 113-148 in *Arabs in America: Building a New Future*, edited by Michael W. Suleiman. Philadelphia: Temple University Press.

ACT. 2007. *ACT National Curriculum Survey 2005-2006*. Retrieved July 29, 2007 (www.act.org).

Adams, Bert N. and R. A. Sydie. 2001. *Sociological Theory*. Thousand Oaks, CA: Pine Forge Press.

Adelman, Clifford. 2006. *The Toolbox Revisited: Paths to Degree Completion from High School Through College*. Washington, DC: U.S. Department of Education.

Adler, Jerry. 2005. "In Search of the Spiritual." *Newsweek*, August 20-September 5, 44 64.

———. 2007. "How to Design a Healthier Planet." *Newsweek*, March 6, 66 70.

Adler, Patricia. A. and Paul Adler. 1994. "Observational Techniques." Pp. 377-392 in *Handbook of Qualitative Research*, edited by Norman K. Denzin and Yvonna S. Lincoln. Thousand Oaks, CA: Sage.

Adler, Susan M. 2003. "Asian-American Families." Pp. 82-91 in *International Encyclopedia Of Marriage And Family*, 2nd ed., volume 2, edited by J. J. Ponzetti, Jr. New York: Macmillan.

Agence France Presse. 2001. "Nigerian Girl to Be Lashed 180 Times." *Christian Science Monitor*, January 4, 7.

Aguirre, Adalberto Jr. and Jonathan H. Turner. 2004. *American Ethnicity: The Dynamics and Consequences of Discrimination*, 4th ed. Boston, MA: McGraw Hill.

Aizenman, N. C. 2005. "In Afghanistan, New Misgivings about an Old but Risky Practice." *Washington Post*, April 17. A16.

Ajrouch, Kristine. 1999. "Family and Ethnic Identity in an Arab-American Community." Pp. 129-139 in *Arabs in America: Building a new future*, edited by Michael W. Suleiman. Philadelphia: Temple University Press.

Akers, Ronald L. 1997. *Criminological Theories: Introduction and Evaluation*, 2nd edition. Los Angeles, CA: Roxbury.

Alder, Christine and Anne Worrall, eds. 2004. *Girls' Violence: Myths and Realities*. Albany: State University of New York Press.

Alesina, Alberto, Edward Glaeser, and Bruce Sacerdote. 2005. "Work and Leisure in the U.S. and Europe: Why So Different?" National Bureau of Economic Research, April. Retrieved June 2, 2007 (www.nber.org).

Ali, Kecia. 2003. "Progressive Muslims and Islamic Jurisprudence: The Necessity for Critical Engagement with Marriage and Divorce Law." Pp. 163-189 in *Progressive Muslims: On Justice, Gender, and Pluralism*, edited by Omid Safi. Oxford, England: Oneworld Publications.

Alini, Erica. 2007. "In Italy, Hard to Get the Kids to Move Out." *Christian Science Monitor*, November 15, 7.

Allegretto, Sylvia A., Sean P. Corcoran, and Lawrence Mishel. 2008. *The Teaching Penalty: Teacher Pay Losing Ground*. Washington, DC: Economic Policy Institute.

Allegretto, Sylvia, and Rob Gray. 2007. "Wealth Inequality is Vast and Growing." Economic Policy Institute. Retrieved March 16, 2007 (www.epi.org).

Allen, Katherine R., and Ben K. Beitin. 2007. "Gender and Class in Culturally Diverse Families." Pp. 63-79 in *Cultural Diversity and Families: Expanding Perspectives*, edited by Bahira Sherif Trask and Raeann R. Hamon. Thousand Oaks, CA: Sage, pp. 63-79.

Alliance for Board Diversity. 2008. "Women and Minorities on Fortune 100 Boards." *Catalyst*. Retrieved April 28, 2008 (www.catalyst.org).

Allport, Gordon. W. 1954. *The Nature of Prejudice*. Reading, MA: Addison-Wesley.

Almeida, R. V., ed. 1994. *Expansions of Feminist Family Theory through Diversity*. New York: Haworth Press.

Almond, Lucinda. 2007. *The Abortion Controversy*. Farmington Hills, MI: Greenhaven Press.

Alt, Betty L. and Sandra K. Wells. 2004. *Fleecing Grandma and Grandpa: Protecting against Scams, Cons, and Frauds*. Westport, CT: Praeger.

Alvarez, Lizette. 2005. "Got 2 Extra Hours for Your E-Mail?" *New York Times* (November 10): E1-E2.

Amato, Paul R., and Tamara D. Afifi. 2006. "Feeling Caught Between Parents: Adult Children's Relations with Parents and Subjective Well-Being." *Journal of Marriage and Family* 68 (February): 222-235.

American Academy of Pediatrics. "Children, Adolescents, and Television." 2001. *Pediatrics* 109 (November): 423-426.

American Association of Suicidology. 2006. "Elderly Suicide Fact Sheet." December 28. Retrieved March 29, 2008 (www.suicidology.org).

American Bar Association, 2006a. "A Current Glance at Women in the Law 2006." Commission on Women in the Profession. Retrieved January 15, 2007 (www.abanet.org/women).

———. 2006b. "Charting Our Progress: The Status of Women in the Profession Today." Commission on Women in the Profession. Retrieved January 15, 2007 (www.abanet.org/women).

American Council on Education. 2008. "Recent Graduates Give Alma Maters High Grades." May 20. Retrieved July 18, 2008 (www.solutions forourfuture.org).

American Federation of Teachers. 2007. "Survey and Analysis of Teacher Salary Trends 2005." Retrieved July 13, 2008 (www.aft.org).

"American Indian and Alaska Native Heritage Month: November 2007." 2007. U.S. Census Bureau, Facts for Features, CB07-FF.18. Retrieved June 10, 2008 (www.census.gov).

American Indian- and Alaska Native-Owned Firms: 2002. 2006. U.S. Census Bureau, SB02-00CS-AIAN (RV). Retrieved May 1, 2007 (www.census.gov).

American Institute of Stress. 2007. "Stress in the Workplace, Job Stress, Occupational Stress, Job Stress Questionnaire." Retrieved June 5, 2007 (www.stress.org).

"American Manners Poll." 2005. *USA Today*, October 14. Retrieved October 15, 2005 (www.usatoday.com).

American Rivers. 2005. "America's Most Endangered Rivers of 2005." Retrieved June 22, 2005 (www.americanrivers.org).

American Sociological Association. 2006. "What Can I Do with a Bachelor's Degree in Sociology?": A National Survey of Seniors Majoring in Sociology: First Glances: What Do They Know and Where Are They Going?" Washington, DC: American Sociological Association.

"Americans Engage in Unhealthy Behaviors to Manage Stress." 2006. American Psychological Association, February 23. Retrieved February 24, 2006 (www.apa.org).

Amnesty International. 2006. *Stonewalled: Police Abuse and Misconduct against Lesbian, Gay, Bisexual, and Transgender People in the U.S.* London: Amnesty International Publications.

Anderson, Craig A., and Brad J. Bushman. 2002. "The Effects of Media Violence on Society." *Science* 295 (March 29): 2377-2379.

Anderson, Craig A., et al. 2003. "The Influence of Media Violence on Youth." *Psychological Science in the Public Interest* 4 (December): 81-110.

Anderson, David A. 1999. "The Aggregate Burden of Crime." *Journal of Law and Economics* 42 (October): 611-42.

Anderson, Elijah. 1999. *Code of the Street: Decency, Violence, and the Moral Life of the Inner City*. New York: W.W. Norton.

Anderson, James F. and Laronistine Dyson. 2002. *Criminological Theories: Understanding Crime in America*. Lanham, MD: University Press of America.

Anderson, John W. 2001. "Iran's Cultural Backlash." *Washington Post*, August 16, A1, A20.

Anderson, Stephen A. and Ronald M. Sabatelli. 2007. *Family Interaction: A Multicultural Developmental Perspective*, 4th ed. Boston, MA: Allyn & Bacon.

Andreescu, Titu, Joseph A. Gallian, Jonathan M. Kane, and Janet E. Mertz. 2008. "Cross-Cultural Analysis of Students with Exceptional Talent in Mathematical Problem Solving." *Notices of the American Mathematical Society* 55 (10): 1248-1260.

Arax, Mark. 2004. "State Money Helped Dairies Dirty the Air." *Los Angeles Times*, October 11. Retrieved October 12, 2004 (www.latimes.com).

Arendt, Hannah. 2004. *The Origins of Totalitarianism*. New York: Schocken.

Ariès, Phillippe. 1962. *Centuries of Childhood*. New York: Vintage.

Armour, Stephanie. 2008. "Hitting Home: New Faces Join Ranks of the Homeless." *USA Today*, June 15. Retrieved June 17, 2008 (www.usatoday.com).

Arnoldy, Ben. 2008. "Battle Not Over with Gay-Marriage Edict." *Christian Science Monitor*, May 19, 3.

Aron-Dine, Aviva, and Isaac Shapiro. 2007. "Share of National Income Going to Wages and Salaries at Record Low in 2006." Center on Budget and Policy Priorities, March 29. Retrieved July 4, 2008 (www.cbpp.org).

Asch, Solomon. 1952. *Social Psychology*. Englewood Cliffs, NJ: Prentice-Hall, Inc.

Ashburn, Elyse. 2007. "A Race to Rescue Native Tongues." *Chronicle of Higher Education*, September 28, B15.

Ashford, Lori. S. 2006. "How HIV and AIDS Affect Populations." Population Reference Bureau. Retrieved February 1, 2008 (www.prb.org).

Asian-Owned Firms: 2002. 2006. U.S. Census Bureau, SB02-00CS-ASIAN (RV). Retrieved May 1, 2007 (www.census.gov).

"Asian/Pacific American Heritage Month: May 2008." 2008. U.S. Census Bureau, Facts for Features, CB08-FF.05. Retrieved June 10, 2008 (www.census.gov).

Aspray, William, Frank Mayadas, and Moshe Y. Vardi, eds. 2006. "Globalization and Offshoring of Software: A Report of the ACM Job Migration Task Force." Association for Computing Machinery. Retrieved December 30, 2006 (www.acm.org).

Astin, Alexander, and Helen Astin. 2005. "Spirituality in Higher Education." Higher Education Research Institute. Retrieved December 15, 2005 (www.gseis.ucla.edu).

Atchley, Robert C., and Amanda S. Barusch. 2004. *Social Forces and Aging: An Introduction to Social Gerontology*, 10th ed. Belmont, CA: Wadsworth.

Attinasi, John J. 1994. "Racism, Language Variety, and Urban U.S. Minorities: Issues in Bilingualism and Bidialectalism." Pp. 319-347 in *Race*, edited by Steven Gregory and Roger Sanjek. New Brunswick, NJ: Rutgers University Press.

Aunola, Kaisa, and Jari-Erik Nurmi. 2005. "The Role of Parenting Styles in Children's Problem Behavior." *Child Development* 76 (November/December): 1144-1159.

Austen, Ian. 2002. "A Leg with a Mind of Its Own." *New York Times*, January 3, G1.

Avellar, Sarah, and Pamela Smock. 2005. "The Economic Consequences of the Dissolution of Cohabiting Unions." *Journal of Marriage and Family* 67 (May): 315-327.

Axtell, Roger E., Tami Briggs, Margaret Corcoran, and Mary Beth Lamb. 1997. *Do's and Taboos Around the World for Women in Business*. New York: John Wiley & Sons, Inc.

Azios, Tony. 2008. "Bottled vs. Tap." *Christian Science Monitor*, January 17, 15-16.

Babbie, Earl. 2002. *The Basics of Social Research*, 2nd edition. Belmont, CA: Wadsworth.

Bader, Christopher, Kevin Dougherty, Paul Froese, Byron Johnson, F. Carson Mencken, Jerry Z. Park, and Rodney Stark. 2006. "American Piety in the 21st Century: New Insights to the Depth and Complexity of Religion in the US." Baylor Institute for Studies of Religion, September. Retrieved August 25, 2007 (www.baylor.edu/isreligion).

Bailey, William, Michael Young, Cliff Knickerbocker, and Tom Doan. 2002. "A Cautionary Tale About Conducting Research On Abstinence Education: How Do State Abstinence Coordinators Define 'Sexual Activity'?" *American Journal of Health Education* 33 (September/October): 290-296.

Bainbridge, William S. 1997. *The Sociology of Religious Movements.* New York: Routledge.

Bakan, Joel. 2004. *The Corporation: The Pathological Pursuit of Profit and Power.* New York: Free Press.

Bakyawa, Jennifer. 2002. "Ugandan Women Seek Right to Own Land." *Women's E-News,* January 30. Retrieved January 30, 2002 (www.womensenews.org).

Baldauf, Scott. 2001. "Afghan Weavers Unravel a Trade Tradition: Opium Use." *Christian Science Monitor* (March 28): 7.

Baldi, Stéphane, Ying Jin, Melanie Skemer, Patricia J. Green, and Deborah Herget. 2007. *Highlights from PISA 2006: Performance of U.S. 15-Year-Old Students in Science and Mathematics Literacy in an International Context* (NCES 2008-016). Washington, DC: U.S. Department of Education.

Bandura, Albert and Richard H. Walters. 1963. *Social Learning and Personality Development.* New York: Holt, Rinehart & Winston.

Banerjee, Neela. 2006. "Clergywomen Find Hard Path to Bigger Pulpit." *New York Times,* August 26, A1, A12.

———. 2007. "For Some Black Pastors, Accepting Gay Members Means Losing Others." *New York Times,* March 27, A12.

Banfield, Edward C. 1974. *The Unheavenly City Revisited.* Boston, MA: Little, Brown.

Banks, Ingrid. 2000. *Hair matters: Beauty, Power, and Black Women's Consciousness.* New York: New York University Press.

Banks, Marian R., Lisa M. Willoughby, and William A. Banks. 2008. "Animal-Assisted Therapy and Loneliness in Nursing Homes: Use of Robotic versus Living Dogs." *Journal of the American Medical Directors Association* 9 (March): 173-177.

Banks, Sandy. 2006. "Firehouse Culture and Ordeal for Women." *Los Angeles Times,* December 3. Retrieved December 4, 2006 (www.latimes.com).

Barabak, Mark Z. 2006. "Guest-Worker Proposal Has Wide Support." *Los Angeles Times,* April 30. Retrieved May 2, 2006 (www.latimes.com).

Barash, David P. 2002. "Evolution, Males, and Violence." *Chronicle of Higher Education,* May 24, B7-B9.

Barbarin, Oscar A., and Terry McCandies. 2003. "African-American Families." Pp. 50-56 in *International Encyclopedia Of Marriage And Family,* 2nd ed., volume 1, edited by James J. Ponzetti Jr., 50-56. New York: Macmillan.

Barber, Bernard, and Lyle S. Lobel. 1952. "Fashion in Women's Clothes and the American Social System." *Social Forces* 31 (December): 124-131.

Bar-Cohen, Yoseph and Cynthia Breazeal. 2003. *Biologically Inspired Intelligent Robots.* Bellingham, WA: Spie Press.

Barker, James B. 1993. "Tightening the Iron Cage: Concertive Control in Self-Managing Teams." *Administrative Science Quarterly* 38 (September): 408-437.

Barnard, Chester. 1938. *The Functions of the Executive.* Cambridge, MA: Harvard University Press.

Barnes, Cynthia. 2006. "China's 'Kingdom of Women'." *Slate,* November 17. Retrieved June 12, 2007 (www.slate.com).

Barnett, Megan. 2004. "Starting Over." *U.S. News & World Report,* May 31, 48-49.

Barnett, Rosalind and Caryl Rivers. 2004. *Same Difference: How Gender Myths Are Hurting Our Relationships, Our Children, and Our Jobs.* New York: Basic Books.

Barro, Robert J., and Rachel M. McCleary. 2003. "Religion and Economic Growth across Countries." *American Sociological Review* 68 (October): 760-781.

Barrows, Sydney B. 1989. *Mayflower Madam: The Secret Life of Sydney Biddle Barrows.* Westminster, MD: Arbor House.

Bart, Pauline B. and Eileen Geil Moran, eds. 1993. *Violence Against Women: The Bloody Footprints.* Thousand Oaks, CA: Sage.

Bartkowski, John P., and Jen'nan Ghazal Read. 2003. "Veiled Submission: Gender, Power, and Identity among Evangelical and Muslim Women in the United States." *Qualitative Sociology* 26 (Spring): 71-92.

Bartlett, Thomas, and Paula Wasley. 2008. "Just Say 'A': Grade Inflation Undergoes Reality Check." *Chronicle of Higher Education,* September 5, A1, A10.

Bartlett, Thomas. 2007. "'I Suffer Not a Woman to Teach'." *Chronicle of Higher Education,* April 13, A10-A12.

Barton, Allen H. 1980. "A Diagnosis of Bureaucratic Maladies." Pp. 27-26 in *Making Bureaucracies Work,* edited by Carol H. Weiss and Allen H. Barton. Beverly Hills, CA: Sage.

Bauer, Mary. 2007. "Close to Slavery: Guestworker Programs in the United States." Southern Poverty Law Center. Retrieved April 2, 2007 (www.splcenter.org).

Baum, Katrina. 2006. "Identity Theft, 2004." Bureau of Justice Statistics Bulletin, April. Retrieved March 10, 2008 (wwwl.ojp.usdoj.gov).

Baumrind, Diana. 1968. "Authoritarian versus Authoritative Parental Control." *Adolescence* 3 (11); 255-72.

———. 1989. "Rearing Competent Children." Pp. 349-78 in *Child Development Today and Tomorrow,* edited by William Damon. San Francisco, CA: Jossey-Bass.

Bayer, Patrick, Hanming Fang, and Robert McMillan. 2005. "Separate When Equal? Racial Inequality and Residential Inequality." National Bureau of Economic Research. Retrieved January 22, 2008 (www.nber.org).

Bean, Frank D. and Gillian Stevens. 2003. *America's Newcomers and the Dynamics of Diversity.* New York: Russell Sage Foundation.

Becker, Bernice. 2008. "Justices List Their Assets; Wide Range of Wealth." *New York Times,* June 7, 12.

Becker, Elizabeth. 2001. "Some Who Vote on Farm Subsidies Get Them as Well." *New York Times,* September 1, 10.

Becker, Howard S. 1963. *Outsiders: Studies in the Sociology of Deviance.* New York: Free Press.

Becker, Jasper. 2002. "China's Workers—No Longer a Privileged Class." *Christian Science Monitor,* October 21, 7-8.

Beeghley, Leonard. 2000. *The Structure of Social Stratification in the United States,* 3rd ed. Boston, MA: Allyn and Bacon.

Behnke, Andrew. 2004. "Latino Dads: Structural Inequalities and Personal Strengths." *Report: National Council on Family Relations* 49 (September): F6-F7.

Belkin, Lisa. 2008. "Smoother Transitions." *New York Times,* September 4, 2.

Belknap, Joanne. 2001. *The Invisible Woman: Gender, Crime, and Justice,* 2nd edition. Belmont, CA: Wadsworth.

———. 2007. *The Invisible Woman: Gender, Crime, and Justice,* 3rd edition. Belmont, CA: Wadsworth Press.

Belsky, Janet K. 1988. *Here Tomorrow: Making the Most of Life after Fifty.* Baltimore, MA: Johns Hopkins University Press.

Belson, Ken. 2005. "Ebbers Mounts an 'I Never Knew' Defense." *New York Times,* March 1, 1.

Bender, Courtney. 2003. *Heaven's Kitchen: Living Religion at God's Love We Deliver.* Chicago, IL: The University of Chicago Press.

Bennhold, Katrin. 2008. "A Veil Closes France's Door to Citizenship." *New York Times,* July 19, 1.

Benokraitis, Nijole V. 2007. *Marriages and Families: Changes, Choices, and Constraints,* 6th ed. New Jersey: Upper Saddle River.

———. 1997. *Subtle Sexism: Current Practices and Prospects for Change.* Thousand Oaks, CA: Sage.

———, ed. 2000. *Feuds about Families: Conservative, Centrist, Liberal, and Feminist Perspectives.* Upper Saddle River, NJ: Prentice Hall.

Benson, Michael L. 2002. *Crime and the Life Course.* Los Angeles, CA: Roxbury.

Benson, Michael L., and Greer Litton Fox. 2004. "When Violence Hits Home: How Economics and Neighborhood Play a Role." National Institute of Justice, September. Retrieved October 1, 2004 (www.ojp.usdoj.gov).

Bergen, Raquel K., Jeffrey L. Edleson, and Claire M. Renzetti, eds. 2005. *Violence against Women: Classic Papers.* Boston, MA: Allyn and Bacon.

Berger, Adam. 2006. "Mind Your Cellphone Manners." July 5. Retrieved March 26, 2008 (www.gadgetell.com).

Berger, Peter L. and Thomas Luckmann. 1966. *The Social Construction of Reality: A Treatise in the Sociology of Knowledge.* New York: Doubleday & Company.

Bergman, Mike. 2004. "Nation Adds 2.2 Million Nonemployer Businesses over Five-Year Period." U.S. Department of Commerce. Retrieved July 19, 2004 (www.census.gov).

Bergquist, Amy. 2006. "Pharmacist Refusals: Dispensing (With) Religious Accommodation under Title VII." *Minnesota Law Review* 90 (April): 1073-1106.

Berk, Richard A. 1974. *Collective Behavior.* Dubuque, Iowa: Wm. C Brown Company.

Berkos, Kristen M., Terre H. Allen, Patricia Kearney, and Timothy G. Plax. 2001. "When Norms are Violated: Imagined Interactions as Processing and Coping Mechanisms." *Communication Monographs* 68 (September): 289-300.

Bernstein, Jared. 2008. "Real Wage Reversal Persists." Economic Policy Institute, February 20. Retrieved July 3, 2008 (www.epi.org).

Bernstein, Jared, Elise Gould, and Lawrence Mishel. 2007. "Poverty, income, and health insurance trends in 2006." Economic Policy Institute, August 28. Retrieved August 29, 2008 (www.epi.org).

Bernstein, Nina. 2008. "Spitzer Wrestles over Response, Paralyzing Albany." *New York Times,* March 12, A21.

Bertrand, Marianne, and Sendhil Mullainathan. 2003. "Are Emily and Greg More Employable than Lakisha and Jamal? A Field Experiment on Labor Market Discrimination." Massachusetts Institute of Technology, Department of Economics. Retrieved May 3, 2004 (http://papers.ssrn.com).

Berube, Alan, Audrey Sinter, Jill H. Wilson, and William H. Frey. 2006. "Finding Exurbia: America's Fast-Growing Communities at the Metropolitan Fringe." Brookings Institution. Retrieved February 2, 2008 (www.brookings.edu).

Best, Joel. 2001. *Damned Lies and Statistics: Untangling Numbers from the Media, Politicians, and Activists.* Berkeley: University of California Press.

———. 2006. *Flavor of the Month: Why Smart People Fall for Fads.* Berkeley: University of California Press.

Betcher, R. William and William S. Pollack. 1993. *In a Time of Fallen Heroes: The Re-Creation of Masculinity.* New York: Atheneum.

Bianchi, Suzanne M., John P. Robinson and Melissa A. Milkie. 2006. *Changing Rhythms of American Family Life.* New York: Russell Sage Foundation.

Billingsley, Andrew. 1992. *Climbing Jacob's Ladder: The Enduring Legacy of African-American Families.* New York: Simon & Schuster.

"Billionaires 2008." 2008. *Forbes,* March 24, 80-86.

Binder, Amy. 1993. "Constructing Racial Rhetoric: Media Depictions of Harm in Heavy Metal and Rap Music." *American Sociological Review* 58 (December): 753-767.

Birch, Douglas. 2006. "Deep Anger, Not Cartoons, Spurred Muslim Protests." *Baltimore Sun,* February 9, 1A, 13A.

Birnbaum, Jeffery H. 2004. "Cost of Congressional Campaigns Skyrockets." *Washington Post,* October 3, A8.

Bishop, George F., Robert W. Oldendick, Alfred J. Tuchfarber, and Stephen E. Bennett. 1980. "Pseudo-opinions on Public Affairs." *The Public Opinion Quarterly* 44 (Summer): 198-209.

Bivens, L. Josh. 2006. "Offshoring." Economic Policy Institute, May. Retrieved June 1, 2007 (www.epinet.org).

———. 2008. "Trade, Jobs, and Wages: Are the Public's Worries about Globalization Justified?" Economic Policy Institute, May 6. Retrieved July 4, 2008 (www.epi.org).

Black-Owned Firms: 2002. 2006. U.S. Census Bureau, SB02-00CS-BLK (RV). Retrieved May 1, 2007 (www.census.gov).

Blanchard, Dallas A. 1994. *The Anti-Abortion Movement and the Rise of the Religious Right: From Polite to Fiery Protest.* New York: Twayne Publishers.

Blankenstein, Andrew, and Tony Barboza. 2007. "More Pot House Busts Revealed." *Los Angeles Times,* April 12. Retrieved April 19, 2007 (www.latimes.com).

Blass, Thomas. 2000. "The Milgram Paradigm After 35 Years: Some Things We Now Know About Obedience to Authority." In Thomas Blass (Ed.), *Obedience to Authority: Current Perspectives on the Milgram Paradigm.* Mahwah, NJ: Lawrence Erlbaum Associates, pp. 35-59.

Blau, Francine B., and Lawrence M. Kahn. 2006. "The Gender Pay Gap: Going, Going . . . but Not Gone." Pp. 37–66 in *The Declining Significance of Gender?* edited by Francine D. Blau, Mary C. Brinton, and David B. Grutsky. New York: Russell Sage Foundation.

Blau, Peter M. 1986. *Exchange and Power in Social Life,* revised ed. New Brunswick, NJ: Transaction.

Blau, Peter M. and Marshall W. Meyer. 1987. *Bureaucracy in Modern Society,* 3rd edition. New York: Random House.

Blauner, Robert. 1969. "Internal Colonialism and Ghetto Revolt." *Social Problems* 16 (Spring): 393–408.

———. 1972. *Racial Oppression in America.* New York: Harper and Row.

Blumer, Herbert. 1946. "Collective Behavior." Pp. 65–121 in *New Outline of the Principles of Sociology,* edited by Alfred M. Lee (Ed.). New York: Barnes & Noble.

———. 1969. *Symbolic Interactionism: Perspective and Method.* Englewood Cliff, NJ: Prentice Hall.

Boase, Jeffrey, John B. Horrigan, Barry Wellman, and Lee Rainie. 2006. "The Strength of Internet Ties." Internet & American Life Project. Retrieved February 12, 2006 (www.pewinternet.org).

Bobroff-Hajal, Anne. 2006. "Why Cousin Marriage Matters in Iraq." *Christian Science Monitor,* December 26, 9.

Boggs, George R. 2004. "Community Colleges in a Perfect Storm." *Change* 36 (November/December): 6–11.

Bohannon, John. 2005. "Disasters: Searching for Lessons from a Bad Year." *Science* 310 (December): 1883.

Bohannon, Lisa. 2000. "Is Your Body Language on Your Side?" *Career World* 29 (November/December): 21–23.

Bohannon, Paul, ed. 1971. *Divorce and After.* New York: Doubleday.

Bollag, Burton. 2007. "Credential Creep." *Chronicle of Higher Education,* June 22, A10–A12.

Bonacich, Edna. 1972. "A Theory of Ethnic Antagonism: The Split Labor Market." *American Sociological Review* 37 (October): 547–559.

Bonger, Willem A. 1916/1969. *Criminality and Economic Conditions.* Bloomington: Indiana University Press.

Bonisteel, Sara. 2006. "Asian Leaders Angered by Rosie O'Donnell's 'Ching Chong' Comments." Fox News, December 11. Retrieved September 10, 2008 (www.foxnews.com).

Bonner, Robert. 2007. "Research—Male Elementary Teachers: Myths and Realities." Men Teach. Retrieved June 12, 2008 (www.menteach.org).

Bonnie, Richard J., and Robert B. Wallace, eds. 2003. *Elder Mistreatment: Abuse, Neglect, and Exploitation in an Aging America.* Washington, DC: The National Academies Press.

Boonstra, Heather D., Rachel B. Gold, Cory L. Richards, and Lawrence B. Finer. 2006. "Abortion in Women's Lives." Guttmacher Institute. Retrieved December 21, 2006 (www.guttmacher.org).

Booza, Jason C., Jackie Cutsinger, and George Galster. 2006. "Where Did They Go? The Decline of Middle-Income Neighborhoods in Metropolitan America." Brookings Institution. Retrieved January 21, 2008 (www.brookings.edu).

Boraas, Stephanie, and William M. Rodgers III. 2003. "How Does Gender Play a Role in the Earnings Gap? An Update." *Monthly Labor Review* 126 (March): 9–15.

Borzekowski, Dina L., and Thomas N. Robinson. 2005. "The Remote, the Mouse, and the No. 2 Pencil." *Archives of Pediatrics & Adolescent Medicine* 159 (July): 607–613.

Bosk, Charles. 1979. *Forgive and Remember: Managing Medical Failure.* Chicago, IL: University of Chicago Press.

"Boss Sells His Company, Shares Proceeds." 1999. *Baltimore Sun,* September 12, 24A.

Bourdieu, Pierre. 1984. *Distinction: A Social Critique of the Judgement of Taste.* Trans. Richard Nice. Cambridge, MA: Harvard University Press.

Bourne, Joel K. Jr. 2006. "Loving Our Coasts to Death." *National Geographic* 210 (July): 60–87.

Bowles, Samuel and Herbert Gintis. 1977. *Schooling in Capitalist America: Educational Reform and the Contradictions of Economic Life.* New York: Basic Books.

Bowles, Samuel, Herbert Gintis and Melissa Osborne Groves. 2005. *Unequal Chances: Family Background and Economic Success.* New York: Russell Sage Foundation.

Bradshaw, York W. and Michael Wallace. 1996. *Global Inequalities.* Thousand Oaks, CA: Pine Forge Press.

Brady, Diane. 2007. "A Little Shame Goes a Long Way." *Business Week,* April 16, 34–35.

Braga, Anthony A. 2003. "Systematic Review of the Effects of Hot Spots Policing on Crime." Unpublished paper. Retrieved May 14, 2005 (www.campbellcollaboration.org/doc-pdf/hotspots.pdf).

Brainard, Jeffery, and J. J. Hermes. 2008. "Colleges' Earmarks Grow, Amid Criticism." *Chronicle of Higher Education,* March 28, A1, A8–A11.

Bramlett, Mosher D., and William D. Mosher. 2002. *Cohabitation, Marriage, Divorce, and Remarriage in the United States.* Centers for Disease Control and Prevention, *Vital and Health Statistics* (Series 23, No. 25). Hyatsville, MD: National Center for Health Statistics.

Braun, Henry, Frank Jenkins, and Wendy Grigg. 2006. *A Closer Look at Charter Schools Using Hierarchical Linear Modeling* (NCES 2006-460). National Center for Education Statistics, U.S. Department of Education. Washington, DC: U.S. Government Printing Office.

Brea, Jorge A. 2003. "Population Dynamics in Latin America." *Population Bulletin* 58 (March): 1–36.

Breazeal, Cynthia L. 2002. *Designing Sociable Robots.* Cambridge, MA: MIT Press.

Brickner, Mindy Kay. 2007. "Vienna Gives Signs a Gender Adjustment." *Christian Science Monitor,* March 14, 13, 16.

Bridgeland, John M., John J. DiIulio, Jr., and Karen Burke Morison. 2006. "The Silent Epidemic: Perspectives of High School Dropouts." Civic Enterprises, March. Retrieved July 21, 2007 (www.civicenterprises.net).

Brittingham, Angela, and G. Patricia de la Cruz. 2005. "We the People of Arab Ancestry in the United States." Census 2000 Special Reports, CENSR-21, March. Retrieved March 26, 2007 (www.census.gov).

Britton, Dana M. 2003. *At Work in the Iron Cage: The Prison as a Gendered Organization.* New York: New York University Press.

Britz, Jennifer D. 2006. "To All the Girls I've Rejected." *New York Times,* March 23, A27.

Broder, John M. 2007. "Industry Flexes Muscle, Weaker Energy Bill Passes." *New York Times,* December 14, 29.

Brodie, Mollyann, Annie Steffenson, Jamie Valdez, and Rebecca Levin. 2002. *2002 National survey of Latinos.* Pew Hispanic Center/Kaiser Family Foundation. Retrieved April 16, 2003 (www.pewhispanic.org).

Brody, Gene H., Shannon Dorsey, Rex Forehand, and Lisa Armistead. 2002. "Unique and Protective Contributions of Parenting and Classroom Processes to the Adjustment of African American Children Living In Single-Parent Families." *Child Development* 73 (January-February): 274–286.

Bronstad, Amanda. 2008. "Is It Creative Style or Harassment?" *National Law Journal* (January 28).

Brookey, Robert A. 2002. *Reinventing the Male Homosexual: The Rhetoric and Power of the Gay Gene.* Bloomington, IN: Indiana University Press.

Brooks, Arthur C. 2007. "I Love My Work." *The American* (online): September-October. Retrieved September 30, 2007 (www.theamericanmag.com).

Brown, David. 2005. "Polio Outbreak Occurs Among Amish Families in Minnesota." *Washington Post,* October 14, A3.

Brown, Matthew Hay. 2005. "Storm Victims Rely on Faith to Help Them Through Crisis." *Baltimore Sun,* September 12, 6A.

Brown, Susan I. 2005. "How Cohabitation Is Reshaping American Families." *Contexts* 4 (Summer): 33–37.

Bruni, Frank. 2001. "Bush Emphasizes Teaching of Values to Children." *New York Times,* August 15, A21.

Brunvand, Jan H. 1999. *Too Good to be True: The Colossal Book of Urban Legends.* New York: W. W. Norton.

———. 2001. *The Truth Never Stands in the Way of a Good Story.* Urbana and Chicago, IL: University of Illinois Press.

Buchmann, Claudia, and Thomas A. DiPrete. 2006. "The Growing Female Advantage in College Completion: The Role of Family Background and Academic Achievement." *American Sociological Review* 71 (August): 515–541.

Buechler, Steven M. 2000. *Social Movements in Advanced Capitalism: The Political Economy and Cultural Construction of Social Activism.* New York: Oxford University Press.

Bullock, Karen. 2005. "Grandfathers and the Impact of Raising Grandchildren." *Journal of Sociology and Social Welfare* 32 (March): 43–59.

Bumiller, Elisabeth. 2005. "In the Struggle over the Iraq War, Women Are On the Front Line." *New York Times,* August 29, 11.

Bumpass, Larry L., R. Kelly Raley, and James A. Sweet. 1995. "The Changing Character of Stepfamilies: Implications of Cohabitation and Nonmarital Childbearing." *Demography* 32 (August): 425–36.

Burns, Barbara J., et al. 2003. "Treatment, Services, and Intervention Programs for Child Delinquents." Washington, DC: U.S. Department of Justice, Office of Juvenile Justice and Delinquency Prevention.

Burr, Chandler. 1996. *A Separate Creation: The Search for the Biological Origins of Sexual Orientation.* New York: Hyperion.

Burrus, Robert T., KimMarie McGoldrick, and Peter W. Schuhmann. 2007. "Self-Reports of Student Cheating: Does a Definition of Cheating Matter?" *Journal of Economic Education* 38 (Winter): 3–16.

Burt, Martha R., John Hedderson, Janine M. Zweig, Mary Jo Ortiz, Laudan Y. Aron, and Sabrina M. Johnson. 2004. "Strategies for Reducing Chronic Street Homelessness." Urban Institute, January 15. Retrieved January 12, 2007 (www.urban.org).

"Bush Defends Domestic Spying." 2006. *Baltimore Sun,* January 2, 2A.

"Bush's Cooks Get Queen Boiling Mad." 2003. *Baltimore Sun,* November 22, 2D.

Bustillo, Miguel. 2006. "Farmers Say They've Got Fruit but No Labor." *Los Angeles Times,* April 17. Retrieved April 28, 2006 (www.latimes.com).

Butterfield, Fox. 2002. "Father Steals Best: Crime in an American Family." *New York Times* (August 15): 1A.

Cadge, Wendy, and Courtney Bender. 2004. "Yoga and Rebirth in America: Asian Religions are Here to Stay." *Contexts* 3 (Winter): 45–51.

Calasanti, Toni M., and Kathleen F. Slevin, eds. 2006. *Age Matters: Realigning Feminist Thinking.* New York: Routledge.

Calvert, Scott. 2001. "The Products They Love to Hate." *Baltimore Sun,* December 6, 2A.

Campbell, Eric G., Russell L. Gruen, James Mountford, Lawrence G. Miller, Paul D. Cleary, and David Blumenthal. 2007. "A National Survey of Physician-Industry Relationships." *New England Journal of Medicine* 356 (April 26): 1742–1750.

Campion-Vincent, Véronique. 2005. "From Evil Others to Evil Elites: A Dominant Pattern in Conspiracy Theories Today." Pp. 103–122 in *Rumor Mills: The Social Impact of Rumor and Legend,* edited by Gary Alan Fine, Véronique Campion-Vincent, and Chip Heath. New Brunswick, NJ: Transaction Publishers.

Campo-Flores, Arian. 2005. "The Battle for Latino Souls." *Newsweek,* March 31, 50–51.

Cantor, Paul A. 2001. *Gilligan Unbound: Pop Culture in the Age of Globalization.* Lanham, MD: Rowman & Littlefield.

Capps, Randy, Rosa Maria Castañeda, Ajay Choudry, and Robert Santos. 2007. *Paying the Price: The Impact of Immigration Raids on America's Children.* Washington, DC: National Council of La Raza.

Carasso, Adam, C. Eugene Steuerle, Gillian Reynolds, Tracy Vericker, and Jennifer Macomber. 2008. "Kids' Share 2008: How Children Fare in the Federal Budget." Urban Institute, June 24. Retrieved July 10, 2008 (www.urban.org).

Card, Josefina J., Marvin B. Eisen, James L. Peterson, and Bonnie Sherman-Williams. 1994. "Evaluating Teenage Pregnancy Prevention and Other Social Programs: Ten Stages of Program Assessment." *Family Planning Perspectives* 26 (May): 116–131.

Carey, Benedict. 2004. "Long After Kinsey, Only the Brave Study Sex." *New York Times,* November 9, F1.

Carey, John. 2007. "Big Strides to Become the Jolly Green Giant." *Business Week,* January 29, 57.

Carey, Kevin. 2005. "One Step from the Finish Line: Higher College Graduation Rates are Within Our Reach." Education Trust. Retrieved January 18, 2005 (www2.edtrust.org).

Carl, Traci. 2002. "Amid Latte, Mocha Craze, Coffee Growers Go Hungry in Paradise." Retrieved September 19, 2002 (http://story.news.yahoo.com).

Carlson, Lewis H. and George A. Colburn, eds. 1972. *In Their Place: White America Defines Her Minorities, 1850–1950.* New York: Wiley.

Carlson, Marcia, Sara McLanahan, Paula England, and Barbara Devaney. 2005. "What We Know about Unmarried Parents: Implications for Building Strong Families Programs." Mathematica Policy Research, Inc., January. Retrieved June 5, 2007 (www.mathematica-mpr.com.

Carmichael, Mary. 2007. "Troubled Waters." *Newsweek*, June 4, 52-56.

Carnagey, Nicholas L., and Craig A. Anderson. 2005. "The Effects of Reward and Punishment in Violent Video Games on Aggressive Affect, Cognition, and Behavior." *Psychological Science* 16 (November): 882-889.

Carney, Ginny. 1997. "Native American Loanwords in American English." *Wicazo SA Review*, 12 (Spring): 189-203.

Carr, J. L. 2005. *American College Health Association Campus Violence White Paper*. Baltimore MD: American College Health Association.

Carroll, Jason S., Laura M. Padilla-Walker, Larry J. Nelson, Chad D. Olson, Carolyn McNamara Barry, and Stephanie D. Madsen. 2008. "Generation XXX: Pornography Acceptance and Use Among Emerging Adults." *Journal of Adolescent Research* 23 (January): 6-30.

Carroll, Jill. 2002. "Getting Good Teaching Evaluations Without Stand-Up Comedy." *Chronicle of Higher Education*, April 15. Retrieved April 15, 2002 (http://chronicle.com).

Carroll, Joseph. 2006. "Americans Prefer Male Boss to a Female Boss." Gallup Organization. Retrieved September 3, 2006 (www.gallup.com).

____. "Most Americans Approve of Interracial Marriages." Gallup News Service, July 6. Retrieved October 10, 2007 (www.galluppoll.com).

____. 2007. "Polluted Drinking Water is Public's Top Environmental Concern." Gallup News Service. Retrieved February 7, 2008 (www.gallup.com).

____. 2007. "Stress More Common Among Younger Americans, Parents, Workers." Gallup News Service, January 24. Retrieved June 2, 2007 (www.gallup-poll.com).

Carter, Jimmy. 2005. *Our Endangered Values: America's Moral Crisis*. New York: Simon & Schuster.

"Cash Countesses." 2007. *Forbes*, October 8, 312, 316-317.

Cassidy, Margaret L., and Gary R. Lee. 1989. "The Study of Polyandry: A Critique and Synthesis." *Journal of Comparative Family Studies* 20 (Spring): 1–11.

Caston, Richard J. 1998. *Life in a Business-Oriented Society: A Sociological Perspective*. Boston, MA: Allyn and Bacon.

Catalano, Shannan M. 2006. "Criminal Victimization, 2005." Washington, DC: U.S. Department of Justice, Bureau of Justice Statistics.

Catalyst. 2006. "Rate of Women's Advancement to Top Corporate Officer Positions Slow, New Catalyst Tenth Anniversary Census Reveals." Retrieved December 29, 2006 (www.catalyst.org).

____. 2007. "2007: Board Directors." Retrieved April 28, 2008 (www.catalyst.org).

Center for American Progress. 2006. "Public Recognizes Debt as a Fast Growing Problem in U.S." July 19. Retrieved June 3, 2007 (www.americanprogress.org).

Center for American Women and Politics. 2008. "Women in Elective Office 2008." Retrieved June 5, 2008 (www.cawp.rutgers.edu).

Center for Responsive Politics, The. 2008a. "Banking on Becoming President." August 31. Retrieved October 14, 2008 (www.opensecrets.org).

____. 2008b. "Most Expensive Races." Retrieved October 14, 2008 (www.opensecrets.org).

____. 2008b. "Presidential Fundraising and Spending, 1976-2008." Retrieved December 2, 2008 (www.opensecrets.org).

____. 2008c. "Top PACs." Retrieved December 2, 2008 (www.opensecrets.org).

Center for the Study of Global Christianity. 2007. "Global Table 5: Status of Global Mission, Presence and Activities, AD 1800-2005." Gordon-Conwell Theological Seminary. Retrieved August 24, 2007 (www.gcts.edu).

Center on Education Policy. 2006. *From the Capital to the Classroom: Year 4 of the No Child Left Behind Act*. Retrieved July 21, 2007 (www.cep-dc.org).

Chamberlain, Andrew, and Gerald Prante. 2007. "Generational Equity: Which Age Groups Pay More Tax, and Which Receive More Government Spending?"

Tax Foundation Special Report, June. Retrieved June 26, 2007 (www.taxfoundation.org).

Chambliss, William J., and Robert B. Seidman. 1982. *Law, Order, and Power*, 2nd edition. Reading, MA: Addison-Wesley.

Chambliss, William J., ed. 1969. *Crime and the Legal Process*. New York: McGraw-Hill.

Chan, Sam. 1997. "Families with Asian Roots." Pp. 251-344 in *Developing Cross-Cultural Competence: a Guide for Working with Children and Their Families*, 2nd edition, edited by Eleanor W. Lynch and Marci J. Hanson. Baltimore, MD: Paul H. Brookes Publishing.

Chandler, Tertius and Gerald Fox. 1974. *3000 Years of Urban Growth*. New York: Academic Press.

Chandra, A., G. A. Martinez, W. D. Mosher, J.C. Abma, and J. Jones. 2005. "Fertility, Family Planning, and Reproductive Health of U.S. Women: Data from the 2002 National Survey of Family Growth." *Vital and Health Statistics* (Series 23, No. 25). Hyatsville, MD: National Center for Health Statistics.

Chapman, Tony, and Jenny Hockey, eds. 1999. *Ideal Homes? Social Change and Domestic Life*. New York: Routledge.

Chasanov, Amy. 2004. "No Longer Getting By: An Increase in the Minimum Wage is Long Overdue." Economic Policy Institute, Briefing Paper #151. Retrieved July 23, 2004 (www.epinet.org).

Chasin, Barbara H. 2004. *Inequality & Violence in the United States: Casualties of Capitalism*, second edition. Amherst, NY: Humanity Books.

Chelala, César. 2002. "World Violence against Women A Great Unspoken Pandemic." *Philadelphia Inquirer*, November 4. Retrieved November 7, 2002 (www.commondreams.org).

Chen, Xianglei. 2005. *First Generation Students in Postsecondary Education: A Look at Their College Transcripts* (NCES 2005-17). U.S. Department of Education, National Center for Education Statistics. Washington, DC: U.S. Government Printing Office.

Chesler, Phyllis. 2006. "The Failure of Feminism." *Chronicle of Higher Education*, February 26, B12.

Chesney-Lind, Meda and Lisa Pasko. 2004. *The Female Offender: Girls, Women, and Crime*, 2nd edition. Thousand Oaks, CA: Sage.

Choice, Pamela, and Leanne K. Lamke. 1997. "A Conceptual Approach to Understanding Abused Women's Stay/Leave Decisions." *Journal of Family Issues* 18 (May): 290–314.

Choudhury, Tufyal, Mohammed Aziz, Duaa Izzidien, Intissar Khreeji, and Dilwar Hussein. 2006. "Perceptions of Discrimination and Islamophobia: Voices from Members of Muslim Communities in the European Union." Retrieved September 1, 2007 (www.eumc.eu).

Christenson, Matthew, Thomas M. McDevitt, and Karen A. Stanecki. 2004. *Global Population Profile: 2002*. U.S. Census Bureau, International Population Reports WP/02. Washington, DC: Government Printing Office.

Christiano, Kevin J. 2000. "Religion and the Family in Modern American Culture." Pp. 43-78 in *Family, Religion, and Social Change in Diverse Societies*, edited by Sharon K. Houseknecht and Jerry G. Pankhurst. New York: Oxford University Press.

Chronicle of Higher Education. 2008. "Students." *Almanac Issue 2008-9* 55 (August 29): 18.

Chu, Henry. 2007. "A Gift for India's Inter-Caste Couples." *Los Angeles Times*, November 4. Retrieved November 6, 2007 (www.latimes.com).

Churchill, Ward. 1997. *A Little Matter of Genocide: Holocaust and Denial in the Americas, 1492 to the Present*. San Francisco: City Lights Books.

Cichocki, Mary K. 1981. "Women's Travel Patterns in a Suburban Development." Pp. 151-163 in *New Space for Women*, edited by Gerda R. Wekerle, Rebecca Peterson, and David Morley. Boulder, CO: Westview Press.

Cintrón-Vélez, Aixa N. 2002. "Custodial Mothers, Welfare Reform, and the New Homeless: A Case Study of Homeless Families in Three Lowell Shelters." Pp. 148-178 in *Laboring Below the Line: The New Ethnography of Poverty, Low-Wage Work, and Survival in the Global Economy*, edited by Frank Munger. New York: Russell Sage Foundation.

Citizens for Responsibility and Ethics in Washington (CREW). 2007. "CREW Releases 'Family Affair' Detailing how 96 House Members Used Campaign Funds to Financially Benefit Family Members."

June 17. Retrieved June 27, 2008 (www.citizens forethics.org).

____. 2008. "CREW Releases New Report Detailing Senators' Use of Positions to Benefit Family Members." February 24. Retrieved June 27, 2008 (www.citizensforethics.org).

Clawson, Dan, Alan Neustadtl and Mark Weller. 1998. *Dollars and Votes: How Business Campaign Contributions Subvert Democracy*. Philadelphia, PA: Temple University Press.

Clayton, Mark. 2004. "Mercury Rising." *Christian Science Monitor*, April 29, 13, 15.

Clayton, Susan L. 2005. "Jail Inmates Bake Their Way to Successful Reentry." *Corrections Today* 67 (April): 78-80, 130.

Cloke, Kenneth and Joan Goldsmith. 2002. *End of Management and the Rise of Organizational Democracy*. San Francisco: Jossey-Bass.

Cohan, Catherine I., and Stacey Kleinbaum. 2002. "Toward a Greater Understanding of the Cohabitation Effect: Premarital Cohabitation and Marital Communication." *Journal of Marriage and Family* 64 (February): 180-192.

Cohen, Adam. 2004. "A Community of Ex-Cons Shows How to Bring Prisoners Back to Society." *New York Times* (January 2): A18.

Cohen, Harry. 1981. *Connections: Understanding Social Relationships*. Ames: Iowa State University Press.

Cohen, Jere. 2002. *Protestantism and Capitalism: The Mechanisms of Influence*. New York: Aldine de Gruyter.

Cohen, Joel E. 2005. "Human Population Grows Up." *Scientific American* 293 (September): 48-56.

Cohn, D'Vera. 2000. "Census Complaints Hit Home." *Washington Post*, May 4, A9.

Cohn, Meredith. 2007. "Green." *Baltimore Sun*, March 11, 1A, 18A.

Colapinto, John. 1997. "The True Story of John/Joan." *Rolling Stone* (December 11): 54 -73, 92 -97.

____. 2001. *As Nature Made Him: The Boy Who Was Raised as a Girl*. New York: Harper Perennial.

____. 2004. "What Were the Real Reasons Behind David Reimer's Suicide?" *Slate,* June 3. Retrieved April 24, 2008 (www.slate.com).

Coley, Rebekah L. 2002. "What Mothers Teach, What Daughters Learn: Gender Mistrust and Self-Sufficiency Among Low-Income Women." Pp. 97-106 in *Just Living Together: Implications of Cohabitation on Families, Children, And Social Policy*, edited by Alan Booth and Ann C. Crouter. Mahwah, NJ: Lawrence Erlbaum Associates.

Coll, Steve. 2008. *The Bin Ladens: An Arabian Family in the American Century*. New York: Penguin.

College Board. 2006. "2006 College-Bound Seniors: Total Group Profile Report." Retrieved July 29, 2007 (www.collegeboard.com).

____. 2008a. "Asian Americans and Pacific Islanders: Facts, not Fiction: Setting the Record Straight." Retrieved July 12, 2008 (www.collegeboard.com).

____. 2008b. "The 4th Annual AP Report to the Nation." February. Retrieved July 15, 2008 (professionals.collegeboard.com).

Collins, Patricia Hill. 1990. *Black Feminist Thought: Knowledge, Consciousness, and the Politics of Empowerment*. New York: Routledge.

Collins, Randall. 2002. "Credential Inflation and the Future of Universities." Pp. 23-46 in *The Future of the City of Intellect: The Changing American University*, edited by Steven Brint. Stanford, CA: Stanford University Press.

Commission on No Child Left Behind, The. 2007. *Beyond NCLB: Fulfilling the Promise to Our Nation's Children*. Retrieved July 21, 2007 (www.aspeninstitute.org).

Computer Industry Almanac Inc. 2007. "Worldwide Internet Users Top 1.2 Billion in 2006." February 12. Retrieved March 10, 2008 (www.c-i-a.com).

Conger, Rand D., and Glen H. Elder, eds. 1994. *Families in Troubled Times: Adapting to Change in Rural America*. New York: Aldine.

Conley, Dalton. 2004. *The Pecking Order: Which Siblings Succeed And Why*. New York: Pantheon.

Connolly, Frank W. 2001. "My Students Don't Know What They're Missing." *Chronicle of Higher Education*, December 21, B5.

Consumer Electronics Association. 2007. "The Energy and Greenhouse Gas Emissions Impact of Telecommuting and e-Commerce." July. Retrieved October 17, 2007 (www.cea.org).

Conway, Kevin P., and Joan McCord. 2005. "Co-Offending and Patterns of Juvenile Crime." Washington, DC: National Institute of Justice. Retrieved October 12, 2006 (www.ojp.usdoj.gov/nij).

Coocy, Paula M., William R. Eakin, and Jay B. McDaniel, eds. 1991. *After Patriarchy: Feminist Transformations of the World Religions*. Maryknoll, NY: Orbis Books.

Cook, Thomas D. and Donald T. Campbell. 1979. *Quasi-Experimentation: Design and Analysis Issues for Field Settings*. Chicago, IL: Rand McNally.

Cooley, Charles Horton. 1909/1983. *Social Organization: A Study of the Larger Mind*. New Brunswick, NJ: Transaction Books.

Coombs, Walter P., and Ralph E. Shaffer. 2008. "Regulating Home Schoolers." *Los Angeles Times*, March 13. Retrieved March 14, 2008 (www.latimes.com).

Corbett, Christianne, Catherine Hill and Andresse St. Rose. 2008. *Where the Girls Are: The Facts about Gender Equity in Education*. Washington, DC: American Association of University Women.

Corporation for National and Community Service, Office of Research and Policy Development. 2007. *Volunteering in America: 2007 State Trends and Rankings in Civic Life*. Washington, DC.

Corso, Regina. 2008. "National Anthem Survey Results Summary." The National Anthem Project, June 6. Retrieved October 17, 2008 (www.thenationalanthemproject.org).

Cose, Ellis. 1999. "Deciphering the Code of the Street." *Newsweek*, August 30, 33.

Coser, Lewis. 1956. *The Functions of Social Conflict*. New York: Free Press.

Council on American-Islamic Relations. 2006. "American Public Opinion about Islam and Muslims." Retrieved August 29, 2007 (www.cair.com).

Cressey, Donald R. 1953. *Other People's Money: A Study in the Social Psychology of Embezzlement*. Glencoe, IL: Free Press.

Crilly, Rob. 2007. "In Sudan, a Blasphemous Teddy Bear." *Christian Science Monitor*, November 29, 7.

Crisis at the Core: Preparing All Students for College and Work. 2004. American College Testing. Retrieved January 5, 2005 (www.act.org).

Crockett, Roger O. 2006. "The Rising Stock of Black Directors." *Business Week*, February 27, 34.

Cross, Terry L. 1998. "Understanding Family Resiliency from a Relational World View." Pp. 143-157 in *Resiliency in Native American and immigrant families*, edited by Hamilton I. McCubbin, Elizabeth A. Thompson, Anne I. Thompson, and Julie E. Fromer. Thousand Oaks, CA: Sage.

Crossley, David. 2008. "The Worst Natural Disasters Ever." LiveScience, May 6. Retrieved August 4, 2008 (www.livescience.com).

Crothers, Charles. 1979. "On the Myth of Rural Tranquility: Comment on Webb and Collette." *American Journal of Sociology*. 84 (May): 1441-1445.

"Crowds Call for Teacher's Death." 2007. *Baltimore Sun*, December 1, 3A.

Curlin, Farr A., Sarah A. Sellergren, John D. Lantos, and Marshall H. Chin. 2007. "Physicians' Observations and Interpretations of the Influence of Religion and Spirituality on Health." *Archives of Internal Medicine* 167 (April 9): 649-654.

Currie, E. 1985. *Confronting Crime: An American Challenge*. New York: Pantheon.

Currie, Janet, and Erdal Tekin. 2006. "Does Child Abuse Cause Crime?" Cambridge, MA: National Bureau of Economic Research, Working Paper 12171. Retrieved June 12, 2006 (http://papers.nber.org).

Curtiss, Susan. 1977. *Genie*. New York: Academic Press.

D'Innocenzio, Anne. 2007. "Working Poor Stretch Paychecks to Breaking Point." *Christian Science Monitor*, October 24, 16.

Dahl, Robert A. 1961. *Who Governs? Democracy and Power in an American City*. New Haven. CT: Yale University Press.

Dalton, Madeline A., et al. 2005. "Use of Cigarettes and Alcohol by Preschoolers While Role-Playing as Adults." *Archives of Pediatrics & Adolescent Medicine* 159 (September): 854-859.

Daly, Mary. 1973. *Beyond God the Father*. Boston, MA: Beacon Press.

Dandaneau, Steven P. 2001. *Taking It Big: Developing Sociological Consciousness in Postmodern Times*. Thousand Oaks, CA: Pine Forge Press.

Dang, Dan Thanh, and Abigail Tucker. 2005. "E-mailers Beware: 'Private' Messages May be Anything But." *Baltimore Sun*, March 15, 1A, 4A.

Dash, Eric. 2006. "Stolen Lives: Protectors, Too, Gather Profits from ID Theft." *New York Times*, December 12, A1, C10.

Davies, James B., Susanna Sandstrom, Anthony Shorrocks, and Edward N. Wolff. 2006. "The World Distribution of Household Wealth." World Institute for Development Economic Research, United Nations University, December 5. Retrieved March 23, 2007 (www.wider.unu.edu).

Davies, James C. 1962. "Toward a Theory of Revolution." *American Sociology Review* 27 (February): 5-19.

———. 1979. "The J-Curve of Rising and Declining Satisfaction as a Cause of Revolution and Rebellion." Pp. 413-36 in *Violence in America: Historical and Comparative Perspectives*, edited by Hugh D. Graham and Ted R. Gurr. Beverly Hills, CA: Sage.

Davis, F. J. 1991. *Who is Black?* University Park: Pennsylvania State University Press.

Davis, James A., Tom W. Smith and Peter V. Marsden. 2005. *General Social Surveys, 1972-2004: Cumulative Codebook*. Chicago, IL: National Opinion Research Center.

Davis, Kingsley, and Wilbert E. Moore. 1945. "Some Principles of Stratification." *American Sociological Review* 10 (April): 242-49.

Day, Jennifer Cheeseman, and Eric C. Newburger. 2002. "The Big Payoff: Educational Attainment and Synthetic Estimates of Work-Life Earnings." U.S. Census Bureau, Current Population Reports, P23-210. Retrieved December 5, 2004 (www.census.gov).

Day, Jennifer Cheeseman, and Kelly Holder. 2004. "Voting and Registration in the Election of November 2002." U.S. Census Bureau, Current Population Reports, P20-552. Retrieved August 2, 2004 (www.census.gov).

de la Cruz, G. Patricia, and Angela Brittingham. 2003. "The Arab Population: 2000." U.S. Census Bureau. Retrieved May 3, 2004 (www.census.gov).

de las Casas, Bartolome. 1992. *The Devastation of the Indies: A Brief Account*. Baltimore, Maryland: Johns Hopkins University Press.

De Long, J. Bradford, and Andrei Shleifer. 1992. "Princes and Merchants: European City Growth Before the Industrial Revolution." National Bureau of Economic Research, December. Retrieved January 10, 2008 (www.nper.org).

De Vries, Raymond, Melissa A. Anderson, and Brian C. Martinson. 2006. "Normal Misbehavior: Scientists Talk about the Ethics of Research." *Journal of Empirical Research on Human Research Ethics* 1 (March): 43-50.

Death Penalty Information Center. 2008. "Deterrence: States without the Death Penalty Have Had Consistently Lower Murder Rates." Retrieved March 26, 2008 (www.deathpenaltyinfo.org).

———. 2008. "Facts about the Death Penalty." May. Retrieved May 10, 2008 (www.deathpenaltyinfo.org).

DeBiaggi, Sylvia D. 2002. *Changing Gender Roles: Brazilian Immigrant Families in the U.S.* New York: LFB Scholarly Publishing LLC.

"Decapitating Appalachia." 2004. *New York Times*, January 13, 24.

"Dents in the Dream." 2008. *Time*, July 28, 40-41.

Dee, Thomas S. 2006. "The Why Chromosome." *Education Next* No 4. (Fall): 69-75.

Dee, Thomas S., and Brian A. Jacob. 2006. "Do High School Exit Exams Influence Educational Attainment or Labor Market Performance?" National Bureau of Economic Research, April. Retrieved, www.nber.org (accessed July 25, 2007).

Deegan, Mary Jo. 1986. *Jane Addams and the Men of the Chicago School, 1892-1918*. New Brunswick, NJ: Transaction Books.

DeFleur, Melvin L. and Margaret H. DeFleur. 2003. *Learning to Hate Americans: How U.S. Media Shape Negative Attitudes among Teenagers in Twelve Countries*. Spokane, WA: Marquette Books.

Degler, Carl. 1981. *At Odds: Women and the Family in America from the Revolution to the Present*. New York: Oxford University Press.

della Porta, Donatella and Mario Diani. 1999. *Social Movements: An Introduction*. Malden, MA: Blackwell Publishers.

DeMaris, Alfred. 2001. "The Influence of Intimate Violence on Transitions Out Of Cohabitation." *Journal of Marriage and Family* 63 (February): 235-246.

Demick, Barbara. 2002. "Where Dogs Are Diners or Dinners." *Baltimore Sun*, January 26, 2A.

Demos, John. 1986. *Past, Present, and Personal: The Family and the Life Course in American History*. New York: Oxford University Press.

DeNavas-Walt, Carmen and Robert Cleveland. 2002. *Money Income in the United States: 2001*. U.S. Census Bureau, Current Population Reports, P60-221. Washington, DC: U.S. Government Printing Office.

DeNavas-Walt, Carmen, Bernadette D. Proctor, and Cheryl Hill Lee. 2006. *Income, Poverty, and Health Insurance Coverage in the United States: 2005*. U.S. Census Bureau, Current Population Reports, P60-23. Washington, DC: U.S. Government Printing Office.

DeNavas-Walt, Carmen, Bernadette D. Proctor, and Jessica Smith. 2007. *Income, Poverty, and Health Insurance Coverage in the United States: 2006*. Current Population Reports, P60-23. Washington, DC: U.S. Government Printing Office, 2007.

———. 2008. *Income, Poverty, and Health Insurance Coverage in the United States: 2007*. U.S. Census Bureau, Current Population Reports, P60-235. Washington, DC: U.S. Government Printing Office.

Denton-Borhaug, Kelly. 2004. "The Complex and Rich Landscape of Student Spirituality: Findings from the Goucher College Spirituality Survey." *Religion & Education* 31 (Fall): 41-61.

Dey, Judy Goldberg, and Catherine Hill. 2007. "Behind the Pay Gap." American Association of University Women Educational Foundation. Retrieved August 1, 2007 (www.aauw.org).

Diamond, Milton, and H. Keith Sigmundson. 1997. "Sex Reassignment at Birth: Long-Term Review and Clinical Implications." *Archives of Pediatrics & Adolescent Medicine* 15 (March): 298-304.

Dickens, William T., and James R. Flynn. 2006. "Black Americans Reduce the Racial IQ Gap: Evidence from Standardization Samples." *Psychological Science* 17 (10): 913-920.

Dickert-Conlin, Stacy and Ross Rubenstein. 2007. *Economic Inequality and Higher Education: Access Persistence and Success*. New York: Russell Sage Foundation.

Dietrich, Shannon. 2007. "Global Sex Trafficking Updates." Project to End Human Trafficking, May. Retrieved November 30, 2008 (www.endhumantrafficking.org).

Dilanian, Ken. 2008. "Lobbyists Find More Ways to Bond with Lawmakers." *USA Today*, January 31, 1A.

Dilworth-Anderson, Peggye, Linda M. Burton, and William L. Turner. 1993. "The Importance of Values in the Study of Culturally Diverse Families." *Family Relations* 42 (July): 238–42.

Dines, Gail. 2008. "Yale Sex Week Glosses over Porn's Dark Side." *Hartford Courant*, February 11. Retrieved June 1, 2008 (www.courant.com).

Dinkes, Rachel, Emily Forrest Cataldi, Wendy Lin-Kelly, and Thomas D. Snyder. 2007. *Indicators of School Crime and Safety: 2007* (NCES 2008-021/NCJ 219553). National Center for Education Statistics, Institute of Education Sciences, U.S. Department of Education, and Bureau of Justice Statistics, Office of Justice Programs, U.S. Department of Justice, Washington, DC.

DiversityInc. 2008. "Fortune 500 Black, Latino, Asian CEOs." July 22. Retrieved September 11, 2008 (www.diversity.com).

Dixon, Marc, and Andrew Martin. 2007. "Can the Labor Movement Succeed Without the Strike?" *Contexts* 6 (Spring): 36-39.

Do, Hien Duc. 1999. *The Vietnamese Americans*. Westport, CT: Greenwood Press.

Dobrzynski, Judith. 2008. "The Highest Paid Women in Corporate America." *ForbesLife Executive Woman*, Fall, 76, 111.

Doeringer, Peter B. and Michael J. Piore. 1971. *Internal Labor Markets and Manpower Analysis*. Lexington, MA: Heath-Lexington Books.

"Domestic Spying Divides U. S." 2006. *Baltimore Sun*, January 8, 2A.

Domhoff, G. William. 1986. *Who Rules America Now? A View for the '80s*. New York: Simon & Schuster.

———. 2006. *Who Rules America? Power, Politics, & Social Change*, 5th ed. Boston, MA: McGraw Hill.

Domosh, Mona and Joni Seager. 2001. *Feminist Geographers Make Sense of the World*. New York: The Guilford Press.

Dorius, Cassandra J., Stephen J. Bahr, John P. Hoffman, and Elizabeth L. Harmon. 2004. "Parenting Practices as Moderators of the Relationship between Peers and Adolescent Marijuana Use." *Journal of Marriage and Family* 66 (February): 163-178.

Douglas, Susan J. and Meredith W. Michaels. 2004. *The Mommy Myth: The Idealization of Motherhood and How It Has Undermined Women.* New York: Free Press.

Downs, Barbara. 2003. "Fertility of American Women: June 2002." U.S. Census Bureau, Current Population Reports, P20-548. Retrieved November 9, 2004 (www.census.gov).

Draffan, George. 2003. "Profile of Boeing." Retrieved September 5, 2004 (www.endgame.org).

Drentea, Patricia and Juan Xi. 2003. "Job Openings in Sociology: What is the Demand?" *Footnotes* (September/October): 9.

Du Bois, W. E. B. 1986. *The Souls of Black Folk.* New York: Library of America.

Dunifon, Rachel and Lori Kowaleski-Jones. 2007. "The Influence of Grandparents in Single-Mother Families." *Journal of Marriage and Family* 69 (May): 465-481.

Dunlap, Riley E. 2007. "The State of Environmentalism in the U.S." Gallup Poll, April 19. Retrieved March 27, 2007 (www.gallup.com).

Durkheim, Emile. 1893/1964. *The Division of Labor in Society.* New York: Free Press.

____. 1897/1951. *Suicide: A Study in Sociology.* (John A. Spaulding and George Simpson, trans,; George Simpson, ed., 1951). New York: Free Press.

____. 1898/1956. *Education and Sociology.* Trans. Sherwood D. Fox. Glencoe, IL: Free Press.

____. 1961. *The Elementary Forms of the Religious Life.* New York: Collier Books.

Durose, Matthew R., and Patrick A. Langan. 2004. "Felony Sentences in State Courts, 2002." Washington, DC: U.S. Department of Justice, Bureau of Justice Statistics.

Dush, Kamp, Catherine Cohan, and Paul Amato. 2003. "The Relationship between Cohabitation and Marital Quality and Stability: Change Across Cohorts?" *Journal of Marriage and Family* (August): 539-549.

Duster, Troy. 2005. "Race and Reification in Science." *Science* 307, February 18, 1050-1051.

Dye, Jane Lawler. 2008. *Fertility of American Women: 2006.* Current Population Reports, P20-558. Washington, DC: U.S. Census Bureau.

Dye, Thomas R. 2002. *Who's Running America? The Bush Restoration,* 7th ed. Upper Saddle River, NJ: Prentice Hall.

Dye, Thomas R. and Harmon Ziegler. 2003. *The Irony of Democracy: An Uncommon Introduction to American Politics,* 12th ed. Belmont, CA: Wadsworth.

Ebaugh, Helen Rose. 2003. "Religion and the New Immigrants." Pp. 225-239 in *Handbook of the Sociology of Religion,* edited by Michele Dillon. New York: Cambridge University Press,

Eckstein, Rick, Rebecca Schoenike, and Kevin Delaney. 1995. "The Voice of Sociology: Obstacles to Teaching and Learning the Sociological Imagination." *Teaching Sociology* 23 (October): 353-363.

Edin, Kathryn and Laura Lein. 1997. *Making Ends Meet: How Single Mothers Survive Welfare And Low-Wage Work.* New York: Russell Sage Foundation.

Education Commission of the States. 2008. "Youth Voting Pattern." Retrieved December 9, 2008 (www.ecs.org).

Eduventures. 2006. *Key Drivers of Educational Value: The Emergence of Educational ROI.* December 12. Retrieved July 19, 2007 (www.eduventures.com).

Egan, Timothy. 2006. "The Rise of Shrinking-Vacation Syndrome." *New York Times,* August 20, A18.

Ehrenreich, Barbara. 2001. *Nickel and Dimed: On (Not) Getting by in America.* New York: Metropolitan Books.

Ehrlich, Paul. 1971. *The Population Bomb,* 2nd ed. San Francisco: Freeman.

Ehrlich, Paul R. and Anne H. Ehrlich. 2008. *The Dominant Animal: Human Evolution and the Environment.* Washington, DC: Island Press.

Eisenberg, Abne M. and Ralph R. Smith, Jr. 1971. *Nonverbal Communication.* Indianapolis: The Bobbs-Merrill Company, Inc.

Eisenberg, Nancy, et al., 2005. "Relations among Positive Parenting, Children's Effortful Control, And Externalizing Problems: A Three-Wave Longitudinal Study." *Child Development* 76 (September/October): 1055-1071.

Eisenbrey, Ross, Mark Levinson, and Lawrence Mishel.

2007. "The Agenda for Shared Prosperity." *EPI Journal* 17 (Winter): 1-8.

Ekman, Paul. 1985. *Telling Lies: Clues to Deceit in the Marketplace, Politics, and Marriage.* New York: W.W. Norton & Company.

Ekman, Paul and Wallace V. Friesen. 1984. *Unmasking the Face: A Guide to Recognizing Emotions from Facial Clues.* Palo Alto, CA: Consulting Psychologists Press, Inc.

Eldridge, R. I., and P. D. Sutton. 2007. "Births, Marriages, Divorces, and Deaths: Provisional Data for November 2006." *National Vital Statistics Reports* 55 (18), National Center for Health Statistics. Retrieved June 3, 2007 (www.cdc.gov/nchs).

Elgin, Ben. 2008. "Green—Up to a Point." *Business Week,* March 3, 25-26.

Ellis, Bill. 2005. "Legend/AntiLegend: Humor as an Integral part of the Contemporary Legend Process." Pp. 123-140 in *Rumor Mills: The Social Impact of Rumor and Legend,* edited by Gary Alan Fine, Véronique Campion-Vincent, and Chip Heath. New Brunswick, NJ: Transaction Publishers.

Engemann, Kristie M. and Michael T. Owyang. 2005. "So Much for That Merit Raise: The Link between Wages and Appearance." *Regional Economist* (April): 10-11.

England, Paula. 2006. "Toward Gender Equality: Progress and Bottlenecks." Pp. 245-264 in *The Declining Significance of Gender?* edited by Francine D. Blau, Mary C. Brinton, and David B. Grutsky. New York: Russell Sage Foundation.

England, Paula and Nancy Folbre. 2005. "Gender and Economic Sociology." Pp. 627-649 in *The Handbook of Economic Sociology,* 2nd edition, edited by Neil J. Smelser and Richard Swedberg. New Jersey: Princeton University Press and New York: Russell Sage Foundation.

Enloe, Cynthia. 2004. *The Curious Feminist: Searching for Women in a New Age of Empire.* Berkeley: University of California Press.

"Environment, The." 2008. Gallup Poll. Retrieved February 15, 2008 (www.gallup.com).

Environmental Integrity Project. 2007. "Paying Less to Pollute: Environmental Enforcement Under the Bush Administration." Retrieved February 12, 2008 (www.enrironmentalintegrity.org).

EPE Research Center. 2007. "More Than 1.2 Million Students Will Not Graduate in 2007; Detailed Graduation Data Available for Every U.S. District and State." *Education Week,* June 12. Retrieved July 23, 2007 (www.edweek.org).

Erikson, Kai T. 1966. *Wayward Puritans: A Study in the Sociology of Deviance.* New York: John Wiley & Sons.

Esposito, John L. and Dalia Mogahed. 2007. *Who Speaks for Islam? What a Billion Muslims Really Think.* New York: Gallup Press.

Essed, Philomena, and David Theo Goldberg, eds. 2002. *Race Critical Theories: Text and Context.* Malden, MA: Blackwell Publishers.

Esty, Daniel C., M. A. Levy, C. H. Kim, A. de Sherbinin, T. Srebojnak, and V. Mara. 2008. *2008 Environmental Performance Index.* New Haven: Yale Center for Environmental Law and Policy.

European Monitoring Centre on Racism and Xenophobia. 2006. "Muslims in the European Union: Discrimination and Islamophobia." Retrieved September 1, 2007 (www.eumc.eu).

Evans, Harold, Gail Buckland, and David Lefer. 2006. *They Made America: From the Steam Engine to the Search Engine: Two Centuries of Innovators.* New York: Little, Brown.

Ewen, Lynda Ann. 1998. *Social Stratification and Power in America: A View from Below.* Six Hills, NY: General Hall, Inc.

Ewers, Justin. 2004. "Drowning in Applications." *U.S. News & World Report* (December 20): 64-65.

____. 2007. "Female Entrepreneurs Describe Growing Pains." *U.S. News & World Report,* November 8. Retrieved September 14, 2008 (www.usnews.com).

"Ex-Merrill Lynch CEO to Walk with $161.5M." 2007. Associated Press, October 30. Retrieved July 3, 2008 (www.msnbc.msn.com).

Expedia.com. 2008. "2008 International Vacation Deprivation Survey Results." Retrieved July 4, 2008 (www.expedia.com).

Fagan, Brian. 2008. *The Great Warming: Climate Change and the Rise and Fall of Civilizations.* New York: Bloomsbury Press.

FairTest. 2007. "Test Scores Do Not Equal Merit." August. Retrieved July 14, 2008 (www.fairtest.org).

"Faith in the System." 2004. *Mother Jones,* September/October, 26-27.

Falah, Ghazi-Wald, and Caroline Nagel, eds. 2005. *Geographies of Muslim Women: Gender, Religion, and Space.* New York: The Guilford Press.

Fallows, Deborah. 2005. "How Women and Men Use the Internet." PEW Internet & American Life Project. Retrieved February 12, 2006 (www.pewinternet.org).

Family Matters: Substance Abuse and the American Family. 2005. The National Center on Addiction and Substance Abuse at Columbia University, March. Retrieved May 3, 2005 (www.casacolumbia.org).

Farley, John E., and Gregory D. Squires. 2005. "Fences and Neighbors: Segregation in 21st-century America." *Contexts* 4 (Winter): 33-39.

Farley, Melissa. 2001. "Prostitution: The Business of Sexual Exploitation." Pp. 879-891 in *Encyclopedia of Women and Gender,* Vol. 2, edited by Judith Worrell. New York: Academic Press.

Farr, Kathryn. 2005. *Sex Trafficking: The Global Market in Women and Children.* New York: Worth.

Feagin, Joe R. 2001. "Social Justice and Sociology: Agendas for the Twenty-First Century." *American Sociological Review* 66 (February): 1-20.

Feagin, Joe R. and Clairece Booher Feagin. 2008. *Racial and Ethnic Relations,* 8th ed. Upper Saddle River, NJ: Prentice Hall.

Feagin, Joe R. and Melvin P. Sikes. 1994. *Living with Racism: The Black Middle-Class Experience.* Boston, MA: Beacon Press.

Feagin, Joe R. and Robert Parker. 1990. *Building American Cities: The Urban Real Estate Game,* 2nd ed. Englewood Cliffs, NJ: Prentice Hall.

Federal Bureau of Investigation. 2005. "Financial Crimes Report to the Public." May. Retrieved October 5, 2006 (www.fbi.gov).

____. 2007. "Hate Crime Statistics, 2006." U.S. Department of Justice, November 19. Retrieved June 4, 2008 (www.fbi.gov).

____. 2007. "Persons Arrested." *Uniform Crime Reports.* Retrieved May 8, 2008 (www.fbi.gov).

Federal Interagency Forum on Aging-Related Statistics. 2006. *Older Americans Update 2006: Key Indicators of Well-Being.* Washington, DC: Government Printing Office.

Federal Trade Commission. 2008. "What is Identity Theft?" Retrieved March 11, 2008 (www.ftc.gov).

Feldmann, Linda. 2005. "Americans Split on Feds Listening In." *Christian Science Monitor,* December 29, 1, 11.

____. 2008. "How Voters May React to Clintons' $109 Million Income." *Christian Science Monitor,* April 7, 4.

____. 2008. "Women Make Modest Gains in Election 2008." *Christian Science Monitor,* November 17, 2.

Ferree, Myra M. 2005. "It's Time to Mainstream Research on Gender." *The Chronicle Review,* August 12, B10.

Fields, Jason. 2004. "America's Families and Living Arrangements: 2003." U.S. Census Bureau, Current Population Reports, P20-553. Retrieved November 30, 2004 (www.census.gov).

Fields, Jason, and Lynne. M. Casper. 2001. "America's Families and Living Arrangements: 2000." U.S. Census Bureau, Current Population Reports, P20-537. Retrieved February 12, 2003 (www.census.gov).

File, Thom. 2008. "Voting and Registration in the Election of November 2006." U.S. Census Bureau, Current Population Reports, P20-557. Retrieved October 13, 2008 (www.census.gov).

Finke, Roger and Rodney Stark. 1992. *The Churching of America, 1776-1990: Winners and Losers in Our Religious Economy.* New Brunswick, NJ: Rutgers University Press.

Finn, Jeremy D. 2006. *The Adult Lives of At-Risk Students: The Roles of Attainment and Engagement in High School* (NCES 2006-328). Washington, DC: U.S. Department of Education.

Fischer, Karin. 2008. "Top Colleges Admit Fewer Low-Income Students." *Chronicle of Higher Education,* May 2, A1, A19-A20.

Flanagan, William G. 1990. *Urban Sociology: Images and Structure.* Boston, MA: Allyn & Bacon.

Flavin, Jeanne. 2001. "Feminism for the Mainstream Criminologist: An Invitation." *Journal of Criminal Justice* 29 (July/August): 271-285.

Fletcher, Douglass Scott and Ian M. Taplin. 2002. *Understanding Organizational Evolution: Its Impact on Management and Performance.* Westport, CT: Quorum Books.

Fletcher, Laurel E., Phuong Pham, Eric Stover, and Patrick Vinck. 2006. "Rebuilding After Katrina: A Population-Based Study of Labor and Human Right in New Orleans." Retrieved June 25, 2006 (www.hrcberkeley.org).

Florida Office of Insurance Regulation. 2007. "The Use of Occupation and Education as Underwriting/Rating Factors for Private Passenger Automobile Insurance." March. Retrieved April 15, 2007 (www.floir.com).

Fogg, Piper. 2005. "Don't Stand So Close to Me." *Chronicle of Higher Education*, April 29, A10-A12.

Fong, Timothy P. 2002. *The Contemporary Asian American Experience: Beyond the Model Minority*, 2nd ed. Upper Saddle River, NJ: Prentice Hall.

Ford, Peter. 2007. "Pollution Puts China Lake Off Limits." *Christian Science Monitor*, June 4, 7.

"Formaldehyde." 2004. Environmental Defense. Retrieved July 10, 2005 (www.scorecard.org).

Forsberg, Hannele. 2005. "Finland's Families." Pp. 262-282 in *Handbook of World Families*, edited by Bert N. Adams and Jan Trost. Thousand Oaks, CA: Sage.

Foster, Andrea L. 2006. "Student Who Sued Operator of Term-Paper Sites Settles Her Case Out of Court." *Chronicle of Higher Education* 52, January 20, A41.

Foust-Cummings, Heather, Laura Sabattini, and Nancy Carter. 2008. "Women in Technology: Maximizing Talent, Minimizing Barriers." *Catalyst*. Retrieved April 28, 2008 (www.catalyst.org).

Fowler, Geoffrey A. 2006. "An Arrest in China Spotlights Limits to Artistic Freedom." *Wall Street Journal*, July 3, A1, A8.

Fox, James Alan, and Marianne W. Zawitz. 2004. "Homicide Trends in the United States: 2002 Update." Washington, DC: U.S. Department of Justice, Bureau of Justice Statistics.

———. 2007. "Homicide Trends in the United States." U.S. Department of Justice, Bureau of Justice Statistics. Retrieved July 10, 2008 (www.ojp.usdoj.gov/bjs).

Fox, Susannah. 2006. "Are 'Wired Seniors' Sitting Ducks?" April, Pew Internet & American Life Project, April. Retrieved September 20, 2007 (www.pewinternet.org).

Fox, Susannah, and Gretchen Livingston. 2007. "Latinos Online." Pew Hispanic Center, March. Retrieved September 20, 2007 (www.pewinternet.org).

Frank, T. A. 2006. "A Brief History of Wal-Mart." CorpWatch. Retrieved February 19, 2007 (www.corpwatch.org).

Franklin, Rachel S. 2003. "Migration of the Young, Single, and College Educated: 1995 to 2000." U.S. Census Bureau, Census 2000 Special Reports CENSR-12. Washington, DC: Government Printing Office.

Freedom House. 2008a. "Freedom in the World 1008: Selected Data from Freedom House's Annual Global Survey of Political Rights and Civil Liberties." Retrieved June 22, 2008 (www.freedomhouse.org).

———. 2008b. "The Worst of the Worst: The World's Most Repressive Societies 2008." Retrieved June 22, 2008 (www.freedomhouse.org).

Freitas, Donna. 2008. *Sex & the Soul: Juggling Sexuality, Spirituality, Romance, and Religion on America's College Campuses*. New York: Oxford University Press.

French, Howard W. 2007. "China Scrambles for Stability as Its Workers Age." *New York Times*, March 22, A1, A8.

Frey, William H. 2006. "Diversity Spreads Out: Metropolitan Shifts in Hispanic, Asian, and Black Populations Since 2000." Brookings Institution. Retrieved February 2, 2008 (www.brookings.edu).

———. 2008. "Race, Immigration, and America's Changing Electorate." Brookings Institution, February 18. Retrieved June 23, 2008 (www.brookings.edu).

Friedan, Betty. 1963. *The Feminine Mystique*. New York: Norton.

Friedrichs, David O. 2004. *Trusted Criminals: White Collar Crime in Contemporary Society*, 2nd edition. Belmont, CA: Wadsworth.

Frosch, Dan. 2007. "18 Air Force Cadets Exit Over Cheating." *New York Times*, May 2, 18.

Fry, Richard. 2005. "The Higher Dropout Rate of Foreign-born Teens: The Role of Schooling Abroad." Pew Hispanic Center, November 1. Retrieved July 20, 2007 (www.pewhispanic.org).

Fukuyama, Francis, Seymour M. Lipset, Orlando Patterson, Marc Plattner, and Fareed Zakaria. 2002. "Democracy's Century: A Survey of Global Political Change in the 20th Century." Freedom House. Retrieved August 11, 2004 (www.freedomhouse.org).

Gamson, William. 1990. *The Strategy of Social Protest*, 2nd ed. Belmont, CA: Wadsworth.

Gans, Herbert J. 1962. "Urbanism and Suburbanism as Ways of Life: A Reevaluation of Definitions." Pp. 625-48 in *Human Behavior and Social Processes: An Interactionist Approach*, edited by Arnold M. Rose. Boston, MA: Houghton Mifflin.

———. 1971. "The Uses Of Poverty: The Poor Pay All." *Social Policy* (July/August): 78–81.

———. 2005. "Race as Class." *Contexts* 4 (Fall): 17-21.

———. 2005. "Wishes for the Discipline's Future." *The Chronicle Review*, August 12, B9.

Garcia, Alma M. 2002. *The Mexican Americans*. Westport, CT: Greenwood Press.

Gardner, Marilyn. 2007. "When a Layoff is the Price of Experience." *Christian Science Monitor*, April 16, 13, 16.

———. 2008. "Happiness is a Warm 'Thank You'." *Christian Science Monitor*, January 28, 13, 16.

Gardyn, Rebecca. 2001. "A League of Their Own." *American Demographics* 23 (March): 12-13.

Garfinkel, Harold. 1967. *Studies in Ethnomethodology*. Englewood Cliffs, NJ: Prentice-Hall, Inc.

Garfinkel, Simson L., and Abhi Shelat. 2003. "Remembrance of Data Passed: A Study of Disk Sanitation Processes." *IEEE Computer Society* 1 (January/February): 17-27.

Garreau, Joel. 2005. "Inventing Our Evolution." *Washington Post*, May 16, A1.

Garrison, Michelle M., and Dimitri A. Christakis. 2005. "A Teacher in the Living Room? Educational Media for Babies, Toddlers and Preschoolers." Henry J. Kaiser Family Foundation, December. Retrieved April 12, 2008 (www.kff.org).

Gavrilos, Dina. 2006. "U. S. News Magazine Coverage of Latinos: 2006 Report." National Association of Hispanic Journalists, June. Retrieved April 2, 2007 (www.nahj.org).

Gaylin, Willard. 1992. *The Male Ego*. New York: Viking.

Geertz, Clifford. 1966. "Religion as a Cultural System." Pp. 1-46 in *Anthropological Approaches to the Study of Religion*, edited by Michael Banton. London: Tavistock.

Gelbard, Alene, Carl Haub, and Mary M. Kent. 1999. "World Population Beyond Six Billion." *Population Bulletin* 54 (March): 1-44.

Gelles, Richard J. 1997. *Intimate Violence in Families*, 3rd ed. Thousand Oaks, CA: Sage.

"Generational Look at the Public: Politics and Policy, A." 2002. *Washington Post*/Kaiser Family Foundation/Harvard University, October. Retrieved August 14, 2004 (www.kff.org).

Gerth, H.H. and C. Wright Mills, eds. 1946. *Max Weber: Essays in Sociology*. Oxford University Press.

Ghosh, Bobby. 2007. "Why They Hate Each Other." *Time*, March 5, 29-40.

Gibbs, Nancy, and Michael Duffy. 2007. "Leveling the Praying Field." *Time*, July 23, 28-34.

Gibson, Gail. 2004. "Women Workers in U.S. Have Bias in Common." *Baltimore Sun* July 16, 1A, 11A.

Gibson-Davis, C. M., K. Edin, and S. McLanahan. 2005. "High Hopes but Even Higher Expectations: The Retreat from Marriage among Low-income Couples." *Journal of Marriage and Family* 67 (December): 1301-1312.

Gilbert, Dennis and Joseph A. Kahl. 1993. *The American Class Structure: A New Synthesis*, 4th ed. Homewood, IL: Dorsey Press.

Gilbert, Dennis. 2008. *The American Class Structure in an Age of Growing Inequality*, 7th ed. Belmont, CA: Wadsworth Press.

Giugni, Marco, Doug McAdam, and Charles Tilley, eds. 1999. *How Social Movements Matter*. Minneapolis: University of Minnesota Press.

Givhan, Robin. 2005. "Dick Cheney, Dressing Down." *Washington Post*, January 28, C1.

Glassman, Ronald M. 2000. *Caring Capitalism: A New Middle-Class Base for the Welfare State*. New York: St. Martin's Press.

Glater, Jonathan D. 2008. "Judge Says Wikileaks Site Can Have Its Web Address Back." *New York Times*, February 29. Retrieved March 2, 2008 (www.nytimes.com).

Glenn, Norval D. 2001. "Social Science Findings and the 'Family Wars'." *Society* 38 (May/June): 13-19.

———. 2005. "With This Ring . . . : A National Survey on Marriage in America." National Fatherhood Initiative. Retrieved April 2, 2006 (www.fatherhood.org).

Glod, Maria, and Bill Turque. 2008. "Report Finds Little Gain from Vouchers." *Washington Post*, June 17, A1.

Goffman, Erving. 1959. *The Presentation of Self in Everyday Life*. New York: Doubleday Anchor Books.

———. 1961. *Asylums: Essays on the Social Situation of Mental Patients and Other Inmates*. Garden City, NY: Anchor Books.

———. 1963. *Stigma: Notes on the Management of Spoiled Identity*. Englewood Cliffs, NJ: Prentice-Hall.

———. 1967. *Interaction Ritual: Essays on Face-to-Face Behavior*. New York: Anchor Books.

———. 1969. *Strategic Interaction*. Philadelphia, PA: University of Pennsylvania Press.

Golden, Daniel. 2006. *The Price of Admission: How America's Ruling Class Buys Its Way into Elite Colleges—and Who Gets Left Outside the Gates*. New York: Crown.

Goldin, Claudia, Lawrence F. Katz, and Ilyana Kuziemko. 2006. "The Homecoming of American College Women: The Reversal of the College Gender Gap." National Bureau of Economic Research. Retrieved July 25, 2007 (www.nber.org/papgers/w12139).

Goldman, Lea, Monte Burke, and Kiri Blakeley. 2007. "Celebrity 100: The Biggest Big Shots." *Forbes*, July 7, 74-84.

Golombok, Susan, and Fiona Tasker. 1996. "Do Parents Influence The Sexual Orientation Of Their Children? Findings from a Longitudinal Study of Lesbian Families." *Developmental Psychology* 32 (1): 3–11.

Goodkin, Kerala. 2005. "Smelling the Roses." *Glimpse* (Winter): 6-13.

Goodman, David. 2004. "Class Dismissed." *Mother Jones* 29, May/June, 41-47.

Gopnik, Alison, Andrew N. Meltzoff and Patricia K. Kuhl. 2001. *The Scientist in the Crib: What Early Learning Tells Us about the Mind*. New York: Perennial.

Gottdiener, Mark and Ray Hutchison. 2000. *The New Urban Sociology*, 2nd ed. New York: McGraw-Hill.

Gottman, John M. 1994. *What Predicts Divorce? The Relationships between Marital Processes and Marital Outcome*. Hillsdale, NJ: Lawrence Erlbaum Associates.

Gouldner, Alvin W. 1962. "Anti-Minotaur: The Myth of a Value-Free Sociology." *Social Problems* 9 (Winter): 199-212.

Graff, E. J. 2007. "The Opt-Out Myth." *Columbia Journalism Review* 45 (March/April): 51-54.

Grall, Timothy S. 2005. "Support Providers: 2002." U.S. Census Bureau, Current Population Reports, P70-99. Retrieved July 9, 2006 (www.census.gov).

Grant, Gerald and Christine E. Murray. 1999. *Teaching in America: The Slow Revolution*. Cambridge, MA: Harvard University Press.

Grauerholz, Liz and Sharon Bouma-Holtrop. 2003. "Exploring Critical Sociological Thinking." *Teaching Sociology* 31 (October): 485-496.

Graves, Joseph. L. Jr. 2001. *The Emperor's New Clothes: Biological Theories of Race at the Millennium*. New Brunswick, NJ: Rutgers University Press.

Gray, Paul S., John B. Williamson, David R. Karp, and John R. Dalphin. 2007. *The Research Imagination: An Introduction to Qualitative and Quantitative Methods*. New York: Cambridge University Press.

Greeley, Andrew M. 1972. *The Denominational Society*. Glenview, IL: Scott, Foresman.

Greenberg, Milton. 2004. "How the GI Bill Changed Higher Education." *Chronicle of Higher Education*, June 18, B9-B11.

Greenhouse, Steven. 2004. "Altering of Worker Time Cards Spurs Growing Number of Suits." *New York Times*, April 4, 1.

Grodsky, Eric, John Robert Warren, and Demetra Kalogrides. 2008. "State High School Exit Examinations and NAEP Long-Term Trends in Reading and Mathematics, 1971-2004." *Education Policy* 22 (June). Retrieved July 17, 2008 (online.sagepub.com).

Gross, Rita M. 1996. *Feminism and Religion: An Introduction*. Boston, MA: Beacon Press.

Gruber, Jonathan. 2005. "Religious Market Structure, Religious Participation, and Outcomes: Is Religion Good for You?" National Bureau of Economic Research, May. Retrieved August 28, 2007 (www.nber.org).

Guo, Guang. 2005. "Twin Studies: What Can They Tell Us about Nature and Nurture?" *Contexts* 4 (Summer): 43-47.

Gurney, Joan M., and Kathleen T. Tierney. 1982. "Relative Deprivation and Social Movements: A Critical Look at Twenty Years of Theory and Research." *Sociological Quarterly* 23 (Winter): 33-47.

Gurwitt, Rob. 2005. "The Nose That Knows." *Mother Jones*, March/April, 24.

Guterman, Lila. 2005. "Lost Count." *Chronicle of Higher Education*, February 4, A10-A13.

Guterman, Stanley S. 1969. "In Defense of Wirth's 'Urbanism as a Way of Life'." *American Journal of Sociology* 74 (March): 492-499.

Guttmacher Institute. 2005. "Facts on Induced Abortion in the United States. Retrieved December 21, 2006 (www.guttmacher.org).

____. 2008. "Facts on Induced Abortion in the United States." July. Retrieved August 15, 2008 (www.guttmacher.org).

Hagan, Frank E. 2008. *Introduction to Criminology: Theories, Methods, and Criminal Behavior*, 6th edition. Thousand Oaks, CA: Sage.

Hagenbaugh, Barbara. 2006. "U.S. Manufacturers Getting Desperate for Skilled People." *USA Today*, December 5, 1.

Hakimzadeh, Shirin, and D'Vera Cohn. 2007. "English Usage among Hispanics in the United States." Pew Hispanic Center, November 29. Retrieved December 11, 2007 (www.pewhispanic.org).

Halfond, Jay A. 2004. "Grade Inflation is Not a Victimless Crime." *Christian Science Monitor*, May 3, 9.

Hall, Edward T. 1959. *The Silent Language*. New York: Doubleday & Company, Inc.

____. 1966. *The Hidden Dimension*. Garden City, NY: Doubleday & Company, Inc.

Halloran, Liz. 2008. "An Uncertain Legacy." *U.S. News & World Report*, September, 34-37.

Hamby, Sherry L. and David Finkelhor. 2001. "Choosing and Using Child Victimization Questionnaires." *OJJDP Juvenile Justice Bulletin*. Washington, DC: U.S. Department of Justice.

Hamermesh, Daniel S. and Amy W. Parker. 2003. "Beauty in the Classroom: Professors' Pulchritude and Putative Pedagogical Productivity." Working Paper 9853, July. Cambridge, MA: National Bureau of Economic Research.

Hamilton, Brady E., Joyce A Martin, and Stephanie J. Ventura. 2007. "Births: Preliminary Data for 2006." *National Vital Statistics Reports*, vol. 56, no 7. Hyattsville, MD: National Center for Health Statistics.

Hamilton, Brady E., Paul D. Sutton, and Stephanie J. Ventura. 2003. "Revised Birth and Fertility Rates for the 1990s and New Rates for Hispanic Populations, 2000 and 2001: United States." *National Vital Statistics Reports*, 51, no 12. Retrieved August 25, 2003 (www.cdc.gov).

Hamilton, Malcolm B. 1995. *The Sociology of Religion: Theoretical and Comparative Perspectives*. New York: Routledge.

____. 2001. *The Sociology of Religion*, 2nd ed. New York: Routledge.

Hamm, Steve. 2007. "The Trouble with India." *Business Week*, March 19, 49-58.

Hampson, Rick. 2006. "Fear 'As Bad As After 9/11'." *USA Today*, December 12, 1A.

Hamre, Bridget K., and Robert C. Pianta. 2001. "Early Teacher-Child Relationships and the Trajectory of Children's School Outcomes Through Eighth Grade." *Child Development* 72 (March/April): 625-38.

Handelsman, Jo et al. 2005. "Careers in Science." *Science* 309 (August 19): 1190-1191.

Haney, Craig, Curtis Banks, and Philip Zimbardo. 1973. "Interpersonal Dynamics in a Simulated Prison." *International Journal of Criminology and Psychology* 1: 69-97.

Hanushek, Eric A., and Ludger Woessman. 2005. "Does Educational Tracking Affect Performance and Inequality? Differences-in-Differences Evidence Across Countries." National Bureau of Economic Research, February. Retrieved July 25, 2007 (www.nber.org).

Harlow, Caroline Wolf. 2003. "Education and Correctional Populations." U.S. Department of Justice, Bureau of Justice Statistics. Retrieved January 16, 2003 (www.ojp.usdoj.gov/bjs).

Harlow, Harry E. and Margaret K. Harlow. 1962. "Social Deprivation in Monkeys." *Scientific American* 206 (November) 136-146.

Harman, Danna. 2007. "Qatar Reformed by a Modern Marriage." *Christian Science Monitor*, March 6, 20.

Harmon, Amy. 2008. "The DNA Age—Insurance Fears Lead Many to Shun DNA Tests." *New York Times*, February 24, 1.

Harrell, Erika. 2007. "Black Victims of Violent Crime." Washington, DC: U.S. Department of Justice, Bureau of Justice Statistics.

Harris, Chauncey D. 1997. "'The Nature of Cities' and Urban Geography in the Last Century." *Urban Geography* 18: 15-35.

Harris, Chauncey D., and Edward L. Ullman. 1945. "The Nature of Cities." *Annals* 242: 7-17.

Harris, Gardiner. 2008. "Cigarette Company Paid for Lung Cancer Study." *New York Times*, March 26, 6.

Harris, Scott R. 2006. *The Meanings of Marital Equality*. Albany: State University of New York Press.

Hart, Timothy C., and Callie Rennison. 2003. "Reporting Crime to the Police, 1992-2000." Washington, DC: U.S. Department of Justice, Bureau of Justice Statistics.

Hartmann, Heidi, Olga Sorokina, and Erica Williams. 2006. *The Best and Worst State Economies for Women*. Washington, DC: Institute for Women's Policy Research.

Harvey, John H., and Mark A. Fine. 2004. *Children of Divorce: Stories of Loss and Growth*. Mahwah, NJ: Lawrence Erlbaum.

Haskins, Rob. 2007a. "Education and Economic Mobility." Pp. 91-104 in *Getting Ahead or Losing Ground: Economic Mobility in America*, edited by Julia B. Isaacs, Isabel V. Sawhill, and Ron Haskins. Washington, DC: The Brookings Institution.

____. 2007b. "Immigration: Wages, Education, and Mobility." Pp. 81-90 in *Getting Ahead or Losing Ground: Economic Mobility in America*, edited by Julia B. Isaacs, Isabel V. Sawhill, and Ron Haskins. Washington, DC: The Brookings Institution.

Hassel, Bryan C. 2005. "Charter School Achievement: What We Know." Charter School Leadership Council, January. Retrieved February 10, 2005 (www.charterschoolleadershipcouncil.org).

Haub, Carl. 2004. "2004 World Population Data Sheet." Washington, DC: Population Reference Bureau, Washington, DC, wall chart.

____. 2007 "2007 World Population Data Sheet." Population Reference Bureau, Washington, DC, wall chart.

Haub, Carl, and Mary Mederios Kent. 2008. "2008 World Population Data Sheet." Population Reference Bureau, Washington, DC, wall chart.

Haub, Carl, and O. P. Sharma. 2006. "India's Population Reality: Reconciling Change and Tradition." *Population Bulletin* 61 (September): 1-19.

Hauser, Robert M. and David L. Featherman. 1977. *The Process of Stratification: Trends and Analysis*. New York: Academic Press.

Hausmann, Ricardo, Laura D. Tyson, and Saadia Zahidi. 2006. "The Global Gender Gap Report 2006." Geneva, Switzerland: World Economic Forum.

____. 2007. "The Global Gender Report 2007." World Economic Forum. Retrieved June 25, 2008 (www.weforum.org).

Hayani, Ibrahim. 1999. "Arabs in Canada: Assimilation or Integration?" Pp. 284-303 in *Arabs in America: Building a New Future*, edited by Michael W. Suleiman. Philadelphia: Temple University Press.

Hayden, Dolores. 2002. *Redesigning the American Dream: Gender, Housing, and Family Life*. New York: W.W. Norton & Company.

Hayward, Steven F. and Amy Kaleita. 2007. *Index of Leading Environmental Indicators: 2007, Twelfth Edition*. Pacific Research Institute. Retrieved February 15, 2008 (www.aconvenientfiction.com).

"Hazardous Air Pollution—A National Overview." 2005. Environmental Defense. Retrieved July 10, 2005 (www.scorecard.org).

He, Wan, Manisha Sengupta, Victoria A. Velkoff, and Kimberly A. DeBarros. 2005. *65+ in the United States: 2005*. U.S. Census Bureau, Current Population Reports, P23-209. Washington, DC: Government Printing Office.

Head, Simon. 2003. *The New Ruthless Economy: Work and Power in the Digital Age*. New York: Oxford University Press.

Heald, Anne, et al. 2003. "A Conversation on School Vouchers." Washington, DC: Economic Policy Institute, June 12. Retrieved January 20, 2005 (www.epinet.org).

Heath, Chip. 2005. "Introduction." Pp. 81-85 in *Rumor Mills: The Social Impact of Rumor and Legend*, edited by Gary Alan Fine, Véronique Campion-Vincent,

and Chip Heath. New Brunswick, NJ: Transaction Publishers.

Hebert, Maeve and Allan Rivlin. 2002. "The Kids Are All Right." *Public Perspective*, 13 (January/February): 31-33.

Hechter, Michael and Karl-Dieter Opp. 2001. "Introduction." Pp. xi-xx in *Social Norms*, edited by Michael Hechter and Karl-Dieter Opp. New York: Russell Sage Foundation.

Heilbroner, R. L. and L. C. Thurow. 1998. *Economics Explained: Everything You Need to Know about How the Economy Works and Where It's Going*. New York: Touchstone.

Heimer, Karen, and Candace Kruttschnitt. 2006. "Introduction: New Insights into the Gendered Nature of Crime and Victimization." Pp. 1-14 in *Gender and Crime: Patterns of Victimization and Offending*, edited by Karen Heimer and Candace Kruttschnitt. New York: New York University Press.

Heimer, Karen, Stacy Wittrock, and Halime Ünal. 2006. "The Crimes of Poverty: Economic Marginalization and the Gender Gap in Crime." Pp. 115-126 in *Gender and Crime: Patterns of Victimization and Offending*, edited by Karen Heimer and Candace Kruttschnitt. New York: New York University Press.

Helgesen, Sally. 2008. "Female Leadership: Changing Business for the Better." *Christian Science Monitor*, January 17, 9.

Helper, Susan. 2008. "Renewing U.S. Manufacturing: Promoting a High-Road Strategy." Economic Policy Institute, February 13. Retrieved July 4, 2008 (www.epi.org).

Henschke, Claudia I., David F. Yankelevitz, Daniel M. Libby, Mark W. Pasmantier, and James Posit. 2006. "Survival of Patients with Stage 1 Lung Cancer Detected on CT Screening." *New England Journal of Medicine* 355 (October 26): 1763-1771.

Heritage, John. 1984. *Garfinkel and Ethnomethodology*. New York: Basil Blackwell, Inc.

Hernandez, Raymond. 2008. "What Would You Drive, if the Taxpayers Paid?" *New York Times*, May 1, 1.

Herod, Andrew. 1993. "Gender Issues in the Use of Interviewing As a Research Method." *The Professional Geographer* 45 (August): 305-17.

Herrnstein, Richard J. and Charles Murray. 1994. *The Bell Curve: Intelligence and Class Structure in American Life*. New York: Free Press.

Hetherington, E. Mavis, Ross D. Parke, and Virginia Otis Locke. 2006. *Child Psychology: A Contemporary Viewpoint*, 6th edition. Boston, MA: McGraw-Hill.

Heubeck, Elizabeth. 2005. "Pressure Grows To Telecommute." *Baltimore Sun*, October 26, K1-K2.

Hilbert, Richard A. 1992. *The Classical Roots of Ethnomethodology: Durkheim, Weber, and Garfinkel*. Chapel Hill: The University of North Carolina Press.

Hinkle, Stephen, and John Schopler. 1986. "Bias in the Evaluation of In-Group and Out-Group Performance." Pp. 196-212 in *Psychology of Everyday Intergroup Relations*, 2nd ed., edited by Stephen Worchel and William. G. Austin. Chicago, IL: Nelson-Hall.

Hira, Ron. 2008. "An Overview of the Offshoring of U.S. Jobs." *Population Bulletin* 63 (June): 14-15.

Hirsch, Amy E., Sharon M. Dietrich, Rue Landau, Peter D. Schneider, and Irv Ackelsberg. 2002. *Every Door Closed: Barriers Facing Parents with Criminal Records*. Washington, DC: Center for Law and Social Policy. Retrieved May 1, 2005 (www.clasp.org).

"Hispanic Heritage Month 2007: Sept. 15-Oct. 15." 2007. U.S. Census Bureau, Facts for Features, CB07-FF.14. Retrieved June 10, 2008 (www.census.gov).

Hispanic-Owned Firms: 2002. 2006. U.S. Census Bureau, SB02-00CS-HISP (RV). Retrieved May 1, 2007 (www.census.gov).

Hochberg, Fred P. 2002. "American Capitalism's Other Side." *New York Times*, July 25, A17.

Hochschild, Arlie Russell. 1983. *The Managed Heart: Commercialization of Human Feeling*. Berkeley: University of California Press.

____. 1989. *The Second Shift: Working Parents and the Revolution At Home*. New York: Penguin.

Hoecker-Drysdale, Susan. 1992. *Harriet Martineau: First Woman Sociologist*. Providence, RI: Berg.

Hoefer, Michael, Nancy Rytina, and Christopher Campbell. 2007. "Estimates of the Unauthorized Immigrant Population Residing in the United States: January 2006." U.S. Department of Homeland Security, Office of Immigration Studies, August. Retrieved September 12, 2008 (www.dhs.gov).

Holder, Kelly. 2006. "Voting and Registration in the

Election of November 2004." U.S. Census Bureau, March, Current Population Reports, P20-556. Retrieved May 15, 2007 (www.census.gov).

Holt, Emily W., Daniel J. McGrath, and Marilyn M. Seastrom. 2006. "Qualifications of Pubic Secondary School History Teachers, 1999-2000." U.S. Department of Education, National Center for Education Statistics, NCES 2006-004. Retrieved July 12, 2007 (http://nces.ed.gov).

Holzer, Harry. 2007. "Better Workers for Better Jobs: Improving Worker Advancement in the Low-Wage Labor Market." Urban Institute, December 12. Retrieved May 18, 2008 (www.urban.org).

Homans, George. 1974. *Social Behavior: Its Elementary Forms*, revised ed. New York: Harcourt Brace Jovanovich.

Hondagneu-Sotelo, Pierrette. 2001. *Doméstica: Immigrant Workers Cleaning and Caring in the Shadows of Affluence*. Berkeley: University of California Press.

hooks, bell. 2000. *Feminism is for Everybody: Passionate Politics*. Cambridge, MA: South End Press.

Horn, Bernie, ed. 2006. *Progressive Agenda for the States 2006: State Policy Leading America*. Washington, DC: Center for Policy Alternatives.

Horn, Laura, Katharin Peter, and C. Dennis Carroll. 2003. *What Colleges Contribute: Institutional Aid to Full-Time Undergraduate Attending 4-Year Colleges and Universities*. U.S. Department of Education, National Center for Education Statistics, April. Retrieved January 6, 2005 (http://nces.ed.gov).

Horrigan, John B. 2004. "How Americans Get in Touch with Government." Pew Internet & American Life, May 24. Retrieved September 9, 2004 (www.pewinternet.org).

Horrigan, John B., and Lee Rainie. 2006. "The Internet's Growing Role in Life's Major Moments." April, Pew Internet & American Life Project, April. Retrieved September 20, 2007 (www.pewinternet.org).

Horwitz, Allan V. and Jerome C. Wakefield. 2006. "The Epidemic in Mental Illness: Clinical Fact or Survey Artifact?" *Contexts* 5 (Winter): 19-23.

Houser, Ari. 2007. "Women & Long-Term Care." AARP Public Policy Institute. Retrieved June 23, 2007 (assets.aarp.org).

Houtenville, Andrew J., and Karen Smith Conway. 2008. "Parental Effort, School Resources, and Student Achievement." *Journal of Human Resources* 43 (Spring): 437-453.

Howard, Jeff. 2003. "Still at Risk: The Causes and Costs of Failure to Educate Poor and Minority Children for the Twenty-First Century." Pp. 81-97 in *A Nation Reformed? American Education 20 Years after A Nation at Risk*, edited by David T. Gordon and Patricia A. Graham. Cambridge, MA: Harvard Education Press.

Howard, Judith A. and Jocelyn A. Hollander. 1997. *Gendered Situations, Gendered Selves: A Gender Lens On Social Psychology*. Thousand Oaks, CA: Sage.

Hoyt, Homer. 1939. *The Structure and Growth of Residential Neighborhoods in American Cities*. Washington, DC: Federal Housing Administration.

Hubbard, Ruth. 1990. *The Politics of Women's Biology*. New Brunswick, NJ: Rutgers University Press.

Hughes, Everett C. 1945. "Dilemmas and Contradictions of Status." *American Journal of Sociology* 50: 353-359.

Hughes, Kristen A. 2006. "Justice Expenditure and Employment in the United States, 2003." Washington, DC: U.S. Department of Justice, Office of Justice Programs.

Hunter, James Davison, and Joshua Yates. 2002. "In the Vanguard of Globalization: The World of American Globalizers." Pp. 323-357 in *Many Globalizations: Cultural Diversity in the Contemporary World*, edited by Peter L. Berger and Samuel P. Huntington. New York: Oxford University Press.

Huntington, Samuel P. 2004. "The Hispanic Challenge." *Foreign Policy* Issue # 141 (March/April): 30-45.

Hurh, Won Moo. 1998. *The Korean Americans*. Westport, CT: Greenwood Press.

Hurst, Charles E. 2001. *Social Inequality: Forms, Causes, and Consequences*, 4th ed. Boston, MA: Allyn and Bacon.

Hutter, Mark. 2007. *Experiencing Cities*. Boston, MA: Allyn & Bacon.

Hyde, Janet S. 2005. "The Gender Similarities Hypothesis." *American Psychologist* 60 (September): 581-592.

Hyde, Janet S., Sara M. Lindberg, Marcia C. Linn, Amy B. Ellis, and Caroline C. Williams. 2008. "Gender

Similarities Characterize Math Performance." *Science* 321 (July): 494-495.

Hymowitz, Carol. 2006. "'Any College Will Do'." *Wall Street Journal*, September 18. Retrieved November 8, 2007 (online.wsj.com).

Independent Sector. 2006. "Value of Volunteer Time." Retrieved September 7, 2006 (www.independentsector.org).

———. 2008. "Giving and Volunteering in the United States." Retrieved May 8, 2008 (www.independentsector.org).

"India Gives Mixed Welcome to Billionth Citizen." 2000. *Baltimore Sun*, May 12, 30A.

Innocence Project. 2008. "Fact Sheet." Retrieved May 12, 2008 (www.innocenceproject.org).

Inside-OUT. 2001. "A Report On The Experiences Of Lesbians, Gays And Bisexuals In American And The Public's Views On Issues And Policies Related To Sexual Orientation." Kaiser Family Foundation. Retrieved June 18, 2003 (www.kff.org).

Institute of Medicine. 2006. *Food Marketing to Children and Youth: Threat or Opportunity?* Washington, DC: National Academy of Sciences.

Insurance Institute for Highway Safety. 2008. "Q&As: Teenagers—General." March. Retrieved March 26, 2008 (www.iihs.org).

Intergovernmental Panel on Climate Change. 2007. "Fourth Assessment Report." Retrieved February 12, 2008 (www.ipcc.ch).

International Institute for Democracy and Electoral Assistance. 2007. "Turnout over Time: Advances and Retreats in Electoral Participation." Retrieved May 24, 2007 (www.idea.int).

———. 2008. "Compulsory Voting." Retrieved June 28, 2008 (www.idea.int).

Internet Crime Complaint Center. 2007. "2007 Internet Crime Report." Retrieved August 25, 2008 (www.ic3.gov).

Internet World Stats. 2008. "World Internet Users December 2007." Retrieved March 10, 2008 (www.internetworldstats.com).

Inter-Parliamentary Union. 2008. "Women in National Parliaments." April 30. Retrieved June 28, 2008 (www.ipu.org).

Ioannidis, John. 2005. "Contradicted and Initially Stronger Effects in Highly Cited Clinical Research." *Journal of the American Medical Association* 294 (July 13): 218-228.

"Iranian Gays Present, Hidden." 2007. *Baltimore Sun*, September 30, 16A.

Irwin, Neil. 2008. "Why We're Gloomier than the Economy." *Washington Post*, June 18, A1.

Isaacs, Julia B. 2007a. "Economic Mobility of Black and White Families." Pp. 71-80 in *Getting Ahead or Losing Ground: Economic Mobility in America*, edited by Julia B. Isaacs, Isabel V. Sawhill, and Ron Haskins. Washington, DC: The Brookings Institution.

———. 2007b. "Economic Mobility of Men and Women." Pp. 61-70 in *Getting Ahead or Losing Ground: Economic Mobility in America*, edited by Julia B. Isaacs, Isabel V. Sawhill, and Ron Haskins. Washington, DC: The Brookings Institution.

———. 2007c. "International Comparisons of Economic Mobility." Pp. 37-46 in *Getting Ahead or Losing Ground: Economic Mobility in America*, edited by Julia B. Isaacs, Isabel V. Sawhill, and Ron Haskins. Washington, DC: The Brookings Institution.

"Island Women Banned From Wearing Trousers." 2002. *Ananova News*, January 4. Retrieved January 5, 2002 (www.ananova.com/news).

Jackson, D. D. 1998. "'This Hole in Our Heart': Urban Indian Identity and the Power of Silence." *American Indian Culture and Research Journal* 22 (4): 227-54.

Jacobs, Jerry A. 2005. "Multiple Methods in ASR." *Footnotes* 33 (December): 1, 4.

Jacobson, Michael. 2005. *Downsizing Prisons: How to Reduce Crime and End Mass Incarceration*. New York: New York University Press.

Jagger, Alison M., and Paula S. Rothenberg, eds. 1984. *Feminist Frameworks*, 2nd ed. New York: McGraw-Hill.

Jandt, Fred E. 2001. *Intercultural Communication: An Introduction*. Thousand Oaks, CA: Sage.

Janis, Irving L. 1972. *Victims of Groupthink: A Psychological Study of Foreign-Policy Decisions and Fiascoes*. Boston, MA: Houghton Mifflin Company.

Janofsky, Michael. 2003. "Amid Acceptance of Gays, a Split on Marriage Issue." *New York Times*, November 23, A12.

———. 2003. "Thurmond Kin Acknowledge Black Daughter." *New York Times*, December 13, A28.

———. 2003. "Young Brides Stir New Outcry on Utah Polygamy." *New York Times*, February 28, 1.

———. 2005. "U.S. Court Backs Bush's Revisions in Clean Air Act." *New York Times*, June 25, A1.

Javers, Eamon. 2007. "Inside the Hidden World of Earmarks." *Business Week*, September 17, 56.

Jeffery, Clara. 2006. "Poor Losers." *Mother Jones*, July/August, 20-21.

Jemal, A., R. Siegel, E. Ward, Y. Hao, J. Xu, T. Murray, and M. J. Thun. 2008. "Cancer Statistics, 2008." *CA: Cancer Journal for Clinicians* 58 (March-April): 71-96.

Jenkins, J. Craig. 1983. "Resource Mobilization Theory and the Study of Social Movements." *Annual Review of Sociology* 9 (August): 527-553.

Jenness, Valerie and Ryken Grattet. 2001. *Making Hate a Crime: From Social Movement to Law Enforcement*. New York: Russell Sage Foundation.

Jeune, Bernard and James W. Vaupel, eds. 1995. *Exceptional Longevity: From Prehistory to the Present*. Denmark: Odense University Press.

Jiménez, Tomás R. 2007. "The Next Americans." *Los Angeles Times*, May 27, M1.

Jo, Moon. H. 1999. *Korean Immigrants and the Challenge of Adjustment*. Westport, CT: Greenwood Press.

Johnson, Allan G. 2008. "Our House is on Fire." Paper presented, March 17. Retrieved June 17, 2008 (http://uhavax.hartford.edu/agjohnson/kellogg.htm).

Johnson, Hank, and Bert Klandermans, eds. 1995. *Social Movements and Culture*. Minneapolis: University of Minnesota Press.

Johnson, Leslie and Justine Lloyd. 2004. *Sentenced to Everyday Life: Feminism and the Housewife*. New York: Berg.

Johnson, Rachel L., Debra Roter, Neil R. Powe, and Lisa A. Cooper. 2004. "Patient Race/Ethnicity and Quality of Patient-Physician Communication during Medical Visits." *American Journal of Public Health* 94 (December): 2084-2090.

Johnson, Sharon. 2008. "U.S. Employees Pushing Women Out of Work Force." Women's E News, August. Retrieved September 1, 2008 (www.womensenews.org).

Johnson, Pamela. 2005. "Dressing the Part." *Chronicle of Higher Education*, August 10. Retrieved Augusts 11, 2005 (http://chronicle.com/jobs).

Jolliffe, Dean. 2006. "The Cost of Living and the Geographic Distribution of Poverty." U.S. Department of Agriculture, Economic Research Service, September. Retrieved November 10, 2006 (www.ers.usda.gov).

Jones, Del. 2005. "Wanted: CEO, no Ivy Required." *USA Today*, April 7. Retrieved November 8, 2007 (www.usatoday.com).

Jones, Jeffrey M. 2005. "Most Americans Approve of Interracial Dating." Gallup News Service, October 7. Retrieved October 10, 2005 (www.galluppoll.com).

———. 2006. "One In Three U.S. Workers Have 'Telecommuted' To Work." Gallup Organization, August 16. Retrieved September 28, 2006 (www.galluppoll.com).

———. 2007. "Public: Family of Four Needs to Earn Average of $52,000 to Get By." Gallup News Service, February 9. Retrieved June 2, 2007 (www.galluppoll.com).

———. 2008. "Confidence in Congress: Lowest Ever for Any U.S. Institution." Gallup, June 20. Retrieved August 20, 2008 (www.gallup.com).

———. 2008. "Fewer Americans Favor Cutting Back Immigration." Gallup poll, July 10. Retrieved September 9, 2008 (www.gallup.com).

———. 2008. "Four in 10 Americans See Their Standard of Living Declining." Gallup News Service, June 9. Retrieved July 4, 2008 (www.gallup.com).

———. 2008. "Trust in Government Remains Low." Gallup, September 18. Retrieved October 14, 2008 (www.gallup.com).

Jones, Nicholas A., and Amy S. Smith. 2001. "*The Two or More Races Population: 2000*." U.S. Census Bureau. Retrieved April 16, 2003 (www.census.gov).

Jones, Steve. 2003. "Let the Games Begin: Gaming Technology and Entertainment among College Students." Pew Internet & American Life Project. Retrieved December 12, 2004 (www.pewinternet.org).

Jones, Susan S., and Hye-Won Hong. 2001. "Onset of Voluntary Communication: Smiling Looks to Mother." *Infancy* 2 (3): 353-70.

Jones, Tim. 2007. "Get Healthy or Get Fired." *Baltimore Sun*, September 26, 4A.

Jonsson, Patrik. 2007. "In US Justice, How Much Bias?" *Christian Science Monitor*, September 21, 1, 10.

Josephson Institute of Ethics. 2006. "2006 Josephson Institute Report Card on the Ethics of American Youth: Part One—Integrity." Retrieved November 16, 2007 (www.josephsoninstitute.org).

Joyce, Amy. 2006. "Vacation Deprivation." *Washington Post*, June 25, D4.

Juergensmeyer, Mark. 2003. "Thinking Globally about Religion." Pp. 3-13 in *Global Religions: An Introduction*, edited by Mark Juergensmeyer. New York: Oxford University Press.

Kahn, Kim Fridkin. 2003. "Assessing the Media's Impact on the Political Fortunes of Women." Pp. 173-189 in *Women and American Politics: New Questions, New Directions*, edited by Susan J. Carroll. Oxford, NY: Oxford University Press.

Kalev, Alexandra, Frank Dobbin, and Erin Kelly. 2006. "Best Practices or Best Guesses? Assessing the Efficacy of Corporate Affirmative Action and Diversity Policies." *American Sociological Review* 71 (August): 589-617.

Kalil, Ariel. 2002. "Cohabitation and Child Development." Pp. 153-160 in *Just Living Together: Implications Of Cohabitation On Families, Children, And Social Policy*, edited by Alan Booth, Ann C. Crouter, and Nancy S. Landale. Mahwah, NJ: Lawrence Erlbaum Associates.

Kalmijn, Matthijs. 1998. "Intermarriage and Homogamy: Causes, Patterns, Trends." *Annual Review of Sociology* 24 (August): 395–421.

Kamber, Richard, and Mary Biggs. 2002. "Grade Conflation: A Question of Credibility." *Chronicle of Higher Education*, April 12, B14.

Kane, Emily. W., and Laura J. Macaulay. 1993. "Interviewer Gender and Gender Attitudes." *Public Opinion Quarterly* 57 (Spring): 1 -28.

Kane, John, and April D. Wall. 2006. *The 2005 National Public Survey on White Collar Crime*. Fairmont, WV: National White Collar Crime Center.

Kapferer, Jean-Noel. 1992. "How Rumors are Born." *Society* 29 (July/August): 53-60.

Karasik, Sherry. 2000. "More Latchkey Kids Means More Trouble: High-Risk Behavior Increases When Parents Are Gone." Retrieved September 28, 2008 (www.apbnews.com).

Karnow, Stanley. 2004. "Keep Your Tired, Poor Stereotypes About Immigrants." *Baltimore Sun*, March 5, 13A.

Katz, Jackson. 2006. *The Macho Paradox: Why Some Men Hurt Women and How All Men Can Help*. Naperville, IL: Sourcebooks, Inc.

Katzenbach, Jon. 2003. *Why Pride Matters More Than Money: The Power of the World's Greatest Motivational Force*. New York: Crown Business.

Kawamoto, Walter T., and Tamara C. Cheshire. 1997. "American Indian Families." Pp. 15-34 in *Families in Cultural Context: Strengths and Challenges in Diversity*, edited by Mary Kay DeGenova. Mountain View, CA: Mayfield Publishing Company.

Keeter, Scott, Juliana Horowitz, and Alec Tyson. 2008. "Young Voters in the 2008 Election." Pew Research Center, November 12. Retrieved November 14, 2008 (www.pewresearch.org).

Kelley, Matt. 1999. "American Indian Boarding Schools: 'That Hurt Never Goes Away'." CNews. Retrieved May 19, 2004 (www.canoe.ca).

____. 2007. "Ex-lawmakers Find Work with Lobbyists." *USA Today*, February 21, 1A.

Kempner, Joanna, Clifford S. Perlis, and Jon F. Merz. 2005. "Ethics: Forbidden Knowledge." *Science* 307 (February 11): 854.

Kendall, Diana. 2002. *The Power of Good Deeds: Privileged Women and the Social Reproduction of the Upper Class*. Lanham, MD: Rowman & Littlefield.

Kennickell, Arthur B. 2006. "Currents and Undercurrents: Changes in the Distribution of Wealth, 1989-2004." Federal Reserve Board. Retrieved May 12, 2008 (www.federalreserve.gov).

Kent, Mary Mederios. 2007. "Immigration and America's Black Population." *Population Bulletin* 62 (December): 1-16.

Kestin, Sally, and Peter Franceschina. 2007. "Federal Watchdog Examines Seminoles' Gambling Profits." *South Florida Sun-Sentinel*, December 8. Retrieved December 10, 2007 (www.sun-sentinel.com).

Kilborn, Peter T. 2001. "Rural Towns Turn to Prisons to Reignite Their Economies." *New York Times*, July 27, 1.

Kilbourne, Jean. 1999. *Deadly Persuasion: Why Women and Girls Must Fight the Addictive Power of Advertising*. New York: Free Press.

Kilson, Martin. 2005. "Black Civil Society's Leadership Burden." *The Black Commentator*, October 13. Retrieved September 24, 2008 (www.blackcommentator.com).

King, C. Wendell. 1956. *Social Movements in the United States*. New York: Random House.

King, Jacqueline E. 2006. *Gender Equity in Higher Education: 2006*. Washington, DC: American Council on Education.

King, Michael, and Annie Bartlett. 2006. "What Same Sex Civil Partnerships May Mean for Health." *Journal of Epidemiology and Community Health* 60 (March): 188-191.

Kinsella, Kevin, and David R. Phillips. 2005. "Global Aging: The Challenge of Success." *Population Bulletin* 60 (March): 1-40.

Kirkpatrick, David D. 2007. "Congress Finds Ways to Avoid Lobbyist Limits." *New York Times*, February 11, 1.

"Kith & Kin Inc." 2004. *New York Times*, February 14, 18.

Kivel, Paul. 2004. *You Call This a Democracy? Who Benefits, Who Pays and Who Really Decides*. New York: Apex Press.

Kivisto, Peter and Dan Pittman. 2001. "Goffman's Dramaturgical Sociology: Personal Sales and Service in a Commodified World." Pp. 311-334 in *Illuminating Social Life: Classical and Contemporary Theory Revisited*, 2nd edition. Thousand Oaks, CA: Pine Forge Press.

Klandermans, Bert. 1984. "Mobilization and Participation: Social Psychological Explanations of Resource Mobilization Theory." *American Sociological Review* 49 (October): 583-600.

Klaus, Marshall H., John H. Kennell, and Phyllis H. Klaus. 1995. *Bonding: Building the Foundations of Secure Attachment and Independence*. Reading, MA: Addison-Wesley.

Klaus, Patsy. 2005. "Crimes against Persons Age 65 or Older, 1993-2002." Washington, DC: U.S. Department of Justice, Bureau of Justice Statistics.

____. 2006. "Crime and the Nation's Households, 2004." Washington, DC: U.S. Department of Justice, Bureau of Justice Statistics.

Klein, Alyson. 2008. "Education Earmarks Get Scrutiny." *Education Week*, February 29. Retrieved March 1, 2008 (www.edweek.org).

Klopfenstein, Kristin, and M. Kathleen Thomas. 2005. "The Advance Placement Performance Advantage: Fact or Fiction?" Retrieved February 10, 2005 (www.utdallas.edu).

Knickerbocker, Brad. 2005. "Fallout of Marijuana Verdict." *Christian Science Monitor*, June 8, 3.

Kochhar, Rakesh. 2007. "1995-2005: Foreign-Born Latinos Make Progress on Wages." Pew Hispanic Center, August 21. Retrieved October 10, 2007 (www.pewhispanic.org).

Koenig, Linda J., Daniel J. Whitaker, Rachel A. Royce, Tracey E. Wilson, Kathleen Ethier, and M. Isabel Fernandez. 2006. "Physical and Sexual Violence during Pregnancy and After Delivery: A Prospective Multistate Study of Women with or At Risk for HIV Infection." *American Journal of Public Health* 96 (June): 1052-1059.

Kohut, Andrew, et al. 2007. "Blacks See Growing Values Gap between Poor and Middle Class." Pew Research Center, November 13. Retrieved December 10, 2007 (www.pewsocialtrends.org).

Kornhauser, William. 1959. *The Politics of Mass Society*. New York: Free Press.

Kozol, Jonathan. 2005. *The Shame of the Nation: The Restoration of Apartheid Schooling in America*. New York: Crown.

Kramer, Laura. 2005. *The Sociology of Gender: A Brief Introduction*, 2nd ed. Los Angeles, CA: Roxbury.

Kreider, Rose M. 2003. "Adopted Children and Stepchildren: 2000." U.S. Census Bureau, Census 2000 Special Reports, CENSR-6RV. Retrieved June 10, 2007 (www.census.gov).

____. 2005. "Number, Timing, and Duration of Marriages and Divorces: 2001." U.S. Census Bureau, Current Population Reports, P70-97. Retrieved July 4, 2006 (www.census.gov).

Kreider, Rose M., and Jason M. Fields. 2002. "Number, Timing, and Duration of Marriages and Divorces: 1996." U.S. Census Bureau, Current Population Reports, P70-80. Retrieved March 1, 2003 (www.census.gov).

Kreps, Gary L. 2006. "Communication and Racial Inequities in Health Care." *American Behavioral Scientist* 49 (February): 760-774.

Kristof, Kathy M., and E. Scott Reckard. 2008. "Ex-Countrywide CEO Angelo Mozilo is Under Formal SEC Investigation." *Los Angeles Times*, August 8. Retrieved August 24, 2008 (www.latimes.com).

Krucoff, Mitchell W, Suzanne W. Crater, and Kerry L. Lee. 2006. "From Efficacy to Safety Concerns: A STEP Forward or a Step Back for Clinical Research and Intercessory Prayer? The Study of Therapeutic Effects of Intercessory Prayer (STEP). *American Heart Journal* 151 (April): 762-4.

Krugman, Paul. 2002. "Crony Capitalism, U.S.A." *New York Times*, January 15, 21.

Kuhn, Manford H., and Thomas. S. McPartland. 1954. "An Empirical Investigation of Self-Attitudes." *American Sociological Review* 19 (February): 68-76.

Kulczycki, Andrei. 2001. "Deepening the Melting Pot: Arab-Americans at the Turn of the Century." *Middle East Journal* 3 (Summer): 459-473.

____. and Arun P. Lobo. 2002. "Patterns, Determinants, and Implications of Intermarriage among Arab Americans." *Journal of Marriage and the Family* 64 (February): 202-210.

Kunkel, D., K. M. Cope, W. J. M. Farinola, E. Biely, E. Rollin and E. D. Donnerstein. 1999. *Sex on TV: A Biennial Report to the Kaiser Family Foundation*. Washington, DC: Henry J. Kaiser Family Foundation.

Kurdek, Lawrence A. 1993. "Predicting Marital Dissolution: A 5-Year Prospective Longitudinal Study of Newlywed Couples." *Journal of Personality and Social Psychology* 64 (February): 221–42.

____. 2004. "Are Gay and Lesbian Cohabiting Couples *Really* Different from Heterosexual Married Couples?" *Journal of Marriage and Family* 66 (November): 880-900.

Kutner, M., E. Greenberg, Y. Jin, B. Boyle, Y. Hsu, and E. Dunleavy. 2007. *Literacy in Everyday Life: Results from the 2003 National Assessment of Adult Literacy* (NCES 2007-480). Washington, DC: U.S. Department of Education.

Lacey, Marc. 2008. "Across Globe, Empty Bellies Bring Rising Anger." *New York Times*, April 18, 1.

LaFraniere, Sharon, and Laurie Goodstein. 2007. "Anglicans Rebuke U.S. Branch on Blessing Same-Sex Unions." *New York Times*, February 20, A1, A11.

Lakoff, Robin. T. 1990. *Talking Power: The Politics of Language*. New York: Basic Books.

Lalasz, Robert. 2006. "Americans Flocking to Outer Suburbs in Record Numbers." Population Reference Bureau. Retrieved January 24, 2008 (www.prg.org).

Lampman, Jane. 2007. "New Fight, Old Foe: Slavery." *Christian Science Monitor*, February 21, 13-14.

Landler, Mark. 2006. "In Munich, Provocation in a Symbol of Foreign Faith." *New York Times*, December 8, A3.

____. 2008. "Austria Stunned by Case of Another Woman Imprisoned in a Cellar." *New York Times*, April 29, 6.

Landsberg, Mitchell. 2007. "They're More Interested in Money Than Mao." *Los Angeles Times*, June 26, A1.

Landsberger, Henry A. 1958. *Hawthorne Revisited*. Ithaca, NY: Cornell University Press.

Lang, Kurt, and Gladys Engel Lang. 1961. *Collective Dynamics*. New York: Thomas Y. Crowell.

Lannutti, Pamela J., Melanie Laliker, and Edward L. Hale. 2001. "Violations of Expectations and Social-Sexual Communication in Student/Professor Interactions." *Communication Education* 50 (January): 69-82.

Laraña, Enrique, Hank Johnston, and Joseph R. Gusfield, eds. 1994. *New Social Movements: From Ideology to Identity*. Philadelphia, PA: Temple University Press.

LaRossa, Ralph. 1986. *Becoming a Parent*. Thousand Oaks, CA: Sage.

LaRossa, Ralph and Donald C. Reitzes. 1993. "Symbolic Interactionism and Family Studies." Pp. 135–63 in *Sourcebook of Family Theories and Methods: A Contextual Approach*, edited by Pauline G. Boss, William J. Doherty, Ralph LaRossa, Walter R. Schumm, and Suzanne K. Steinmetz. New York: Plenum Press.

Larsen, Janet. 2004. "Sterilization is Most Popular Contraceptive." *Population Press* 10 (Fall): 18-19.

Larson, Dale G., and Daniel R. Tobin. 2000. "End-of-Life Conversations: Evolving Practice and Theory." 2000. *JAMA* 284 (September 27): 1573-78.

Larson, Jeffry and Rachel Hickman. 2004. "Are College

Marriage Textbooks Teaching Students the Premarital Predictors of Marital Quality?" *Family Relations* 53 (July): 385-392.

Laskow, Sarah. 2006. "Hired Guns: State Lobbying Becomes Billion-Dollar Business." Center for Public Integrity, December 20. Retrieved May 23, 2007 (www.publicintegrity.org).

Lasswell, Harold. 1936. *Politics: Who Gets What, When and How*. New York: McGraw-Hill.

Lathrop, Richard G., and John Hasse. 2007. "Tracing New Jersey's Dynamic Landscape: A Municipal Report Card on Urban Grown and Open Space Loss." Retrieved January 28, 2008 (www.crssa.rutgers.edu).

Lavelle, Marianne. 2007. "Water Woes." *U.S. News & World Report*, June 4, 37-46.

Lawless, Jennifer L. and Richard L. Fox. 2005. *It Takes a Candidate: Why Women Don't Run for Office*. New York: Cambridge University Press.

Lazo, Alejandro. 2008. "Hispanics Hit Hard as Workers Lose Hours." *Washington Post*, September 1, E11.

Leaper, Campbell, and Melanie M. Ayres. 2007. "A Meta-Analytic Review of Gender Variations in Adults' Language Use: Talkativeness, Affiliative Speech, and Assertive Speech." *Personality and Social Psychology Review* 11 (November): 328-363.

Lee, Don. 2007. "In China, Income Disparity Takes a Great Leap." *Los Angeles Times*, June 10, C1.

Lee, J., W. Grigg, and G. Dion. 2007. *The Nation's Report Card: Mathematics 2007* (NCES 2007-494). Washington, DC: National Center for Education Statistics, Institute of Education Sciences, U.S. Department of Education.

Lemann, Nicholas, 2000. *The Big Test: The Secret History of the American Meritocracy*. New York: Farrar, Straus, and Giroux.

Lemert, Edwin M. 1951. *Social Pathology: A Systematic Approach to the Theory of Sociopathic Behavior*. New York: McGraw-Hill.

————. 1967. *Human Deviance, Social Problems and Social Control*. Englewood Cliffs, NJ: Prentice Hall.

Lengermann, Patricia Madoo and Jill Niebrugge-Brantley. 1992. "Contemporary Feminist Theory." Pp. 308-357 in *Contemporary Sociological Theory*, 3rd edition, edited by George Ritzer. New York: McGraw-Hill.

Lenhart, Amanda. 2007. "Cyberbullying and Online Teens." June, Pew Internet & American Life Project, April. Retrieved September 20, 2007 (www.pewinternet.org).

Leonardsen, Dag. 2004. *Japan as a Low-Crime Nation*. New York: Palgrove Macmillan.

Lessig, Lawrence. 2001. "The Internet under Siege," *Foreign Policy*, Issue 127 (November/December): 56-66.

Levine, Arthur. 1994. *Educating School Teachers*. The Education Schools Project. Retrieved August 2, 2007 (http://edschools.org).

Levine, Marc V. 1994. "A Nation of Hamburger Flippers?" *Baltimore Sun*, July 31, 1E, 4E.

"Levi's Set to Close Last U.S. Factory." *Baltimore Sun*, October 19, A14.

Lewis, David Levering. 1993. *W. E. B. Du Bois: Biography of a Race, 1868-1919*. New York: Henry Holt.

Lewis, Oscar. 1966. "The Culture of Poverty." *Scientific American* 115 (October): 19-25.

Lewis, Peter. 2005. "Taking a Bite Out of Identity Theft." *Fortune*, May 2, 36, 38.

LexisNexis Corporate Affiliations. 2003. New Providence, NJ: LexisNexis Group.

Lichter, Daniel T., Deborah R. Graefe, and J. Brian Brown. 2003. "Is Marriage a Panacea? Union Formation among Economically Disadvantaged Mothers." *Social Problems* 50 (February): 60-86.

Lichter, Daniel T., Zhenchao Qian, and Leanna M. Mellott. 2006. "Marriage or Dissolution? Union Transitions among Poor Cohabiting Women." *Demography* 43 (May): 223-240.

Light, Paul C. 2002. "The Troubled State of the Federal Public Service," The Brookings Institution. Retrieved November 1, 2002 (www.brookings.edu).

Lilly, J. Robert, Francis T. Cullen and Richard A. Ball. 1995. *Criminological Theory: Context and Consequences*, 2nd edition. Thousand Oaks, CA: Sage.

Lindsay, D. Michael. 2007. *Faith in the Hall of Power: How Evangelicals Joined the American Elite*. New York: Oxford University Press.

Lindsey, Linda L. 2005. *Gender Roles: A Sociological Perspective*, 4th ed. Upper Saddle River, NJ: Prentice Hall.

Lino, Mark. 2008. "Expenditures on Children by Families, 2007." U.S. Department of Agriculture,

Center for Nutrition Policy and Promotion, March. Retrieved April 15, 2008 (www.cnpp.usda.gov).

Linton, Ralph. 1936. *The Study of Man*. New York: Appleton-Century-Crofts.

————. 1964. *The Study of Man: An Introduction*. New York: Appleton-Century-Crofts.

Lipka, Sara. 2005. "Winning Streaks." *Chronicle of Higher Education*, March 18, A6.

Liu, Goodwin. 2006. "How the Federal Government Makes Rich States Richer." The Education Trust, *Funding Gaps 2006*, pp. 2-4. Retrieved July 23, 2007 (www2.edtrust.org).

"Living Humbled, Unhappy Lives." 1996. *Baltimore Sun*, May 23, 2A.

Llana, Sara Miller. 2005. "Can a $103 Fine Stop Students from Swearing?" *Christian Science Monitor*, December 7, 1-2.

Locher, David A. 2002. *Collective Behavior*. Upper Saddle River, NJ: Prentice Hall.

Lofland, John. 1993. "Collective Behavior: The Elementary Forms." Pp. 70-75 in *Collective Behavior and Social Movements*, edited by Russell L. Curtis, Jr. and Benigno E. Aguirre. Boston, MA: Allyn & Bacon.

————. 1996. *Social Movement Organizations: Guide to Research on Insurgent Realities*. New York: Aldine De Gruyter.

Logan, John and Harvey Molotch. 1987. *Urban Fortunes: The Political Economy of Place*. Berkeley: University of California Press.

Long, Lynette. 2006. "Stamp Out Gender Bias in Postal Service Commemoratives." *Baltimore Sun*, January 13, A11.

————. 2008. "Painful Lessons." *Baltimore Sun*, May 18, 11A.

Longman, Phillip. 2004. "Which Nations Will Go Forth and Multiply?" *Fortune*, April 12, 60-61.

Lorber, Judith. 2005. *Gender Inequality: Feminist Theories and Politics*, 3rd ed. Los Angeles, CA: Roxbury.

Lorber, Judith and Lisa Jean Moore. 2007. *Gendered Bodies: Feminist Perspectives*. Los Angeles, CA: Roxbury.

Louw, P. Eric. 2001. *The Media and Cultural Production*. Thousand Oaks, CA: Sage.

Loveless, Tom, Steve Farkas, and Ann Duffett. 2008. "High-Achieving Students in the Era of NCLB." Thomas B. Fordham Institute. Retrieved July 15, 2008 (www.edexcellence.net).

Lowenkamp, Christopher T., and Edward J. Latessa. 2005. "Developing Successful Reentry Programs: Lessons Learned from the 'What Works' Research." *Corrections Today* 67 (April): 72-77.

Lugaila, Terry. A. 1998. "Marital Status and Living Arrangements: March 1998 (Update)." Current Population Reports, P20-514. Retrieved August 8, 2000 (www.census.gov).

Lukemeyer, Anna, Marcia K. Meyers, and Timothy Smeeding. "Expensive Children In Poor Families: Out-Of-Pocket Expenditures for the Care of Disabled and Chronically Ill Children in Welfare Families." *Journal of Marriage and the Family* 62 (May): 399-415.

Lund, Elizabeth. 2004. "Breakfast, Lunch, and a Brighter Future." *Christian Science Monitor*, November 3, 16-17.

Lundberg, Shelly, and Robert A. Pollak. 2007. "The American Family and Family Economics." *Journal of Economic Perspectives* 21 (Spring): 3-26.

Lutz, William. 1989. *Doublespeak*. New York: Harper & Row.

Lynch, Eleanor W., and Marci J. Hanson, eds. 1999. *Developing Cross-Cultural Competence: A Guide for Working with Children and Their Families*, 2nd ed. Baltimore, MD: Paul H. Brookes Publishing Co.

Lynn, David B. 1969. *Parental and Sex Role Identification: A Theoretical Formulation*. Berkeley, CA: McCutchen.

Lynn, M. and M. Todoroff. 1995. "Women's Work and Family Lives." Pp. 244-71 in *Feminist Issues: Race, Class, and Sexuality*, edited by Nancy Mandell. Scarborough, Ontario: Prentice Hall Canada.

Lyons, Linda. 2005. "Tracking U.S. Religious Preferences Over the Decades." The Gallup Poll, May 24. Retrieved August 25, 2007 (www.galluppoll.com).

Maccoby, Eleanor E., and John A. Martin. 1983. "Socialization in the Context of the Family: Parent-Child Interaction." Pp. 1-101 in *Socialization, Personality, and Social Development: Vol. 4. Handbook of Child Psychology*, edited by E. Mavis Hetherington. New York: Wiley.

Macionis, John J. and Vincent N. Parrillo. 2007. *Cities*

and Urban Life, 4th ed. New Jersey: Upper Saddle River.

MacKenzie, Donald, and Judy Wajcman, eds. 1999. *The Social Shaping of Technology*, 2nd ed. Philadelphia, PA: Open University Press.

MacKinnon, Catharine. 1982. "Feminism, Marxism, Method, and the State: An Agenda for Theory." *Signs* 7 (Spring): 515-544.

MacLean, Kim. 2003. "The Impact of Institutionalization on Child Development." *Development and Psychopathology* 15 (December): 853-884.

Macomber, Jennifer. 2006. "An Overview of Selected Data on Children in Vulnerable Families." Urban Institute and Child Trends, August 10. Retrieved October 1, 2006 (www.urban.org).

MacPhee, David, Janet Fritz, and Jan Miller-Heyl. 1996. Ethnic variations in personal social networks and parenting. *Child Development* 67 (December): 3278-95.

Madden, Mary, and Amanda Lenhart. 2006. "Online Dating." March, Pew Internet & American Life Project, April. Retrieved September 20, 2007 (www.pewinternet.org).

Madigan, Nick. 2003. "Suspect's Wife Is Said To Cite Polygamy Plan." *New York Times*, March 15, 1.

————. 2005. "Sheehan's Ad Says Bush Lied About Iraq War." *Baltimore Sun*, August 27, 1D, 4D-5D.

Mahon, Joe. 2004. "Banking on the Fringe." The Federal Reserve Bank of Minneapolis, July. Retrieved September 10, 2008 (www.minneapolisfed.org).

Maines, David R. 2001. *The Faultline of Consciousness: A View of Interactionism in Sociology*. New York: Aldine De Gruyter.

Major, Brenda, Mark Appelbaum, Linda Beckman, Mary Ann Dutton, Nancy Felipe Russo, and Carolyn West. 2008. "Report of the APA Task Force on Mental Health and Abortion." August 13, American Psychological Association. Retrieved September 5, 2008 (www.apa.org).

"Major Religions of the World Ranked by Number of Adherents." 2007. Retrieved August 24, 2007 (www.adherents.com).

Malle, Bertram F., and Gale E. Pearce. 2001. "Attention to Behavioral Events During Interaction: Two Actor-Observer Gaps and Three Attempts to Close Them." *Journal of Personality and Social Psychology* 81 (2): 278-294.

Malley-Morrison, Kathleen and Denise A. Hines. 2004. *Family Violence in a Cultural Perspective: Defining, Understanding, and Combating Abuse*. Thousand Oaks, CA: Sage.

Malthus, Thomas Robert. 1798/1965. *An Essay on Population*. New York: Augustus Kelley.

————. 1872/1991. *An Essay on the Principle of Population*, 7th ed. London: Reeves & Turner.

Mananzan, Mary John. 2002. "Theological Reflections on Violence against Women (A Catholic Perspective)." Pp. 205-212 in *Gendering the Spirit: Women, Religion & the Post-Colonial Response*, edited by Durre S. Ahmed. New York: Zed Books.

Manz, Charles C., and Henry P. Sims. 1987. "Leading Workers to Lead Themselves: The External Leadership of Self-Managing Work Teams." *Administrative Science Quarterly* 32: 106-108.

Margolis, Jane, and Allan Fisher. 2002. *Unlocking the Clubhouse: Women in Computing*. Cambridge, MA: MIT Press.

Marks, Alexandra. 2006. "Prosecutions Drop for White-Collar Crime." *Christian Science Monitor*, August 31, 3.

————. 2008. "Troubled Economy Hits Women Hard." *Christian Science Monitor*, January 2, 2.

————. 2008. "Voter Turnout Historical in Numbers and Diversity." *Christian Science Monitor*, November 6, 2.

Marks, Joseph L. 2007. *Fact Book on Higher Education 2007*. Atlanta, GA: Southern Regional Education Board.

Marshall, Tyler. 2004. "In Hong Kong, a Demand for Democracy." *Los Angeles Times*, July 2. Retrieved July 3, 2004 (www.latimes.com).

Martens, Jens. 2005. "A Compendium of Inequality." Global Policy Forum. Retrieved October 25, 2006 (www.globalpolicy.org).

Martin, Paige D., Don Martin, and Maggie Martin. 2001. "Adolescent Premarital Sexual Activity, Cohabitation, and Attitudes toward Marriage." *Adolescence* 36 (Fall): 601-609.

Martin, Philip, and Elizabeth Midgley. 2006.

"Immigration: Shaping and Reshaping America." *Population Bulletin* 61 (December): 1-28.

Martin, Philip, and Gottfried Zürcher. 2008. "Managing Migration: The Global Challenge." *Population Bulletin* 63 (March): 1-20.

Martin, Philip, and Jonas Widgren. 2002. "International Migration: Facing the Challenge." *Population Bulletin* 57 (March): 1-40.

Martin, Suzanne. 2006. "Advertising to Youth: What Youth Want and What Advertisers Need to Know." *Trends and Tudes* 5 (August): 1-5.

Martinez, G. M., A. Chandra, J. C. Abma, J. Jones, and W. D. Mosher. 2006. "Fertility, Contraception, and Fatherhood: Data on Men and Women from Cycle 6 (2002) of the National Survey of Family Growth. *Vital Health Statistics* 23 (26). Retrieved June 3, 2007 (www.cdc.gov/nchs).

Marty, Martin E. 2005. *When Faiths Collide*. Malden, MA: Blackwell.

Marvasti, Amir and Karyn D. McKinney. 2004. *Middle Eastern Lives in America*. Lanham, MD: Rowman & Littlefield.

Marx, Gary T. and Douglas McAdam. 1994. *Collective Behavior and Social Movements: Process and Structure*. Upper Saddle River, NJ: Prentice Hall.

Marx, Karl. 1844/1964. *Economic and Philosophic Manuscripts of 1844*. New York: International Publishers.

———. 1845/1972. "The German Ideology." Pp. 110-164 in *The Marx-Engels Reader*, edited by Robert C. Tucker. New York: W. W. Norton.

———. 1867/1967. *Capital*. Friedrich Engels, ed. New York: International Publishers.

———. 1934. *The Class Struggles in France*. New York: International Publishers.

———. 1964. *Karl Marx: Selected Writings in Sociology and Social Philosophy*. T. B. Bottomore, trans. New York: McGraw-Hill.

Masci, David. 2007. "How the Public Resolves Conflicts Between Faith and Science." Pew Forum on Religion & Public Life, August 27. Retrieved September 3, 2007 (www.pewforum.org).

Massey, Douglas S. 2005. "From Social Sameness, a Fascination with Differences." *The Chronicle Review*, August 12, B11-B12.

———. 2006. "Blackballed by Bush." *Contexts* 5 (Winter): 40-42.

———. 2007. *Categorically Unequal: The American Stratification System*. New York: Russell Sage Foundation.

Mather, Mark. 2008. "Population Losses Mount in U.S. Rural Areas." March, Population Reference Bureau. Retrieved July 29, 2008 (www.prb.org).

Mattera, Philip, and Anna Purinton. 2004. "Shopping for Subsidies: How Wal-Mart Uses Taxpayer Money to Finance Its Never-Ending Growth." Good Jobs First, May. Retrieved May 24, 2008 (www.goodjobsfirst.org).

Matthews, Warren. 2004. *World Religions*, 4th ed. Belmont, CA: Wadsworth.

Mauss, Armand. 1975. *Social Problems as Social Movements*. Philadelphia, PA: Lippincott.

Mayer, Caroline E. 2002. "Stuck Under a Load of Debt." *Washington Post*, November 9, E01.

Mayo, Elton. 1945. *The Problems of an Industrial Civilization*. Cambridge, MA: Harvard University Press.

McAdam, Doug, and Ronnelle Paulsen. 1994. "Specifying the Relationship between Social Ties and Activism." *American Journal of Sociology* 99 (November): 640-67.

McAdoo, Harriette P. 2002. "African American Parenting." Pp. 47-58 in *Handbook of Parenting*, 2nd ed., Volume 4: *Social Conditions and Applied Parenting*, edited by Marc H. Bornstein. Mahwah, NJ: Lawrence Erlbaum Associates.

McCabe, Donald. 2005. "Levels of Cheating and Plagiarism Remain High. Honor Codes and Modified Codes are Shown to be Effective in Reducing Academic Misconduct." Center for Academic Integrity, June. Retrieved October 5, 2006 (www.academicintegrity.org).

McCabe, Donald L., Kenneth D. Butterfield, and Linda K. Treviño. 2006. "Academic Dishonesty in Graduate Business Programs: Prevalence, Causes, and Proposed Action." *Academy of Management Learning Education* 5 (September): 294-305.

McCarthy, Ellen. 2004. "Md. Professor Archives History of Dot-Com Bombs." *Washington Post*, October 28, E1.

McCarthy, John, and Mayer Zald. 1977. "Resource Mobilization and Social Movements: A Partial Theory." *American Journal of Sociology* 82 (May): 1212–1241.

McCoy, J. Kelly, Gene H. Brody, and Zolinda Stoneman. 2002. "Temperament and the Quality of Best Friendships: Effect of Same-Sex Sibling Relationships." *Family Relations* 51 (July): 248-255.

McDonald, Renee, Ernest N. Jouriles, Suhasini Ramisetty-Mikler, Raul Caetano, and Charles E. Green. 2006. "Estimating the Number of American Children Living in Partner-Violent Families." *Journal of Family Psychology* 20 (March): 137-142.

McFalls, Joseph A., Jr. 2007. "Population: A Lively Introduction," 5th ed. *Population Bulletin* 62 (March): 1-31.

McGinn, Daniel. 2004. "Help Not Wanted." *Newsweek*, March, 31-33.

McGregor, Jena, and Steve Hamm. 2008. "Managing the Workforce." *Business Week*, January 28, 34-43.

McHale, Susan M. 2001. "Free-Time Activities in Middle Childhood: Links with Adjustment in Early Adolescence." *Child Development* 76 (November/December): 1764-78.

McIntosh, Peggy. 1995. "White Privilege and Male Privilege: A Personal Account of Coming to See Correspondences through Work in Women's Studies." Pp. 76-87 in *Race, Class, and Gender: An Anthology*, 2nd ed., edited by Margaret L. Andersen and Patricia Hill Collins. Belmont, CA: Wadsworth.

McLaughlin, Kathleen E. 2005. "With Woofs and Wet Noses, Dogs Help Heal in China." *Christian Science Monitor,* June 1, 7.

McMurtrie, Beth. 2001. "For Many Muslim Students, College Is a Balancing Act." *Chronicle of Higher Education*, November 9, A55-A57.

McNamee, Stephen J., and Robert K. Miller Jr. 1998. "Inheritance and Stratification." Pp. 193-213 in *Inheritance and Wealth in America*, edited by Robert K. Miller Jr. and Stephen J. McNamee. New York: Plenum Press.

McNeely, Clea, et al. 2002. "Mothers' Influence on the Timing of First Sex Among 14- and 15-year-olds." *Journal of Adolescent Health*, 31 (September): 256-265.

McPhail, Clark, and Ronald T. Wohlstein. 1983. "Individual and Collective Behavior within Gatherings, Demonstrations, and Riots." *Annual Review of Sociology*, 9 (August): 579-600.

McRae, Susan. 1999. "Cohabitation or Marriage?" Pp. 172-190 in *The Sociology of the Family*, edited by Graham Allan. Malden, MA: Blackwell Publishers.

McRoberts, Omar M. 2003. *Streets of Glory: Church and Community in a Black Urban Neighborhood*. Chicago, IL: The University of Chicago Press.

Mead, George Herbert. 1934. *Mind, Self, and Society*. Chicago, IL: University of Chicago Press.

———. 1964. *On Social Psychology*. Chicago, IL: University of Chicago Press.

Mead, Margaret. 1935. *Sex and Temperament in Three Primitive Societies*. New York: Morrow.

Mead, Lawrence M. 2008. "Comment on 'Helping Poor Working Parents Get Ahead'." Urban Institute, July 17. Retrieved August 25, 2008 (www.urban.org).

Mehl, Matthias R., Simine Vazire, Nairán Ramírez-Esparza, Richard B. Slatcher, and James W. Pennebaker. 2007. "Are Women Really More Talkative Than Men?" *Science* 317 (July): 82.

Mehta, Seema. 2008. "Parents May Home-School Children Without Teaching Credential, California Court Says." *Los Angeles Times*, August 9. Retrieved August 14, 2008 (www.latimes.com).

Mello, Michelle M., Brian R. Clarridge, and David M. Studdert. 2005. "Academic Medical Centers' Standards for Clinical-Trial Agreements with Industry." *New England Journal of Medicine* 352 (May 26): 2202-2210.

Melucci, Alberto. 1995. "The New Social Movements Revisited: Reflections on a Sociological Misunderstanding." Pp. 107-119 in *Social Movements and Social Classes: The Future of Collective Action*, edited by Louis Maheu. Thousand Oaks, CA: Sage.

Mendelsohn, Oliver and Maria Vicziany. 1998. *The Untouchables, Subordination, Poverty and the State in Modern India*. New York: Cambridge University Press.

Menifield, Charles E., Winfield H. Rose, John Homa, and Anita B. Cunningham. 2001. "The Media's Portrayal of Urban and Rural School Violence: A Preliminary Analysis." *Deviant Behavior: An Interdisciplinary Journal*, 22 (September/October): 447-64.

Menissi, Fatima. 1991. *The Veil and the Male Elite: A Feminist Interpretation of Women's Rights in Islam*. Reading, MA: Addison-Wesley.

———. 1996. *Women's Rebellion and Islamic Memory*. Atlantic Highlands: Zed Books.

"Mercury Pollution Endangers Public Health." 2005. Environmental Defense. Retrieved July 10, 2005 (www.environmentaldefense.org).

"The Merits of Gay Marriage." 2003. Editorial, *Washington Post*, November 20, A40.

Mermin, Gordon B. T. 2007. "Are Employers Willing to Hire and Retain Older Workers?" Urban Institute, December 7. Retrieved December 30, 2007 (www.urban.org).

Merton, Robert K. 1938. "Social Structure and Anomie." *American Sociological Review* 3 (December): 672-682.

———. 1948/1996. "The Self-Fulfilling Prophecy." Pp. 183-201 in *Robert K. Merton: On Social Structure and Science*, edited by Piotr Sztompka. Chicago, IL: University of Chicago Press.

———. 1949. "Discrimination and the American Creed." Pp. 99-126 in *Discrimination and National Welfare*, edited by Robert M. MacIver. New York: Harper.

———. 1968. *Social Theory and Social Structure*. New York: Free Press.

Merton, Robert K., and Alice K. Rossi. 1950. "Contributions to the Theory of Reference Group Behavior." Pp. 40-105 in *Continuities in Social Research*, edited by Robert K. Merton and Paul L. Lazarsfeld. New York: Free Press.

Metz, Mary Haywood. 2003. *Different by Design: The Context and Character of Three Magnet Schools*. New York: Teachers College, Columbia University.

Meyer, Cheryl L. and Michelle Oberman. 2001. *Mothers Who Kill Their Children: Understanding the Acts of Moms from Susan Smith to the "Prom Mom."* New York: New York University Press.

Michels, Robert. 1911/1949. *Political Parties*. Glencoe, IL: Free Press.

Mikkelson, Barbara, and David Mikkelson. 2005. "Does KFC Use Real Chickens?" Retrieved September 15, 2005 (www.scambusters.org).

Milanovic, Branko. 2006. "Global Income Inequality: What It Is and Why It Matters." World Bank.Retrieved October 20, 2006 (siteresources.worldbank.org).

Milbank, Dana. 2003. "Bush Delivers Religious Address." *Washington Post*, February 10, A2.

———. 2004. "Leaked Salary List Shows Bush's Highest-Paid Staff is Mostly Male." *Washington Post*, July 13, A13.

Milgram, Stanley. 1963. "Behavioral Study of Obedience." *Journal of Abnormal and Social Psychology* 67: 4 (371-378).

———. 1965. "Some Conditions of Obedience and Disobedience to Authority." *Human Relations* 18 (February): 57-76.

Miller, D. W. 2001. "DARE Reinvents Itself—With Help From Its Social-Scientist Critics." *Chronicle of Higher Education*, October 16, A12-A-14.

Miller, Matthew. 2005. "The (Porn) Player." *Forbes*, July 4, 124, 126, 128.

Miller, S. M. 2001. "My Meritocratic Rise." *Tikkun* (March/April): 1-3. Retrieved January 20, 2003 (www.tikkun.org).

Mills, C. Wright. 1956. *The Power Elite*. New York: Oxford University Press.

———. 1959. *The Sociological Imagination*. New York: Oxford University Press.

Min, Pyong Gap. 2002. "Introduction." Pp. 1-14 in *Religions in Asian America: Building Faith Communities*, edited by Pyong Gap Min and Jung Ha Kim. Walnut Creek, CA: AltaMira Press.

Mincy, Ronald, ed. 2006. *Black Males Left Behind*. Washington, DC: Urban Institute Press.

Miniño, A. M., M. P. Heron, S. L. Murphy, and K. D. Kochanek. 2007. "Deaths: Final Data for 2004." *National Vital Statistics Reports*, vol. 55, no 19. Hyattsville MD: National Center for Health Statistics

Mirowsky, John, and Catherine E. Ross. 2007. "Creative Work and Health." *Journal of Health and Social Behavior* 48 (December): 385-403.

Mischel, Walter. 1966. "A Social Learning View of Sex Differences." Pp. 57-81 in *The Development of Sex Differences*, edited by Eleanor E. Maccoby. Stanford, CA: Stanford University Press.

Mishel, Lawrence, Jared Bernstein, and Sylvia Allegretto. 2007. *The State of Working America 2006-2007*. Ithaca, NY: Cornell University Press.

Mitchell, Josh. 2002. "Hot Rod as Hobby, and Obsession." *Baltimore Sun*, July 24, 1A, 6A.

Monastersky, Richard. 2004. "'The Day After Tomorrow'." *Chronicle of Higher Education*, June 18, A8.

Money, John and Anke A. Ehrhardt. 1972. *Man & Woman, Boy & Girl: The Differentiation and Dimorphism of Gender Identity from Conception to Maturity*. Baltimore, MD: Johns Hopkins University Press.

Montagu, Ashley, ed. 1999. *Race and IQ: Expanded Edition*. New York: Oxford University Press.

Montenegro, Xenia P. 2004. "The Divorce Experience: A Study of Divorce at Midlife and Beyond." *AARP the Magazine* (May): 40-79.

Montlake, Simon. 2006. "China Reins in Reach of Foreign News." *Christian Science Monitor*, September 13, 6.

Moore, Elizabeth S. 2006. "It's Child's Play: Advergaming and the online Marketing of Food to Children." Henry J. Kaiser Family Foundation, July. Retrieved April 12, 2008 (www.kff.org).

Moore, Kathleen. 2008. "Low-income Homes Green—and Affordable." *Daily Gazette*, July 1. Retrieved September 21, 2008 (www.dailygazette.com).

Moore, Lisa. 2007. "The Secret to Smarter Schools." *U.S. News & World Report*, March 26-April 7, 54-55.

Moore, Martha T. 2006. "Cows Power Plan for Alternative Fuel." *USA Today*, December 3, 11A.

Morais, Richard C. 2007. "Desperate Arrangements." *Forbes*, January 29, 72-79.

Morales, Gail. 2008. "Record Numbers of Women to Serve in Senate and House." Center for American Women and Politics, November 10. Retrieved November 20, 2008 (www.cawp.rutgers.edu).

Moreau, Ron, and Sami Yousafzai. 2006. "A War on Schoolgirls." *Newsweek*, June 26, 34-35.

Morgenson, Gretchen. 2004. "No Wonder C.E.O.'s Love Those Mergers." *New York Times*, July 18, C1.

Mori, Kyoko. 1997. *Polite Lies: On Being a Woman Caught Between Cultures*. New York: Henry Holt and Company.

Morris, Desmond. 1994. *Bodytalk: The Meaning of Human Gestures*. New York: Crown Trade Paperbacks.

Morris, Jim. 2006. "Power Trips: Privately Sponsored Trips Hot Tickets on Capitol Hill." Center for Public Integrity, June 5. Retrieved May 23, 2007 (www.publicintegrity.org).

Morrison, Denton E. 1971. "Some Notes toward Theory on Relative Deprivation, Social Movements, and Social Change." *The American Behavioral Scientist* 14 (May-June): 675-690.

Mosher, William D., Anjani Chandra, and Jo Jones. 2005. "Sexual Behavior and Selected Health Measures: Men and Women 15-44 Years of Age, United States, 2002." National Center for Health Statistics, *Vital and Health Statistics*, No. 362, September 15. Retrieved January 10, 2006 (www.cdc.gov/nchs).

Motivans, Mark. 2004. "Intellectual Property Theft, 2002." Washington, DC: U.S. Department of Justice, Bureau of Justice Statistics.

Moyer, Imogene L. 2001. *Criminological Theories: Traditional and Nontraditional Voices and Themes*. Thousand Oaks, CA: Sage.

_____. 2003. "Jane Addams: Pioneer in Criminology." *Women & Criminal Justice* 14 (3/4): 1-14.

Mulac, Anthony. 1998. "The Gender-Linked Language Effect: Do Language Differences Really Make a Difference?" Pp. 127-153 in *Sex Differences and Similarities in Communication: Critical Essays and Empirical Investigations of Sex and Gender in Interaction*, edited by Daniel J. Canary and Kathryn Dindia. Mahwah, NJ: Lawrence Erlbaum Associates.

Mulligan, Thomas S. 2004. "Beanie Babies Rode Own Bubble." *Baltimore Sun*, August 31, 1C, 3C.

Mulvihill, Keith. 2001. "US Kids' Distaste for Obese Peers Has Grown." Retrieved November 15, 2001 (http://news.excite.com).

Mumford, Lewis. 1961. *The City in History: Its Origins, Transformations, and Its Prospects*. New York: Harcourt, Brace.

Mumola, Christopher J. 2000. "Incarcerated Parents and Their Children." U.S. Department of Justice, November. Retrieved July 14, 2002 (www.ojp.usdoj.gov/bjs).

"Municipal Solid Waste." 2005. U.S. Environmental Protection Agency. Retrieved June 30, 2005 (www.epa.gov).

Murdock, George P. 1940. "The Cross-Cultural Survey." *American Sociological Review*, 5: 361-70.

_____. 1945. "The Common Denominator of Cultures." Pp. 123-142 in *The Science of Man in the World Crisis*, edited by Ralph Linton. New York: Columbia University Press.

_____. 1967. "Ethnographic Atlas: A Summary." *Ethnology* 6, 109–236.

Murphy, Cait. 2005. "Fast-Forward to the Future." *Fortune*, September 19, 271.

Murphy, Evelyn and E. J. Graff. 2005. *Getting Even: Why Women Don't Get Paid Like Men—And What To Do About It*. New York: Simon & Schuster.

Murphy, John. 2004. "S. Africa's New Goal: Economic Equality." *Baltimore Sun*, April 27, 1A, 4A.

Murray, Charles. 2008. "College Daze." *Forbes*, September 1, 32.

Myers, John P. 2007. *Dominant-Minority Relations in America: Convergence in the New World*, 2nd ed. Boston, MA: Allyn and Bacon.

Naisbitt, John, Nana Naisbitt, and Douglas Philips. 1999. *High Tech/High Touch: Technology and Our Search for Meaning*. New York: Broadway Books.

Nakao, Keiko and Judith Treas. 1992. "The 1989 Socioeconomic Index of Occupations: Construction from the 1989 Occupational Prestige Scores," GSS Methodological Report No. 74. Chicago, IL: National Opinion Research Center.

Nakashima, Ellen. 2003. "Debating Polygamy's Resurgence." *Washington Post*, November 28, A1.

_____. 2008. "Prescription Data Used to Assess Consumers." *Washington Post*, August 4, A1.

Nathan, Rebekah. 2005. *My Freshman Year: What a Professor Learned by Becoming a Student*. Ithaca, NY: Cornell University Press.

National Academy of Sciences. 2007. *Treatment of PTSD: An Assessment of the Evidence*. Washington, DC: National Academies Press.

_____. 2008. *Science, Evolution, and Creationism*. Washington, DC: National Academy of Sciences.

National Center for Education Statistics. 2007. *Mapping 2005 State Proficiency Standards onto the NAEP Scales* (NCES 2007-482). Washington, DC: U.S. Department of Education.

National Center for Health Statistics. 2005. *Health, United States, 2005, With Chartbook on Trends in the Health of Americans*. Hyattsville, Maryland.

_____. 2007. *Health, United States, 2007 with Chartbook on Trends in the Health of Americans*. Washington, DC: U.S. Government Printing Office.

National Center on Addiction and Substance Abuse. 2003. "National Survey of American Attitudes on Substance Abuse VIII: Teens and Parents." Retrieved August 25, 2003 (www.casacolumbia.org).

National Center on Elder Abuse. 2005. "Elder Abuse Prevalence and Incidence." Retrieved June 23, 2007 (www.elderabusecenter.org).

National Coalition for the Homeless. 2007a. "How Many People Experience Homelessness?" August. Retrieved May 17, 2008 (www.nationalhomeless.org).

_____. 2007b. "Why are People Homeless?" June. Retrieved May 17, 2008 (www.nationalhomeless.org).

National Council of Teachers of English Committee on Public Doublespeak. 2005. Retrieved February 12, 2006 (www.ncte.org).

National Eating Disorders Association. 2008. "Facts and Statistics." Retrieved September 6, 2008 (www.nationaleatingdisorders.org).

National Election Study. 2003. University of Michigan Codebook. Retrieved August 21, 2004 (ftp://ftp.nes.isr.umich.edu).

National Geographic Education Foundation. 2006. "Final Report: 2006 Geographic Literacy Study." Washington, DC. Retrieved July 21, 2007 (www.nationalgeographic.com).

National Healthcare Disparities Report. 2003. U.S. Department of Health and Human Services. Retrieved April 20, 2004 (www.qualitytools.ahrq.gov).

National Institutes of Health. 2008. "Stem Cell Information." Retrieved August 6, 2008 (www.stemcells.nih.gov/info).

National Latino Alliance for the Elimination of Domestic Violence. 2005. "Domestic Violence Affects Families of All Racial, Ethnic, and Economic Backgrounds." Retrieved June 20, 2007 (www.dvalianza.org).

National Research Council. 2001. *Informing America's Policy on Illegal Drugs: What We Don't Know Keeps Hurting Us*. Washington, DC: National Academy Press.

National Survey of Student Engagement. 2006. "Engaged Learning: Fostering Success for All Students, Annual Report 2006." National Survey of Student Engagement. Retrieved August 5, 2007 (http://nsse.iub.edu).

"Nation's Population One-Third Minority." 2006. U.S. Census Bureau News, May 10. Retrieved July 28, 2006 (www.census.gov).

Natural Resources Defense Council. 2007. "Beach Pollution." Retrieved February 8, 2008 (www.ndrc.org).

Neelakantan, Shailaja. 2006. "In India, Conservatives Want Women Under Wraps." *Chronicle of Higher Education*, May 26, A47-A48.

_____. 2008. "India Expands Quota System for Lower-Caste Students." *Chronicle of Higher Education* (April 25): A31.

Neider, Linda L. and Chester A. Schriesheim, eds. 2005. *Understanding Teams*. Greenwich, CT: Information Age.

Netherlands Environmental Assessment Agency. 2007. "China Now No. 1 in CO_2 Emissions; USA in Second Position." Retrieved February 12, 2008 (www.mnp.nl).

Newman, David M. 2005. *Identities and Inequalities: Exploring the Intersections of Race, Class, Gender, and Sexuality*. New York: McGraw-Hill.

Newman, Katherine S. 1999. *No Shame in My Game: The Working Poor in the Inner City*. New York: Russell Sage Foundation.

Newman, Katherine S., and Victor Tan Chen. 2007. "The Crisis of the Near Poor." *Chronicle of Higher Education*, October 5, B10-B11.

Newport, Frank. 2001. "Americans See Women as Emotional and Affectionate, Men as More Aggressive." *Gallup Poll Monthly* No. 425 (February, 2001): 34-38.

_____. 2006a. "Religion Most Important to Blacks, Women, and Older Americans." The Gallup Poll, November 29. Retrieved August 27, 2007 (www.galluppoll.com).

_____. 2006b. "Who Believes in God and Who Doesn't?" The Gallup Poll, June 23. Retrieved June 24, 2006 (www.galluppoll.com).

_____. 2007. "Just Why Do Americans Attend Church?" The Gallup Poll, April 6. Retrieved August 27, 2007 (www.galluppoll.com).

_____. 2007. "Sixty-Nine Percent of Americans Support Death Penalty." Gallup News Service, October 12. Retrieved May 12, 2008 (www.gallup.com).

_____. 2008. "Bush Job Approval at 25%, His Lowest Yet." Gallup, October 6. Retrieved October 14, 2008 (www.gallup.com).

_____. 2008. "Personal Financial Assessments Remain at 32-Year Low." Gallup News, September 29. Retrieved October 16, 2008 (www.gallup.com).

_____. 2008. "Wives Still Do Laundry, Men Do Yard Work." Gallup News Service, April 4. Retrieved June 2, 2008 (www.gallup.com).

Newport, Frank, and Lydia Saad. 2006. "Religion, Politics Inform Americans' Views on Abortion." Gallup Organization. Retrieved April 5, 2005 (www.gallup.com).

"New words in the Concise Oxford Dictionary." 2001. Retrieved January 20, 2002 (www.askoxford.com).

Nichols, Austin. 2006. "Understanding Recent Changes in Child Poverty." Urban Institute, August 25. Retrieved February 15, 2007 (www.urban.org).

Nie, Norman H., and Saar D. Golde. 2008. "Does Education Really Make You Smarter?" *Miller-McCune* 1 (June/July): 56-64.

Niebuhr, H. Richard. 1929. *The Social Sources of Denominationalism*. New York: Meridian.

Nielsen Company, The. 2008. "College Spring Break Study." February 27. Retrieved April 10, 2008 (www.alcoholstats.com).

Nitkin, David. 2008. "Hopkins' Carson to Get Medal of Freedom." *Baltimore Sun*, June 12, 7B.

"No Action on Greenhouse Gases." 2008. *Baltimore Sun*, July 12, 2A.

Noguchi, Yuki. 2005. "Good Samaritans Turn to Web to Help Victims." *Washington Post*, September 2, D4.

Norris, Pippa. 2003. "The Gender Gap: Old Challenges, New Approaches." Pp. 146-170 in *Women and American Politics: New Questions, New Directions*, edited by Susan J. Carroll. Oxford, NY: Oxford University Press.

Nunberg, Geoffrey. 2007. "The Language of Death." *Los Angeles Times*, February 12. Retrieved February 14 (www.latimes.com).

Nussbaum, Martha C. 2003. "Women's Education: A Global Challenge." *Signs: Journal of Women in Culture and Society* 29 (Winter): 325-355.

O'Brien, Jodi and Peter Kollock. 2001. *The Production of Reality: Essays and Readings on Social Interaction*, 3rd ed. Thousand Oaks, CA: Pine Forge.

O'Donnell, Victoria. 2007. *Television Criticism*. Thousand Oaks, CA: Sage.

O'Hare, William P. 2002. "Tracking the Trends in Low-Income Working Families," *Population Today* 30 (August/September): 1-3.

O'Harrow, Jr., Robert. 2008. "Earmark Spending Makes a Comeback." *Washington Post*, June 13, A1.

O'Neill, Brendan. 2007. "Gaelic-Only Laws: Linguistic Apartheid?" *Christian Science Monitor*, February 6, 20.

Oakes, Jeannie. 1985. *Keeping Track: How Schools Structure Inequality*. New Haven, CT: Yale University Press.

Obach, Brian K. 2004. *Labor and the Environmental Movement: The Quest for Common Ground*. Cambridge, MA: MIT Press.

Oberschall, Anthony. 1973. *Social Conflict and Social Movements*. Englewood Cliffs, NJ: Prentice Hall.

Office of Applied Studies. 2004. "Alcohol and Drug Services Study (ADSS) Cost Study." Substance Abuse and Mental Health Services Administration (SAMSA), June 18. Retrieved May 14, 2005 (oas.samhsa.gov).

Office of National Drug Control Policy. 2007. "Teens and Prescription Drugs: An Analysis of Recent Trends on the Emerging Drug Threat." Retrieved May 10, 2008 (www.mediacampaign.org).

Office of the Deputy Chief of Staff for Intelligence. 2006. "Arab Cultural Awareness: 58 Factsheets." U.S. Army Training and Doctrine Command, Ft. Leavenworth, Kansas, January. Retrieved February 15, 2006 (www.fas.org).

Ogburn, William F. 1922. *Social Change with Respect to Culture and Original Nature*. New York: Dell.

Ogden, Cynthia L., Margaret D. Carroll, and Katherine M. Flegal. 2008. "High Body Mass Index for Age among US Children and Adolescents, 2003-2006." *JAMA* 299 (May 28): 2401-2405.

Ogunwole, Stella U. 2006. "We the People: American Indians and Alaska Natives in the United States." Census 2000 Special Reports, CENSR-28. Retrieved March 26, 2007 (www.census.gov).

Ore, Tracy E. 2006. *The Social Construction of Difference and Inequality*, 3rd ed. New York: McGraw-Hill.

Ornstein, Allan C. 2003. *Pushing the Envelope: Critical Issues in Education*. Upper Saddle River, NJ: Prentice Hall.

Orrenius, Pia. 2004. "Immigrant Assimilation: Is the U.S. Still a Melting Pot?" *Southwest Economy*, Federal Reserve Bank of Dallas, May/June. Retrieved January 3, 2005 (www.dallasfed.org).

Orum, Anthony M. 2001. *Introduction to Political Sociology*, 4th ed. Upper Saddle River, NJ: Prentice Hall.

Packard, Vance. 1959. *The Status Seekers*. New York: David McKay.

Padgett, Tim. 2007. "Pilfering Priests." *Time*, February 26, 46-47.

Park, Lisa Sun-Hee. 2002. "Asian Immigrant Entrepreneurial Children." Pp. 161-174 in *Contemporary Asian American Communities: Intersections and Divergences*, edited by Linda Trinh Võ and Rick Bonus. Philadelphia: Temple University Press.

Park, Robert and Ernest Burgess. 1921. *Introduction to the Science of Sociology*. Chicago, IL: University of Chicago Press.

Parkinson, C. Northcote. 1962. *Parkinson's Law*, 2nd ed. Boston, MA: Houghton Mifflin.

Parsons, Talcott. 1954. *Essays in Sociological Theory*, revised edition. New York: Free Press.

———. 1959. "The School Class as a Social System: Some of its Functions in American Society." *Harvard Educational Review* 29 (Fall): 297-313.

———. 1960. *Structure and Process in Modern Societies*. New York: Free Press.

Parsons, Talcott, and Robert F. Bales, eds. 1955. *Family, Socialization and Interaction Process*. New York: Free Press.

Passel, Jeffrey S. 2006. "The Size and Characteristics of the Unauthorized Migrant Population in the U.S." Pew Hispanic Center, March 7. Retrieved May 2, 2007 (www.pewhispanic.org).

Pasupathi, Monisha. 2002. "Arranged Marriages: What's Love Got To Do With It?" Pp. 211-235 in *Inside the American Couple: New Thinking/New Challenges*, edited by Marilyn Yalom and Laura L. Carstensen. Berkeley: University of California Press.

Paul, Noel. C. 2002. "The Birth of a Would-Be-Fad." *Christian Science Monitor*, September 23, 11, 14-16.

Paul, Pamela. 2005. *Pornified: How Pornography is Transforming Our Lives, Our Relationships, and Our Families*. New York: Henry Holt.

Paul, Richard and Linda Elder. 2007. *The Miniature Guide to Critical Thinking: Concepts and Tools*. Dillon Beach, CA: The Foundation for Critical Thinking.

Pauley, Bruce F. 1997. *Hitler, Stalin, and Mussolini: Totalitarianism in the Twentieth Century*. Wheeling, IL: Harlan Davidson, Inc.

Pearce, Diana. 1978. "The Feminization of Poverty: Women, Work, and Welfare." *Urban and Social Change Review* 11, 28–36.

Pedersen, Paul. 1995. *The Five Stages of Culture Shock: Critical Incidents around the World*. Westport, CT: Greenwood Press.

Pelton, Tom. 2007. U.S. Blocks States on Emissions." *Baltimore Sun*, February 20, A7.

Pennington, Bill. 2006. "Small Colleges, Short of Men, Embrace Football." *New York Times*, July 10, 1.

Perie, Marianne, Rebecca Moran, and Anthony D. Lutkus. 2005. *NAEP 2004 Trends in Academic Progress: Three Decades of Student Performance in Reading and Mathematics* (NCES 2005-464). U.S. Department of Education, Institute of Education Sciences, National Center for Education Statistics. Washington, DC: U.S. Government Printing Office.

Perlmutter, David D. 2001. "Students are Blithely Ignorant; Professors are Bitter." *Chronicle of Higher Education*, July 27, B20.

Perloff, Richard M., Bette Bonder, George B. Ray, Eileen B. Ray, and Laura A. Siminoff. 2006. "Doctor-Patient Communication, Cultural Competence, and Minority Health." *American Behavioral Scientist* 49 (February): 835-852.

Perrucci, Robert and Earl Wysong. 1999. *The New Class Society*. Lanham, MD: Rowman & Littlefield Publishers.

Pershing, Ben. 2008. "Arizona's Rep. Renzi is Indicted in Land Deal." *Washington Post*, February 23, A1.

Peter, Lawrence J., and Raymond Hull. 1969. *The Peter Principle*. New York: Morrow.

Peterson, Jonathan. 2006. "Many Forced To Retire Early." *Los Angeles Times*, May 15. Retrieved May 18, 2006 (www.latimes.com).

Pew Center on the States. 2008. "One in 100: Behind Bars in America 2008." February. Retrieved May 11, 2008 (www.pewcenteronthestates.org).

Pew Forum on Religion & Public Life. 2008. "U.S. Religious Landscape Survey: Religious Beliefs and Practices: Diverse and Politically Relevant." June. Retrieved June 26, 2008 (religions.pewforum.org).

Pew Global Attitudes Project. 2006. "The Great Divide: How Westerners and Muslims View Each Other." June 22. Retrieved August 15, 2007 (www.pew-global.org).

———. 2007. "World Publics Welcome Global Trade—But Not Immigration." October 4. Retrieved November 25, 2007 (www.pewglobal.org).

Pew Initiative on Food and Biotechnology. 2004. "Genetically Modified Crops in the United States." August. Retrieved September 30, 2005 (www.pewagbiotech.org).

Pew Internet & American Life Project Surveys. 2007. "Percentage of U.S. Adults Online." February 15. Retrieved April 23, 2008 (www.pewinternet.org).

Pew Internet & American Life Project. 2008. "Demographics of Internet Users." February 15. Retrieved April 23, 2008 (www.pewinternet.org).

Pew Research Center. 2006. "Pragmatic Americans Liberal *and* Conservative on Social Issues." Washington, DC: Pew Research Center for the People & the Press. Retrieved January 8, 2007 (www.people-press.org).

———. 2008. "Inside the Middle Class: Bad Times Hit the Good Life." April 9. Retrieved April 10, 2008 (pewresearch.org).

———. 2008. "Public Support Falls for Religion's Role in Politics." August 21. Retrieved September 18, 2008 (www.pewresearch.org).

Pewewardy, Cornel. 1998. "Fluff and Feathers: Treatment of American Indians in the Literature and the Classroom." *Equity & Excellence in Education* 31 (April): 69–76.

Phillips, Kevin. 2002. *Wealth and Democracy: A Political History of the American Rich*. New York: Broadway Books.

Phinney, Jean S., Irma Romero, Monica Nava, and Dan Huang. 2001. "The Role of Language, Parents, and Peers in Ethnic Identity among Adolescents in Immigrant Families." *Journal of Youth and Adolescence* 30 (April): 135-153.

Pianta, Robert C. 2002. *School Readiness: A Focus on Children, Families, Communities, and Schools*. Arlington, VA: Educational Research Service.

Pianta, Robert C., Jay Belsky, Renate Houts, Fred Morrison, and The National Institute of Child Health and Human Development (NICHD) Early Child Care Research Network. 2007. "Teaching: Opportunities to Learn in America's Elementary Classrooms." *Science* 315 (March 30): 1795-6.

Piore, Adam. 2005. "A Higher Frequency." *Mother Jones*, December, 47-51, 80-81.

Pipher, Mary. 1994. *Reviving Ophelia: Saving the Selves of Adolescent Girls*. New York: Putnam.

Pittz, Will. 2005. "Closing the Gap: Solutions to Race-Based Health Disparities." Applied Research Center & Northwest Federation of Community Organizations, June. Retrieved April 20, 2007 (www.arc.org).

"Plan to Build Texas Mosque Sends Neighbors Into Uproar." 2006. *Baltimore Sun*, December 8, 2A.

Planty, M. W. Hussar, T. Snyder, S. Provasnik, G. Kena, R. Dinkes, A. KewalRamani, and J. Kemp. 2008. *The Condition of Education 2008* (NCES 2008-031). Washington, DC: National Center for Education Statistics, Institute of Education Sciences, U.S. Department of Education.

Planty, Michael, et al. 2007. *The Condition of Education 2007* (NCES 2007-064). Washington, DC: U.S. Government Printing Office.

Plateris, Alexander A. 1973. *100 Years of Marriage and Divorce Statistics: 1867-1967*. Rockville, MD: National Center for Health Statistics.

Political Money Line. 2004. "Money in Politics Databases." Retrieved August 18, 2004 (www.fecinfo.com).

"Polling Prayers." 2004. *U.S. News & World Report*, December 20, 56.

Polsby, Nelson W. 1959. "Three Problems in the Analysis of Community Power." *American Sociological Review* 24 (December): 796-803.

Pong, Suet-ling, Lingxin Hao, and Erica Gardner. 2005. "The Roles of Parenting Styles and Social Capital in the School Performance of Immigrant Asian and Hispanic Adolescents." *Social Science Quarterly* 86 (December): 928-950.

Popenoe, David and Barbara D. Whitehead. 2002. *Should We Live Together? What Young Adults Need To Know About Cohabitation Before Marriage—A Comprehensive Review Of Recent Research*, 2nd ed. New Brunswick, NJ: The National Marriage Project, Rutgers University. Retrieved July 12, 2003 (marriage.rutgers.edu).

———. 2006. *The State of Our Unions: 2006*. Rutgers, the State University of New Jersey, The National Marriage Project. Retrieved July 28, 2006 (http://marriage.rutgers.edu).

Population Division, U.S. Census Bureau. 2008. "Projections of the Population by Race and Hispanic Origin for the United States: 2208 to 2050" and "Percent of the Projected Population by Race and Hispanic Origin for the United States: 2008 to 2050." August 14. Retrieved September 8, 2008 (www.census.gov).

Population Reference Bureau. 2007. "World Population Highlights." *Population Bulletin* 62 (September): 1-12.

Postman, Neil and Steve Powers. 1992. *How to Watch TV News*. New York: Penguin Books.

Powell, Lisa M., Glen Szczypka, Frank J. Chaloupka, and Carol L. Braunschweig. 2007. "Nutritional Content of Television Food Advertisements Seen by Children and Adolescents in the United States." *Pediatrics* 120 (September): 576-583.

Powers, Charles H. 2004. *Making Sense of Social Theory: A Practical Introduction*. Lanham, MD: Rowman & Littlefield.

Price, Barbara Raffel, and Natalie J. Sokoloff, eds. 2004. *The Criminal Justice System and Women: Offenders, Prisoners, Victims, & Workers*, 3rd edition. New York: McGraw-Hill.

Price, J. A. 1981. "North American Indian Families." Pp. 245–68 in *Ethnic Families in America: Patterns and Variations*, 2nd ed., edited by Charles H. Mindel and Robert W. Habenstein. New York: Elsevier.

Princiotta, Daniel, and Stacey Bielick. 2006. *Homeschooling in the United States: 2003* (NCES 2006-042). National Center for Education Statistics, U.S. Department of Education. Washington, DC: U.S. Government Printing Office.

Prothrow-Stith, Deborah and Howard R. Spivak. 2005. *Sugar and Spice and No Longer Nice: How We Can Stop Girls' Violence*. San Francisco, CA: Jossey-Bass.

Provasnik, Stephen and Scott Dorfman. 2005. *Mobility in the Teacher Workforce* (NCES 2005-114). U.S. Department of Education, National Center for Education Statistics. Washington, DC: U.S. Government Printing Office.

Pryor, John H., Sylvia Hurtado, Victor B. Saenz, José Louis Santos, and William S. Korn. 2007. *The American Freshman: Forty Year Trends*. Los Angeles: Higher Education Research Institute, UCLA.

Public Education Finances 2005. 2007. U.S. Census Bureau. Retrieved July 20, 2007 (www.census.gov).

Putnam, Frank W. 2006. "The Impact of Trauma on Child Development." *Juvenile and Family Court Journal* 57 (Winter): 1-11.

Qian, Zhenchao. 2005. "Breaking the Last Taboo: Interracial Marriage in America." *Contexts* 4 (Fall): 33-37.

Qian, Zhenchao, and Daniel T. Lichter. 2007. "Social Boundaries and Marital Assimilation: Interpreting Trends in Racial and Ethnic Intermarriage." *American Sociological Review* 72 (February): 68-94.

"Quality Counts 2007: From Cradle to Career." 2007. *Education Week*, January 4. Retrieved July 23, 2007 (www.edweek.org).

Quinney, Richard. 1980. *Class, State, and Crime*. Boston, MA: Little, Brown.

Radosh, Polly F. 1993. "Women and Crime in the United States: A Marxian Explanation." Pp. 263-289 in *Female Criminality: The State of the Art*, edited by Concetta C. Culliver. New York: Garland.

Raintree Nutrition, Inc. 2008. "Rainforest Facts." Retrieved February 12, 2008 (www.rain-tree.com).

Ratner, Edward R., and John Y. Song. 2002. "Education for the End of Life." *Chronicle of Higher Education*, June 7, B12.

Ray, Rebecca, and John Schmitt. 2007. "No-Vacation Nation." Center for Economic and Policy Research, May. Retrieved July 4, 2008 (www.cepr.net).

Rayasam, Renuka. 2007. "Immigrants: The Unsung Heroes of the U.S. Economy." *U.S. News & World Report*, February 26, 58.

Reay, Campbell A. M., and K. D. Browne. 2001. "Risk Factor Characteristics for Carers Who Physically Abuse or Neglect Their Elderly Dependents." *Aging and Mental Health* 5 (February): 56-62.

Reed, Betsy. 2008. "Race to the Bottom." *The Nation*, May 1. Retrieved June 25, 2008 (www.thenation.com).

Reeves, Mary E. 1999. "School Is Hell: Learning with (and from) *The Simpsons*." Pp. 55-82 in *Popular Culture and Critical Pedagogy: Reading, Constructing, Connecting*, edited by Toby Daspit and John A. Weaver. New York: Garland Publishing.

Reindl, Travis. 2007. *Hitting Home: Quality, Cost, and Access Challenges Confronting Higher Education Today*. Making Opportunity Available. Retrieved August 2, 2007 (www.makingopportunityaffordable.org).

Reiss, Steven. 2004. "The Sixteen Strivings for God." *Zygon* 39 (June): 303-320.

"Religions of the World: Number of Adherents, Names of Houses of Worship . . .". 2007. Religious Tolerance. Retrieved August 24, 2007 (www.religioustolerance.org).

Rennison, Callie M. 2003. *Intimate Partner Violence, 1993-2001*. Washington, DC: U.S. Department of Justice.

Renzulli, Linda A., and Vincent J. Roscigno. 2007. "Charter Schools and the Public Good." *Contexts* 6 (Winter): 31-36.

Reynolds, Christopher. 2007. "Without Foreign Workers, U.S. Parks Struggle." *Los Angeles Times*, May 27. Retrieved October 10, 2007 (www.latimes.com).

Reynolds, Maura, Michelle Quinn, and Ronald D. White. 2008. "$4 Gasoline? It's News to Bush." *Los Angeles*

Times, February 29. Retrieved March 1, 2008 (www.latimes.com).

Rheingold, Harriet L. 1969. "The Social and Socializing Infant." Pp. 779-90 in *Handbook of Socialization Theory and Research*, edited by David A. Goslin. Chicago, IL: Rand McNally.

Rhoads, Steven E. 2004. *Taking Sex Differences Seriously*. San Francisco, CA: Encounter Books.

Rideout, Victoria. 2007. "Parents, Children & Media." Henry J. Kaiser Family Foundation, June. Retrieved April 12, 2008 (www.kff.org).

Rideout, Victoria, Donald F. Roberts and Ulla G. Foehr. 2005. *Generation M: Media in the Lives of 8-18 Year-Olds*. Menlo Park, CA: Kaiser Family Foundation.

Rice, George, Carol Anderson, Neil Risch, and George Ebers. 1999. "Male Homosexuality: Absence of Linkage to Microsatellite Markers at Xq28." *Science* 284 (April 23): 665–67.

Riche, Martha Farnsworth. 2000. "America's Diversity and Growth: Signposts for the 21st Century." *Population Bulletin* 55 (June): 1-41.

Richtel, Matt, and Alexei Barrionuevo. 2005. "Wendy's Restaurants." *New York Times*, April 22, A9.

Rickles, Jordan, Paul M. Ong, and Doug Houston. 2002. "The Integrating (and Segregating) Effect of Charter, Magnet, and Traditional Elementary Schools: The Case of Five California Metropolitan Areas." UCLA School of Public Policy and Social Research, October 11. Retrieved January 2, 2005 (http://lewis.sppsr.ucla.edu).

Ridge, Mian. 2008. "Anti-Christian Attacks Flare in India." *Christian Science Monitor*, September 24, 7.

Riesman, David. 1953. *The Lonely Crowd*. New York: Doubleday.

Riordan, Cornelius. 1997. *Equality and Achievement: An Introduction to the Sociology of Education*. New York: Addison-Wesley Longman.

"Rising to the Challenge: Are High School Graduates Prepared for College and Work? A Study of Recent High School Graduates, College Instructors, and Employers." 2005. Achieve, Inc., February. Retrieved February 10, 2005 (www.achieve.org).

Ritter, John. 2007. "Pot Growing Moves to Suburbs." *USA Today*, February 7, 3A.

Ritzer, George. 1992. *Contemporary Sociological Theory*, 3rd ed. New York: McGraw-Hill.

————. 1996. *The McDonaldization of Society: An Investigation into the Changing Character of Contemporary Social Life*. Thousand Oaks, CA: Pine Forge Press.

————. 2008. *The McDonaldization of Society*, 5th edition. Los Angeles, CA: Pine Forge Press.

Roan, Shari. 2008. "Cut in Paid Sick Days Leave Unhealthy Employees Stuck in the Workplace." *Los Angeles Times*, July 7. Retrieved July 7, 2008 (www.latimes.com).

Roberts, Keith A. 2004. *Religion in Sociological Perspective*, 4th ed. Belmont, CA: Wadsworth.

Roberts, Les, Riyadh Lafta, Richard Garfield, and Jamal Khudhairi. 2004. "Mortality Before and After the 2003 Invasion of Iraq: Cluster Sample Survey." *Lancet* 364 (October 29): 1857-64.

Robey, Elizabeth B., Daniel J. Canary, and Cynthia S. Burggraf. 1998. "Conversational Maintenance Behaviors of Husbands and Wives: An Observational Analysis." Pp. 373-92 in *Sex Differences and Similarities in Communication: Critical Essays and Empirical Investigations of Sex and Gender in Interaction*, edited by Daniel J. Canary and Kathryn Dindia. Mahwah, NJ: Lawrence Erlbaum Associates.

Robinson, John P., Geoffrey Godbey and Robert D. Putnam. 1999. *Time for Life: The Surprising Ways Americans Use Their Time*, 2nd edition. University Park: The Pennsylvania State University Press.

Roethlisberger, F. J., and William J. Dickson. 1939/1942. *Management and the Worker: An Account of a Research Program Conducted at the Western Electric Company, Hawthorne Works, Chicago*. Cambridge, MA: Harvard University Press.

Rogers, Stacy J. 2004. "Dollars, Dependency, and Divorce: Four Perspectives on the Role of Wives' Income." *Journal of Marriage and Family* 66 (February): 59-74.

Roof, Wade Clark. 1993. *A Generation of Seekers*. New York: HarperCollins.

Roosevelt, Margot. 2008. "An Elusive Billionaire Gives

Away His Good Fortune." *Los Angeles Times*, March 8. Retrieved March 9, 2008 (www.latimes.com).

Rose, Fred. 1997. "Toward a Class-Cultural Theory of Social Movements: Reinterpreting New Social Movements." *Sociological Forum* 12 (September): 461-494.

Rose, Peter I. 1997. *They and We: Racial and Ethnic Relations in the United States*, 5th edition. New York: McGraw-Hill.

Rosen, Jacob, and Blake Hannaford. 2006. "Doc at a Distance." *IEEE Spectrum* (October): 34-39.

Rosenberg, Janice. 1993. "Just the Two of Us." Pp. 301-307 in *Reinventing Love: Six Women Talk About Lust, Sex, And Romance*, edited by Laurie Abraham, Laura Green, Magda Krance, Janice Rosenberg, Janice Somerville, and Carroll Stoner. New York: Plume.

Rosenfeld, Michael J. 2002. "Measures of Assimilation in the Marriage Market: Mexican Americans 1970-1990." *Journal of Marriage and the Family* 64 (February): 152-162.

Ross, Jeffrey Ian and Stephen C. Richards. 2002. *Behind Bars: Surviving Prison*. Indianapolis, IN: Alpha Books.

Ross, Joseph S., Kevin P. Hill, David S. Egilman, and Harlan M. Krumholz. 2008. "Guest Authorship and Ghostwriting in Publications Related to Rofecoxib." *JAMA* 299 (April 15): 1800-1812.

Rothenberg, Paula S. 2008. *White Privilege: Essential Reading on the Other Side of Racism*, 3rd ed. New York: Worth.

Rothman, Sheila M. 1978. *Women's Proper Place: A History of Changing Ideals and Practices, 1870 To the Present*. New York: Basic Books.

Rowe-Finkbeiner, Kristin. 2004. *The F-Word: Feminist in Jeopardy, Women, Politics, and the Future*. Emeryville, CA: Seal Press.

Roylance, Frank D., and Michael Hill. 2008. "Scientists Say Evolution Fits." *Baltimore Sun*, January 5, 1A, 14A.

Royster, Deirdre A. 2003. *Race and the Invisible Hand: How White Networks Exclude Black Men from Blue-Collar Jobs*. Berkeley: University of California Press.

Rubin, Kenneth, William Bukowski, and Jeffrey G. Parker. 1998. "Peer Interactions, Relationships, and Groups." Pp. 619-700 in *Handbook of Child Psychology: Vol. 3. Social, Emotional, and Personality Development*, edited by William Damon and Nancy Eisenberg. New York: Wiley.

Rubin, Rita. 2004. "'Smart Pills' Make Headway." *USA Today*, July 7, 1D.

Rubin, Trudy. 2006. "Reaction to Cartoons Shows Lack of Understanding." *Baltimore Sun*, February 7, 13A.

Rutenberg, Jim, and Patrick D. Healy. 2005. "Bloomberg Spends $46.6 Million on Race." *New York Times*, October 8, A1.

Ryan, Missy. 2008. "In US, Record Numbers Seeking Food Stamps." *Christian Science Monitor*, May 8, 4.

Ryerson, William. 2004. "Responses to: Demographic 'Bomb' May Only Go 'Pop!'" *Population Press* 10 (Fall): 21.

Saad, Lydia. 2001. "Majority Considers Sex Before Marriage Morally Okay." *The Gallup Poll Monthly*, No. 428 (May): 46-48.

————. 2006. "Families of Drug and Alcohol Abusers Pay an Emotional Toll." August 25, Gallup News Service. Retrieved August 27, 2006 (www.galluppoll.com).

————. 2007. "Americans Rate the Morality of 16 Social Issues." Gallup News Service, June 4. Retrieved June 8, 2007 (www.gallup.com).

————. 2007. "Environmental Concern Holds Firm During Past Year." Gallup Poll, March 26. Retrieved March 27, 2007 (www.gallup.com).

————. 2007. "Women Slightly More Likely to Prefer Working to Homemaking." Gallup News Service, August 31. Retrieved September 3, 2007 (www.gallup.com).

————. 2008. "CA Ruling on Same-Sex Marriage Bucks Majority View." Gallup News Service, May 15. Retrieved June 2, 2008 (www.gallup.com).

————. 2008. "Cultural Tolerance for Divorce Grows to 70%." Gallup News Service, May 19. Retrieved July 8, 2008 (www.gallup.com).

————. 2008. "Telecommuting Still a Rare Perk." Gallup, August 15. Retrieved August 17, 2008 (www.gallup.com).

Sabattini, Laura, Nancy M. Carter, Jeanine Prime, and

David Megathlin. 2007. "The Double-Bind Dilemma for Women in Leadership: Damned if You Do, Doomed if You Don't." *Catalyst*. Retrieved April 28, 2008 (www.catalyst.org).

Sachs, Jeffrey S. 2005. "Confusion over Population: Growth or Dearth?" *Pop!ulation Press* 11 (Winter/Spring): 17.

Sacks, Peter. 2007. *Tearing Down the Gates: Confronting the Class Divide in American Education*. Berkeley: University of California Press.

Saegert, Susan, and Gary Winkel. 1981. "The Home: A Critical Problem for Changing Sex Roles." Pp. 41-63 in *New Space for Women*, edited by Gerda R. Wekerle, Rebecca Peterson, and David Morley. Boulder, CO: Westview Press.

Saenz, Victor B. and Douglas S. Barrera. 2007. "Findings from the 2005 College Student Survey (CSS): National Aggregates." UCLA Graduate School of Education Information Studies. Retrieved August 5, 2007 (www.gseis.ucla.edu).

Safi, Omid. 2003. "Introduction: The Times They Are A-Changin'—A Muslim Quest for Justice, Gender, Equality, and Pluralism." Pp. 1-29 in *Progressive Muslims: On Justice, Gender, and Pluralism*, edited by Omid Safi. Oxford, England: Oneworld Publications.

Sagarin, Edward. 1975. *Deviants and Deviance*. New York: Praeger.

Salladay, Robert. 2006. "Gov.'s Candid Moments Caught on Audiotape." *Los Angeles Times*, September 8. Retrieved September 10, 2008 (www.latimes.com).

Samples, John, ed. 2005. *Welfare for Politicians? Taxpayer Financing of Campaigns*. Washington, DC: Cato Institute.

Sandstrom, Kent L., Daniel D. Martin and Gary Alan Fine. 2006. *Symbols, Selves, and Social Reality: A Symbolic Interactionist Approach to Social Psychology and Sociology*, 2nd edition. Los Angeles, CA: Roxbury.

Santos, Fernanda. 2008. "After the War, a New Battle to Become Citizens." *New York Times*, February 24, 1.

Sarmiento, Soccorro T. 2002. *Making Ends Meet: Income-Generating Strategies among Mexican Immigrants*. New York: LFB Scholarly Publishing LLC.

Sasseen, Jane. 2005. "White-Collar Crime: Who Does It?" *Business Week*, February 6, 60-61.

"Saudi Arabia: Women without the Vote." 2008. International Museum of Women, September. Retrieved September 17, 2008 (www.imow.org).

Saul, Stephanie. 2007. "Doctors and Drug Makers: A Move to End Cozy Ties." *New York Times*, February 11, C10.

Savage, David G. 2008. "California Gay Marriage Ruling Isn't Seen as a Trend." *Los Angeles Times*, May 26. Retrieved May 27, 2008 (www.latimes.com).

Scarlett, W. George, Sophie Naudeau, Dorothy Salonius-Pasternak and Iris Ponte. 2005. *Children's Play*. Thousand Oaks, CA: Sage.

Scelfo, Julie. 2007. "Come Back, Mr. Chips." *Newsweek*, September 17, 44.

Schachter, Jason P. 2004. "Geographical Mobility: 2002 to 2003." U.S. Census Bureau, March, Current Population Reports, P20-549. Retrieved May 15, 2007 (www.census.gov).

Schaeffer, Robert K. 2003. *Understanding Globalization: The Social Consequences of Political, Economic, and Environmental Change*, 2nd edition. Lanham, MD: Rowman & Littlefield Publishers.

Schehr, Robert C. 1997. *Dynamic Utopia: Establishing Intentional Communities as a New Social Movement*. Westport, CT: Bergin & Garvey.

Schirber, Michael. 2007. "Why Desalination Doesn't Work (Yet)." June 25, *Live Science*. Retrieved February 7, 2008 (www.livescience.com).

Schmidt, Steffen W., Mack C. Shelley and Barbara A. Bardes. 2001. *American Government and Politics Today*. Belmont, CA: Wadsworth.

Schorr, Melissa. 2001. "Gay and Lesbian Couples Do Well At Parenting." Excite News, Aug. 28. Retrieved Aug. 29, 2001 (news.excite.com).

Schuckit, Marc A. 1999. "New Findings in the Genetics of Alcoholism." *JAMA* 281 (May 26): 1875-1876.

Schulz, Dorothy M. 2004. *Breaking the Brass Ceiling: Women Police Chiefs and Their Paths to the Top*. Westport, CT: Praeger.

Schur, Edwin M. 1968. *Law and Society: A Sociological View*. New York: Random House.

Schurman-Kauflin, Deborah. 2000. *The New Predator: Women Who Kill*. New York: Algora.

Schutz, Alfred. 1967. *The Phenomenology of the Social World*. Evanston, IL: Northwestern University Press.

Schwalbe, Michael. 2001. *The Sociologically Examined Life: Pieces Of the Conversation*, 2nd ed. Mountain View, CA: Mayfield.

Schwartz, John. 2004. "Always on the Job, Employees Pay With Health." *New York Times*, September 5, A1, A22.

Schwartz, Marlene B. Lenny R. Vartanian, Brian A. Nosek, and Kelly D. Brownell. 2006. "The Influence of One's Own Body Weight on Implicit and Explicit Anti-fat Bias." *Obesity* 14 (July): 440-447.

Sciolino, Elaine. 2003. "France Makes Major Changes in Pension Law." *New York Times*, July 25, 7.

Scommegna, Paola. 2005. "Clean Water's Historic Effect on U.S. Mortality Rates Provides Hope for Developing Countries." Population Reference Bureau. Retrieved June 25, 2005 (www.prb.org).

Scott, Katherine V. 1995. *Gender and Development: Rethinking Modernization and Dependency Theory*. Boulder, CO: Lynne Rienner.

Seager, Richard, et al. 2007. "Model Projections of an Imminent Transition to a More Arid Climate in Southwestern North America." *Science* 316 (May): 1181–1184.

Seccombe, Karen. 2007. *Families in Poverty*. New York: Allyn and Bacon.

"Second National Report on Human Exposure to Environmental Chemicals." 2003. Centers for Disease Control and Prevention, January. Retrieved June 25, 2005 (www.cdc.gov).

Sedlacek, William E. 2004. *Beyond the Big Test: Noncognitive Assessment in Higher Education*. San Francisco, CA: Jossey-Bass.

Segura, Denise. A. 1994. "Working at Motherhood: Chicana and Mexican Immigrant Mothers and Employment." Pp. 211-33 in *Mothering: Ideology, Experience, And Agency*, edited by Evelyn N. Glenn, Grace Chang, and Linda R. Forcey. New York: Routledge.

Seligson, Hannah. 2007. "Colleges Go Light on Women's Pay Inquiry." Women's e-news, April 25. Retrieved April 26, 2007 (www.womensenews.org).

Seltzer, Judith A. 2004. "Cohabitation and Family Change." Pp. 57-78 in *Handbook of Contemporary Families: Considering The Past, Contemplating the Future*, edited by Marilyn Coleman and Lawrence H. Ganong. Thousand Oaks, CA: Sage.

Sengupta, Somini. 2006. "Report Shows Muslims Near Bottom of Social Ladder." *New York Times*, November 29, A4.

Sequist, Thomas D., Garrett M. Fitzmaurice, Richard Marshall, Shimon Shaykevich, Dena Gelb Safran, and John Z. Ayanian. 2008. "Physical Performance and Racial Disparities in Diabetes Mellitus Care." *Archives of Internal Medicine* 168 (June 9): 1145-1151.

Sernau, Scott. 2001. *Worlds Apart: Social Inequalities in a New Century*. Thousand Oaks, CA: Pine Forge Press.

Sharifzadeh, Virginia-Shirin. 1997. "Families with Middle Eastern Roots." Pp. 441-482 in *Developing Cross-Cultural Competence: a Guide for Working with Children and Families*, edited by Eleanor W. Lynch and Marci J. Hanson. Baltimore, MD: Paul H. Brookes Publishing.

Shaw-Taylor, Yoku and Nijole V. Benokraitis. 1995. "The Presentation of Minorities in Marriage and Family Textbooks." *Teaching Sociology* 23 (April): 122-35.

Shenkman, Rick. 2008. *Just How Stupid Are We? Facing the Truth about the American Voter*. New York: Basic Books.

Sherman, Lawrence. 1992. *Policing Domestic Violence: Experiment and Dilemmas*. New York: Free Press.

Shibutani, Tamotsu. 1986. *Social Process: An Introduction to Sociology*. Berkeley, CA: University of California Press.

Shinagawa, Larry H., and Gin Y. Pang. 1996. "Asian American Panethnicity and Intermarriage." *Amerasia Journal* 22 (Spring): 127–52.

Shore, Michael. 2003. "Out of Control and Close to Home: Mercury Pollution from Power Plants." Environmental Defense. Retrieved July 10, 2005 (www.environmentaldefense.org).

Shorto, Russell. 2008. "No Babies?" *New York Times*, June 29, 34.

Siegman, Aron W., and Stanley Feldstein, eds. 1987. *Nonverbal Behavior and Communication*. Hillsdale, NJ: Lawrence Erlbaum Associates.

Silverman, Rachel E. 2003. "Provisions Boost Rights of Couples Living Together." *Wall Street Journal*, March 5, D1.

Silverstein, Ken. 2002. "Unjust Rewards." *Mother Jones*, May/June, 69-86.

Simmons, Tavia, and Martin O'Connell. 2003. "Married-couple and Unmarried-partner Households: 2000." U.S. Census Bureau. Retrieved April 20, 2003 (www.census.gov).

Simon, Rita J. and Jean Landis. 1991. *The Crimes Women Commit, the Punishments They Receive*. Lexington, MA: Lexington Books.

Simon, Robin W., and Leda E. Nath. 2004. "Gender and Emotion in the United States: Do Men and Women Differ in Self-Reports of Feelings and Expressive Behavior?" *American Journal of Sociology* 109 (March): 1137-76.

Simpson, Sally S., and Carole Gibbs. 2006. "Making Sense of Intersections." Pp. 269-302 in *Gender and Crime: Patterns of Victimization and Offending*, edited by Karen Heimer and Candace Kruttschnitt. New York: New York University Press.

Sjoberg, Gideon. 1960. *The Preindustrial City: Past and Present*. Glencoe, IL: Free Press.

Skerry, Peter. 2002. "Beyond Sushiology: Does Diversity Work?" *Brookings Review* 20 (Winter): 20-23.

Slackman, Michael. 2007. "In Arab Hub, Poor are Left in Dire Straits." *New York Times*, March 1, A1, A4.

Slevin, Peter. 2005. "In Heartland, Stem Cell Research Meets Fierce Opposition." *Washington Post*, August 10, A1.

Smedley, Audrey. 2007. *Race in North America: Origin and Evolution of a Worldview*, 3rd ed. Boulder, CO: Westview Press.

Smelser, Neil J. 1962. *Theory of Collective Behavior*. New York: Free Press.

_____. 1988. "Social Structure." Pp. 103-129 in *Handbook of Sociology*, edited by Neil J. Smelser. Newbury Park, CA: Sage.

Smith, Adam. 1937. *An Inquiry into the Nature and Causes of the Wealth of Nations*. New York: The Modern Library, orig. 1776.

Smith, Craig S. 2002. "Kandahar Journal; Shh, It's an Open Secret: Warlords and Pedophilia." *New York Times*, February 21, A4.

Smith, Dorothy E. 1987. *The Everyday World as Problematic: A Feminist Sociology*. Toronto: University of Toronto Press.

Smith, Grant F., and Tanya Carina Hsu. 2007. "Visa Denied: How Anti-Arab Visa Policies Destroy US Export, Jobs and Higher Education." Institute for Research: Middle Eastern Policy. Retrieved May 7, 2007 (www.irmep.org).

Smith, Jane I. 1994. "Women in Islam." Pp. 303-325 in *Today's Woman in World Religions*, edited by Arvind Sharma. Albany: State University of New York Press.

Smith, Tom W. 2007. "Job Satisfaction in the United States." April 17, NORC/University of Chicago. Retrieved June 1, 2007 (www-news.uchicago.edu).

Smith, Tom W., and Seokho Kim. 2004. "The Vanishing Protestant Majority." NORC/University of Chicago, July. Retrieved March 3, 2005 (www-news.uchicago.edu).

Snell, Marilyn B. 2007. "The Talking Way." *Mother Jones*, January/February, 30-35.

Snipp, C. Matthew. 1996. "A Demographic Comeback for American Indians." *Population Today* 24 (November): 4-5.

Snyder, Howard N. and Melissa Sickmund. 2006. *Juvenile Offenders and Victims: 2006 National Report*. Washington, DC: U.S. Department of Justice, Office of Justice Programs, Office of Juvenile Justice and Delinquency Prevention.

Snyder, Leslie B., Frances Fleming Milici, Michael Slater, Helen Sun, and Yulija Strizhakova. 2006. "Effects of Alcohol Advertising Exposure on Drinking among Youth." *Archives of Pediatrics & Adolescent Medicine* 160 (January): 18-24.

Snyder, Thomas D., Sally A. Dillow, and Charlene M. Hoffman. 2008. *Digest of Education Statistics 2007* (NCES 2008-022). Washington, DC: National Center for Education Statistics, Institute of Education Sciences, U.S. Department of Education.

Soares, Joseph A. 2007. *The Power of Privilege: Yale and America's Elite Colleges*. Stanford, CA: Stanford University Press.

Sokol, Ronald P. 2008. "Why France Can't See Past the Burqa." *Christian Science Monitor*, July 21, 9.

Sontag, Deborah. 2002. "Fierce entanglements." *New York Times Magazine*, November 17, 52.

Sorkin, Andrew R. 2002. "When a Trusted Secretary Takes More Than a Letter." *New York Times*, November 27, 1.

Sowell, Thomas. 2008. "A Prestigious Degree Doesn't Always Equal Success." *Chronicle of Higher Education*, January 18, A34.

Spalter-Roth, Roberta, and Nicole Van Vooren. 2008. "What Are They Doing with a Bachelor's Degree in Sociology? Data Brief on Current Jobs." January, American Sociological Association, Department of Research and Development. Retrieved March 27, 2008 (www.asanet.org).

Spano, John. 2007. "Abuse Deal Reached in L.A." *Baltimore Sun*, July 15, 3A.

Spector, Malcolm and John Kitsuse. 1977. *Constructing Social Problems*. Menlo Park, CA: Cummings.

Spectorsky, A. C. 1955. *The Exurbanites*. Philadelphia: J. B. Lippincott.

Spencer, Herbert. 1862/1901. *First Principles*. New York: P. F. Collier & Son.

Sperry, Shelley. 2008. "Ozone Defense." *National Geographic* 214 (October): no page given.

Spiess, Michele. 2003. "Juveniles and Drugs." Rockville, MD: ONDCP Drug Policy Information Clearinghouse.

Spinner, Jackie. 2004. "Head Scarves Now a Protective Accessory in Iraq." *Washington Post*, December 30, A15.

Spotts, Peter N. 2004. "Blowing in the Wind: Transatlantic Pollution." *Christian Science Monitor*, August 5, 14, 17.

Spradley, J.P. and M. Phillips. 1972. "Culture and Stress: A Quantitative Analysis." *American Anthropologist* 74 (3): 518-529.

Squires, Gregory D., and Charis E. Kubrin. 2006. *Privileged Places: Race, Residence, and the Structure of Opportunity*. Boulder, CO: Lynne Rienner.

Stafford, Frank. 2008. "Exactly How Much Housework Does a Husband Create?" University of Michigan News Service, April 3. Retrieved June 2, 2008 (www.ns.umich.edu).

Stanczak, Gregory C. 2006. *Social Change and American Religion*. New Brunswick, NJ: Rutgers University Press.

Stanley, Scott M., and Gary Smalley. 2005. *The Power of Commitment: A Guide to Active, Lifelong Love*. San Francisco, CA: Jossey Bass.

Stanley, Thomas J. and William D. Danko. 1996. *The Millionaire Next Door: The Surprising Secrets of America's Wealthy*. Atlanta, GA: Longstreet Press.

Staples, Brent. 2004. "Why Some Politicians Need Their Prisons to Stay Full." *New York Times*, December 27, 16.

Starr, Tama. 1991. *The "Natural Inferiority" of Women*. New York: Poseidon Press.

Stauss, Joseph H. 1995. "Reframing and Refocusing American Indian Family Strengths." Pp. 105-18 in *American Families: Issues in Race and Ethnicity*, edited by Cardell K. Jacobson. New York: Garland

Stearns, Matt. 2006. "Organ Transplants Called Biased for the Well-To-Do." *Philadelphia Inquirer*, June 7, A10.

Steffensmeier, Darrell, Jennifer Schwartz, Hua Zhong, and Jeff Ackerman. 2005. "An Assessment of Recent Trends in Girls' Violence Using Diverse Longitudinal Sources: Is the Gender Gap Closing?" *Criminology* 43 (May): 355-405.

Steinhauer, Jennifer. 2005. "When the Joneses Wear Jeans." *New York Times*, May 29, 1.

Stephen, J. 2004. *State Prison Expenditures, 2001*. Bureau of Justice Statistics Special Report. Washington, DC: U.S. Department of Justice.

Stephenson, John B. 2007. "Environmental Protection: EPA-State Enforcement Partnership Has Improved, but EPA's Oversight Needs Further Enhancement." United States Government Accountability Office, July, GAO—07-883. Retrieved February 12, 2008 (www.gao.gov).

Sternheimer, Karen. 2007. "Do Video Games Kill?" *Contexts* 6 (Winter): 13-17.

Stewart, Pearl, and Katia Paz Goldfarb. 2007. "Historical Trends in the Study of Diverse Families." Pp. 3-19 in *Cultural Diversity and Families: Expanding Perspectives*, edited by Bahira Sherif Trask and Raeann R. Hamon. Thousand Oaks, CA: Sage, pp. 3-19.

Stillars, Alan L. 1991. "Behavioral Observation." Pp. 197-218 in *Studying Interpersonal Interaction*, edited by B. M. Montgomery and S. Duck. New York: Guilford Press.

Stolley, Giordano, and Somchai Taphaneeyapan. 2002. "Stomaching Bugs in Thailand." *Baltimore Sun*, June 20, 2A.

Stoops, Nicole. 2004. "Educational Attainment in the United States: 2003." U.S. Census Bureau, Current Population Reports. Retrieved December 7, 2004 (www.census.gov).

Story, Louise. 2008. "To Aim Ads, Web is Keeping Closer Eye on You." *New York Times*, March 10, 1.

Strasburger, Victor C. and Barbara J. Wilson. 2002. *Children, Adolescents, & the Media*. Thousand Oaks, CA: Sage Publications.

Strasburger, Victor C., et al. 2006. "Children, Adolescents, and Advertising." *Pediatrics* 118 (December 6): 2563-2569.

Strauss, Valerie. 2005. "More and More, Kids Say the Foulest Things." *Washington Post*, April 12, 12A.

Strazdins, Lyndall, Mark S. Clements, Rosemary J. Korda, Dorothy H. Broom, and Rennie M. D'Souza. 2006. "Unsociable Work? Nonstandard Work Schedules, Family Relationships, And Children's Well-Being." *Journal of Marriage and Family* 68 (May): 393-410.

"Streaking." 2005. Wikipedia Encyclopedia. Retrieved September 8, 2005 (www.en.wikipedia.org/wiki).

Sturken, Marita and Lisa Cartwright. 2001. *Practices of Looking: An Introduction to Visual Culture*. New York: Oxford University Press.

Substance Abuse and Mental Health Services Administration. 2007. "Results from the 2006 National Survey on Drug Use and Health: National Findings." U.S. Department of Health and Human Services. Retrieved May 12, 2008 (www.samhsa.gov).

Sugg, Diana K. 2002. "Baltimore Drug Programs are Effective, Study Says." *Baltimore Sun*, January 31, 7A.

Sullivan, Andrew, ed. 1997. *Same-Sex Marriage, Pro and Con: A Reader*. New York: Vintage.

———. 2003. "The Conservative Case for Gay Marriages." *Time*, June 30, 76.

———. 2004. "Showdown at the Communion Rail." *Time*, May 24, 94.

Sullivan, Evelin. 2001. *The Concise Book of Lying*. New York: Farrar, Straus and Giroux.

Sullivan, Will. 2007 "Road Warriors." *US News & World Report*, May 7, 40-49.

Sultana, Ronald G. 2008. "The Girls' Education Initiative in Egypt." UNICEF, January. Retrieved September 18, 2008 (www.ungei.org).

Sumner, William G. 1906. *Folkways*. New York: Ginn.

Sutherland, Edwin H. 1949. *White Collar Crime*. New York: Holt, Rinehart, and Winston.

Sutherland, Edwin H., and D. R. Cressey. 1970. *Criminology*, 8th edition. Philadelphia, PA: Lippincott.

Sutton, Charles T., and Mary A. Broken Nose. 1996. "American Indian Families: An Overview." Pp. 31-54 in *Ethnicity and Family Therapy*, 2nd ed., edited by Monica McGoldrick, Joe Giordano, and John K. Pearce. New York: The Guilford Press.

Sutton, Philip W. 2000. *Explaining Environmentalism: In Search of a New Social Movement*. Burlington, VT: Ashgate Publishing Company.

Swanberg, Jennifer E. 2005. "Job-Family Role Strain Among Low-Wage Workers." *Journal of Family and Economic Issues* 26 (Spring): 143-158.

Swers, Michele L. 2002. *The Difference Women Make: The Policy Impact of Women in Congress*. Chicago, IL: University of Chicago Press.

Switzer, Jacqueline V. 2003. *Disabled Rights: American Disability Policy and the Fight for Equality*. Washington, DC: Georgetown University Press.

Symonds, William C. 2003. "College Admissions: The Real Barrier is Class." *Business Week* (April 14): 66, 68.

Szinovacz, Maximiliane E., ed. 1998. *Handbook on Grandparenthood*. Westport, CT: Greenwood Press.

Tajfel, Henri. 1982. "Social Psychology of Intergroup Relations." *Annual Review of Psychology*. Palo Alto, CA: Annual Reviews, 1-39.

Tannen, Deborah. 1990. *You Just Don't Understand: Women And Men In Conversation*. New York: Ballantine.

Tannenbaum, Frank. 1938. *Crime and the Community*. New York: Columbia University Press.

Tanur, Judith M. 1994. "The Trustworthiness of Survey Research." *Chronicle of Higher Education*, May 25, B1–B3.

Tatara, Toshio. 1998. "The National Elder Abuse Incidence Study." The National Center on Elder Abuse and the American Public Humane Services Association. Retrieved October 19, 2000 (www.aoa.gov).

Taylor, Frederick W. 1911/1967. *The Principles of Scientific Management*. New York: W. W. Norton & Company.

Taylor, Jay. 1993. *The Rise and Fall of Totalitarianism in the Twentieth Century*. New York: Paragon House.

Taylor, Jonathan B. and Joseph P. Kalt. 2005. *American Indians on Reservations: A Databook of Socioeconomic Change between the 1990 and 2000 Censuses*. Harvard Project on American Indian Economic Development, January. Retrieved April 20, 2007 (www.ksg.harvard.edu).

Taylor, Susan C. 2003. *Brown Skin: Dr. Susan Taylor's Prescription for Flawless Skin, Hair, and Nails*. New York: HarperCollins.

Teaster, Pamela B., Tyler A. Dugar, Marta S. Mendiondo, Erin L. Abner, Kara A. Cecil, and Joanne M. Otto. 2006. "The 2004 Survey of State Adult Protective Services: Abuse of Adults 60 Years of Age and Older." National Center on Elder Abuse. Retrieved June 27, 2007 (www.elderabusecenter.org).

Teicher, Martin. 2000. "Wounds That Time Won't Heal: The Neurobiology of Child Abuse." *Cerebrum* 2 (Fall), The Dana Foundation. Retrieved July 7, 2002 (www.dana.org).

Teicher, Stacy A. 2007. "Closing the Gaps." *Christian Science Monitor*, April 5, 16-17.

Terhune, Chad. 2008. "They Know What's in Your Medicine Cabinet." *Business Week*, August 4, 48-52.

Teske, Paul, Mark Schneider, Jack Buckley, and Sara Clark. 2001. "Can Charter Schools Change Traditional Public Schools?" Pp. 188-214 in *Charters, Vouchers, and Public Education*, edited by Paul E. Peterson and David E. Campbell. Washington, D.C.: Brookings Institution Press.

Tessier, Marie. 2001. "Women's Appointments Plummet Under Bush." Women's E-News, July 6. Retrieved September 30, 2004 (www.womensenews.org).

Tétreault, Mary Ann. 2001. "A State of Two Minds: State Cultures, Women, and Politics in Kuwait." *International Journal of Middle East Studies* 33 (May): 203-220.

Thee, Megan. 2007. "Cellphones Challenge Poll Sampling." *New York Times*, December 7, 29.

Thibaut, John W. and Harold H. Kelley. 1959. *The Social Psychology of Groups*. New York: Wiley.

Thomas, W. I. and Dorothy Swaine Thomas. 1928. *The Child in America*. New York: Alfred A. Knopf.

Thompson, Gabriel. 2007. *There's No José Here: Following the Hidden Lives of Mexican Immigrants*. New York: Nation Books.

Thompson, Robert S., et al. 2006. "Intimate Partner Violence: Prevalence, Types, and Chronicity in Adult Women." *American Journal of Preventive Medicine* 30 (June): 447-57.

Thottam, Jyoti. 2004. "Is Your Job Going Abroad?" *Time*, March 1, 26-36.

Thumma, Scott, Dave Travis, and Warren Bird. 2005. "Megachurches Today 2005." Hartford Institute for Religion Research. Retrieved August 14, 2007 (http://hirr.hartsem.edu).

"Ties that Bind, The." 2000. *Public Perspective* 11 (May–June): 10.

Tilly, Charles. 1978. *From Mobilization to Revolution*. Reading, MA: Addison-Wesley.

Tizon, Tomas. 2008. "An Alaskan Village Prepares to Move." *Christian Science Monitor*, January 10, 17.

Toch, Thomas. 2007. "In Testing, the Infrastructure is Buckling." *Education Week*, July 24. Retrieved July 11, 2007 (www.edweek.org).

Toder, Eric J. 2005. "What Will Happen To Poverty Rates Among Older Americans In The Future And Why?" The Urban Institute, November. Retrieved August 16, 2002 (www.urban.org).

Tormey, Simon. 1995. *Making Sense of Tyranny: Interpretations of Totalitarianism*. New York: Manchester University Press.

Tornatzky, Louis G., Richard Cutler, and Jongho Lee. 2002. "College Knowledge: What Latino Parents Need to Know and Why They Don't Know It." The Tomás Rivera Policy Institute. Retrieved January 2, 2005 (www.trpi.org).

Toth, John F., and Xiaohe Xu. 2002. "Fathers' Child-Rearing Involvement in African American, Latino, and White Families." Pp. 130-140 in *Contemporary Ethnic Families in the United States: Characteristics, Variations, And Dynamics*, edited by Nijole V. Benokraitis. Upper Saddle River, NJ: Prentice Hall.

Touraine, Alain. 1981. *The Voice and the Eye: An Analysis of Social Movements*. Cambridge: Cambridge University Press.

____. 2002. "The Importance of Social Movements." *Social Movement Studies* 1 (April): 89-96.

Treiman, Donald. 1977. *Occupational Prestige in Comparative Perspective*. New York: Academic Press.

"Trends in Political Values and Core Attitudes: 1987-2007." 2007. Pew Research Center for the People & the Press, March 22. Retrieved May 10, 2007 (www.people-press.org).

Trumbull, Mark. 2007. "Life at America's Bottom Wage." *Christian Science Monitor* January 9, 1, 10-11.

Tucker, Cynthia. 2007. "Lingering Sexism Impedes Women's Path to Highest Level of Power." *Baltimore Sun*, January 8, A9.

Tumin, Melvin M. 1953. "Some Principles of Stratification: A Critical Analysis." *American Sociological Review* 18 (August): 387-393.

Turk, Austin T. 1969. *Criminality and the Legal Order*. Chicago, IL: Rand-McNally.

____. 1976. "Law as a Weapon in Social Conflict." *Social Problems* 23 (February): 276-291.

Turner, Ralph H., and Lewis M. Killian. 1987. *Collective Behavior*, 3rd ed. Englewood Cliffs, NJ: Prentice Hall.

Twitchell, Geoffrey R., Gregory L. Hanna, Edwin H. Cook, Scott F. Stoltenberg, Hiram E. Fitzgerald, and Robert A. Zucker. 2001. "Serotonin Transporter Promoter Polymorphism Genotype Is Associated With Behavioral Disinhibition and Negative Affect in Children of Alcoholics." *Alcoholism: Clinical and Experimental Research* 25 (July): 953-959.

Tyre, Peg. 2007. "Learning Takes Time." *Newsweek*, January 22, 12.

U.S. Bureau of Labor Statistics, 2007. "Women in the Labor Force: A Databook." September. Retrieved July 2, 2008 (www.bls.gov).

____. 2008. Table 39: Median Weekly Earning of Full-Time Wage and Salary Workers by Detailed Occupation and Sex. Retrieved July 2, 2008 (www.bls.gov).

U.S. Census Bureau. 1999. *Statistical Abstract of the United States, 1999*, 119th ed. Washington, DC: U.S. Government Printing Office.

____. 2006. "2006 American Community Survey." Table C04001: First Ancestry Reported, Total Population. Retrieved October 13, 2008 (factinder.census.gov).

____. 2007. *Statistical Abstract of the United States*, 126th ed. Washington, DC: U.S. Government Printing Office.

____. 2007a. "Table MS-1. Marital Status of the Population 15 Years Old and Over, by Sex and Race: 1950 to Present." Current Population Survey, Annual Social and Economic Supplements. Retrieved July 1, 2007 (www.census.gov).

____. 2007b. "Table MS-2. Estimated Median Age at First Marriage, by Sex: 1890 to the Present." Current Population Survey, Annual Social and Economic Supplements. Retrieved July 1, 2007 (www.census.gov).

____. 2007c. "Table UC1. Opposite Sex Unmarried Partner Households, by Labor Force Status of Both Partners, and Race and Hispanic Origin of the Householder: 2006." Current Population Survey, Annual Social and Economic Supplements. Retrieved July 1, 2007 (www.census.gov).

____. 2007d. "Table MS-2. Estimated Median Age at First Marriage, by Sex: 1890 to the Present." Current Population Survey, Annual Social and Economic Supplements. Retrieved July 1, 2007 (www.census.gov).

____. 2008. *Statistical Abstract of the United States*, 127th ed. Washington, DC: U.S. Government Printing Office.

U.S. Census Bureau News. 2008. "An Older and More Diverse Nation by Midcentury." August 14. Retrieved August 18, 2008 (www.census.gov).

U.S. Census Bureau Press Release. 2008. "Unmarried and Single Americans Week: Sept. 21-27, 2008." July 21. Retrieved August 1, 2008 (www.census.gov).

U.S. Department of Education. 2006. *Digest of Education Statistics 2005*, National Center for Education Statistics, Table 246. Retrieved August 8, 2006 (nces.ed.gov/programs).

U.S. Department of Health and Human Services.1999. *Mental Health: A Report of the Surgeon General*. Rockville, MD: U.S. Department of Health and Human Services, National Institutes of Health, National Institute of Mental Health.

____. 2007. *Child Maltreatment 2005*. Washington, DC: U.S. Government Printing Office.

U.S. Department of Health and Human Services, Administration on Children, Youth and Families. 2007. *Child Maltreatment 2005*. Washington, DC: U.S. Government Printing Office.

U.S. Department of Housing and Urban Development. 2007. *The Annual Homeless Assessment Report to Congress*. February. Retrieved March 24, 2007 (www.huduser.org).

U.S. Department of Justice. 1996. "Policing Drug Hot Spots." National Institute of Justice, January. Retrieved May 3, 2005 (www.ncjrs.org).

U. S. Department of Labor. 2002. *Working in the 21st Century*. Washington, DC: U.S. Government Printing Office.

____. 2005. "Characteristics of Minimum Wage Workers: 2005." Retrieved May 31, 2006 (www.bls.gov).

____. 2005. "Occupational Employment Statistics." Bureau of Labor Statistics. Retrieved October 15, 2006 (www.bls.gov).

____. 2005. "Workers on Flexible and Shift Schedules in 2004 Summary." July 1. Retrieved June 4, 2007 (www.bls.gov).

____. 2007. "Women in the Labor Force: A Databook." U.S. Bureau of Labor Statistics, September. Retrieved July 2, 2008 (www.bls.gov).

____. 2008. "Employment and Earning Online." Bureau of Labor Statistics. Retrieved January, 2008.

____. 2008. "Volunteering in the United States, 2007." Bureau of Labor Statistics. Retrieved May 5, 2008 (www.bls.gov/cps).

U.S. Department of State. 2006. *Trafficking in Persons Report*. June. Retrieved November 1, 2006 (www.state.gov).

U.S. Equal Employment Opportunity Commission. 2008a. "Pregnancy Discrimination Charges EEOC & FEPAs Combined: FY 1997-FY 2007." Retrieved June 3, 2008 (www.eeoc.gov).

____. 2008b. "Sexual Harassment Charges EEOC & FEPAs Combined: FY 1997-FY 2007." Retrieved June 3, 2008 (www.eeoc.gov).

U.S. General Accounting Office. 2004. "Defense of Marriage Act: Update to Prior Report." www.gao.gov (accessed July 5, 2007).

U.S. Senate Special Committee on Aging, American Association of Retired Persons, Federal Council on the Aging, and U.S. Administration on Aging. 1991. *Aging America: Trends and projections, 1991*. Washington, DC: Department of Health and Human Services.

UNESCO World Report. 2005. *Towards Knowledge Societies*. New York: UNESCO Publishing.

Union of Concerned Scientists. 2008. "Interference at the EPA: Science and Politics at the U.S. Environmental Protection Agency." April. Retrieved April 25, 2008 (ucsusa.org).

United Nations Development Fund for Women. 2007. "Violence against Women—Facts and Figures." March. Retrieved March 11, 2007 (www.unifem.org).

United Nations Development Programme. 2006. *Human Development Report 2006: Beyond Scarcity: Power, Poverty, and the Global Water Crisis*. New York: United Nations Development Programme.

United Nations Population Division. 2008. "An Overview of Urbanization, Internal Migration, Population Distribution and Development in the World." Population Division. Retrieved January 20, 2008 (www.un.org).

United Nations World Water Development Report 2. 2006. "Water: A Shared Responsibility." Retrieved February 3, 2008 (www.unesco.org).

United States Conference of Mayors. 2006. *Hunger and Homelessness Survey 2006*. December. Retrieved March 24, 2007 (www.mayors.org).

United States Government Manual, 2005/2006, The. 2006. Washington, DC: U.S. Government Printing Office.

United States Senate. 2004. *Report on the U.S. Intelligence Community's Prewar Intelligence Assessments on Iraq*. Retrieved September 13, 2006 (www.gpoaccess.gov).

University of Connecticut, Center for Survey Research & Analysis. 2007. "Two Americas but One American Dream." July 6. Retrieved May 25, 2008 (www.csra.uconn.edu).

Urbina, Ian. 2007. "Court Rejects Law Limiting Pornography on Internet." *New York Times*, March 23, A11.

Uzzi, Brian, and Ryon Lancaster. 2004. "Embeddedness and Price Formation in the Corporate Law Market." *American Sociological Review* 69 (June): 319-344.

Valenti, Jessica. 2007. "How the Web Became a Sexists' Paradise." *The Guardian*, April 6. Retrieved April 9, 2007 (www.guardian.co.uk).

Van Biema, David. 2004. "Rising Above the Stained-Glass Ceiling." *Time*, June 28, 58-61.

Van Doorn-Harder, Nelly. 2006. "Behind the Cartoon War: Radical Clerics Competing for Followers." *Christian Science Monitor*, February 23, 9.

van Dyk, Deirdre. 2005. "*Parlez-vous* Twixter?" *Time*, January 24, 49.

Van Ginneken, Jaap. 2003. *Collective Behavior and Public Opinion: Rapid Shifts in Opinion and Communication*. Mahwah, NJ: Lawrence Erlbaum Associates.

VandeHei, Jim, and Dan Eggen. 2006. "Cheney Cites Justifications for Domestic Eavesdropping." *Washington Post*, January 5, A2.

Vanderpool, Tim. 2002. "Tribes Move Beyond Casinos to Malls and Concert Halls." *Christian Science Monitor*, October 22, 2-3.

Vandewater, Elizabeth A., Victoria J. Rideout, Ellen A. Wartella, Xuan Huang, June H. Lee, and Mi-suk Shim. 2007. "Digital Childhood: Electronic Media and Technology Use Among Infants, Toddlers, and Preschoolers." *Pediatrics* 119 (May): e106-e1015.

Veblen, Thorstein. 1899/1953. *The Theory of the Leisure Class*. New York: New American Library.

____. 1899/1953. *The Theory of the Leisure Class*. New York: New American Library.

Vedantam, Shankar. 2002. "Negative View? It May Be Brain 'Knob'." *Seattle Times*, February 12. Retrieved February 13, 2002 (http://seattletimes.nwsource.com).

Venkatesh, Sudhir. 2008. *Gang Leader for a Day: A Rogue Sociologist Takes to the Streets*. New York: Penguin Press.

Ventura, S. J., J. A. Martin, S. C. Curtin, T. J. Mathews, and M. M. Park. 2000. "Births: Final Data for 1998." *National Vital Statistics Reports* 48, March 28, Centers for Disease Control and Prevention. Retrieved September 22, 2000 (www.cdc.gov).

Victory, Joy. 2005. "Face Transplant Patient Signed Movie Deal." ABC News, December 8. Retrieved September 18, 2008 (www.abcnews.go.com).

Vieth, Warren. 2005. "In Rare Tale of War Casualties, Bush Says 30,000 Iraqis Killed." December 13, *Los Angeles Times*. Retrieved December 14 (www.latimes.com).

Vigdor, Jacob L. 2008. "Measuring Immigrant Assimilation in the United States." Center for Civic Innovation, May. Retrieved June 6, 2008 (www.manhattan-institute.org).

Villani, Susan. 2001. "Impact of Media on Children and Adolescents: A 10-Year Review of the Research." *Journal of the American Academy of Child and Adolescent Psychology*, 40 (April): 392-401.

Vogel, Nancy. 2007. "Donor Money Talks, Often in a Whisper." *Los Angeles Times*, December 27. Retrieved December 28, 2007 (www.latimes.com).

Vold, George B. 1958. *Theoretical Criminology*. New York: Oxford University Press.

WAGE. 2006. "Occupational segregation." WAGE: Women are Getting Even. Retrieved May 9, 2006 (www.wageproject.org).

Wakeman, Jessica. 2008. "Misogyny's Greatest Hits." *Extra!* Fairness & Accuracy and Reporting. May/June, pp. 6-7.

Wakin, Neil J. 2004. "A Divisive Issue for Catholics: Bishops, Politicians and Communion." *New York Times*, May 31, 12.

"Wal-Mart Subsidy Watch." 2007. Good Jobs First, June 5. Retrieved May 24, 2008 (www.goodjobsfirst.org).

Wald, Kenneth D. 2003. *Religion and Politics in the United States*, 4th ed. Lanham, MD: Rowman & Littlefield.

Walker, Lenore E. 2000. *The Battered Woman Syndrome*, 2nd ed. New York: Springer.

Wallace, Bruce. 2006. "Japanese Schools to Teach Patriotism." *Los Angeles Times*, December 16. Retrieved December 18, 2006 (www.latimes.com).

____. 2008. "Debate Grows with Philippine Population." *Los Angeles Times*, May 7. Retrieved May 9, 2008 (www.latimes.com).

Wallechinsky, David. 2008. "The World's 10 Worst Dictators." *Parade*, February 17, 4-5.

Walsh, Bryan. 2007. "Back to the Tap." *Time*, August 9, 56.

Walton, Richard E., and J. Richard Hackman. 1986. "Groups under Contrasting Management Strategies." Pp. 168-201 in *Designing Effective Work Groups*, edited by Paul S. Goodman and Associates. San Francisco, CA: Jossey-Bass.

Wasley, Paula. 2006. "Review Blasts Professors for Plagiarism by Graduate Students." *Chronicle of Higher Education*, June 16, A13.

———. 2007a. "46 Students are Disciplined for Cheating at Indiana University's Dental School." *Chronicle of Higher Education*, May 9. Retrieved May 10, 2007 (www.chronicle.com).

———. 2007b. "Home-Schooled Students Rise in Supply and Demand." *Chronicle of Higher Education*, October 12, A1, A33.

———. 2008. "The Syllabus Becomes a Repository of Legalese." *Chronicle of Higher Education* 54, March 14, A1, A8-A11.

Water Quality & Health Council. 2005. "Facts About Chlorine and Drinking Water." Retrieved July 3, 2005 (www.waterandhealth.org).

Watson, C. W. 2000. *Multiculturalism*. Philadelphia, PA: Open University Press.

Weber, Lynn, Tina Hancock, and Elizabeth Higginbotham. 1997. "Women, Power, and Mental Health." Pp. 380-96 in *Women's Health: Complexities And Differences*, edited by Sheryl B. Ruzek, Virginia L. Olesen, and Adele E. Clarke. Columbus: Ohio State University Press.

Weber, Max. 1904-1905. *The Protestant Ethic and the Spirit of Capitalism*. (Talcott Parsons, trans., 1958). New York: Charles Scribner's Sons.

———. 1920. *The Protestant Ethic and the Spirit of Capitalism*. 1904 1905. (Talcott Parsons, Trans., 1958). New York: Charles Scribner's Sons.

———. 1925/1947. *The Theory of Social and Economic Organization*. New York: Free Press.

———. 1925/1978. *Economy and Society*. Guenther Roth and Claus Wittich, eds. Berkeley: University of California Press.

———. 1946. *From Max Weber: Essays in Sociology*, translated and edited by H. H. Gerth and C. Wright Mills. Berkeley: University of California Press.

Wedner, Diane. 2008. "Housing for the Growing 55-plus Crowd." *Los Angeles Times*, April 6. Retrieved April 7, 2008 (www.latimes.com).

Weil, Elizabeth. 2006. "What if It's (Sort of) a Boy and (Sort of) a Girl?" *New York Times Magazine* (September 24): 48-53.

Weinstock, Jeffrey A. 2004. "Respond Now! E-mail, Acceleration, and a Pedagogy of Patience." *Pedagogy* 4 (Fall): 365-383.

Weiss, Carol H. 1998. *Evaluation: Methods for Studying Programs and Policies*, 2nd edition. Upper Saddle River, NJ: Prentice Hall.

Weiss, Harold B. 2003. "Pregnancy-Associated Assault Hospitalizations in Selected U.S. States, 1997: Exploring the Incidence and Risk for Hospitalized Assault against Women during Pregnancy." Unpublished research report submitted to the U.S. Department of Justice. Retrieved December 1, 2004 (www.ncjrs.org).

Weiss, Michael J. 2001. "The New Summer Break." *American Demographics* 23 (August): 49-55.

———. 2004. "Online America." *American Demographics* 23 (March): 53-60.

Weiss, Rick. 2004. "Nanomedicine's Promise is Anything but Tiny." *Washington Post*, January 31, A8.

———. 2005. "The Power to Divide." *National Geographic* 208 (July): 3-27.

Welch, Susan, John Gruhl, John Comer, and Susan Rigdon. 2004. *Understanding American Government*, 7th ed. Belmont, CA: Wadsworth.

Wellins, Richard S., William C. Byham, and Jeanne M. Wilson. 1991. *Empowered Teams: Creating Self-Directed Work Groups that Improve Quality, Productivity, and Participation*. San Francisco, CA: Jossey-Bass.

Wellner, Alison Stein. 2003. "The Wealth Effect." *American Demographics* 24 (January): 35-47.

Wenneras, Christine and Agnes Wold. 1997. "Nepotism and Sexism in Peer Review." *Nature* 387 (May 22): 341–43.

West, Candace, and Don H. Zimmerman. 1987. "Doing Gender." *Gender and Society* 1 (June): 125–51.

West, Martha S., and John W. Curtis. 2006. *AAUP Faculty Gender Equity Indicators 2006*. Washington, DC: American Association of University Professors.

Western, Bruce, Vincent Schiraldi, and Jason Ziedenberg. 2003. "Education & Incarceration." Washington, DC: Justice Policy Institute, August 28. Retrieved January 4, 2005) (www.justicepolicy.org).

"What's That Commandment against Stealing?" 2006. *Forbes*, December 25, 36.

Wheeler, Larry. 2007. "Power Plants are Focus of Drive to Cut Mercury." *USA Today*, October 29. Retrieved October 30, 2007 (www.usatoday.com).

Whelan, David. 2007. "The Sentencing Game." *Forbes*, February 12, 40.

Whisnant, Rebecca, and Christine Stark, eds. 2004. *Not for Sale: Feminists Resisting Prostitution and Pornography*. North Melbourne, Australia: Spinifex Press.

White, James W. 2005. *Advancing Family Theories*. Thousand Oaks, CA: Sage.

Whitelaw, Kevin. 2000. "But What to Call It?" *U.S. News & World Report*, October 16, 42.

Whoriskey, Peter. 2008. "Skilled-Worker Visa Demand Expected to Far Exceed Supply." *Washington Post*, April 1, D3.

"Who Votes, Who Doesn't, and Why: Regular Voters, Intermittent Voters, and Those Who Don't." 2006. Pew Research Center for the People & the Press, October 18. Retrieved May 3, 2007 (www.people-press.org).

Wickenden, Dorothy. 2006. "Top of the Class." *The New Yorker*, October 2. Retrieved July 26, 2007 (www.newyorker.com).

Wiehe, Vernon R. 1997. *Sibling Abuse: Hidden Physical, Emotional, And Sexual Trauma*, 2nd edition. Thousand Oaks, CA: Sage Publications.

Wiener, Ross, and Eli Pristoop. 2006. "How States Shortchange the Districts That Need the Most Help." The Education Trust, *Funding Gaps 2006*, pp. 5-9. Retrieved July 23, 2007 (www2.edtrust.org).

Wiig, Janet and Cathy Spatz Widom. 2003. *Understanding Child Maltreatment & Juvenile Delinquency*. Washington, DC: Child Welfare League of America.

Wilcox, W. Bradford. 2006. "Religion and the Domestication of Men." *Contexts* 5 (Fall): 42-46.

Wilkinson, Charles. 2006. *Blood Struggle: The Rise of Modern Indian Nations*. New York: W.W. Norton & Company.

Williams III, Frank P. and Marilyn D. McShane. 2004. *Criminological Theory*, 4th edition. Upper Saddle River, NJ: Prentice Hall.

Williams, David E., Sean Kennedy, and Thomas A. Schatz. 2008. "2008 Congressional Pig Book Summary." Citizens Against Government Waste. Retrieved June 25, 2008 (www.cagw.org).

Williams, Dmitri, and Marko Skoric. 2005. "Internet Fantasy Violence: A Test of Aggression in an Online Game." *Communication Monographs* 72 (June): 217-33.

Williams, Krissah. 2005. "Tenacity Drives Immigrant's Dream." *Washington Post*, August 7, A1.

Williams, Robin M. Jr. 1970. *American Society: A Sociological Interpretation*, 3rd edition. New York: Knopf.

Willie, Charles Vert, and Richard J. Reddick. 2003. *A New Look at Black Families*, 5th ed. Walnut Creek, CA: AltaMira Press.

Wilson, Robin. 2005. "Second Sex." *Chronicle of Higher Education*, October 7, A10.

Wilson, Thomas C. 1993. "Urbanism and Kinship Bonds: A Test of Four Generalizations." *Social Forces* 71 (March): 703-712.

Wilson, William Julius. 1996. *When Work Disappears: The World of the New Urban Poor*. New York: Knopf.

Wilt, G. E. 1998. "Vote Early And Often." *American Demographics* 20 (December): 23.

Wiltenburg, Mary. 2002. "Minority." *Christian Science Monitor*, January 31, 14.

Winseman, Albert L. 2004. "Women in the Clergy: Perception and Reality." Gallup Organization, March 30. Retrieved March 30, 2004 (www.gallup.com).

Winston Group, The. 2006. "Math and Science Education and United States Competitiveness: Does the Public Care?" Retrieved July 15, 2007 (www.solutionsforourfuture.org).

Wirt, John, et al. 2004. *The Condition of Education 2004*. U.S. Department of Education, National Center for Education Statistics, June. Retrieved January 3, 2005 (http://nces.ed.gov).

Wirth, Louis. 1938. "Urbanism as a Way of Life." *American Journal of Sociology* 44 (July): 1-24.

Wolfe, Alan. 2008. "Pew in the Pews." *Chronicle of Higher Education*, March 21, B5-B6.

Wolff, Edward N. 2007. "Recent Trends in Household Wealth in the United States: Rising Debt and the Middle-Class Squeeze." The Levy Economics Institute, June. Retrieved May 20, 2008 (www.levy.org).

Wolfson, Mark. 2001. *The Fight Against Big Tobacco: The Movement, the State, and the Public's Health*. New York: Aldine de Gruyter.

Wood, Julia T. 2003. *Gendered Lives: Communication, Gender, and Culture*, fifth edition. Belmont, CA: Wadsworth.

Woolf, Alex. 2004. *Fundamentalism*. Chicago, IL: Raintree.

World Bank. 2007. "World Development Indicators." April 23. Retrieved June 2, 2007 (siteresources.worldbank.org).

———. 2008. *World Development Indicators 2008*. Washington, DC: The International Bank.

World Food Program. 2006. *World Hunger Series 2006: Hunger and Learning*. United Nations. Retrieved November 3, 2006 (www.wfp.org).

World Health Organization. 2005. *WHO Multi-country Study on Women's Health and Domestic Violence against Women: Summary Report of Initial Results on Prevalence, Health Outcomes and Women's Responses*. Geneva: World Health Organization.

Wright, John C., Aletha C. Huston, and Kimberlee C. Murphy. 2001. "The Relations Of Early Television Viewing To School Readiness And Vocabulary Of Children From Low-Income Families: The Early Window Project." *Child Development* 72 (September/October): 1347-66.

Wulthorst, Ellen. 2006. "US Mothers Deserve $134,121 In Salary." May 3. Retrieved May 10, 2006 (http://today.rcuters.com).

Yagelski, Robert, et al. 2005. "The Impact of the SAT and ACT Timed Writing Tests." National Council of Teachers of English, April 16. Retrieved July 4, 2007 (www.ncte.org).

Yang, Fenggang. 2002. "Religious Diversity among the Chinese in America." Pp. 71-98 in *Religions in Asian America: Building Faith Communities*, edited by Pyong Gap Min and Jung Ha Kim. Walnut Creek, CA: AltaMira Press.

Yardley, Jim. 2006. "Rules Ignored, Toxic Sludge Sinks Chinese Village." *New York Times*, September 4, A1, A8.

Yellowbird, Michael, and C. Matthew Snipp. 2002. "American Indian Families." Pp. 227-249 in *Minority Families in The United States: A Multicultural Perspective*, 3rd ed., edited by Ronald L. Taylor. Upper Saddle River, NJ: Prentice Hall.

Yokota, Fumie and Kimberly M. Thompson. 2000. "Violence in G-rated Animated Films." *JAMA: Journal of the American Medical Association* 283 (May 24/31): 2716-20.

Young, Katherine K. 1999. "Introduction." Pp. 1-24 in *Feminism and World Religions*, edited by Arvind Sharma and Katherine K. Young. Albany: State University of New York Press.

Yu, Roger. 2007. "Indian-Americans Book Years of Success." *USA Today*, April 18, 1B-2B.

Zajonc, Robert B., and Gregory B. Markus. 1975. "Birth Order and Intellectual Development." *Psychological Review* 82 (January): 74-88.

Zellner, Wendy. 2002. "How Wal-Mart Keeps Unions at Bay." *Business Week*, October 28, 94, 96.

Zhou, Min, Carl L. Bankston III, and Rebecca Y. Kim. 2002. "Rebuilding Spiritual Lives in the New Land: Religious Practices among Southeast Asian Refugees in the United States." Pp. 37-70 in *Religions in Asian America: Building Faith Communities*, edited by Pyong Gap Min and Jung Ha Kim. Walnut Creek, CA: AltaMira Press.

Zimbardo, Philip G., Christina Maslach, and Craig Haney. 2000. "Reflections on the Stanford Prison Experiment: Genesis, Transformations, Consequences." Pp. 193-237 in *Obedience to Authority: Current Perspectives on the Milgram Paradigm*, edited by Thomas Blass. Mahwah, NJ: Lawrence Erlbaum Associates.

Zimbardo, Philip. G. 1975. "Transforming Experimental Research into Advocacy for Social Change." Pp. 33-66 in *Applying Social Psychology: Implications for Research, Practice, and Training*, edited by Morton Deutsch and Harvey A. Hornstein. Hillsdale, NJ: Erlbaum.

Zweigenhaft, Richard L. and G. William Domhoff. 1998. *Diversity in the Power Elite: Have Women and Minorities Reached the Top?* New Haven: Yale University Press.

———. 2006. *Diversity in the Power Elite: How It Happened, Why It Matters*. Lanham, MD: Rowman & Littlefield.

NAME INDEX

A

Aberle, David, 321
Abma, J. C., 241
Abner, Erin L., 246
Abudabbeh, Nuha, 194
Abu-Laban, Baha, 244
Abu-Laban, Sharon M., 244
Ackelsberg, Irv, 131
Ackerman, Jeff, 128
Adams, Bert N., 9, 10, 15
Addams, Jane, 9, 14
Adelman, Clifford, 258
Adler, Jerry, 277, 308
Adler, Patricia A., 30
Adler, Paul, 30
Adler, Susan M., 191
Afifi, Tamara D., 238
Aguirre, Adalberto, Jr., 191
Aizenman, N. C., 234
Ajrouch, Kristine, 194
Akers, Ronald L., 125
Alder, Christine, 128
Alesina, Alberto, 226
Ali, Kecia, 288
Alini, Erica, 235
Allegretto, Sylvia, 222, 224, 268
Allegretto, Sylvia A., 139
Allen, Katherine R., 251
Allen, Terre H., 88
Allen, Woody, 78, 163
Allport, Gordon W., 183, 188
Almeida, R. V., 18
Almond, Lucinda, 167
Alt, Betty L., 121
Alvarez, Lizette, 97
Amato, Paul R., 238, 240
Anderson, Carol, 165
Anderson, Craig A., 75
Anderson, David A., 127
Anderson, Elijah, 30
Anderson, James F., 125, 131
Anderson, John W., 48
Anderson, Melissa A., 37
Anderson, Stephen A., 245
Andreescu, Titu, 267
Appelbaum, Mark, 168
Arax, Mark, 307
Arendt, Hannah, 201
Aries, Philippe, 76
Aristotle, 91
Armistead, Lisa, 72
Armour, Stephanie, 221
Arnoldy, Ben, 169
Aron, Laudan Y., 147
Aron-Dine, Aviva, 139, 220
Asch, Solomon, 102–103, 113
Ashburn, Elyse, 192
Ashford, Lori S., 295
Aspray, William, 222
Astin, Alexander, 276
Astin, Helen, 276
Atchley, Robert C., 249, 252
Attinasi, John J., 44
Augustine, 287
Aunola, Kaisa, 72
Austen, Ian, 58
Austen, Jane, 138
Avellar, Sarah, 171
Axtell, Roger E., 48
Ayanian, John Z., 184
Ayres, Melanie M., 90
Azios, Tony, 307
Aziz, Mohammed, 288

B

Babbage, Charles, 326
Babbie, Earl, 33
Bacon, Lynn, 164
Bacon, Meredith, 164
Bader, Christopher, 278
Bahr, Stephen J., 72
Bailey, William, 37
Bainbridge, William S., 277
Bakan, Joel, 111, 219
Bakyawa, Jennifer, 171
Baldauf, Scott, 153
Baldi, Stephane, 267
Bales, Robert F., 170, 234
Ball, Richard A., 126
Bandura, Albert, 67
Banerjee, Neela, 166, 288
Banfield, Edward C., 146
Banks, Curtis, 103
Banks, Ingrid, 316
Banks, Marian R., 327
Banks, Sandy, 162
Banks, William A., 327
Bankston, Carl L., III, 281
Barabak, Mark Z., 179
Barash, David P., 64
Barbarin, Oscar A., 242
Barber, Bernard, 317
Barboza, Tony, 122
Bar-Cohen, Yoseph, 327
Bardes, Barbara A., 203
Barker, James B., 110
Barnard, Chester, 108
Barnes, Cynthia, 235
Barnett, Megan, 221
Barnett, Rosalind, 6
Barrera, Douglas S., 6, 266
Barrionuevo, Alexei, 315
Barro, Robert J., 284
Barrows, Sydney B., 172
Barry, Carolyn McNamara, 169
Bart, Pauline B., 172
Bartkowski, John P., 289
Bartlett, Annie, 169
Bartlett, Thomas, 271, 288
Barton, Allen H., 107
Barusch, Amanda S., 249, 252
Baum, Katrina, 123
Baumrind, Diana, 72
Bayer, Patrick, 302
Bean, Frank D., 180
Becker, Bernice, 213
Becker, Elizabeth, 213
Becker, Howard S., 83, 118, 131
Becker, Jasper, 218
Beckman, Linda, 168
Beeghley, Leonard, 144, 171
Behnke, Andrew, 242
Beitin, Ben K., 251
Belkin, Lisa, 166
Belknap, Joanne, 127, 128, 129, 132
Belsky, Janet K., 66
Belsky, Jay, 265
Belson, Ken, 126
Bender, Courtney, 284, 322
Bennett, Stephen E., 29
Bennhold, Katrin, 282
Benokraitis, Nijole V., 32, 159, 204,
     236, 245, 251
Benson, Michael L., 131, 245
Bergen, Raquel K., 90
Berger, Adam, 3
Berger, Peter L., 87, 289

Bergman, Mike, 224
Bergquist, Amy, 282
Berk, Richard A., 314
Berkos, Kristen M., 88
Bernstein, Jared, 147, 220, 222, 224
Bernstein, Nina, 49
Bertrand, Marianne, 177
Berube, Alan, 302
Best, Joel, 29, 317, 318
Betcher, R. William, 170
Bianchi, Suzanne M., 160, 242
Bielick, Stacey, 272
Biely, E., 75
Biggs, Mary, 271
Billingsley, Andrew, 242
Binder, Amy, 32
bin Laden, Osama, 236
Birch, Douglas, 289
Bird, Warren, 280
Birnbaum, Jeffery H., 205
Bishop, George F., 29
Bivens, L. Josh, 221, 222
Blakeley, Kiri, 152
Blanchard, Dallas A., 324
Blankenstein, Andrew, 122
Blass, Thomas, 113
Blau, Francine B., 162
Blau, Peter M., 89, 107
Blauner, Robert, 181
Bloomberg, Michael R., 205
Blumenthal, David, 115
Blumer, George G., 19, 87
Blumer, Herbert, 19, 316, 320
Boase, Jeffrey, 96
Bobroff-Hajal, Anne, 234
Boggs, George R., 255
Bogle, Dale "Rooster," 129, 130
Bogle, Tracey, 129, 130
Bohannon, John, 57
Bohannon, Lisa, 93
Bohannon, Paul, 86
Bollag, Burton, 261
Bonacich, Edna, 186
Bonder, Bette, 81
Bonger, Willem A., 125
Bonisteel, Sara, 183
Bonner, Robert, 161
Bonnie, Richard J., 246
Boonstra, Heather D., 166, 167
Booza, Jason C., 302
Boraas, Stephanie, 162
Borzekowski, Dina L., 74
Bosk, Charles, 231
Bouma-Holtrop, Sharon, 7
Bourdieu, Pierre, 149, 258, 317
Bourne, Joel K., Jr., 306
Bowles, Samuel, 259, 260
Boyle, B., 258
Bradshaw, York W., 155
Brady, Diane, 231
Braga, Anthony A., 133
Brainard, Jeffery, 199
Bramlett, Mosher D., 238, 239
Braun, Henry, 272
Braunschweig, Carol L., 75
Brea, Jorge A., 298
Breazeal, Cynthia L., 327
Brickner, Mindy Kay, 173
Bridgeland, John M., 270
Briggs, Tami, 48
Brittingham, Angela, 194, 244
Britton, Dana M., 129
Broder, John M., 309
Brodie, Mollyann, 184
Brody, Gene H., 72

Broken Nose, Mary A., 243
Bronstad, Amanda, 162
Brookey, Robert A., 165
Brooks, Arthur C., 229
Broom, Dorothy H., 223
Brown, David, 58
Brown, Matthew Hay, 284
Brown, Susan I., 239
Browne, K. D., 246
Brownwell, Kelly D., 101
Bruni, Frank, 44
Brunvand, Jan H., 315, 316
Buchmann, Claudia, 263
Buckland, Gail, 57
Buckley, Jack, 272
Buechler, Steven M., 323, 324
Bukowski, William, 73
Bullock, 27
Bumiller, Elisabeth, 314
Bumpass, Larry L., 238
Burgess, Ernest, 302, 303
Burggraf, Cynthia S., 91
Burke, Monte, 152
Burns, Barbara J., 132
Burr, Chandler, 165
Burrus, Robert T., 271
Burt, Martha R., 147
Burton, Linda M., 242
Bush, George W., 35, 40, 41, 44, 52,
     93, 111, 205, 209, 213, 261, 269,
     284, 308, 314
Buslman, Brad J., 75
Bustillo, Miguel, 185
Butterfield, Fox, 130
Butterfield, Kenneth D., 271
Byham, William C., 110

C

Cadge, Wendy, 322
Caetano, Raul, 246
Calasanti, Toni M., 173
Calvert, Scott, 42
Calvin, John, 12
Campbell, Christopher, 179
Campbell, Donald T., 33
Campbell, Eric G., 115
Campion-Vincent, Veronique, 315
Campo-Flores, Arian, 281
Canary, Daniel J., 91
Cantor, Paul A., 51
Capps, Randy, 179
Carasso, Adam, 250
Card, Josefina J., 34
Carey, Benedict, 35
Carey, John, 310
Carl, Traci, 155
Carlson, Lewis H., 4
Carlson, Marcia, 240
Carmichael, Mary, 307
Carnagey, Nicholas L., 75
Carney, Ginny, 44
Carr, J. L., 123
Carroll, C. Dennis, 264
Carroll, Jason S., 169
Carroll, Jill, 271
Carroll, Joseph, 186, 195, 225, 288,
     307
Carroll, Margaret D., 54
Carson, Ben, 149
Carter, Jimmy, 201
Carter, Nancy M., 113
Cartwright, Lisa, 76
Casper, Lynne M., 77, 195

SUBJECT INDEX

Note: Italic page numbers indicate material in tables or figures. Bold page numbers indicate definitions.

social organization, 12
social policies, 6
Social Register, 142
social research, **23**, 23–24
    ethical research, 35, 37
    outside influences, 35–37
    scientific method. *See* Scientific
      method
social sciences, 7–9
social service agencies, 132
social solidarity, **10**
social statics, 10
social strain, 124
social stratification, **137**, 137
    closed systems, 137–138
    conflict perspectives, 151, *151*
    dimensions of, 138–141
    ethnic inequality, 185–188
    feminist perspectives, *151*, 152
    functionalist perspectives, 150,
      151
    gender. *See* Gender inequality
    global economic inequality, *153*,
      153–155
    open systems, 138
    racial inequality, 185–188
    religious justification for, 286
    symbolic interactionist perspec-
      tives, *151*, 152–153
social structure, **82**, 82
social workers, 8–9
society, 39
sociobiology, **64**
    and socialization, 64–65
socioeconomic status (SES), **141**. *See
also* Social stratification
sociological imagination, 4–5, *5*
sociology, **3**
spam, 96
special-interest group, **204**
special-interest groups, 204–207
spirituality, 276–277
spousal abuse, 245
standardized testing, 259–260
Stanford Prison Experiment, 34,
    103–104
status, **82**, 82–84. *See also* Social
    stratification
status inconsistency, **83**, 83–84
status set, **82**, 82, *83*
status symbols, 142
stem cell research, 327
stepfamily, **238**
stereotypes, **183**, 183
    gender, 158–159
stigma, **118**
strain theory, **124**
stratification. *See* Social stratification
structural functionalism. *See* Func-
    tionalism
structural strain theory, 314
student diversity, 256
student engagement, 265–266
subcultures, **51**, 51–52
subjective research, 23
subjective understanding, 12–13
suburbs, 301
Suicide, 10
suicide rates, 10–11, *11*
surveys, **28**, 28–29, *36*
    credibility, 29
    victim survey, 120
sustainable development, **309**,
    309–310, *310*
symbolic interactionism, **19**, 19–20, *21*

crime and deviance, 129–132, *130*
culture, 59, *59*
education, 264–266, *265*
ethnic-racial inequality, *186, 187,*
    187–188
families, 251–253, *252, 253*
gender inequality and sexuality,
    173–174, *174*
groups and organizations, 113,
    114
religion, 289–291, *291*
social interaction theories, 87–89,
    90
socialization theories, 68–71
social stratification, *151,* 152–153
urbanization, 304–305, *305*
work and the economy, *230,*
    231, 231
symbols, 20, **41**
    cultural, 41–42
    religious, *288–289,* 289
    sex, *174*

T

teachers
    children's development, impact
      on, 74
    effectiveness of, 267–268
technology, **326**
    benefits/costs of, 328–329
    and cultural change, 54–57
    and cultural lag, 57–58
    ethical issues, 330
    recent advances in, 326–328
    and social change, 326–330
teenage drivers, 4
telecommuting, 97
telephone interviews, 28–29
television's role in socialization, 74–75
Ten Commandments in public places,
    276
theories, **9**
    conflict theory. *See* Conflict theory
    feminist theories. *See* Feminist
      theories
    functionalism. *See* Functionalism
    social exchange theory, 89–90, *90*
    social learning theories, 66–68, *67*
    symbolic interactionism. *See*
      Symbolic interactionism
    what are, 9
thinking critically, 7
Thomas Theorem, 87
"timid bigots," 185
total institutions, **79**
totalitarianism, **201**, 201–202
touch, 93–94
tracking students, **264**, 264
traditional authority, **199**, *199, 199*
transgendered people, **165**, 165
transnational conglomerates, **220**, 220
transnational corporations, **220**, 220
transsexuals, 165
transvestites, 165
two-income families, 241–242, *241*

U

underclass, **144**, 144–145. *See also*
    Poverty
underemployed, **224**
underemployed people, 224

uninvolved parenting, 72, *72*
universal education, 255
universities. *See* Public universities
unprejudiced discriminators, 184–185
unprejudiced nondiscriminators, 184
upper class, 141–142
upper-middle class, 142–143
upper-upper class, 141–142
urban ecology, **302**
urbanization, **299**
    consequences of, 302
    global, 299–300, *300*
    sociological explanations of,
      302–305, *305*
    in the U.S., 300–302
urban legends, **315**, 315–316
urban sprawl, **302**, 302
U.S. education
    attainment statistics, *256*
    changes in, 255–256
    charter schools, 271–272
    college preparedness, *268*
    curricula control, 269–270
    dropout students, *269,* 270
    grade inflation, *270,* 270–271
    home schooling, 272
    magnet schools, 272
    new directions in, 271–272
    problems with, 266–271
    school funding, 267
    school vouchers, 271
    teacher effectiveness, 267–268
U.S. English, *44*
U.S. families
    changes in, 237–242
    cohabitation, 239–240, *240*
    diversity in, 242–244
    divorce and, *237,* 237–238
    singlehood, 238–239
    two-income families, 241–242,
      241
    unmarried parents, 240–241, *241*
U.S. government, *198*
U.S. politics
    democratic system, 201, *202*
    political parties, 203–204, *204*
    sociological perspectives on,
      210–213
    special-interest groups, 204–207
    voting in America, 207–210, *208*
U.S. urbanization, 300–302
U.S. values, 44–45
USA PATRIOT Act, 40–41
uses of sociology, 6

V

vacation days, *225,* 225–226
validity, **25**, 25
value free, **13**
value-free sociology, 13–14
values, **44**
    across cultures, 46
    emotion-laden nature of, 46
    gender differences, *46*
    U.S. values, 44–45
variable, **24**
verstehen, 13
vertical mobility, **148**
victimless crimes, **120**, 120
victims of crimes, 120–121
    women as, 127–128
victim survey, **120**, 120
voluntary associations, **105**, *105*

volunteering in the U.S., *105*
voting in America, 207–210, *208*
vouchers, **271**

W

wage gap between men and women,
    161–162, *162*
water, 306–308
water pollution, 307, *307*
wealth, **138**, 138–139
Weber, Max, 12–14
white-collar crime, **125**, 125
wife beating, 245
women. *See also* Gender
    as criminal offenders, 128–129
    feminist theories. *See* Feminist
      theories
    and poverty, 146–147
    as victims of crimes, 127–128
    violence against, 245
    in the workplace, 226, 226–227,
      227, 228
work, **215**, 215–216
    sociological explanation of,
      227–231, *231*
    in the U.S., 220–227
working class, 143–144
working poor, **144**, 144
workplace issues
    contingent workers, 223–224
    deindustrialization, 221
    downsizing, 223, *223*
    ethnic inequality, 185–188
    gender pay gap, 161–162, *162,*
      226–227, *227, 228*
    gender-segregated work, 160–161
    globalization, 221
    "guest workers," *180*
    job satisfaction and stress, *224,*
      224–226
    labor unions, 221
    low-wage jobs, 222
    minorities in the workplace, 227,
      228
    offshoring, 221–222
    part-time work, 223
    pregnancy discrimination, 162
    racial inequality, 185–188
    sexual harassment, 162
    shift work, 222–223
    underemployed people, 224
    vacation days, *225,* 225–226
    women in the workplace, *226,*
      226–227, *227, 228*
work roles, 77–78
work teams, 110
world religions, 278, *279*
world-system theory, 155

Y

young adulthood, 77

Z

zero population growth (ZPG), **299**,
    299
Zimbardo's research, 103–104

To help you succeed, we've designed a review card for each chapter.

CHAPTER 2 | IN REVIEW

Examining Our Social World

Chapter 2 Topics

1 What Is Social Research?

Social research is the ongoing study of human behavior. [It requires] and imagination, but also an understanding of the rules and [it's use]ful scientific study. The process entails choosing a socially-re[levant] [resear]ch question, developing a hypothesis, testing that hypothe[sis] [finding]s. Unbiased research is fundamental to research, while per[sonal, self-] help literature, often ignore the scientific method. Beca[use] researchers examine past studies and modify research designs to better understand a social phenomenon.

> In this column, you'll find summary points supported by diagrams when relevant to help you better understand important concepts.

2 The Scientific Method

The *scientific method* is an established research process which incorporates careful data collection, exact measurement, accurate recording and analysis of findings, thoughtful interpretation of results, and, when appropriate, a generalization of the findings to a larger group. Sociologists, like other researchers, use the scientific method to measure the relationships between *variables*—characteristics that can change in value or magnitude under different conditions. A research question or a *hypothesis* examines the association between an *independent variable* (a characteristic that determines or has an effect) and the *dependent variable* (the result or outcome). Sociologists use both qualitative and quantitative approaches to determine if there is a relationship between variables and are always concerned about the *reliability* and *validity* of their measures.

3 Some Major Data Collection Methods

There are many data collection methods, but six are especially common in social research, including sociology.

- Sociologists use *surveys* to systematically collect data from respondents using questionnaires, face-to-face or telephone interviews, or a combination of these.

someone else, such as

in their natural

ta are categorized,
nd other characteristics.
en two or more variables

l techniques to assess the
e sectors.

ollection method in

FIGURE 2.2

Steps in the Scientific Method Using the Inductive Approach

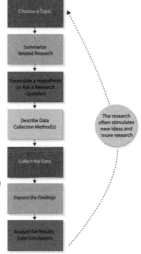

How to Use This Card

1. Look over the card to preview the new concepts you'll be introduced to in the chapter.
2. Read your chapter to fully understand the material.
3. Go to class (and pay attention).
4. Review the card one more time to make sure you've registered the key concepts.
5. Don't forget, this card is only one of many SOC learning tools available to help you succeed in your sociology course.

Key Terms

social research research that examines human behavior.

scientific method the steps in the research process that include careful data collection, exact measurement, accurate recording and analysis of the findings, thoughtful interpretation of results, and, when appropriate, a generalization of the findings to a larger group.

variable a characteristic that can change in value or magnitude under different conditions.

hypothesis a statement of a relationship between two or more variables that researchers want to test.

independent variable a characteristic that determines or has an effect on the dependent variable.

dependent variable the outcome, which may be affected by the independent variable.

reliability the consistency with which the [same measure produces the same] resu[lt]

> Here you'll find key terms and definitions in the order in which they appear in the chapter.

val[idity] [the degree to which a m]easure is ac[curate or measures wha]t it clai[ms to measure.]

de[...] [...]ng that [...] [...]n, or g[...]ted through data collection.

inductive reasoning reasoning that begins with a specific observation, followed by data collection and the development of a general conclusion or theory.

population any well-defined group of people (or things) about whom researchers want to know something.

sample a group of people (or things) that are representative of the population that researchers wish to study.

probability sample a sample for which each person (or thing, such as an e-mail address) has an equal chance of

> When it's time to prepare for exams, use the review card and the technique to the left to ensure successful study sessions.

[...] examines nonnumerical material and interprets it.

quantitative research research that focuses on a numerical analysis of people's responses or specific characteristics.

surveys a systematic method for collecting data from respondents, including questionnaires, face-to-face or telephone interviews, or a combination of these.

secondary analysis examination of data that have been collected by someone else.

field research data collection by systematically observing people in their natural surroundings.

content analysis data collection method that systematically examines examples of some form of communication.

experiment a carefully controlled artificial situation that allows researchers to manipulate variables and measure the effects.

experimental group the group of subjects in an experiment who are exposed to the independent variable.

control group the group of subjects in an experiment who are not exposed to the independent variable.

4 Ethics, Politics, and Sociological Research

To avoid exploitation and maltreatment of participants, sociological research demands a strict code of ethics throughout every research step. For example, participants must give informed consent and must not be harmed, humiliated, abused, or coerced; researchers must honor their guarantees of privacy, confidentiality, and/or anonymity. Still, sociologists, like other researchers, often encounter considerable pressure from policy makers and others to limit their research to topics that won't generate controversy on sensitive issues.

Example: *Sociological Research in Action*

The Human Terrain Team is an experimental Pentagon scientists to U.S. combat units in Afghanistan and Iraq social scientists to gather information about local pop understanding of cultural differences and then to neg encourage the growth of agitators.

Source: Rohde, David. 2007. "Army Enlists Anthropology in V

evaluation research research that relies on all of the standard data collection techniques to assess the effectiveness of social programs in both the public and the private sectors.

> Some review cards have extra examples to help you better connect the theoretical concepts from the chapter to events and situations that occur or exist in the real world.

> Each review card contains some tables. Most summarize the different sociological perspectives on the chapter topics; this one, from early in the book, summarizes the different data collection methods used in sociological research.

TABLE 2.2
Some Data Collection Methods in Sociological Research

METHOD	EXAMPLE	ADVANTAGES	DISADVANTAGES
Surveys	Sending questionnaires and/or interviewing students on why they succeeded in college or dropped out	Questionnaires are fairly inexpensive and simple to administer; interviews have high response rates; findings are often generalizable	Mailed questionnaires may have low response rates; respondents may be self-selected; interviews are usually expensive
Secondary analysis	Using data from the National Center for Education Statistics (or similar organizations) to examine why students drop out of college	Usually accessible, convenient and inexpensive; often longitudinal and historical	Information may be incomplete; biases
Field research	Observing first-year college students with high and low grade-point averages (GPAs) regarding their classroom participation and other activities	Flexible; offers deeper understanding of social behavior; usually inexpensive	
Content analysis	Comparing the transcripts of college graduates and dropouts on variables such as gender, race/ethnicity, and social class	Usually inexpensive; can recode errors easily; unobtrusive; permits comparisons over time	
Experiments	Providing tutors to some students with low GPAs to find out if such resources increase college graduation rates	Usually inexpensive; plentiful supply of subjects; can be replicated	Volunteers and paid subjects aren't representative of a larger population; the laboratory setting is artificial
Evaluation research	Examining student records; interviewing administrators, faculty, and students; observing students in a variety of settings (such as classroom and extracurricular activities); and using surveys to determine students' employment and family responsibilities.	Usually inexpensive; valuable in real-life applications	Often political; findings might be rejected

r 7 Topics

eviance?

violation of social norms that is usually accompanied by a *stigma*, a negative lues a person and changes her or his self-concept and social identity. deviance vary across and within societies, and can change over time. Those power decide what's right or wrong.

iance and College Drinking

any college presidents, alcohol abuse is the most serious problem on e it results in alcohol poisoning and blackouts and leads to sexual assault, , injuries, and academic problems. Because drinking laws are rarely college presidents have proposed that the drinking age be lowered from argue that changing the law would increase deviant behavior, including mong young people and drinking problems. Young people can get a 16 and vote and enlist in the military at 18. Should they be the ones, then, er drinking laws should be changed?

2   ne?

Cri n of societal norms and rules for which punishment is specified by law. Many
soc *iminologists*, researchers who use scientific methods to study the nature,
ext control of criminal behavior. Violent crimes are most likely to be covered by
the ericans are much more likely to be victimized by theft or burglary than to be
mu obbed, or assaulted with a deadly weapon. Most offenders are never caught,
but ow that offenders are usually young white and African American males from
lowe ic levels. Most victims of crime are male, black, under age 25, poor, and live
in u *imless crimes* are acts that violate laws but the parties involved don't consider
them

3 c Deviance and Crime

Socia to the techniques and strategies that regulate people's behavior in
societ se of social control is to eliminate, or at least reduce, deviance and crime.
Forma ol is administered by those in authority or power and exists outside of
the in mal social control is internalized from childhood. Most people conform
becau , punishments or rewards for obeying or violating a norm. Positive
sanctio ood" behavior. Negative sanctions are punishments for violating a norm.

differ **ciation** people learning
devianc eraction, especially with
significa

labelin a perspective that holds
that soci to behavior is a major
factor in self or others as deviant.

primar the initial violation of a
norm or l

second ce rule-breaking
behavior dopt in response to the
reactions

criminal justice system the government agencies—including the police, courts, and prisons—that are charged with enforcing laws, passing judgment on offenders, and changing criminal behavior.

crime control model an approach that holds that crimes rates increase when offenders don't fear apprehension or punishment

rehabilitation a social control approach that holds that appropriate treatment can change offenders into productive, law-abiding citizens.

Key Terms

deviance behavior that violates expected rules or norms.

stigma a negative label that devalues a person and changes her or his self-concept and social identity.

crime a violation of societal norms and rules for which punishment is specified by public law.

criminologists researchers who use scientific methods to study the nature, extent, cause, and control of criminal behavior.

victim survey a method of gathering data that involves interviewing people about their experiences as crime victims.

victimless crimes acts that violate laws but involve individuals who don't consider themselves victims.

social control the techniques and strategies that regulate people's behavior in society.

sanctions punishments or rewards for obeying or violating a norm.

anomie the condition in which people are unsure of how to behave because of absent, conflicting, or confusing social norms.

strain theory the idea that people may engage in deviant behavior when they experience a conflict between goals and the means available to obtain the goals.

white-collar crime illegal activities committed by high-status individuals in the course of their occupation.

occupational crimes crimes committed in the workplace by individuals acting solely in their own personal interest.

corporate crimes white-collar crimes committed by executives to benefit themselves and their companies (also known as *organizational crimes*).

cybercrime white-collar crimes that are conducted online.

organized crime activities of individuals and groups that supply illegal goods and services for profit.

Sociological Perspectives on Deviance and Crime

As with previous chapters, this chapter examines deviance and crime through the lens of four sociological approaches. The following table organizes the approaches for key topics 4 through 7.

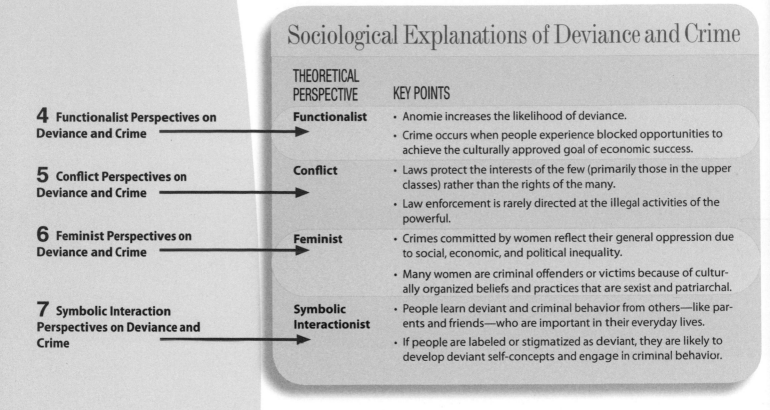

Sociological Explanations of Deviance and Crime

THEORETICAL PERSPECTIVE	KEY POINTS
4 Functionalist Perspectives on Deviance and Crime → **Functionalist**	• Anomie increases the likelihood of deviance. • Crime occurs when people experience blocked opportunities to achieve the culturally approved goal of economic success.
5 Conflict Perspectives on Deviance and Crime → **Conflict**	• Laws protect the interests of the few (primarily those in the upper classes) rather than the rights of the many. • Law enforcement is rarely directed at the illegal activities of the powerful.
6 Feminist Perspectives on Deviance and Crime → **Feminist**	• Crimes committed by women reflect their general oppression due to social, economic, and political inequality. • Many women are criminal offenders or victims because of culturally organized beliefs and practices that are sexist and patriarchal.
7 Symbolic Interaction Perspectives on Deviance and Crime → **Symbolic Interactionist**	• People learn deviant and criminal behavior from others—like parents and friends—who are important in their everyday lives. • If people are labeled or stigmatized as deviant, they are likely to develop deviant self-concepts and engage in criminal behavior.

8 The Criminal Justice System and Social Control

The *criminal justice system* refers to government agencies that are charged with enforcing laws, passing judgment on offenders, and changing criminal behavior, and relies on three major approaches in controlling crime: prevention and intervention, punishment, and rehabilitation. A *crime control model* emphasizes protecting society and supports a tough approach toward criminals in sentencing, imprisonment, and capital punishment. In contrast, many people believe that *rehabilitation* can change offenders into productive and law-abiding citizens, especially if offenders are provided with educational opportunities, job training, and crisis intervention programs.

Chapter 8 Topics

1 What Is Social Stratification?

Social stratification is a hierarchical ranking of people who have different access to valued resources such as property, prestige, power, and status. In a *closed stratification system*, movement from one social position to another is very limited due to ascribed statuses such as gender, skin color, and family background. An *open stratification system* is based on a person's individual achievement and allows for movement up or down. A *social class* is a category of people who have a similar standing or rank based on wealth, education, power, prestige, and other valued resources.

2 Dimensions of Stratification

In explaining stratification, sociologists use a multidimensional approach that includes wealth, prestige, and power. *Wealth* is the money and economic assets that a person or family owns, including property and income. *Prestige* is the respect, recognition, or regard attached to particular social positions, and is based on many criteria, including wealth, family background, fame, leadership, occupation, skills, and power. *Power* is the ability of individuals or groups to achieve goals, control events, and maintain influence over others despite opposition.

3 Social Class in America

A good indicator of social class is *socioeconomic status (SES)*, an overall rank of people's social position based on their income, education, and occupation. Using SES and other variables (such as values, power, social networks, lifestyles, and *conspicuous consumption*), most sociologists agree that there are at least four general social classes in the United States—upper, middle, working, and lower. These groups can be divided further into upper-upper, lower-upper, upper-middle, lower-middle, and the working class. The lower class includes the *working poor* and the *underclass*. A major outcome of social stratification is *life chances*, the extent to which people have positive experiences and can secure good things in life.

Example: *Restaurant Menus and Stratification*

What do our everyday, taken-for-granted establishments tell us about social class? Two sociologists—in Iowa and Virginia—asked their students in introductory sociology classes to do a content analysis (see Chapter 1) of 10 menus that represented a cross-sampling of restaurants by social class. What did the students find? The restaurants that catered to upper-class clientele had higher-than-average entrée prices, described the entrées in foreign languages, used fancy sauces, recommended expensive wines, and had few illustrations. Middle-class menus emphasized "value for the dollar," presented photos of entrées with "bountiful plates overflowing with appetizing food," and popular items such as quesadillas. Menus at lower-class restaurants featured low prices ($3 to $10 entrées), the items were numbered, none of the entrées had "pretentious names," and the typesetting was simple. In effect, then, even menus denote social class and social status.

Source: Wright, Wynne, and Elizabeth Ransom, 2005, "Stratification on the menu: Using restaurant menus to examine social class," *Teaching Sociology* 33 (July): 310–316.

4 Poverty in America

Absolute poverty is not having enough money to afford the basic necessities of life such as food, clothing, and shelter. *Relative poverty* is not having enough money to maintain an

Key Terms

social stratification the hierarchical ranking of people in a society who have different access to valued resources, such as property, prestige, power, and status.

open stratification system a system that is based on individual achievement and allows movement up or down.

closed stratification system a system in which movement from one social position to another is limited by ascribed statuses such as one's sex, skin color, and family background.

social class a category of people who have a similar standing or rank in a society based on wealth, education, power, prestige, and other valued resources.

wealth the money and other economic assets that a person or family owns, including property and income.

prestige respect, recognition, or regard attached to social positions.

power the ability of individuals or groups to achieve goals, control events, and maintain influence over others despite opposition.

socioeconomic status (SES) an overall ranking of a person's position in the class hierarchy based on income, education, and occupation.

conspicuous consumption lavish spending on goods and services to display one's social status and to enhance one's prestige.

working poor people who work at least 27 weeks a year but receive such low wages that they live in or near poverty.

underclass people who are persistently poor and seldom employed, segregated residentially, and relatively isolated from the rest of the population.

life chances the extent to which people have positive experiences and can secure the good things in life because they have economic resources.

absolute poverty not having enough money to afford the most basic necessities of life.

relative poverty not having enough money to maintain an average standard of living.

poverty line the minimal level of income that the federal government considers necessary for basic subsistence.

feminization of poverty the higher likelihood that female heads of households will be poor.

social mobility a person's ability to move up or down the class hierarchy.

horizontal mobility moving from one position to another at the same class level.

vertical mobility moving up or down the class hierarchy.

intragenerational mobility moving up or down the class hierarchy over a lifetime.

intergenerational mobility moving up or down the class hierarchy relative to the position of one's parents.

Davis-Moore thesis the functionalist view that social stratification has beneficial consequences for a society's operation.

meritocracy a belief that individuals are rewarded for what they do and how well rather than on the basis of their ascribed status.

bourgeoisie those who own the means of production and can amass wealth and power.

proletariat workers who sell their labor for wages.

corporate welfare an array of direct subsidies, tax breaks, and assistance that the government has created for businesses.

average standard of living. The *poverty line* is the minimal level of income that the federal government considers necessary for basic subsistence. Explanations for poverty vary, but two general perspectives propose that individual characteristics lead to poverty or that a society's organization creates and sustains poverty.

5 Social Mobility

Social mobility is a person's movement up or down the stratification hierarchy. Social mobility can be *horizontal, vertical, intragenerational,* or *intergenerational.* Structural, demographic, and individual factors affect a person's social mobility. Structural factors include the economy, changes in the number of available job positions, and immigration. Demographic factors include education, gender, and race and ethnicity. Individual factors include family origin, socialization and *habitus,* and connections and chance.

6 Why There Are Haves and Have-Nots

Sociological Explanations of Social Stratification

PERSPECTIVE	KEY POINTS
Functionalist	• Fills social positions that are necessary for a society's survival • Motivates people to succeed and ensures that the most qualified people will fill the most important positions
Conflict	• Encourages workers' exploitation and promotes the interests of the rich and powerful • Ignores a wealth of talent among the poor
Feminist	• Constructs numerous barriers in patriarchal societies that limit women's achieving wealth, status, and prestige • Requires most women, not men, to juggle domestic and employment responsibilities that impede upward mobility
Symbolic Interactionist	• Shapes stratification through socialization, everyday interaction, and group membership • Reflects social class identification through symbols, especially products that signify social status

7 Inequality across Societies

Global inequality is widespread, but some societies are much wealthier than others. There are 53 low-income countries, 96 middle-income countries, and 60 high-income countries. Sociologists often use *modernization theory, dependency theory*, and *world-system theory* to explain why inequality is universal.

Chapter 9 Topics

1 Female-Male Similarities and Differences

Sex refers to biological characteristics apparent at birth and include chromosomes, anatomy, hormones, and other physical and physiological attributes. *Gender* refers to learned attitudes and behaviors that characterize people of one sex or the other. Many people use the terms sex and gender interchangeably, but sex is a biological designation, whereas gender and gender roles are social creations. *Gender identity* is the individual's perception of himself or herself as either masculine or feminine. *Gender roles* are the characteristics, attitudes, feelings, and behaviors that society expects of females and males. Many Americans still have *gender stereotypes*, expectations about how people will look, act, think, and feel based on their sex.

2 Contemporary Gender Stratification and Inequality

Sexism is an attitude or behavior that discriminates against one sex, mostly women, based on the assumed superiority of the other sex. Sexism is widespread due to *gender stratification*, people's unequal access to wealth, power, status, prestige, and other valued resources on the basis of sex. Gender stratification can lead to inequality in the family, education, the workplace (as in the case of a *gender pay gap*), and politics. *Sexual harassment* and *pregnancy discrimination* are also common in the workplace.

3 Sexual Orientation

Our sexual identity incorporates a *sexual orientation*—a preference for sexual partners of the same sex (*homosexuality*), of the opposite sex (*heterosexuality*), of both sexes (*bisexuality*), and of neither sex (*asexuality*). *Transgendered* people include those living on the boundaries of the sexes. Many biological theories maintain that sexual orientation has a strong genetic basis, but social constructionists argue that sexual behavior is largely the result of socialization, and that culture, not biology, plays a large role in forming sexual identity.

4 Some Current Controversies about Gender and Sexuality

Abortion is the expulsion of an embryo or fetus from the uterus. The abortion rate, which has decreased steadily since 1980, is controversial because almost equal percentages of Americans support or condemn the practice. Those who favor same-sex marriage argue that people should have the same legal rights regardless of sexual orientation and that marriage may increase the stability of same-sex couples and lead to better physical and mental health for gays and lesbians. Those who oppose same-sex marriage contend that such unions are immoral, weaken our traditional notions of marriage, and are contrary to religious beliefs. *Pornography* is the graphic depiction of images—including photographs and videos, especially those on the Internet—that cause sexual arousal. Some people view pornography as erotic recreation, but others denounce it as obscene and as debasing women.

Key Terms

sex the biological characteristics with which we are born.

gender learned attitudes and behaviors that characterize people of one sex or the other.

gender identity a perception of oneself as either masculine or feminine.

gender roles the characteristics, attitudes, feelings, and behaviors that society expects of females and males.

gender stereotypes expectations about how people will look, act, think, and feel based on their sex.

sexism an attitude or behavior that discriminates against one sex, usually females, based on the assumed superiority of the other sex.

gender stratification people's unequal access to wealth, power, status, prestige, and other valued resources as a result of their sex.

gender pay gap the overall income difference between women and men in the workplace (also called the *wage gap*).

sexual harassment any unwanted sexual advance, request for sexual favors, or other conduct of a sexual nature that makes a person uncomfortable and interferes with her or his work.

sexual orientation a preference for sexual partners of the same sex, of the opposite sex, or of both sexes.

homosexuals those who are sexually attracted to people of the same sex.

heterosexuals those who are sexually attracted to people of the opposite sex.

bisexuals those who are sexually attracted to members of both sexes.

asexuals those who lack any interest in or desire for sex.

transgendered people those who are transsexuals, intersexuals, or transvestites.

heterosexism the belief that heterosexuality is superior to and more natural than homosexuality or bisexuality.

homophobia the fear and hatred of homosexuality.

abortion the expulsion of an embryo or fetus from the uterus.

pornography the graphic depiction of images that cause sexual arousal.

5 Sociological Explanations of Gender Inequality and Sexuality

Sociological Explanations of Gender Inequality and Sexuality

THEORETICAL PERSPECTIVE	KEY POINTS
Functionalist	• Gender roles are complementary and equally important. • Agreed-upon sexual norms contribute to a society's order and stability.
Conflict	• Gender roles give men power to control women's lives instead of allowing the sexes to be complementary and equally important. • Most societies regulate women's, not men's, sexual behavior.
Feminist	• Women's inequality reflects their historical and current domination by men, especially in the workplace. • Many men use violence—including sexual harassment, rape, and global sex trafficking—to control women's sexuality.
Symbolic Interactionist	• Gender inequality is a social construction that emerges through day-to-day interactions and reflects people's gender role expectations. • The social construction of sexuality varies across cultures because of societal norms and values.

Example: *Gender Roles and Hooking Up—Are Women the Losers?*

Hooking up (or "hookin' up") refers to physical encounters, no strings attached, and can mean anything from kissing and genital fondling to oral sex and sexual intercourse. Several studies at colleges have found that between 60 to 84 percent of the students had hooked up at one time or another (cited in McGuinn, 2004). Hooking up has its advantages because it's cheaper than dating. Also, because no one knows for sure what, if anything, happened, women can avoid getting a bad reputation for being "loose" or "easy." Most importantly, it's assumed that hooking up requires no commitment of time or emotion: "A girl and a guy get together for a physical encounter and don't necessarily expect anything further" (Wolcott, 2004: 11). Hooking up also has disadvantages. For example, in a study of 4,000 undergraduates at five large U.S. campuses, sociologists Paula England and her colleagues (2007) found that hook ups are gendered in three important ways: (1) men initiate most of the sexual action; (2) men have orgasms more frequently than women and see pleasure, rather than an enduring relationship, as a high priority; and (3) a sexual double standard persists because women are more at risk than men of getting a bad reputation for hooking up with multiple partners.

Sources: McGinn, Daniel, 2004, "Mating Behavior 101," *Newsweek* (October 4): 44–45; Wolcott, Jennifer, 2004, "Is Dating Dated on College Campuses?" *Christian Science Monitor*, March 2: 11, 14; England, Paula, Emily Fitzgibbons Shafer, and Alison C. K. Fogarty, 2007, "Hooking Up and Forming Romantic Relationships on Today's College Campuses," pp. 531–547 in *The Gendered Society Reader*, 3rd edition, edited by Michael Kimmel, New York: Oxford University Press.

Chapter 10 Topics

1 Racial and Ethnic Diversity in America

One in five Americans is either foreign-born or a first-generation resident. Perhaps the most multicultural country in the world, the United States includes about 150 distinct ethnic or racial groups among more than 305 million inhabitants. By 2025, only 58 percent of the U.S. population will be white—down from 86 percent in 1950.

2 The Significance of Race and Ethnicity

Race refers to a group of people who share physical characteristics, such as skin color and facial features, that are passed on through reproduction. An *ethnic group* is a set of people who identify with a common national origin or cultural heritage, such as Puerto Ricans and Hungarians. A *racial-ethnic group* is a category of people that has both distinctive physical and cultural characteristics. For example, Japanese-American designates a specific region of origin, language, and customs.

3 Our Immigration Mosaic

In 1900, almost 85 percent of immigrants came from Europe; now immigrants come primarily from Asia (mainly China and the Philippines) and Latin America (mainly Mexico). Many Americans are ambivalent about immigrants, especially those who are in the country illegally, but most scholars argue, that on balance and in the long run, both legal and undocumented immigrants bring more benefits than costs.

4 Dominant and Minority Groups

A *dominant group* is any physically or culturally distinctive group that has the most economic and political power, the greatest privileges, and the highest social status in a society. A *minority group*, which may be larger in number than a dominant group, is a group of people who may be subject to differential and unequal treatment because of their physical, cultural, or other characteristics, such as gender, sexual orientation, religion, ethnicity, or skin color. The pattern of dominant-minority group relations includes *genocide, internal colonialism, segregation, assimilation,* and *pluralism.*

FIGURE 10.2

Continuum of Some Dominant-Minority Group Relations

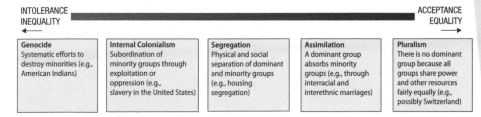

INTOLERANCE INEQUALITY ←				ACCEPTANCE EQUALITY →
Genocide Systematic efforts to destroy minorities (e.g., American Indians)	**Internal Colonialism** Subordination of minority groups through exploitation or oppression (e.g., slavery in the United States)	**Segregation** Physical and social separation of dominant and minority groups (e.g., housing segregation)	**Assimilation** A dominant group absorbs minority groups (e.g., through interracial and interethnic marriages)	**Pluralism** There is no dominant group because all groups share power and other resources fairly equally (e.g., possibly Switzerland)

5 Sources of Racial-Ethnic Friction

Racism is a set of beliefs that one's own racial group is naturally superior to other groups, and justifies and preserves the social, economic, and political interests of dominant groups. *Prejudice* is an attitude, usually negative, toward people because of their group membership. All of us can be prejudiced, but minorities are typically targets of *stereotypes* and *ethnocentrism* that often lead to *scapegoating. Discrimination* is any act that treats people unequally or unfairly because of their group membership, and occurs at both the individual and institutional level.

Key Terms

race a group of people who share physical characteristics, such as skin color and facial features, that are passed on through reproduction.

ethnic group a set of people who identify with a common national origin or cultural heritage that includes language, geographic roots, food, customs, traditions, and/or religion.

racial-ethnic group a group of people who have both distinctive physical and cultural characteristics.

dominant group any physically or culturally distinctive group that has the most economic and political power, the greatest privileges, and the highest social status.

apartheid a formal system of racial segregation.

minority group a group of people who may be subject to differential and unequal treatment because of their physical, cultural, or other characteristics, such as gender, sexual orientation, religion, ethnicity, or skin color.

genocide the systematic effort to kill all members of a particular ethnic, religious, political, racial, or national group.

internal colonialism the unequal treatment and subordinate status of groups within a nation.

segregation the physical and social separation of dominant and minority groups.

assimilation the process of conforming to the culture of the dominant group (by adopting its language and values) and intermarrying with that group.

pluralism minority groups retain their culture but have equal social standing in a society.

racism a set of beliefs that one's own racial group is naturally superior to other groups.

prejudice an attitude, positive or negative, toward people because of their group membership.

stereotype an oversimplified or exaggerated generalization about a category of people.

ethnocentrism the belief that one's own culture, society, or group is inherently superior to others.

scapegoats individuals or groups whom people blame for their own problems or shortcomings.

discrimination any act that treats people unequally or unfairly because of their group membership.

individual discrimination harmful action directed intentionally, on a one-to-one basis, by a member of a dominant group against a member of a minority group.

institutional discrimination unequal treatment and opportunities that members of minority groups experience as a result of the everyday operations of a society's laws, rules, policies, practices, and customs.

gendered racism the combined and cumulative effects of inequality due to racism and sexism.

contact hypothesis the idea that the more people get to know members of a minority group personally, the less likely they are to be prejudiced against that group.

miscegenation marriage or sexual relations between a man and a woman of different races.

6 Sociological Explanations of Racial-Ethnic Inequality

THEORETICAL PERSPECTIVE	KEY POINTS
Functionalist	Prejudice and discrimination can be dysfunctional, but they provide benefits for dominant groups and stabilize society.
Conflict	Powerful groups maintain their advantages and perpetuate racial-ethnic inequality primarily through economic exploitation.
Feminist	Minority women suffer from the combined effects of racism and sexism.
Symbolic Interactionist	Hostile attitudes toward minorities, which are learned, can be reduced through cooperative interracial and interethnic contacts.

7 Major Racial and Ethnic Groups in the United States

America is home to a multitude of ethnic groups. European Americans, who settled the first colonies, are declining while Latinos are the fastest-growing minority group and now comprise 15 percent of the population. African Americans, who make up 13 percent of the population, have diverse roots. Asian Americans, who make up 5 percent of the population, come from at least 26 countries. American Indians make up less than 2 percent of the population, but are heterogeneous and growing in number. Middle Eastern Americans, who comprise the smallest minority group, come from over 30 countries, and reflect a multitude of ethnic and linguistic groups with very different native languages and customs. All of these groups have experienced prejudice and discrimination, but they have enhanced U.S. society and culture.

Example: *Stuff White People Like*

The popular blog, *Stuff White People Like,* has generated clones (like *Stuff Educated Black People Like* and *Stuff Asian People Like*). Why are these sites so popular? Many fans say that the descriptions are funny because they're true. According to some critics, however, by poking fun at privileged upper-middle class people, the sites fuel stereotypes and in frivolous ways instead of having painfully frank discussions about race and racism in the United States. In addition, sites such as *Stuff White People Like* allow readers to feel superior because the entries don't reflect their own lifestyles or because it's comforting to recognize oneself as a member of a comfortable middle class (Jones 2008; Sternbergh 2008). Do you agree? Or not?

Sources: Jones, Vanessa J. 2008. "Coffee and Yoga and Prius and 'Juno'." *Boston Globe,* March 24. Retrieved November 6, 2008 (www.boston.com); Sternbergh, Adam. 2008. "Why White People Like 'Stuff White People Like'." *The New Republic*, March 17. Retrieved November 6, 2008 (www.tnr.com).

8 Interracial and Interethnic Relationships

Almost 99 percent of Americans report being only one race, but the numbers of biracial children are rising due to an increase of interracial dating and marriage. The rise of racial-ethnic intermarriage reflects many micro and macro factors that include greater acceptance of integration and interethnic and interracial contact.

Chapter 11 Topics

1 Government

A *government* is a formal organization that has the authority to make and enforce laws. Governments maintain order, provide welfare services, regulate the economy, and establish educational systems. The U.S. government, like many other democracies, is also affected by a civic society, a group of citizens that includes community-based organizations, the mass media, lobbyists, and voters.

2 Politics, Power, and Authority

Politics is the social process through which individuals and groups acquire and exercise power and authority. *Power* is the ability of a person or group to affect the behavior of others despite resistance and opposition. *Authority*, the legitimate use of power, can be based on tradition, charisma, rational-legal power, or a combination of these sources.

Example: *Rational-legal authority*

In societies based on rational-legal authority, people obey the rules even when they disagree. Americans don't revolt, for example, if candidates they support lose an election. Nor do they question the authority of police, social workers, judges, and other state employees even if they dislike them. On the other hand, leaders who violate laws can lose their authority and office. For example, President Richard Nixon was forced to resign during the 1970s when it became apparent that he had approved a break-in at the Democratic Party's National Committee offices in Washington, D.C.

3 Types of Political Systems

A *democracy* is a political system in which, ideally, citizens have a high degree of control over the state and its actions. Citizens can participate in governmental decisions and elect leaders, and the government recognizes individual rights, such as freedom of speech, press, and assembly. *Totalitarianism* is a political system in which the government controls every aspect of people's lives. The society is controlled by a single party and led by one person, a dictator, who stays in office indefinitely. *Authoritarianism* is a political system in which the state controls the lives of its citizens, but generally permits some degree of individual freedom. A *monarchy*, in which power is allocated solely on the basis of heredity, is the oldest type of authoritarian regime.

Example: *Is the Internet Promoting Democracy in a Totalitarian Country?*

In 1999, there were just 4 million Internet users in China; by early 2008, the number had increased to 210 million and is expected to surge even higher in the future. The government has tried to control Internet use in many ways: blocking politically outspoken blogs, denying access to international Web sites like Wikipedia, and censoring content on sites containing topics like corruption among government officials, the independence movements in Taiwan and Tibet, and citizens' uprisings. The Chinese Communist Party (CCP) justifies its control of all media to avoid damaging China's culture or traditions and its unity and sovereignty. However, some skilled online users are evading the content filters with specially designed Web browsers.

Sources: Ford, Peter. 2007. "Web Opens World for Young Chinese, but Erodes Respect." *Christian Science Monitor*, May 14, 1, 12; Demick, Barbara. 2008. "China Cracks Down on Irreverent Websites." *Los Angeles Times*, February 5. Retrieved February 6, 2008 (www.latimes.com).

Key Terms

government a formal organization that has the authority to make and enforce laws.

politics a social process through which individuals and groups acquire and exercise power and authority.

power the ability of a person or group to affect the behavior of others despite resistance and opposition.

authority the legitimate use of power.

traditional authority authority based on customs that justify the position of the ruler.

charismatic authority authority based on exceptional individual abilities and characteristics that inspire devotion, trust, and obedience.

rational-legal authority authority based on the belief that laws and appointed or elected political leaders are legitimate.

democracy a political system in which, ideally, citizens have control over the state and its actions.

totalitarianism a political system in which the government controls every aspect of people's lives.

authoritarianism a political system in which the state controls the lives of citizens but generally permits some degree of individual freedom.

monarchy a political system in which power is allocated solely on the basis of heredity and passes from generation to generation.

political party an organization that tries to influence and control government by recruiting, nominating, and electing its members to public office.

special-interest group (sometimes called an *interest group*) a voluntary and organized association of people that attempts to influence public policy and policymakers on a particular issue.

lobbyist a representative of a special-interest group who tries to influence political decisions on the group's behalf.

political action committee (PAC) a special-interest group set up to raise money to elect a candidate to public office.

pluralism a political system in which power is distributed among a variety of competing groups in a society.

power elite a small group of influential people who make a nation's major political decisions.

4 Power and Politics in U.S. Society

A *political party* is an organization that tries to influence and control government by recruiting, nominating, and electing its members to public office. Unlike the two-party system in the United States, many democratic countries around the world have numerous parties. A *special-interest group* is a voluntary and organized association of people who attempt to influence policymakers on a particular issue. Some of the most influential special-interest groups include *lobbyists*, wealthy campaign contributors, and *political action committees (PACs)*.

Example: *How Corporations Influence Politics*

In 2008, corporations spent at least $112 million on the Democratic and Republican conventions. Some of the largesse included reduced-fare tickets by United Airlines; use of "plush new vehicles" by General Motors; state-of-the-art technology underwritten by Microsoft, Google, AT&T, and other companies; and corporate-funded events for convention members. Donors who contributed $250,000 or more enjoyed private meetings with top government officials. A major benefit is that the biggest donors, regardless who wins an election, will have greater access to elected officials than does the average citizen.

Source: Campaign Finance Institute. 2008. "Inside Fundraising for the 2008 Party Conventions: Party Surrogates Gather Soft Money While Federal Regulators Turn a Blind Eye." Retrieved June 26, 2008 (www.cfinst.org).

5 Who Votes, Who Doesn't, and Why

Typically, only about half of eligible Americans vote in national elections and only 25 percent vote in local elections. The voting rate is much higher among older than younger people, and increases with age. Married people are more likely to vote than those who are divorced, never married, or widowed. The voting rates of those with a college degree are almost twice as high as those who have not completed high school, and voting rates increase with income levels. Whites are typically the most likely to vote because, among other reasons, they are more likely to be citizens, registered to vote, and are more optimistic than racial-ethnic groups about government and politics. Religion often affects who votes and for whom. Situational and structural factors, such as convenience, can also encourage or discourage voting.

6 Who Rules America?

Sociological Explanations of Political Power

	FUNCTIONALISM: A PLURALIST MODEL	CONFLICT THEORY: A POWER ELITE MODEL	FEMINIST THEORIES: A PATRIARCHAL MODEL
Who has political power?	The people	Rich upper-class people—especially those at top levels in business, government, and the military	White men in Western countries; most men in traditional societies
What is the source of political power?	Citizens' participation	Wealthy people in government, business corporations, the military, and the media	Being white, male, and very rich
Does one group dominate politics?	No	Yes	Yes
Do political leaders represent the average person?	Yes, the leaders speak for a majority of the people.	No, the leaders are most concerned with keeping or increasing their personal wealth and power.	No, the leaders—who are typically white, elite men—are most concerned with protecting or increasing their personal wealth and power.

Chapter **12** Topics

1 The Social Significance of Work

The *economy* is a social institution that determines how a society produces, distributes, and consumes goods and services. *Work* is physical or mental activity that accomplishes or produces something, either goods or services. Work provides a sense of stability, accomplishment, and social identity, but it is also a major source of stress.

2 Global Economic Systems

Capitalism is an economic system in which wealth is in private hands and is invested and reinvested to produce profits. Capitalistic systems frequently spawn *monopolies* and *oligopolies*, which dominate the market and discourage competition. *Socialism* is an economic and political system based on the principle of the public ownership of the production of goods and services. Ideally, socialistic systems emphasize cooperation, a collective ownership of property, and forbid private profits; in reality, there is considerable economic inequality. The late twentieth century experienced *globalization*, the growth and spread of investment, trade, production, communication, and new technology around the world.

3 Corporations and Capitalism

A *corporation* is a social entity that has legal rights, privileges, and liabilities apart from those of its members. Today, there are more than 5 million, most created for profit. Many have formed *conglomerates*, giant corporations that own a collection of companies in different industries. Both corporations and conglomerates are governed by *interlocking directorates*, in which the same people serve on the boards of directors of several companies or corporations. Interlocking directorates have become more powerful than ever because of the growth of *transnational corporations* and *transnational conglomerates*, both of which own and operate a variety of companies in a number of countries.

Example: *Runaway CEO Pay*

U.S. CEOs, even during hard economic times, enjoy huge pay packages. In 2007, the average CEO pay was almost $11 million, 344 times the pay of the typical American worker. The top 50 private fund managers averaged $588 million each, more than 19,000 times as much as typical U.S. workers earned (Anderson et al. 2008b). Management consultant Peter Drucker has proposed that top CEOs shouldn't get more than 20 times the average salary in the company, but some Americans see such proposals as "nothing but Communist rhetoric" (Wartzman 2008). In 2008, after the federal government's $100 billion bailout of a number of corporations with taxpayer money, the new legislation for executive pay did not set any monetary limits on the pay of top executives at bailed out companies (Anderson et al. 2008a).

Sources: Anderson, Sarah, John Cavanagh, Chuck Collins, Dedrick Muhammad, and Sam Pizzigati. 2008a. "Analysis of Treasury Department Rules on Executive Compensation for Bailout Firms." Institute for Policy Studies, October 15. Retrieved November 7, 2008 (www.ips-dc.org); Anderson, Sarah, John Cavanagh, Chuck Collins, Sam Pizzigati, and Mike Lapham. 2008b. "Executive Excess 2008." Institute for Policy Studies and United for a Fair Economy, August 25. Retrieved November 7, 2008 (www.faireconomy.org); Wartzman, Rick. 2008 "Put a Cap on CEO Pay." *Business Week*, September 12. Retrieved November 7, 2008 (www.businessweek.com).

Key Terms

economy a social institution that determines how a society produces, distributes, and consumes goods and services.

work physical or mental activity that accomplishes or produces something, either goods or services.

capitalism an economic system in which wealth is in private hands and is invested and reinvested to produce profits.

monopoly domination of a particular market or industry by one person or company.

oligopoly a market dominated by a few large producers or suppliers.

socialism an economic and political system based on the principle of the public ownership of the production of goods and services.

communism a political and economic system in which all members of a society are equal.

globalization the growth and spread of investment, trade, production, communication, and new technology around the world.

corporation a social entity that has legal rights, privileges, and liabilities apart from those of its members.

conglomerate a giant corporation that owns a collection of companies in different industries.

interlocking directorate a situation in which the same people serve on the boards of directors of several companies or corporations.

transnational corporation (sometimes called a *multinational corporation* or an *international corporation*) a large company that is based in one country but operates across international boundaries.

transnational conglomerate a corporation that owns a collection of different companies in various industries in a number of countries.

deindustrialization a process of social and economic change due to the reduction of industrial activity, especially manufacturing.

offshoring sending work or jobs to another country to cut a company's costs at home.

downsizing a euphemism for firing large numbers of employees at once.

contingent workers people who don't expect their jobs to last or who say that their jobs are temporary.

underemployed people who have part-time jobs but want full-time work or whose jobs are below their experience and education level.

4 Work in U.S. Society Today

Many American workers have been casualties of *deindustrialization*, a process of social and economic change due to the reduction of industrial activity, especially manufacturing. Others have lost their jobs to *offshoring*, sending work or jobs to another country to cut a company's costs at home. Because globalization, deindustrialization, and offshoring have decreased job security, many Americans have had to take low-wage jobs and work shifts. The widespread occurrence of *downsizing*, firing large numbers of employees at once, has created a large pool of *contingent workers* who can find only temporary jobs and part-time work. In addition, millions of Americans are *underemployed*—they have part-time jobs but want full-time work or their jobs are below their experience and education level.

Example: *Underemployment: "I'll Take Any Job"*

Kim Tolivar, 37, who has a master's degree in organizational development, was recently laid off and accepted a temporary job as a clerical assistant, making 20 percent of her previous salary. She is depressed but says that any job is better than nothing. Tolivar is one of over 7 million Americans (up from about 4.5 million in 2007 and representing 13 percent of the workforce) who are overqualified for her job. Some economists contend that by counting the unemployed but not the underemployed, the government skews the unemployment picture because underemployment "is a much more accurate measure of what the economy is really like for people." Underemployment is usually cyclical, and many people return to jobs in their chosen occupation when the economy bounces back. This is little consolation to underemployed workers, however, who often exhaust their savings and investments to pay for mortgages and everyday expenses during the period of underemployment. Being unemployed is worse, but underemployment takes an emotional toll. Says Tolivar, "It makes you wonder if you got the right degree, if you messed up or something."

Source: Rosenwald, Michael S. 2008. "Rising Underemployment Contributes to Pain of Jobs Slump." *Washington Post*, December 6, D1.

5 Sociological Explanations of Work and the Economy

THEORETICAL PERSPECTIVE	KEY POINTS
Functionalist	Capitalism benefits society; work provides an income, structures people's lives, and gives them a sense of accomplishment.
Conflict	Capitalism enables the rich to exploit other groups; most jobs pay little and are monotonous and alienating, creating anger and resentment.
Feminist	Gender roles structure women's and men's work experiences differently and inequitably.
Symbolic Interactionist	How people define and experience work in their everyday lives affects their workplace behavior and relationships with co-workers and employers.

Chapter **13** Topics

1 What Is a Family?

A *family* is an intimate group consisting of two or more people who live together in a committed relationship, care for one another and any children, and share close emotional ties and functions. Worldwide, families are alike in fulfilling similar functions, encouraging *marriage*, and trying to ensure that people select appropriate mates. There are also considerable worldwide variations in many family characteristics such as whether the family structure is a *nuclear* or an *extended* family, living arrangements (*patrilocal*, *matrilocal*, or *neolocal*), who has authority (*matriarchal*, *patriarchal*, or *egalitarian*), and how many marriage mates a person can have (*monogamy* or *polygamy*).

2 How U.S. Families Are Changing

The American family has changed dramatically since the 1960s. Couples of all ages experience *divorce*, the legal dissolution of a marriage, and divorce is easier to obtain than in the past because all states have enacted *no-fault divorce* laws so that neither partner need establish wrongdoing on the part of the other. The number of single people has also risen greatly, primarily because many people are postponing marriage. There has also been a striking increase in *cohabitation* and out-of-wedlock births. In *dual-earner couples*, both partners are employed outside the home. Median income can be twice as high when wives work full time, but the couples must also cope with conflicts between domestic and employment responsibilities.

3 Diversity in American Families

There is considerable diversity across American families because of the increase of racial-ethnic populations. For many Latinos, African Americans, American Indians, and Asian Americans, extended families are common and provide considerable emotional and economic support. Among some of these groups, family structures vary widely depending on the members' time of arrival to the United States and socioeconomic status. Middle Eastern American families tend to have fewer children than the average American family, often because the parents postpone childbearing until they have attained a college or professional degree. Gay and lesbian families are very similar to heterosexual families, but often lack the legal rights and benefits that married couples enjoy.

4 Family Conflict and Violence

Families can be warm, loving, and nurturing, but we are more likely to experience violence with an intimate partner or family member than with a stranger. Nationally, 20 percent of women and 3 percent of men say that a current or former spouse, cohabiting partner, or girlfriend/boyfriend has physically assaulted them at some time. Millions of American children experience abuse and neglect on a daily basis: Almost 80 percent of the perpetrators are parents, and almost a third of all children live in homes where parents or other adults engage in violence. In the case of elder abuse and neglect, similarly, most of the offenders are adult children, spouses, or other family members. Across all families, low income is a major contributing factor for the stress that leads to conflict and violence.

Key Terms

family an intimate group consisting of two or more people who: (1) live together in a committed relationship, (2) care for one another and any children, and (3) share close emotional ties and functions.

incest taboo cultural norms and laws that forbid sexual intercourse between close blood relatives, such as brother and sister, father and daughter, or uncle and niece.

marriage a socially approved mating relationship that people expect to be stable and enduring.

endogamy (sometimes called *homogamy*) the practice of selecting mates from within one's group.

exogamy (sometimes called *heterogamy*) the practice of selecting mates from outside one's group.

nuclear family a form of family consisting of married parents and their biological or adopted children.

extended family a family consisting of parents and children as well as other kin, such as uncles and aunts, nieces and nephews, cousins, and grandparents.

patrilocal residence pattern newly married couples live with the husband's family.

matrilocal residence pattern newly married couples live with the wife's family.

neolocal residence pattern each newly married couple sets up its own residence.

boomerang generation young adults who move back into their parents' home after living independently for a while or who never leave it in the first place.

matriarchal family system the oldest females (usually grandmothers and mothers) control cultural, political, and economic resources and, consequently, have power over males.

patriarchal family system the oldest men (grandfathers, fathers, and uncles) control cultural, political, and economic resources and, consequently, have power over females.

egalitarian family system both partners share power and authority fairly equally.

marriage market a process in which prospective spouses compare the assets and liabilities of eligible partners and choose the best available mate.

monogamy one person is married exclusively to another person.

serial monogamy individuals marry several people, but one at a time.

polygamy a marriage in which a man or woman has two or more spouses.

divorce the legal dissolution of a marriage.

no-fault divorce state laws that do not require either partner to establish guilt or wrongdoing on the part of the other to get a divorce.

stepfamily a household in which two adults are married or living together and at least one of them has a child.

cohabitation an arrangement in which two unrelated people are not married but live together and have a sexual relationship.

dual-earner couples both partners are employed outside the home (also called *dual-income, two-income, two-earner,* or *dual-worker* couples).

fictive kin nonrelatives who are accepted as part of an African American family.

gerontologists scientists who study the biological, psychological, and social aspects of aging.

life expectancy the average length of time people of the same age will live.

sandwich generation people in a middle generation who care for their own children as well as their aging parents.

activity theory proposes that many older people remain engaged in numerous roles and activities, including work.

exchange theory contends that people seek through their interactions with others to maximize their rewards and to minimize their costs.

continuity theory posits that older adults can substitute satisfying new roles for those they've lost.

5 Our Aging Society

How people define "old" varies across societies depending on *life expectancy*. For the most part, however, people in industrialized societies are deemed old at age 65 because they can retire and become eligible for pensions and governmental benefits. *Gerontologists* emphasize that the aging population should not be lumped into one group because, for example, there are significant differences between the young-old and the oldest-old. The United States, like many other countries, is rapidly graying. This means that there will be more debates in the future on how to provide health care, social security, and other resources, especially if schools don't turn out enough skilled and well-educated workers to replace those who are retiring.

Example: *"He Gets Prettier; I Get Older"*

When comparing her own public image with that of her actor husband, the late Paul Newman, actress Joanne Woodward once remarked, "He gets prettier; I get older." Was she right? When Dove started marketing Pro Age, a new line of skin and hair care products for older women, with ads on billboards and elsewhere, many women (including some feminists) praised the ads for "celebrating older women." Why, however, aren't any of the women at least 50 pounds overweight or in their 70s and 80s? If aging gracefully is acceptable, why does Dove tout anti-aging products? And, where are the men in these and similar anti-aging ads?

6 Sociological Explanations of Family and Aging

THEORETICAL PERSPECTIVE	KEY POINTS
Functionalist	• Families are important in maintaining societal stability and meeting family members' needs. • Older people who are active and engaged are more satisfied with life.
Conflict	• Families promote social inequality because of social class differences. • Many corporations view older workers as disposable.
Feminist	• Families both mirror and perpetuate patriarchy and gender inequality. • Women have an unequal burden in caring for children as well as older family members and relatives.
Symbolic Interactionist	• Families construct their everyday lives through interaction and subjective interpretations of family roles. • Many older family members adapt to aging and often maintain previous activities.

Chapter 14 Topics

1 How Education in the United States Has Changed

Education is a social institution whereas *schooling*, a narrower term, refers to the formal training and instruction provided in a classroom setting. In the United States, both education and schooling have changed in four important ways: Education has expanded and mass schooling is universal, community colleges have flourished, public higher education has burgeoned, and student diversity has greatly increased.

Educational Attainment of the U.S. Population, 1947–2007

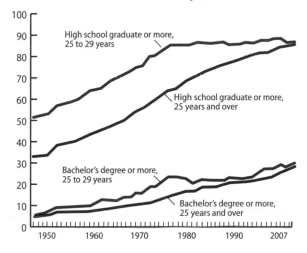

2 Sociological Perspectives on Education

THEORETICAL PERSPECTIVE	VIEW OF EDUCATION	SOME MAJOR QUESTIONS
Functionalist	Contributes to society's stability, solidarity, and cohesion	What are the manifest and latent functions of education?
Conflict	Reproduces and reinforces inequality and maintains a rigid social class structure	How does education limit equal opportunity?
Feminist	Produces inequality based on gender	How does gender inequality in education limit women's upward mobility?
Symbolic interactionist	Teaches roles and values through everyday face-to-face interaction and practices	How do tracking, labeling, self-fulfilling prophecies, and engagement affect students' educational experiences?

Key Terms

education a social institution that transmits attitudes, knowledge, beliefs, values, norms, and skills to its members through formal, systematic training.

schooling formal training and instruction provided in a classroom setting.

intelligence quotient (IQ) an index of an individual's performance on a standardized test relative to the performance level of others of the same age.

hidden curriculum school practices that transmit nonacademic knowledge, values, attitudes, norms, and beliefs which legitimize economic inequality and fill unequal work roles.

credentialism an emphasis on certificates or degrees to show that people have certain skills, educational attainment levels, or job qualifications.

literacy the ability to read and write in at least one language.

tracking (also called *streaming* or *ability grouping*) assigning students to specific educational programs and classes on the basis of test scores, previous grades, or perceived ability.

vouchers publicly funded payments that parents can apply toward tuition or fees at a public or private school of their choice.

charter schools self-governing public schools that have signed an agreement with their state government to improve students' education.

magnet school a public school that is typically small and offers students a distinctive program and specialized curriculum in a particular area, such as business, science, the arts, or technology.

home schooling teaching children in the home as an alternative to enrolling them in a public or private elementary, middle, or high school.

3 Some Problems with U.S. Education

Despite numerous strengths, the U.S. educational system suffers from serious problems. Compared with many other countries, large numbers of students are performing poorly in elementary and high schools—especially in mathematics and the sciences—because they experience a lower quality of instruction and spend considerably fewer hours and days in school. Most public schools, particularly in low-income communities, are struggling to survive financially because of inadequate funding. Compared with countries in Europe and elsewhere, American teachers' entry salaries are low, many teachers are out-of-field, and they have less control over curricula than ever before. And despite widespread grade inflation, high school and college dropout rates are high.

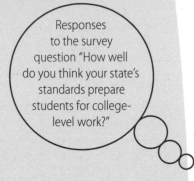

Responses to the survey question "How well do you think your state's standards prepare students for college-level work?"

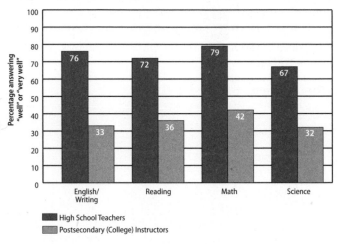

4 New Directions in U.S. Education

Because many traditional public schools are failing to educate students adequately, parents and legislators have turned to a variety of alternatives, some more controversial than others. Some states offer vouchers, but there is no significant overall difference in achievement between the children participating in voucher programs and those who remain in public schools. *Charter schools* promise to improve students' education, but students perform about the same as in traditional public schools if the teachers are certified and have had at least 5 years' experience and if the charter school is not new. A *magnet school* offers students a distinctive program and specialized curriculum—such as business, science, the arts, or technology—but few students can be accepted because enrollments are limited. *Home schooling* has grown, but there are no national data on whether this alternative is more successful in improving the quality of students' schooling than attending a traditional public or private school.

Example: *Paying Students to Excel*

There's an emerging trend at some public schools to reward student progress on standardized test scores with cash, certificates, gift certificates, McDonald's meals, and so on. And at Baylor University, a private Baptist university in Texas, the administration said that it would give a $1,000 merit scholarship each year to each first-year student (who had already been admitted) whose total SAT score rose by at least 50 points when they retook the exam, presumably to improve the university's place in national rankings, which then attracts more students. According to some high school counselors, other colleges have the same goal by informing admitted students that they could qualify for better scholarships if they raise their grade-point averages or standardized-test scores. Are such practices a good idea in motivating high school students to excel and in enhancing a college's reputation?

Sources: Medina, Jennifer. 2008. "Next Question: Can Students Be Paid to Excel." *New York Times*, March 5, 1; O'Brien, Rourke. 2008. "Paying City Students Is a Wise Investment." *Baltimore Sun*, June 27, p. 19A; Supiano, Beckie, and Eric Hoover. 2008. "Baylor U. Rewarded Freshmen Who Retook the SAT." *Chronicle of Higher Education*, October 24, A20.

Chapter 15 Topics

1 What Is Religion?

Religion, an important social institution, unites believers into a community that shares similar beliefs, values, and practices about the supernatural. Every known society distinguishes between *sacred* (spiritual) and *secular* (nonspiritual) activities. For sociologists, religion, *religiosity*, and spirituality differ because, for example, people who describe themselves as religious may not attend services and those who see themselves as spiritual may not belong to an organized group.

2 Types of Religious Organization

People express their religious beliefs most commonly through organized groups, including *cults* (also called *new religious movements*, NRMs), *sects*, *denominations*, and *churches*. Some NRMs, which usually organize around a *charismatic leader*, have been short-lived while others have become established religions with highly organized bureaucracies. Like NRMs, some sects have disappeared but others have persisted. Sometimes sects develop into denominations, which typically accommodate themselves to the larger society rather than try to dominate or change it. In both denominations and churches, the members are born into the group but may decide to withdraw later, the institutions are typically bureaucratically organized, and they often maintain some control over educational and political institutions.

3 Some Major World Religions

Worldwide the largest religious group is Christians, followed by Muslims, but there is no religious group that comes close to being a global majority. The third largest group consists of nonbelievers. The number of followers varies, but five religions—Christianity, Islam, Hinduism, Buddhism, and Judaism—have had an impact on economic, political, and social issues since their origin.

RELIGION	DATE OF ORIGIN	FOUNDER	PREVALENCE	NUMBER OF FOLLOWERS (IN MID-2007)	CORE BELIEFS
Christianity	0 C.E.	Jesus Christ	All continents, with largest numbers in Latin America and Europe	2.2 billion	Jesus, the son of God, sacrificed his life to redeem humankind. Those who follow Christ's teachings and live a moral life will enter the Kingdom of Heaven. Sinners who don't repent will burn in hell for eternity.
Islam	600 C.E.	Muhammad	Mainly Asia (including Indonesia), but also parts of Africa, China, India, and Malaysia	1.4 billion	God is creator of the universe, omnipotent, omniscient, just, forgiving, and merciful. Those who sincerely repent and submit (the literal meaning of *islam*) to God will attain salvation, while the wicked will burn in hell.
Hinduism	Between 4000 and 1500 B.C.E.	No specific founder	Mainly India, Nepal, Malaysia, and Sri Lanka, but also Africa, Europe, and North America	887 million	Life in all its forms is an aspect of the divine. The aim of every Hindu is to use pure acts, thoughts, and devotion to escape a cycle of birth and rebirth (*samsara*) determined by the purity or impurity of past deeds (*karma*).
Buddhism	525 B.C.E.	Siddhartha Gautama	Throughout Asia, from Sri Lanka to Japan	386 million	Life is misery and decay with no ultimate reality. Meditation and good deeds will end the *cycle* of endless birth and rebirth, and the person will achieve *nirvana*, a state of liberation and bliss.
Judaism	2000 B.C.E.	Abraham	Mainly Israel and the United States	15 million	God is the creator and the absolute ruler of the universe. God established a particular relationship with the Hebrew people. By obeying the divine law God gave them, Jews bear special witness to God's mercy and justice.

Key Terms

religion a social institution that involves shared beliefs, values, and practices based on the supernatural and unites believers into a community.

sacred anything that people see as mysterious, awe-inspiring, extraordinary and powerful, holy, and not part of the natural world.

profane anything that is not related to religion.

secular the term sociologists use (instead of *profane*) to characterize worldly rather than spiritual things.

religiosity the ways people demonstrate their religious beliefs.

cult a religious group that is devoted to beliefs and practices that are outside of those accepted in mainstream society.

new religious movement (NRM) term used instead of *cult* by most sociologists.

charismatic leader a religious leader whom followers see as having exceptional or superhuman powers and qualities.

sect a religious group that has broken away from an established religion.

denomination a subgroup within a religion that shares its name and traditions and is generally on good terms with the main group.

church a large established religious group that has strong ties to mainstream society.

secularization a process of removing institutions such as education and government from the dominance or influence of religion.

fundamentalism the belief in the literal meaning of a sacred text.

civil religion (sometimes called *secular religion*) practices in which citizenship takes on religious aspects.

Protestant ethic a belief that hard work, diligence, self-denial, frugality, and economic success will lead to salvation in the afterlife.

false consciousness an acceptance of a system that prevents people from protesting oppression.

ritual (sometimes called a *rite*) a formal and repeated behavior in which the members of a group regularly engage.

4 Religion in the United States

About 92 percent of Americans believe in God, but religion in the United States is complex and diverse because people may change their faith and those who describe themselves as religious identify with one of over 150 groups. The number of Americans who say that religion is "very important in their lives" has decreased since the 1950s. More than 40 percent of U.S. adults have changed their religion since childhood, many opting for no religion at all. A major change has been the decline of the so-called mainline Protestant groups and the surge of evangelicals. Overall, many Americans are more likely to believe in a religion than to practice it by attending formal services regularly. Religious participants vary by gender, age, race and ethnicity, and social class. For example, women and those aged 65 and older tend to be more religious than men or younger people, whites are more likely than racial-ethnic groups to have no religious affiliation, and people with lower levels of educational attainment are generally more religious than those with higher educational levels.

5 Secularization: Is Religion Declining?

Many European countries are undergoing *secularization*, but such trends are less clear in the United States. Some sociologists maintain that secularization is increasing rapidly in the United States, but others contend that this claim is greatly exaggerated, especially as witnessed by the growth of *fundamentalism* and the prevalence of *civil religion*.

6 Sociological Perspectives on Religion

THEORETICAL PERSPECTIVE	VIEW OF RELIGION	SOME MAJOR QUESTIONS
Functionalist	Religion benefits society by providing a sense of belonging, identity, meaning, emotional comfort, and social control over deviant behavior.	How does religion contribute to social cohesion?
Conflict	Religion promotes and legitimates social inequality, condones strife and violence between groups, and justifies oppression of poor people.	How does religion control and oppress people, especially those at lower socioeconomic levels?
Feminist	Religion subordinates women, excludes them from decision-making positions, and legitimizes patriarchal control of society.	How is religion patriarchal and sexist?
Symbolic Interactionist	Religion provides meaning and sustenance in everyday life through symbols, rituals, and beliefs and binds people together in a physical and spiritual community.	How does religion differ within and across societies?

Example: Religion Can Be Liberating or Constraining

S. Truett Cathy, the founder and chairman of Chick-fil-A (a franchise of stores that prepares chicken-breast sandwiches), believes that serving chicken is God's work. The corporate mission, as stated on a plaque at company headquarters, is "to glorify God." Chick-fil-A is the only national fast-food chain that closes on Sunday so employees can go to church and prospective employees are asked about their religious activities. Many franchise operators are delighted with the religious emphasis because, according to one operator, "I'm not working for Chick-fil-A; I'm working for the Lord." Others feel that a business should stay out of its workers' personal lives. For instance, a Muslim who was a Chick-fil-A manager in Texas settled a lawsuit after being fired a day after he refused to participate in a group prayer to Jesus Christ at a company training program.

Source: Schmall, Emily. 2007. "The Cult of Chick-fil-A." *Forbes*, July 23, 80, 83.

Chapter 16 Topics

1 Population Dynamics

Demography is the scientific study of human *population* that looks at the interplay between *fertility, mortality*, and *migration*. There are several ways to measure fertility, but one of the most common is the *crude birth rate*. The *crude death rate* and the *infant mortality rate* measure a population's life expectancy and health. Push and pull factors affect international migration and internal migration. Demographers also use *sex ratios* and *population pyramids* to understand a population's composition and structure.

Has population growth gotten out of hand? Demographers who believe that population growth is a ticking bomb subscribe to *Malthusian theory*, which argues that the world's food supply will not keep up with population growth. As a result, masses of people will live in poverty or die of starvation. *Demographic transition theory*, in contrast, maintains that population growth is kept in check and stabilizes as countries experience greater economic and technological development.

FIGURE 16.3
The Classical Demographic Transition Model

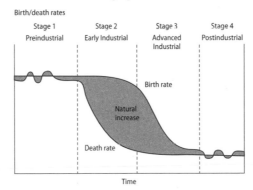

2 Urbanization

Globally and in the United States, *cities* and *urbanization* mushroomed during the twentieth century and are expected to increase. As more people move from rural to urban areas, many of the world's largest cities are becoming *megacities*. In the United States, urban growth has lead to *suburbanization, edge cities, exurbs, gentrification*, and *urban sprawl*.

Sociologists offer several perspectives on how and why cities change, and how these changes affect people:

THEORETICAL PERSPECTIVE	KEY POINTS
Functionalist	People create urban growth by moving to cities to find jobs and to suburbs to enhance their quality of life.
Conflict	Driven by greed and profit, large corporations, banks, developers, and other capitalistic groups determine the growth of cities and suburbs.
Feminist	Whether they live in cities or suburbs, women generally experience fewer choices and more constraints than do men.
Symbolic interactionist	City people are more tolerant of different lifestyles, but they tend to interact superficially and are generally socially isolated.

Key Terms

demography the scientific study of human populations.

population a collection of people who share a geographic territory.

fertility the number of babies born during a specified period in a particular society.

crude birth rate (also known as the *birth rate*) the number of live births per 1,000 people in a population in a given year.

mortality the number of deaths during a specified period in a population.

crude death rate (also called the *death rate*) the number of deaths per 1,000 people in a population in a given year.

infant mortality rate the number of deaths of infants (under 1 year of age) per 1,000 live births in a population.

migration the movement of people into or out of a specific geographic area.

sex ratio the proportion of men to women in a population.

population pyramid a visual representation of the makeup of a population in terms of the age and sex of its members at a given point in time.

Malthusian theory the idea that population is growing faster than the food supply needed to sustain it.

demographic transition theory the idea that population growth is kept in check and stabilizes as countries experience economic and technological development.

zero population growth (ZPG) a stable population level that occurs when each woman has no more than two children.

city a geographic area where a large number of people live relatively permanently and secure their livelihood primarily through nonagricultural activities.

urbanization population movement from rural to urban areas.

megacities metropolitan areas with at least 10 million inhabitants.

suburbanization population movement from cities to the areas surrounding them.

edge cities business centers that are within or close to suburban residential areas.

exurbs areas of new development beyond the suburbs that are more rural but on the fringe of urbanized areas.

urban sprawl the rapid, unplanned, and uncontrolled spread of development into regions adjacent to cities.

gentrification the process in which middle-class and affluent people buy and renovate houses and stores in downtown urban neighborhoods.

urban ecology the study of the relationships between people and urban environments, originated by sociologists at the University of Chicago.

new urban sociology urban changes are largely the result of decisions made by powerful capitalists and other groups in the dominant social class.

ecosystem a system in which all forms of life live in relation to one another and a shared physical environment.

greenhouse effect the heating of the earth's atmosphere due to the presence of certain atmospheric gases.

climate change a change of overall temperatures and weather conditions over time.

global warming the increase in the average temperature of the earth's atmosphere.

sustainable development economic activities that meet the needs of the present without threatening the environmental legacy of future generations.

FIGURE 16.4
Four Models of City Growth and Change

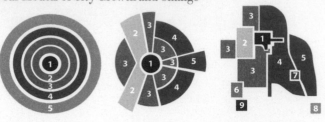

Concentric Zone Theory   Sector Theory   Multiple Nuclei Theory   Peripheral Theory

3 Environmental Issues

Population growth and urbanization are changing the planet's *ecosystem* and, many argue, endangering the close connections between plants, animals, and humans that affect survival. Water and air pollution and global warming—two interrelated environmental problems—are good examples of threats to the ecosystem in the United States and globally.

Water, which has an enormous impact on all life, is not as abundant as in the past. Industrialized nations not only have greater access to clean water than the developing world, they use more and pay less for it. In contrast, for millions of people in developing nations, clean water is an expensive and scarce luxury. Clean water has been depleted for many reasons, including pollution, privatization, and mismanagement.

Four of the most common sources and causes of air pollution are the burning of fossil fuels, manufacturing plants that spew pollutants into the air, winds that carry contaminants across borders and oceans, and lax governmental policies. Air pollution, which can lead to the *greenhouse effect*, is a major cause of *climate change* and *global warming*.

The rise of environmental problems has sparked a concern about *sustainable development*. Those who are pessimistic about achieving sustainable development show that, worldwide, the United States has one of the worst records on environmental performance, largely because of the close ties between government officials and corporations. Others are optimistic about achieving sustainable development and point to examples such as decreases of the emissions of major air pollutants and some large U.S. corporations' switching to practices that decrease pollution and energy consumption.

Example: *The Privatization of Water and Its Impact on the Environment*

In the United States alone, bottled water sales have surged—from 1 billion liters in the mid-1970s to almost 32 billion liters in 2006. In 2006, the bottled-water industry saw sales triple to almost $11 billion from a decade earlier. Up to 40 percent of bottled water is actually tap water, and a study of 103 brands of bottled water found that about one-third contained varying levels of contamination, including chemicals, bacteria, parasites, and fecal matter. In fact, one bottler with a picture of a lake surrounded by mountains on the label was actually using water from an industrial parking lot next to a hazardous waste site. For the United States alone,

- producing the bottles requires the equivalent of more than 17 million barrels of oil, not including the energy for transportation;
- bottling water produces almost 3 million tons of carbon dioxide (a toxin);
- it takes 3 liters of water to produce 1 liter of bottled water; and
- fewer than a quarter of the plastic bottles are recycled every year, leaving 2 billion pounds that clog landfills.

Sources: Azios, Tony. 2008. "Bottled vs. Tap." *Christian Science Monitor*, January 17, 15–16. "Bottled Water: Pure Drink or Pure Hype?" 1999. National Resources Defense Council. Retrieved June 19, 2005 (www.nrdc.org); Food & Water Watch. 2007. "Take Back the Tap: Why Choosing Tap Water over Bottled Water Is Better for Your Health, Your Pocketbook, and the Environment." Retrieved July 21, 2008 (www.foodandwaterwatch.org); Pacific Institute. 2007. "Bottled Water and Energy: A Fact Sheet." Retrieved February 7, 2008 (www.pacinst.org); Walsh, Bryan. 2007. "Back to the Tap." *Time*, August 20, 55–56.

Chapter 17 Topics

1 Collective Behavior

Collective behavior, a major source of social change, is the spontaneous and unstructured behavior of a large number of people. According to an influential sociological theory, six macro-level conditions can encourage or discourage collective behavior—structural conduciveness, structural change, the growth and spread of a generalized belief, precipitating factors, mobilization, and social control. There are many types of collective behavior, some more short-lived or harmful than others. *Rumors, gossip,* and *urban legends* are typically untrue, but many people believe and pass them on for a number of reasons, such as anxiety and fear about living in an unpredictable world, to reinforce a community's moral standards, or simply because they're fun to tell. In contrast, *panic* and *mass hysteria* can have dire consequences, including death.

Fashions, fads, and *crazes* are harmless because they usually last only a short time and change over time. However, these forms of collective behavior can deplete consumers' wallets while corporations profit. People choose to participate in fashions, fads, and crazes, but a *disaster* is an unexpected event that causes widespread damage, destruction, distress, and loss. Disasters are due to social, technological, and natural causes, some of which are more costly in human lives than others.

Publics, public opinion, and *propaganda* also affect large numbers of people, and some of these types of collective behavior are more harmful than others. In contrast to publics and public opinion, for example, propaganda can be especially dangerous because it purposely manipulates people with misinformation that can even result in riots. *Crowds* vary in their motives, interests, and emotional level. A casual crowd, for example, has little, if any, interaction, the gathering is temporary, and there is little emotion. On the other hand, protest crowds, especially *mobs* and those involved in a *riot*, can wreak considerable havoc on property and result in death.

Example: *Crowds Can Be Deadly*

On Thanksgiving, 2008, crowds started gathering at 9:00 P.M. outside of the Wal-Mart store in Valley Stream, New York, for a bargain-hunting ritual known as Black Friday, the day after Thanksgiving. By 4:55 A.M. the next morning, the crowd had grown to more than 2,000 people and could no longer be held back: "Fists banged and shoulders pressed on the sliding-glass double doors, which bowed in with the weight of the assault. Six to 10 workers inside tried to push back, but it was hopeless." Suddenly, according to witnesses, the door shattered and "the shrieking mob surged through in a blind rush for holiday bargains." A 34-year-old male temporary worker, who had been hired for the holiday season, was trampled to death, and four other people, including a 28-year-old woman who was eight months pregnant, were treated for injuries (McFadden and Macropoulos, 2008). Review the types of crowds on pp. 319–321 of textbook. Which type of crowd do you think is most representative of this Wal-Mart incident?

Source: McFadden, Robert D. and Angela Macropoulos. 2008. "Wal-Mart Employee Trampled to Death." *New York Times,* November 29. Retrieved November 29, 2008 (www.times.com).

2 Social Movements

Unlike collective behavior, *social movements* are typically organized, have long-lasting effects in promoting or resisting a particular social change, and may be perceived as threatening because they challenge the existing status quo. Some of the most common social movements are alternative, redemptive, reformative, resistance, and revolutionary. Sociologists have offered

Key Terms

collective behavior the spontaneous and unstructured behavior of a large number of people.

rumor unfounded information that is spread quickly.

gossip rumors, often negative, about other people's personal lives.

urban legends (also called *contemporary legends* and *modern legends*) a type of rumor consisting of stories that supposedly happened.

panic a collective flight, typically irrational, from a real or perceived danger.

mass hysteria an intense, fearful, and anxious reaction to a real or imagined threat by large numbers of people.

fashion a standard of appearance that enjoys widespread but temporary acceptance within a society.

fad a form of collective behavior that spreads rapidly and enthusiastically but lasts only a short time.

craze a fad that becomes an all-consuming passion for many people for a short period of time.

disaster an unexpected event that causes widespread damage, destruction, distress, and loss.

public a collection of people, not necessarily in direct contact with each other, who are interested in a particular issue.

public opinion widespread attitudes on a particular issue.

propaganda the presentation of information in a manner deliberately designed to influence people's opinions or actions.

crowd a temporary gathering of people who share a common interest or participate in a particular event.

mob a highly emotional and disorderly crowd that uses force or violence against a specific target.

riot a violent crowd that directs its hostility at a wide and shifting range of targets.

social movement a large and organized activity to promote or resist some particular social change.

relative deprivation a gap between what people have and what they think they should have based on what others in a society have.

technology the application of scientific knowledge for practical purposes.

several explanations for the emergence of social movements that include *mass society theory, relative deprivation theory, resource mobilization theory*, and *new social movements theory*. Each perspective has strengths and weaknesses in helping us understand social movements.

TABLE 17.1
Five Types of Social Movements

MOVEMENT	GOAL	EXAMPLES
Alternative	Change some people in a specific way	Alcoholics Anonymous, transcendental meditation
Redemptive	Change some people, but completely	Jehovah's Witnesses, born-again Christians
Reformative	Change everyone, but in specific ways	gay rights advocates, Mothers Against Drunk Driving (MADD)
Resistance	Change everyone but in specific ways	anti-abortion groups, white supremacists
Revolutionary	Change everyone completely	right-wing militia groups, Communism

Most social movements are short-lived because some never really get off the ground and others meet their goals and disband. Social movements generally go through four stages—emergence, organization, institutionalization, and decline. Decline is most likely when a social movement is successful and becomes a part of society's fabric; when the members, especially the leaders, are co-opted; when the members become distracted because the group loses sight of it its original goals and/or their enthusiasm diminishes; the membership fragments because the participants disagree about goals, strategies, or tactics; and when a government quashes dissent through repression. Social movements are important because they can either create or resist change on the individual, institutional, and societal level.

3 Technology and Social Change

Technology also generates changes. Some of the most important technological advances have included computer technology, biotechnology, and nanotechnology—all of which have changed our lives dramatically. Nanotechnology, especially, may increase longevity by finding early signs of disease and destroying unhealthy cells. Technology has both benefits and costs, however. For example, the Internet and other forms of telecommunication technology can bring people together but can also intrude on our privacy. In addition, technological advances raise numerous ethical questions, such as its greater availability to the wealthy, educated, and computer-literate.

Example: *Is Revealing Clothing a Fad or a Social Movement?*

© Hiroko Masuike/
The New York Times/Redux

In 2007, several towns in Louisiana outlawed sagging pants that revealed underwear. The penalties included fines up to $500 or up to a six-month jail sentence. Similar laws were considered or proposed but defeated in Connecticut, Georgia, Maryland, New Jersey, New York, Oklahoma, and Virginia because, according to opponents, politicians shouldn't legislate fashion preferences and towns "have more pressing issues." Still, in 2008, the police chief of Flint, Michigan, ordered his officers "to start arresting saggers on sight, threatening them with jail time and hefty fines for a fad he calls 'immoral self-expression'" (Bennett and Chapman 2008: 9). Do low-hanging pants that expose underwear indicate delinquency and "moral decay" in a society? A temporary fad? Or a social movement that's here to stay? Also, are bans against baggy pants, but not other revealing clothing (such as thongs and women's apparel showing cleavage), racially motivated because many of the wearers are young black men?

Sources: Bennett, Jessica, and Mary Chapman. 2008. "An Equal-Opportunity Crackdown?" *Newsweek*, July 28, 9; Koppel, Niko. 2007. "Are Your Jeans Sagging? Go Directly to Jail." *New York Times*, August 30, 1; White, Tanika. 2007. "Pants Safe from Laws." *Baltimore Sun*, November 28, A1, A20.

iron law of oligarchy the tendency of a bureaucracy to become increasingly dominated by a small group of people.

glass ceiling a collection of attitudinal or organizational biases in the workplace that prevent women from advancing to leadership positions.

social institution an organized and established social system that meets one or more of a society's basic needs.

3 Sociological Perspectives on Groups and Organizations

TABLE 6.3
Sociological Perspectives on Groups and Organizations

THEORETICAL PERSPECTIVE	LEVEL OF ANALYSIS	MAIN POINTS	KEY QUESTIONS
Functionalist	Macro	Organizations are made up of inter-related parts and rules and regulations that produce cooperation in meeting a common goal.	• Why are some organizations more effective than others? • How do dysfunctions prevent organizations from being rational and effective?
Conflict	Macro	Organizations promote inequality that benefits elites, not workers.	• Who controls an organization's resources and decision making? • How do those with power protect their interests and privileges?
Feminist	Macro and micro	Organizations tend not to recognize or reward talented women and regularly exclude them from decision-making processes.	• Why do many women hit a glass ceiling? • How do gender stereotypes affect women in groups and organizations?
Symbolic Interactionist	Micro	People aren't puppets but can determine what goes on in a group or organization.	• Why do people ignore or change an organization's rules? • How do members of social groups influence workplace behavior?

4 Social Institutions

A *social institution* is an organized and established social system that meets a society's basic needs to survive. Functionalists identify five core social institutions: family, economy, political institutions, education, and religion. The family replaces citizens through procreation, socializes its members, and legitimizes sexual activity between adults; economic institutions organize a society's development, production, distribution, and consumption of goods and services; and political systems maintain law and order, pass legislation, and form military groups to safeguard its members from internal and external violence.

Chapter 6 Topics

1 Social Groups

A *social group* consists of two or more people who interact with one another and who share a common identity and a sense of belonging (such as friends or work groups). A *primary group* is a relatively small group of people (such as a family) who engage in intimate face-to-face interaction over an extended period of time. A *secondary group*, in contrast (such as the students in your sociology class), is a large, usually formal, impersonal, and temporary collection of people who pursue a specific goal or activity. Members of an *in-group* share a sense of identity and "we-ness" that typically excludes and devalues outsiders. *Out-groups* consist of people who are viewed and treated negatively because they are seen as having values, beliefs, and other characteristics different from those of the in-group. We also have *reference groups*, collections of people who shape our behavior, values, and attitudes as well as influence who we are, what we do, and who we'd like to be in the future. Groups often form a *social network*, a web of social ties that links an individual to others (such as members of a local hiking group).

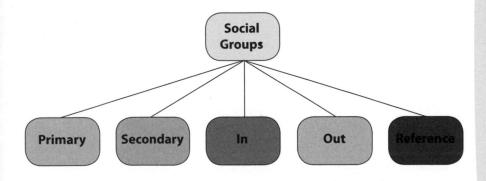

Example: *Secondary Groups Can Replace Primary Groups*

In 1864, an alcoholic who had ruined a promising career on Wall Street because of his constant drunkenness, co-founded Alcoholics Anonymous (AA), a program that would enable people to stop drinking by undergoing a spiritual awakening and seeking help from a buddy to stay sober. Initially, AA was a secondary group that tried to beat alcoholism by encouraging its members to attend regular meetings where alcoholics talked about their accomplishments in staying sober. Over the years, however, AA has become a primary group for many members because it offers a relatively small group of people who engage in face-to-face interaction over an extended period of time, especially when their family and friends have rejected them.

2 Formal Organizations

A *formal organization* is a complex and structured secondary group that is deliberately created to achieve specific goals in an efficient manner. Two of the most widespread and important types of formal organizations in the United States are voluntary associations and bureaucracies. A *voluntary association* is a formal organization created by people who share a common set of interests and who are not paid for their participation, such as members of charitable groups. In contrast, a *bureaucracy* is a large formal organization, such as your college, that is designed to accomplish goals and tasks in the most efficient and rational way possible.

Key Terms

social group two or more people who interact with one another and who share a common identity and a sense of belonging or "we-ness."

primary group a relatively small group of people who engage in intimate face-to-face interaction over an extended period of time.

secondary group a large, usually formal, impersonal, and temporary collection of people who pursue a specific goal or activity.

ideal types general traits that describe a social phenomenon rather than every case.

in-groups sets of people who share a sense of identity and "we-ness" that typically excludes and devalues outsiders.

out-groups people who are viewed and treated negatively because they are seen as having values, beliefs, and other characteristics different from those of an in-group.

reference group a collection of people who shape our behavior, values, and attitudes.

groupthink a tendency of in-group members to conform without critically testing, analyzing, and evaluating ideas, that results in a narrow view of an issue.

social network a web of social ties that links an individual to others.

formal organization a complex and structured secondary group that has been deliberately created to achieve specific goals in an efficient manner.

voluntary association a formal organization created by people who share a common set of interests and who are not paid for their participation.

bureaucracy a formal organization that is designed to accomplish goals and tasks through the efforts of a large number of people in the most efficient and rational way possible.

alienation a feeling of isolation, meaninglessness, and powerlessness that may affect workers in a bureaucracy.

social exchange theory the perspective whose fundamental premise is that any social interaction between two people is based on each person's trying to maximize rewards (or benefits) and minimize punishments (or costs).

nonverbal communication messages that are sent without using words.

stage, partners inform friends, family, the children's teachers, and others, that they are no longer married. Finally, the couple goes through a "psychic divorce," in which the partners separate from each other emotionally. In many cases, one or both spouses never complete this stage because they can't let go of their pain, anger, and resentment—even if they remarry.

4 Explaining Social Interaction

TABLE 5.2
Sociological Explanations of Social Interaction

PERSPECTIVE	KEY POINTS
Symbolic Interactionist	• People create and define their reality through social interaction. • Our definitions of reality, which vary according to context, can lead to self-fulfilling prophecies.
Social Exchange	• Social interaction is based on a balancing of benefits and costs. • Relationships involve trading a variety of resources, such as money, youth, and good looks.
Feminist	• The sexes act similarly in many interactions but often differ in communication styles and speech patterns. • Men are more likely to use speech that's assertive (to achieve dominance and goals), while women are more likely to use language that connects with others.

5 Nonverbal Communication

Our *nonverbal communication,* messages that are sent without using words, includes gestures, facial expressions, eye contact, and silence. Touching and how we use space are also important forms of nonverbal communication because they send powerful messages about our feelings and power.

6 Online Communication

Many people interact in *cyberspace,* an online world of computer networks. Internet usage varies by gender, age, ethnicity, and social class. Cyberspace can be impersonal and socially isolating, but it can also save time, foster closer relationships among family members and friends, and facilitate working from home.

Zones of Personal Space

public — 12+ feet

social — 4 feet

private — >1½–2 feet

Chapter 5 Topics

1 Social Structure

Social interaction is the process by which we act toward and react to people around us. Our interaction is part of the *social structure,* an organized pattern of behavior that governs people's relationships. Every society has a social structure that encompasses statuses, roles, groups, organizations, and institutions.

2 Status

A *status* is a social position that an individual occupies in a society. Every person has many statuses that form her or his *status set,* a collection of social positions that a person occupies at a given time. Status sets include both ascribed and achieved statuses. An *ascribed status* is a social position that a person is born into and can't control, change, or choose (such as, age, race, family relationships, and being male or female). An *achieved status* is a social position that a person attains through personal effort or assumes voluntarily (such as college student or wife). Because we hold many statuses, some clash. *Status inconsistency* refers to the conflict that arises from occupying social positions that are ranked differently (such as being a low-paid college professor).

3 Role

A *role* is the behavior expected of a person who has a particular status. Roles define how we are expected to behave in a particular status, but people vary considerably in their fulfillment of the responsibilities associated with their roles. These differences reflect *role performance,* the actual behavior of a person who occupies a status. A *role set* refers to the different roles attached to a single status (such as a parent who also plays the roles of teacher, chauffeur, and PTA member). Playing many roles often leads to *role conflict,* the frustrations and uncertainties a person experiences when confronted with the requirements of two or more statuses, and *role strain,* the stress that arises due to incompatible demands among roles within a single status.

Ways to Resolve Role Conflict

- compromise
- negotiate
- set priorities
- compartmentalize
- not take on more roles
- exit the roles

Example: *Exiting a Marriage*

Divorce is a good example of role exit, but often involves a long process of five stages that may last several decades (Bohannon 1971). The "emotional divorce" begins when one or both partners feel disillusioned, unhappy, or rejected. The "legal divorce" is the formal dissolution of the marriage during which the partner who does not want the divorce may try to stall the end of the marriage. During the "economic divorce" stage, the partners may argue about who should pay past debts, property taxes, and unforeseen expenses (such as moving costs). The "coparental divorce" stage involves the agreements between the parents regarding the legal responsibility for financial support of the children and the rights of both parents to spend time with the children. During the "community divorce"

Key Terms

social interaction the process by which we act toward and react to people around us.

social structure an organized pattern of behavior that governs people's relationships.

status a social position that a person occupies in a society.

status set a collection of social statuses that an individual occupies at a given time.

ascribed status a social position that a person is born into.

achieved status a social position that a person attains through personal effort or assumes voluntarily.

master status an ascribed or achieved status that determines a person's identity.

status inconsistency the conflict or tension that arises from occupying social positions that are ranked differently.

role the behavior expected of a person who has a particular status.

role performance the actual behavior of a person who occupies a status.

role set the different roles attached to a single status.

role conflict the frustrations and uncertainties a person experiences when confronted with the requirements of two or more statuses.

role strain the stress arising from incompatible demands among roles within a single status.

self-fulfilling prophecy a situation where if we define something as real and act upon it, it can, in fact, become real.

ethnomethodology the study of how people construct and learn to share definitions of reality that make everyday interactions possible.

dramaturgical analysis a technique that examines social interaction as if occurring on a stage where people play different roles and act out scenes for the audiences with whom they interact.

Chapter 3 Topics

1 Culture and Society

Culture refers to the learned and shared behaviors, beliefs, attitudes, values and material objects that characterize a particular group or society. Culture is learned, transmitted from one generation to another, adaptive, and always changing. *Material culture* consists of the tangible objects that members of a society make, use and share. *Nonmaterial culture* includes the shared set of meanings that people use to interpret and understand the world such as symbols and values. A *society* is an organized population that shares a culture and sees itself as a social unit.

2 The Building Blocks of Culture

The following are some of the fundamental building blocks of culture.

- *Symbols* are anything that hold particular meaning for people who share a culture. Symbols take many forms, can change over time, can unify or divide a society, and can affect cross-cultural views.
- *Language* is a system of shared symbols that enables people to interact with others, can change over time, and can affect perceptions of gender, race, class and ethnicity.
- *Values* are cultural standards that provide general guidelines for behavior. Values are usually emotion laden, vary across cultures, and change over time.
- *Norms*—whether they are folkways, mores, or laws—are a society's specific rules that regulate our behavior. Like the other building blocks, norms vary across cultures, can change over time, and are subject to sanctions ranging from mild to severe.

Example: *Sanctions for Violating the Dead*

Sanctions are more severe for violating laws than folkways. Legacy.com, which carries a death notice or obituary for virtually all of the roughly 2.4 million Americans who die each year, dedicates at least 30 percent of its budget to weeding out comments (a relatively mild punishment) that "diss the dead" (Urbina, 2006). In contrast, when there's no prior criminal record, the penalties in many states for vandalizing a tombstone, a property crime, can result in a fine up to $1,000, up to a year in jail, or both.

Source: Urbina, Ian. 2006. "In Online Mourning, Don't Speak Ill of the Dead." *New York Times*, November 5, 1.

3 Cultural Similarities

While many cultural characteristics vary across countries, *cultural universals* are customs and practices that are common to all societies, such as some form of food taboo. When an individual is exposed to an unfamiliar way of life or environment, she or he may experience *culture shock*, a state of confusion, anxiety, and uncertainty on how to behave.

4 Cultural Diversity

Subcultures and countercultures account for some of the complexity within a society. A *subculture* is a group or category of people whose distinctive ways of thinking, feeling, and acting differ somewhat from those of the larger society. In contrast, a *counterculture* deliberately opposes and rejects some of the dominant culture's basic beliefs, values, and norms. *Ethnocentrism* is the belief that one's culture and way of life are superior to those of other groups. *Cultural relativism,* the opposite of ethnocentrism, is a belief that no culture is better than another and that all cultures should be judged by their own standards.

Key Terms

culture the learned and shared behaviors, beliefs, attitudes, values, and material objects that characterize a particular group or society.

society a group of people that has lived and worked together long enough to become an organized population and to think of themselves as a social unit.

material culture the tangible objects that members of a society make, use, and share.

nonmaterial culture the shared set of meanings that people in a society use to interpret and understand the world.

symbol anything that stands for something else and has a particular meaning for people who share a culture.

language a system of shared symbols that enables people to communicate with one another.

values the standards by which members of a particular culture define what is good or bad, moral or immoral, proper or improper, desirable or undesirable, beautiful or ugly.

norms a society's specific rules concerning right and wrong behavior.

folkways norms that members of a society (or a group within a society) look upon as not being critical and that may be broken without severe punishment.

mores norms that members of a society consider very important because they maintain moral and ethical behavior.

laws formal rules about behavior that are defined by a political authority that has the power to punish violators.

sanctions rewards for good or appropriate behavior and/or penalties for bad or inappropriate behavior.

ideal culture the beliefs, values, and norms that people in a society say they hold or follow.

real culture the actual everyday behavior of people in a society.

cultural universals customs and practices that are common to all societies.

Multiculturalism occurs when many cultures coexist in the same geographic area without any one culture dominating another.

5 Popular Culture

Popular culture refers to the beliefs, practices, activities, and products that are widely shared among a population in everyday life. Popular culture includes television, music, radio, advertising, sports, hobbies, fads, fashions, and movies as well as the food we eat, the people with whom we spend time, the gossip we share, and the jokes we pass along. Popular culture is typically disseminated through *mass media,* including television and the Internet, and has enormous power in shaping our perceptions and opinions.

6 Cultural Change and Technology

Some societies are relatively stable because of *cultural integration,* but all societies change over time because of diffusion, innovation and invention, discovery, external pressures, and changes in the physical environment. A *cultural lag* occurs when a culture's material side changes more rapidly than its nonmaterial side.

7 Sociological Perspectives on Culture

TABLE 3.3
Sociological Explanations of Culture

THEORETICAL PERSPECTIVE	FUNCTIONALIST	CONFLICT	FEMINIST	SYMBOLIC INTERACTIONIST
Level Of Analysis	Macro	Macro	Macro and Micro	Micro
Key Points	• Similar beliefs bind people together and create stability. • Sharing core values unifies a society and promotes cultural solidarity.	• Culture benefits some groups at the expense of others. • As powerful economic monopolies increase worldwide, the rich get richer and the rest of us get poorer.	• Women and men often experience culture differently. • Cultural values and norms can increase inequality because of gender, race/ethnicity, and social class.	• Cultural symbols forge identities (that change over time). • Culture (such as norms and values) helps people merge into a society despite their differences.

culture shock a sense of confusion, uncertainty, disorientation, or anxiety that accompanies exposure to an unfamiliar way of life or environment.

subculture a group or category of people whose distinctive ways of thinking, feeling, and acting differ somewhat from those of the larger society.

counterculture a group or category of people who deliberately oppose and consciously reject some of the basic beliefs, values, and norms of the dominant culture.

ethnocentrism the belief that one's culture and way of life are superior to those of other groups.

cultural relativism the recognition that no culture is better than another and that a culture should be judged by its own standards.

multiculturalism *(cultural pluralism)* the coexistence of several cultures in the same geographic area, without any one culture dominating another.

popular culture the beliefs, practices, activities, and products that are widely shared among a population in everyday life.

mass media forms of communication designed to reach large numbers of people.

cultural imperialism the influence or domination of the cultural values and products of one society over those of another.

cultural integration the consistency of various aspects of society, which promotes order and stability.

cultural lag the gap when nonmaterial culture changes more slowly than material culture.

Chapter 2 Topics

1 What Is Social Research?

Social research is the ongoing study of human behavior. Research requires curiosity and imagination, but also an understanding of the rules and procedures that govern careful scientific study. The process entails choosing a socially-relevant topic, asking a research question, developing a hypothesis, testing that hypothesis, and analyzing the findings. Unbiased research is fundamental to research, while personal opinions, as seen in self-help literature, often ignore the scientific method. Because knowledge is cumulative, researchers examine past studies and modify research designs to better understand a social phenomenon.

2 The Scientific Method

The *scientific method* is an established research process which incorporates careful data collection, exact measurement, accurate recording and analysis of findings, thoughtful interpretation of results, and, when appropriate, a generalization of the findings to a larger group. Sociologists, like other researchers, use the scientific method to measure the relationships between *variables*—characteristics that can change in value or magnitude under different conditions. A research question or a *hypothesis* examines the association between an *independent variable* (a characteristic that determines or has an effect) and the *dependent variable* (the result or outcome). Sociologists use both qualitative and quantitative approaches to determine if there is a relationship between variables and are always concerned about the *reliability* and *validity* of their measures.

FIGURE 2.2

Steps in the Scientific Method Using the Inductive Approach

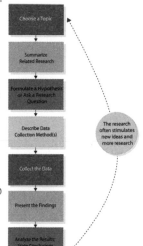

3 Some Major Data Collection Methods

There are many data collection methods, but six are especially common in social research, including sociology.

- Sociologists use *surveys* to systematically collect data from respondents using questionnaires, face-to-face or telephone interviews, or a combination of these.
- *Secondary analysis* examines data that have been collected by someone else, such as historical material and official statistics.
- *Field research* is a method of systematically observing subjects in their natural surroundings, using participant or nonparticipant observation.
- *Content analysis* examines written and oral communication. Data are categorized, coded, sorted, and analyzed in terms of frequency, intensity, and other characteristics.
- *Experiments* allow researchers to study cause and effect between two or more variables in a controlled setting.
- *Evaluation research* relies on all of the standard methodological techniques to assess the effectiveness of social programs in both the public and private sectors.

Sociologists weigh the advantages and limitations of each data collection method in designing their studies.

Key Terms

social research research that examines human behavior.

scientific method the steps in the research process that include careful data collection, exact measurement, accurate recording and analysis of the findings, thoughtful interpretation of results, and, when appropriate, a generalization of the findings to a larger group.

variable a characteristic that can change in value or magnitude under different conditions.

hypothesis a statement of a relationship between two or more variables that researchers want to test.

independent variable a characteristic that determines or has an effect on the dependent variable.

dependent variable the outcome, which may be affected by the independent variable.

reliability the consistency with which the same measure produces similar results time after time.

validity the degree to which a measure is accurate and really measures what it claims to measure.

deductive reasoning reasoning that begins with a theory, prediction, or general principle that is then tested through data collection.

inductive reasoning reasoning that begins with a specific observation, followed by data collection and the development of a general conclusion or theory.

population any well-defined group of people (or things) about whom researchers want to know something.

sample a group of people (or things) that are representative of the population that researchers wish to study.

probability sample a sample for which each person (or thing, such as an e-mail address) has an equal chance of being selected because the selection is random.

nonprobability sample a sample for which little or no attempt is made to get a representative cross section of the population.

qualitative research research that examines nonnumerical material and interprets it.

quantitative research research that focuses on a numerical analysis of people's responses or specific characteristics.

surveys a systematic method for collecting data from respondents, including questionnaires, face-to-face or telephone interviews, or a combination of these.

secondary analysis examination of data that have been collected by someone else.

field research data collection by systematically observing people in their natural surroundings.

content analysis data collection method that systematically examines examples of some form of communication.

experiment a carefully controlled artificial situation that allows researchers to manipulate variables and measure the effects.

experimental group the group of subjects in an experiment who are exposed to the independent variable.

control group the group of subjects in an experiment who are not exposed to the independent variable.

4 Ethics, Politics, and Sociological Research

To avoid exploitation and maltreatment of participants, sociological research demands a strict code of ethics throughout every research step. For example, participants must give informed consent and must not be harmed, humiliated, abused, or coerced; researchers must honor their guarantees of privacy, confidentiality, and/or anonymity. Still, sociologists, like other researchers, often encounter considerable pressure from policy makers and others to limit their research to topics that won't generate controversy on sensitive issues.

Example: *Sociological Research in Action*

The Human Terrain Team is an experimental Pentagon program that assigns social scientists to U.S. combat units in Afghanistan and Iraq. The purpose of the program is for social scientists to gather information about local populations to improve the military's understanding of cultural differences and then to negotiate tribal disputes that sometimes encourage the growth of agitators.

Source: Rohde, David. 2007. "Army Enlists Anthropology in War Zones." *New York Times,* October 5, 1.

evaluation research research that relies on all of the standard data collection techniques to assess the effectiveness of social programs in both the public and the private sectors.

TABLE 2.2
Some Data Collection Methods in Sociological Research

METHOD	EXAMPLE	ADVANTAGES	DISADVANTAGES
Surveys	Sending questionnaires and/or interviewing students on why they succeeded in college or dropped out	Questionnaires are fairly inexpensive and simple to administer; interviews have high response rates; findings are often generalizable	Mailed questionnaires may have low response rates; respondents may be self-selected; interviews are usually expensive
Secondary analysis	Using data from the National Center for Education Statistics (or similar organizations) to examine why students drop out of college	Usually accessible, convenient and inexpensive; often longitudinal and historical	Information may be incomplete; some documents may be inaccessible; some data can't be collected over time
Field research	Observing first-year college students with high and low grade-point averages (GPAs) regarding their classroom participation and other activities	Flexible; offers deeper understanding of social behavior; usually inexpensive	Difficult to quantify and to maintain observer/subject boundaries; the observer may be biased or judgmental; findings are not generalizable
Content analysis	Comparing the transcripts of college graduates and dropouts on variables such as gender, race/ethnicity, and social class	Usually inexpensive; can recode errors easily; unobtrusive; permits comparisons over time	Can be labor-intensive; coding is often subjective (and may be distorted); may reflect social class biases
Experiments	Providing tutors to some students with low GPAs to find out if such resources increase college graduation rates	Usually inexpensive; plentiful supply of subjects; can be replicated	Volunteers and paid subjects aren't representative of a larger population; the laboratory setting is artificial
Evaluation research	Examining student records; interviewing administrators, faculty, and students; observing students in a variety of settings (such as classroom and extracurricular activities); and using surveys to determine students' employment and family responsibilities.	Usually inexpensive; valuable in real-life applications	Often political; findings might be rejected

conflict theory an approach that examines the ways in which groups disagree, struggle over power, and compete for scarce resources (such as property, wealth, and prestige).

feminist theories approaches that try to explain the social, economic, and political position of women in society with a view to freeing women from traditionally oppressive expectations, constraints, roles, and behavior.

symbolic interactionism (*interactionism*) a micro-level perspective that looks at individuals' everyday behavior through the communication of knowledge, ideas, beliefs, and attitudes.

interaction action in which people take each other into account in their own behavior.

Example: Critical Thinking & "Common Sense"

When thinking critically about a topic, it's important to differentiate between "common sense" myths and fact. Following are a few examples:

Myth: The elderly make up the largest group of those who are poor.
Fact: Children under six years of age, and not the elderly, make up the largest group of those who are poor (see Chapters 8 and 11).

Myth: Divorce rates are higher today than ever before.
Fact: Divorce rates are lower today than they were between 1975 and 1990 (see Chapter 13).

Myth: Living together decreases the chance of divorce after marriage.
Fact: Living with someone increases, rather than decreases, the chance of divorce after marriage. Some of the reasons include ongoing personal and communication problems and viewing cohabitation as a more desirable alternative than working on a marriage (see Chapter 13).

Leading Contemporary Perspectives in Sociology*

THEORETICAL PERSPECTIVE	FUNCTIONALIST	CONFLICT	FEMINIST	SYMBOLIC INTERACTIONIST
Level Of Analysis	Macro	Macro	Macro and Micro	Micro
Key Points	• Society is composed of interrelated, mutually dependent parts. • Structures and functions maintain a society's or group's stability, cohesion, and continuity. • Dysfunctional activities that threaten a society's or group's survival are controlled or eliminated.	• Life is a continuous struggle between the "haves" and the "have-nots." • People compete for limited resources that are controlled by a small number of powerful groups. • Society is based on inequality in terms of ethnicity, race, social class, and gender.	• Women experience widespread inequality in society because, as a group, they have little power. • Gender, ethnicity, race, age, sexual orientation, and social class—rather than a person's intelligence and ability—explain many of our social interactions and lack of access to resources. • Social change is possible only if we change our institutional structures and our day-to-day interactions.	• People act on the basis of the meanings they attribute to others. • Meaning grows out of the social interaction that we have with others. • People continuously reinterpret and reevaluate their knowledge and information in their everyday encounters.

*For full table, see Table 1.1 on page 21.

Chapter 1 Topics

1 What Is Sociology?

Sociology is the systematic study of social interaction—processes by which we act toward and react to people around us—at a variety of levels. Sociologists use scientific research to discover patterns and create theories about who we are, how we interact with others, and why we do what we do. Sociology goes beyond common sense in the pursuit of knowledge about the entire scope of our social world, including small groups (like families and friends), large organizations and institutions (like your school), and entire societies (like the United States and other countries).

2 What Is the Sociological Imagination?

While psychology examines how the mind works, anthropology examines the structures of developing countries, and social work endeavors to better the lives of others, sociology is unique in its examination of contemporary, evolving social dynamics. Sociology helps us understand diversity within the community, make socially-conscious decisions, evaluate public policy, and understand how we fit into the big picture. Broadly speaking, sociology can be used to describe the intersection between individual lives and external social influences. Normal, micro-level interactions and conversations with other people subtly shape our everyday lives and habits, while large, macro-level systems and institutions shape the greater society, sometimes limiting our personal options on the micro level.

Some Origins of Sociological Thinking

Social theory sits at the heart of sociology. Sociologists use theories, also called theoretical perspectives, to explain the diverse, and sometimes baffling and bizarre, patterns that people, institutions, and societies follow. Why is society structured like it is? What holds society together? What pulls it apart? These questions and many others can be answered through theories. Theories not only produce knowledge—they can also offer solutions to real social problems. Because society is constantly evolving, so too are social theories. New theories build upon old ones and are tested through ongoing research and observation. Some of the most influential theorists in the discipline of sociology have included Auguste Comte, Harriet Martineau, Émile Durkheim, Karl Marx, Max Weber, Jane Addams, Georg Simmel, and W.E.B. Du Bois. Each brought to sociology a new level of understanding and perspective about our world.

4 Contemporary Sociological Theories

Today, there are four major frameworks of sociological perspective under which new theories are proposed. These frameworks, often used in conjunction with each other, present points of view for thinking about society and social interactions.

- *Functionalism grew out of the work of Auguste Comte and Émile Durkheim, and explains the world in terms of interconnected social systems.* Critics claim that it ignores inequality and social liquidity.
- *Conflict theory developed as opposition to functionalism grew, and sees disagreement and the resulting changes in society as natural, inevitable, and even desirable.* Critics claim that it presents a negative view of humanity.
- *The feminist perspective builds on conflict theory, proposing in addition that gender inequality is central to all conflict.* Critics claim that feminism is far too narrowly focused.
- *Symbolic interactionism focuses on the symbolic meanings of micro-level interactions.* Critics claim that it ignores the impact of macro-level factors on our everyday behavior.

Key Terms

sociology the systematic study of social interaction at a variety of levels.

sociological imagination the intersection between individual lives and larger social influences.

microsociology the study of small-scale patterns of individuals' social interaction in specific settings.

macrosociology the study of large-scale patterns and processes that characterize society as a whole.

theory a set of statements that explains why a phenomenon occurs.

empirical information that is based on observations, experiments, or experiences rather than on ideology, religion, or intuition.

social facts aspects of social life, external to the individual, that can be measured.

social solidarity social cohesiveness and harmony.

division of labor an interdependence of different tasks and occupations, characteristic of industrialized societies, that produce social unity and facilitate change.

capitalism an economic system in which the ownership of the means of production—like land, factories, large sums of money, and machines—is in private hands.

alienation the feeling of separation from one's group or society.

value free separating one's personal values, opinions, ideology, and beliefs from scientific research.

functionalism (*structural functionalism*) an approach that maintains that society is a complex system of interdependent parts that work together to ensure a society's survival.

dysfunctional social patterns that have a negative impact on a group or society.

manifest functions functions that are intended and recognized; they are present and clearly evident.

latent functions functions that are unintended and unrecognized; they are present but not immediately obvious.

Chapter 4 Topics

1 Socialization: Its Purpose and Importance

Socialization, a lifelong process, teaches us attitudes, values, and behavior that are essential for effective participation in a society. Through social contact and interaction, we learn to be human. Socialization fulfills four key purposes: It establishes our social identity, teaches us role taking, controls our behavior (through *internalization*), and transmits culture to the next generation.

2 Nature and Nurture

Biologists tend to focus on the role of heredity (or genetics) in human development. In contrast, most social scientists, including sociologists, underscore the role of learning, socialization, and culture. This difference of opinion is often called the "nature-nurture debate." *Sociobiologists* argue that genetics (nature) can explain much of our behavior. Most social scientists, including sociologists, maintain that socialization and culture (nurture) shape even biological inputs.

TABLE 4.1
The Nature-Nurture Debate

NATURE Human development is . . .
Innate
Biological, physiological
Due largely to heredity
Fairly fixed

NURTURE Human development is . . .
Learned
Psychological, social, cultural
Due largely to environment
Fairly changeable

Example: *Is Male Violence Genetic?*

Many people believe that, because of their genetic makeup, men are innately more aggressive than women. Homicide rates where the assailant is male vary considerably across societies, however: 62 per 100,000 population in Colombia, 22 in Russia, 7 in the United States, and 1 in practically all of the European countries (Krug et al. 2002). Such variations suggest that the environment (nurture), rather than nature, affects violence, including factors such as attitudes about crime, socialization, law enforcement policies, and the extent of poverty in a population.

Source: Etienne J. Krug, Linda L. Dahlberg, James A. Mercy, Anthony B. Zwi, and Rafael Lozano, eds. 2002. *World Report on Violence and Health*. Geneva: World Health Organization.

Sociological Explanations of Socialization

Sociologists have offered many explanations of socialization, but two of the most influential, both at the micro-level, have been social learning and symbolic interaction theories.

Key Terms

socialization the lifelong process of social interaction in which the individual acquires a social identity and ways of thinking, feeling, and acting that are essential for effective participation in a society.

internalization the process of learning cultural behaviors and expectations so deeply that we assume they are correct and accept them without question.

sociobiology a theoretical approach that applies biological principles to explain the behavior of animals, including human beings.

social learning theories approaches whose central notion is that people learn new attitudes, beliefs, and behaviors through social interaction, especially during childhood.

looking-glass self a self-image based on how we think others see us.

self an awareness of one's social identity.

role taking learning to take the perspective of others.

significant others the people who are important in one's life, such as parents or other primary caregivers and siblings.

anticipatory socialization the process of learning how to perform a role one doesn't yet occupy.

generalized other a term used by George Herbert Mead to refer to people who do not have close ties to a child but who influence the child's internalization of society's norms and values.

impression management the process of providing information and cues to others to present oneself in a favorable light while downplaying or concealing one's less appealing qualities.

reference groups groups of people who shape an individual's self-image, behavior, values, and attitudes in different contexts.

agents of socialization the individuals, groups, or institutions that teach us what we need to know to participate effectively in society.

peer group any set of people who are similar in age, social status, and interests.

resocialization the process of unlearning old ways of doing things and adopting new attitudes, values, norms, and behavior.

total institutions places where people are isolated from the rest of society, stripped of their former identities, and required to conform to new rules and behavior.

TABLE 4.2
Key Elements of Socialization Theories

SOCIAL LEARNING THEORIES	SYMBOLIC INTERACTION THEORIES
• Social interaction is important in learning appropriate and inappropriate behavior. • Socialization relies on direct and indirect reinforcement. *Example:* Children learn how to behave when they are scolded or praised for specific behaviors.	• The self emerges through social interaction with significant others. • Socialization includes role taking and controlling the impression we give to others. *Example:* Children who are praised are more likely to develop a strong self-image than those who are always criticized.

4 Primary Socialization Agents

Agents of socialization are the persons, groups, and institutions that teach us how to participate effectively in society. Parents are the first and most important socialization agents, but siblings, grandparents, and other family members also play important roles. Other important socialization agents include play and peer groups—those who are similar in age, social status, and interests—as well as teachers and schools, and popular culture and the media. Advertising is an especially powerful force in socialization.

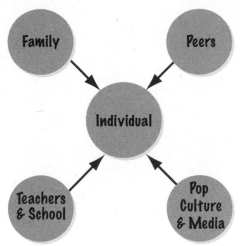

5 Socialization Throughout Life

As we progress through the life course—from infancy to death—we learn culturally-approved norms, values, and roles. Infants are born with an enormous capacity for learning that parents and other caregivers can enrich and shape. In adolescence, these and other socialization agents teach children how to form relationships on their own, to get along with others, and to develop their social identity through play and peer groups. In adulthood, people must learn new roles that include singlehood, marriage, parenthood, divorce, work, buying a house, and experiencing the death of a loved one. Socialization continues in later life when many people learn still new roles such as grandparents, retirees, older workers, and being widowed.

6 Resocialization

Resocialization—which can be voluntary or involuntary—is the process of unlearning old ways of doing things and adopting new attitudes, values, norms, and behavior. In *total institution* people are isolated from the rest of society, stripped of their former identities, and required conform to new rules and behavior.